Families and Society

Classic and Contemporary Readings

SCOTT COLTRANE
University of California, Riverside

Australia • Canada • Mexico • Singapore • Spain
United Kingdom • United States

THOMSON
WADSWORTH

Sociology Editor: *Robert Jucha*
Assistant Editor: *Stephanie Monzon*
Editorial Assistant: *Melissa Walter*
Technology Project Manager: *Dee Dee Zobian*
Marketing Manager: *Matthew Wright*
Marketing Assistant: *Michael Silverstein*
Advertising Project Manager: *Linda Yip*
Project Manager, Editorial Production: *Kirk Bomont*
Print/Media Buyer: *Karen Hunt*
Permissions Editor: *Kiely Sexton*
Production Service: *Carlisle Communications*
Copy Editor: *Kathy Pruno*
Cover Designer: *Bill Stanton*
Cover Image: *Planet Art*
Cover Printer: *Webcom*
Compositor: *Carlisle Communications*
Printer: *Webcom*

Printed in Canada
4 5 6 7 07

For more information about our product, contact us at:
Thomson Learning Academic Resource Center
1-800-423-0563
For permission to use material from this text, contact us by:
Phone: 1-800-730-2214
Fax: 1-800-730-2215
Web: http://www.thomsonrights.com

Library of Congress Control Number: 2003106236

ISBN- 13: 978-0-534-59130-4
ISBN- 10: 0-534-59130-2

Wadsworth/Thomson Learning
10 Davis Drive
Belmont, CA 94002-3098
USA

Asia
Thomson Learning
5 Shenton Way #01-01
UIC Building
Singapore 068808

Australia/New Zealand
Thomson Learning
102 Dodds Street
Southbank, Victoria 3006
Australia

Canada
Nelson
1120 Birchmount Road
Toronto, Ontario M1K 5G4
Canada

Europe/Middle East/Africa
Thomson Learning
High Holborn House
50/51 Bedford Row
London WC1R 4LR
United Kingdom

Latin America
Thomson Learning
Seneca, 53
Colonia Polanco
11560 Mexico D.F.
Mexico

Spain/Portugal
Paraninfo
Calle/Magallanes, 25
28015 Madrid, Spain

Contents

CHAPTER 13
Family Policy 459

Preface

The original idea for this book was launched in conversations with Publisher Eve Howard of Wadsworth and took shape in consultation with colleagues at the University of California, Riverside and across the country. I am particularly indebted to Lynet Uttal, of the University of Wisconsin, for her thoughtful collaboration in the formative stages of the project and to Randall Collins, now of the University of Pennsylvania, for inviting me to be co-author of *Sociology of Marriage and the Family: Gender, Love, and Property.* The work of assembling, reviewing, and evaluating possible selections for this book was facilitated by the expert research assistance of Michele Adams, now Assistant Professor at Tulane University, who also contributed pedagogical exercises and served as a valued editorial consultant and contributor. Selecting from among the best classic and contemporary readings in family studies was a daunting task, but I was inspired by plentiful innovative scholarship in the field. For every article included, there were several worthy contenders set aside because of duplication or space limitations. I thank Nancy Chodorow for introducing me to the work of Rose Laub Coser, Margaret Mead, Dorothy Smith and Lillian Rubin, and to sociologists Marcia Millman, Candace West, Barrie Thorne, and Arlie Hochschild for helping me to discover what was important about family studies while I was still a graduate student. Fellow members of the Council on Contemporary Families provided inspiration for this and other endeavors, and colleagues at the University of California nurtured my imagination and sustained my enthusiasm for the project. Senior editor Bob Jucha of Wadsworth aided me in broadening the book's appeal. The following people provided helpful outside reviews: Anne Barrett, Florida State University; Tim Biblarz, University of Southern California; Ann Beutel, University of Oklahoma; Jill E. Fuller, University of North Carolina at Greensboro; Jennifer E. Glick, Arizona State University; Amy Holzgang, Cerritos College; Trina Hope, University of

Oklahoma; Terry Huffman, University of North Dakota; Barbara Richardson, Eastern Michigan University; Allison L. Vetter, University of Central Arkansas; Peggy Walsh, Kenne State College; and June Youatt, Michigan State University.

Students in my Sociology of Family classes at the University of California, Riverside gave me valuable feedback by reading and commenting on countless articles. I also want to acknowledge the influence of my now adult children, Colin and Shannon, who have taught me more about life than they realize. Finally, I dedicate this book to my lifelong partner, Wendy Wheeler, whose support and encouragement have allowed me to undertake such endeavors, and whose patience, understanding, and critical reflection made this book possible.

About the Author

Scott Coltrane is a professor of Sociology at the University of California, Riverside. He has written about the interrelationships among fatherhood, motherhood, marriage, parenting, domestic labor, popular culture, ethnicity, and structural inequality. Results of his research have been published in various scholarly journals, including the *American Journal of Sociology, Social Problems, Journal of Marriage and Family, Journal of Family Issues, Family Relations, Sex Roles,* and *Gender & Society.* He is author of *Gender and Families, Family Man,* and co-author (with Randall Collins) of *Sociology of Marriage and the Family: Gender, Love, and Property.* He serves as Associate Director of the UCR *Center for Family Studies* and is principal Investigator (with Ross D. Parke) on an NIH-funded longitudinal study of Mexican-American and European-American families. He is the recipient of the University of California, Riverside, Distinguished Teaching Award and was the 2000/2001 President of the Pacific Sociological Association.

Introduction

The family is a paradox. On the one hand, families are about as ordinary and mundane as anything can get. Parent and child, husband and wife, brother and sister, sitting in front of the TV, arguing over what program to watch, eating dinner, washing dishes, going to bed. On the other hand, families are among the most engaging and dramatic subjects the world has ever known. Famous plays and novels describe families, television comedies revolve around family life, movies chronicle romantic encounters or dark family histories, and the news is full of stories that portray family successes or tragedies. Whether about sex, romance, love, loyalty, or betrayal, most popular media refer to couples or families, and most people cherish ideal images of families, even when they know such images are unrealistic.

What should we make of this paradox? It is not just that media images are different from real lives. Rather, the contrast between mundane and dramatic is evident in virtually everyone's family life. Dating, falling in love, getting married, and having children are among the most exciting things in life. Family life can also be routine and taken for granted. What could be more boring than staying at home all the time, away from the "action" of going out with friends? Family life presents us with profoundly negative experiences as well: parents' and children's struggles with each other, couples' quarrels, jealousies, extramarital affairs, violence, abuse, separation, divorce, and death. Families thus account for our personal highs and our worst lows. Moreover, these two sides are connected. The excitement of love and sex, the ideal image of the bride, the joy of a new baby—all of these lead to the routine, mundane side of family life. The transition from dramatic to mundane is often a surprise and the resulting disgruntlement and conflict can lead to difficult times for families.

The family is thus a site of powerful ideals—or illusions—but as family life gets underway, the reality of the everyday work asserts itself—there are diapers to be changed, meals to prepare, bills to be paid. The fantasies, joys, and high points of family life can reassert themselves through the years, but the most successful families are those that can balance the mundane with the ideal. That balancing act is profoundly influenced by an individual's placement in the larger society. For example, a child who grows up in a family with polo ponies and a house at the beach is likely to feel differently about being asked to scrub the floor or wipe a runny nose than a child who grows up eating corn flakes for dinner and having to sleep with two siblings in the same bed. It is not simply that being upper-class makes family life easier—though there is ample evidence that having more material resources improves one's health, marital prospects, career opportunities, and longevity. Rather, it is important to understand how one's race, ethnicity, income, education, occupation, gender, age, and other social and demographic characteristics can shape the conditions under which family life is negotiated. Such an approach to studying families is fundamentally sociological.

Sociology is based on the idea that society organizes people's lives and that individual experiences both reflect and reproduce cultural ideals and patterns of inequality in the larger society. Nowhere is this more evident than in the study of families. It is impossible to describe families without understanding how larger social processes shape them and it is equally impossible to describe individual family experiences without reference to the ways that those experiences shape the social world. Although these key assumptions have guided sociological inquiry for over a century, they are often overlooked when students and scholars turn their attention toward intimate personal topics such as love, sex, marriage, parenting, and a wide array of family-related joys and sorrows. That observation prompted Randall Collins and me to write a family textbook that drew on the most important theoretical insights and critical perspectives developed in sociology and other social sciences. In *Sociology of Marriage and the Family: Gender, Love and Property,* we wrote about how families and society are mutually produced and how one cannot understand one without knowing about the other. In particular, we drew on concepts such as stratification, social construction, and social comparison (described below) to show how a sociological approach to families is more illuminating than the more individualistic approaches used in most marriage and family textbooks.

Families and Society is a collection of readings organized into thirteen chapters covering the most important topics in the sociology of the family, including defining families; sex and romance; marriage and cohabitation; reproduction, birth, and babies; childhood socialization; motherhood; fatherhood; families and work; divorce; diversity of family forms; domestic violence; families over the life course, and family policy. The forty-five articles in this collection were selected because they represent some of the main concepts contained in our earlier textbook, but they also update and expand that work. This reader can thus be used with our textbook or another family textbook, but can also be used as a stand-alone text. Each chapter opens with a theoretical introduction to the topic and articles, and each topic area includes a classic article reflecting historical scholarship on the subject. The classic readings show how families and society have both changed and retained some key features. These historical selections are intended to help you develop an understanding of how society, the economy and politics influence the ways we think about families and intimate relationships. The classic readings also show how scholarship about families has changed over time as families and society have undergone various transformations. The combination of classic and contemporary readings provides students with a broad introduction to some of the most important conceptual and empirical issues in family studies. Articles were selected because they exemplify the creative critical scholarship that has expanded the focus of family sociology and helped it to address current social issues. Whether you use the book as a supplemental reader or a main text, I hope that it will stimulate your thinking about families and help you to see families in a new light.

A sociological approach to studying families relies on three major concepts: stratification, social construction, and social comparison. Sociologists argue that without knowing how families fit into patterns of **stratification** in the larger society, it is impossible to understand family structure, functions, or patterns of interaction. Sociologists analyze stratification in terms of three dimensions: (1) **Economic class** is the material dimension, ranging from poverty to wealth, and includes the occupations and careers by which people get their incomes. (2) **Power** determines who controls whom. It ranges from the actions of political elites at the most macro level of the state down to the most intimate interpersonal matters, such as who has the power to decide when to have sex. (3) **Status** is the dimension concerning who is held in most respect. It is manifested in the way people talk about each other and in the way they display themselves, their homes, and their possessions. It is a realm of symbols and feelings, ranging all the way from admiring smiles to not even noticing someone is there (Coltrane and Collins, 2001, p. 8). Most of the selections in this book examine some aspect of stratification as it relates to the family, especially how families place individuals into the class structure of society and how gender, race, ethnicity, and age stratification affect family interaction and family dynamics.

Another key concept used to understand families is **social construction.** Depending on personal histories, individual characteristics, social experiences, and environmental contexts, people tend to "see" the world differently. For example, a child who grew up near the Arctic Circle has learned many different names for "snow" and can recognize subtle distinctions among them. In contrast, a child who grew up in a rain forest near the equator and sees snow for the first time might call it "rain" or perhaps not know what to call it at all. The rain forest child would "see" the snow differently from the arctic child because the cognitive framework and language for making sense of it would be missing. Naming physical objects (like snow) encourages us to think and talk about them in specific ways, which, in turn, helps us to perceive them in culturally appropriate ways (Coltrane, 1998, p. 2). The same process applies to understanding families and other aspects of the social world. How we interpret our own or other people's thoughts, feelings, and behaviors depends on shared understandings, which, in turn, are produced through interaction using common language and symbols. Many articles in this book use a social constructionist approach to focus on how cultural understandings and everyday practices shape people's views of what a family is and how family members should behave. Adopting a social constructionist approach reveals how families are constantly changing and provides insight into what kinds of changes we are witnessing today.

Another tool for understanding families is **social comparison.** Many articles in this book describe practices common in other parts of the world or in past times. Comparisons between family ideals and practices of today and those of different cultures or historical eras allow us to identify the conditions associated with different types of family systems and to understand the processes producing different family structures. Selections in the book also focus on variation in patterns of marriage and the relationship between spouses within marriage or other types of couple relationships. Comparison among different types of contemporary families (including same-sex couples, single-parent families, childless couples, and blended families) allows us to see how institutions like the law or religion contribute to differing views of sex, marriage, divorce, and the normative roles of husbands and wives. Many of the family practices and ideals described in these selections are likely to differ from your own experiences. By using social comparison, however, you should be able to see beyond your own circumstances and avoid assuming that a particular family practice you are familiar with is more permanent or fundamental than it actually is. The InfoTrac® College Edition exercises are also designed for you to conduct social comparisons of your own. Social comparison is an essential tool for building theories in family sociology, for only by placing families within their social contexts can we understand how various social forces produce differing family practices and ideals.

The words **sex** and **gender** are used in a special way in this book and in the writing of most social scientists. Sex refers to relatively distinct biological differences between males and females such as genitals, hormones, and chromosomes. Sex can also refer to erotic behavior, or sexuality. Gender, on the other hand, describes how, in a particular culture, the typical man is supposed to present himself (as masculine) and how the typical female is supposed to present herself (as feminine). Whereas sex refers to biological characteristics (and erotic behavior), gender refers to the social and cultural aspects of being a man or woman. The classical and contemporary readings in this book reveal how the meaning of gender changes in response to differing cultural and historical contexts. The readings also show how families play a key role in defining and enforcing gender ideals.

The articles in this book and the exercises at the end of every chapter introduction are designed to challenge you and to stimulate you to think for yourself. As I tell students when I teach classes on family, gender, and social psychology, if something I say doesn't offend you, then you are probably not paying enough attention. As a sociologist, it is my job to question the obvious, to look for patterns of power and domination, and to challenge conventional wisdom. This makes many students uncomfortable, but discomfort should not stop you from thinking and talking about the underlying issues involved. This collection of readings is therefore intended to stimulate discussion and debate. Only by questioning received wisdom will you be able to build an understanding of families that resonates with your own life. And only by questioning received wisdom and thinking for yourself will you move closer to a true understanding of the links between family and society.

REFERENCES

Coltrane, Scott, and Randall Collins. 2001. *Sociology of Marriage and the Family: Gender, Love, and Property.* 5th ed. Belmont, CA: Wadsworth/ITP.

Coltrane, Scott. 1998. *Gender and Families.* Thousand Oaks, CA: Pine Forge Press.

CHAPTER 1

Defining Family

What are the basic features of a family? Most of us know a family when we see one, but what are its essential parts? Although this seems a simple question, scholars have been debating it for centuries. The United States Census Bureau defines a **family** as "two or more persons who are related by birth, marriage, or adoption who live together as one household." This definition allows census takers to decide who to count as a "living together" family, but most of us understand family to be much more than this. As divorce, remarriage, same-sex marriage, nonmarital birth, cohabitation, and related practices are becoming more accepted, both scholars and average citizens are asking what should count as a "real" family. Should we consider a man, wife, and children who live together in the same house as the only legitimate form of family, or should we broaden the definition of family to include single parents, committed partners, and other arrangements? And how should we handle the question of family members who do not live together? Is there a core set of features that defines a family, and how should we decide what those features are? The selections in this chapter provide some answers to these questions and raise new ones about the implications of selecting one definition or another. As these three readings suggest, definitions of families carry important consequences for cultural ideals, family practices, and individual thoughts and actions.

The first reading is a classic theoretical statement in the sociology of the family that focuses on the functions that families serve for society. In an excerpt from *The Family: Its Structure and Functions* (1964), Rose Coser suggests that we can only explain families in terms of the larger social order and that we can identify some universal features of families. Coser wrote this piece before the second wave of the women's movement in the United States, at a time when most women married in their early twenties to become mothers and stay-at-home housewives. Her analysis of family life reflects assumptions about how families are similar to one another; how all family members benefit from the family; and how families, as she describes them, are necessary to the functioning of society. Invoking the **principle of legitimacy,** she suggests that fathers secure a place for children in a system of social hierarchy. This follows Bronislaw Malinowski's (1929, 1964) conclusion that marriage legitimates childbearing, based on his observation that all cultures (at least the ones he knew of) disapproved of women bearing children out of wedlock. (Historically in England and America, such children might be called "bastards," or more politely "illegitimate children.") The principle of legitimacy implies that all societies demand that children have a socially designated father to serve as an official male link between the child and the larger community. In later selections, other authors challenge the universality of this pattern, but Coser succinctly states the case for considering this one of the common features of families. Another universal feature of families described by Coser is the **principle of reciprocity.** In discussing incest taboos and other cultural customs that regulate family formation, Coser relies on the principle of reciprocity to suggest that families must regulate sexual and marital relations to maintain the larger social order.

Coser, Malinowski, and other family theorists adopt a comparative perspective to examine how family systems are similar or different depending on the cultural and historical contexts in which they exist. For example, Coser compares how a capitalist economic system like that of the United States promotes individualistic child-rearing practices and rewards personal achievement, sometimes at the expense of group solidarity. She also suggests that periods of rapid social change undermine the authority of parents in families, because they are less able to provide for or regulate the behavior of their offspring. Coser uses theories from the 1950s and 1960s that assume a harmony of interests within the family, and between families and the larger society; however, in this selection she also implies that families both reflect and help to perpetuate inequalities along social class lines. She hints that marriage and family life are not always beneficial for women, thereby laying the groundwork for later critical analyses of families as sites of gender struggle by contemporary feminists. As you read this excerpt, ask yourself how Coser's theoretical approach makes assumptions about similarities or differences among families.

Writing three decades later, Dorothy Smith reflects on her experiences as a single parent, and presents a different opinion on the idea that families have universal features. To Smith, family practices and family ideology are increasingly at odds, and she is unhappy with the confining stereotypes produced by popular culture and reinforced by social science researchers. She describes how the **standard North American family** (SNAF) consisting of middle-class breadwinner father, homemaker mother, and children represents an "ideological code" that patterns our thinking even when we don't consciously agree with it. Part of this ideology is assuming that "intact" two-parent families are the ideal and any deviations from this pattern will be detrimental for children and adults. She critically examines her own methods of research in studying families, as well as those of sociologists studying African American families, especially those who assume that current problems of African American households would be reduced if we reinstated men as rightful heads of families. Here and in her other writings, Smith wonders why we apply different standards to men and women when it comes to holding jobs, being loving partners, and caring for children. She assumes that the dominant popular view of what constitutes a family (SNAF) is part of the problem, rather than part of the solution. According to Dorothy Smith, family researchers often reproduce this narrow normative ideology without examining how it might be detrimental to American families from all race and ethnic backgrounds. Do you agree with Smith that assumptions about normal families can be especially harmful to women and people of color?

The third selection in this chapter examines how popular cultural images of families are not as simple or straightforward as they first appear. Paul Cantor analyzes family imagery and political messages from the popular prime-time cartoon, *The Simpsons.* Although this show is often criticized for undermining "traditional" family values, Cantor suggests that the show actually presents an enduring and endearing image of the **nuclear family.** He sees the show as reflecting an anti-authoritarianism that is an American tradition, and he also suggests that family authority has always been problematic in democratic America. It is not hard to see how *The Simpsons* satirizes and makes fun of family life, but as the author notes, the show also makes fun of everything else. Ultimately Cantor argues that the show promotes family loyalty and commitment. Do you agree? Do you think that television shows like this one influence how people feel about their own families?

Taken together, the three readings in this chapter provide comparative evidence to challenge some taken-for-granted assumptions about family life and to show how family definitions and family ideology matter. They matter to social theorists, family researchers, and everyone who is a consumer of popular culture. These selections also support the idea that families cannot be understood without reference to larger political and social issues, a theme continued throughout the book.

REFERENCES

Coser, Rose Laub. 1964. *The Family: Its Structure and Functions.* New York: St. Martin's Press.

Malinowski, Bronislaw. 1929. *The Sexual Life of Savages in North Western Melanesia.* New York: Harcourt.

Malinowski, Bronislaw. 1964. Parenthood: The Basis of Social Structure. In *The Family: Its Structure and Functions,* edited by R. L. Coser. New York: St. Martin's Press.

SUGGESTED READINGS

Bittman, Michael, and Jocelyn Pixley. 1997. *The Double Life of the Family: Myth, Hope & Experience.* St. Leonards, Australia: Allen & Unwin.

Carrington, Victoria. 2002. *New Times: New Families.* Dordrecht, The Netherlands: Kluwer Academic Publishers.

Casper, Lynne, and Suzanne Bianchi. *Continuity and Change in the American Family.* Thousand Oaks, CA: Sage Publications.

Coltrane, Scott. 1998. *Gender and Families.* Thousand Oaks, CA: Pine Forge Press.

Coontz, Stephanie. 1997. *The Way We Really Are: Coming to Terms with America's Changing Families.* New York: BasicBooks.

Gubrium, Jaber, and James A. Holstein. 1990. *What Is Family?* Mountain View, CA: Mayfield Publishing.

Hawes, Joseph M. and Elizabeth I. Nybakken, eds. 2001. *Family and Society in American History.* Urbana: University of Illinois Press.

Weston, Kath. 1997. *Families We Choose: Lesbians, Gays, Kinship.* Rev. ed. New York: Columbia University Press.

INFOTRAC® COLLEGE EDITION EXERCISES

The exercises that follow allow you to use the InfoTrac® College Edition online database of scholarly articles to explore the sociological implications of the selections in this chapter.

Search Keyword: Family. Draw on the InfoTrac® College Edition database to find articles that discuss different definitions of family. In what ways has the definition of family changed in the last decade or so? In what ways has it stayed the same?

Search Keyword: Family ideology. What is meant by the term *family ideology*? What family ideologies are described in the articles you find? How does family ideology in the larger culture affect the types of social policies developed to solve family problems?

Search Keyword: Nuclear family. Use InfoTrac® College Edition to find articles that discuss the nuclear family. How is this type of family portrayed? What traits are characteristic of this type of family? How does the nuclear family compare to other family types? Is the number of nuclear families increasing or decreasing?

Search Keyword: Family values. In the articles your search turns up, examine the phrase *family values.* What is meant by this term? How do family values connect to politics and political debates in the United States?

Search Keyword: Familism. Using InfoTrac® College Edition, search for articles that deal with the concept of *familism*. How do the articles you find define this concept? Are some groups described as more familistic than others? Explain. How does familism relate to caregiving in families?

1

The Family

Its Structure and Functions

ROSE LAUB COSER
1964

A study of the family as an institution must relate it to the context of the particular society in which it functions, because relationships within the family and its pattern of life must be to some extent congruent with the demands that the community makes upon its members. For example, if there is strong emphasis on individual achievement as the basis of social status—as in our own society, especially in the middle class—the values of individualism and achievement will guide the family's concerns in relation to its young members. For it is in the family that the next generation is prepared for the roles to be occupied in the society at large.

The basic assumption that patterns of family life must be explained in terms of the larger social order is contrary to the view that the political and economic life of the society must be explained by reference to the character and personality of individuals, as they are shaped in the family through its training practices. While it may be true that from the point of view of the individual's life span the family is the first institution to which he is exposed, it is important to remember that from the point of view of society the family is a mediator of social values. As Erich Fromm has aptly stated: "In spite of individual differences that exist in different families [of the same society], the family represents primarily the content of the society; the most important social function of the family is to transmit this content, not only through formation of opinions and points of view but through the creation of a socially desirable attitudinal structure" (Fromm, 1936).

Following the French anthropologist Claude Levi-Strauss, (Levi-Strauss, 1960) the family may be defined as a group manifesting the following organizational attributes: it finds its origin in marriage; it consists of husband, wife, and children born in their wedlock, though other relatives may find their place close to this nuclear group; and the group is united by moral, legal, economic, religious, and social rights and obligations (including sexual rights and prohibitions as well as such socially patterned feelings as love, attraction, piety, and awe).

That the family is a universal institution cannot be explained simply by its manifest functions—such as reproduction, economic activities, socialization of the young—all of which could conceivably be fulfilled outside the institutionalized family. Moreover, the family does not serve exactly the same functions in every society or at every time: it may be an economic enterprise or have no economic function other than consumption; religious worship may center around the family or may be regulated by other social institutions. Nevertheless, though the manifold characteristics of families are as diverse in structure as the cultures in which they are embedded, two features stand out universally: the family serves as an agent of social placement for the new members of society, and by acting as an agent of control of marital relations, it regulates social alliances between family units and helps to place individuals into a patterned network of interweaving social relationships. The first of these features is known as the *Principle of Legitimacy,* the second as the *Principle of Reciprocity.*

Though some earlier social theorists—of the evolutionist school, for example—assumed that "in the beginning" (whatever this may mean) there was no family, that men and women lived in a "free" or "promiscuous" state and that their children belonged to the "collectivity," contemporary anthropologists, sociologists, and historians agree that no known society, present or past, exists without groups of two or more adults, who are responsible for the reproduction and maintenance of new members as well as for their socialization. The first discoveries of previously unknown territories peopled by "savages" (like earlier contacts with Eastern civilization) led Westerners to the realization that their social arrangements were not universal or divinely ordained. Early travelers and missionaries regarded patterns of sexual life that they were unfamiliar with as lacking sexual mores and rules. Societies where men had sexual relations with more than one woman, where promiscuity seemed to be the common pattern, or where sexual relations between boys and girls before marriage appeared to be unrestricted seemed to lack any order at all. Yet we now know that in every society that has been discovered so far the relations between men and women, the processes of reproduction and of socialization, and the social exchange of women have been subject to some social control.

It is well established, for example, that polygamous societies institutionalized either polyandry, when a woman has more than one husband, or polygyny, when a man has more than one wife (the latter being much more frequent), but never random mating. Whether the culture practices polyandry or polygyny, there is usually role differentiation and division of labor among the various partners of the same sex, and assignment of offspring to the proper lineage. Furthermore, polygamy is limited primarily, if not exclusively, to those individuals who occupy a high place in the social structure, chiefs or nobles, for example, so that marital selection tends to be related to the status order of the larger society.

One important aspect of the interrelation between the family and the society is the "placement" function the family has for its members. The family, more specifically the father, gives the children a social identity; that is, it places them in a specific pattern of social relationships. Indeed, the distinctly social nature of the family is characterized by the universal insistence on fatherhood. If the function of the family were simply reproduction, the link between man and woman could be severed after conception, because no necessary ties arise from the fact of biological descent. The Latin distinction between *genitor* (father as procreator) and *pater* (social father) indicates well the dual nature of fatherhood. The institution of adoption demonstrates that it is the *social father* who assigns status to the children that are said to belong to the family, whether the father, or both parents, are the physiological agents or not. It is in this way that every society, no matter what its sexual mores, establishes the legitimacy of its new members.

The dyad mother-child, self-sufficient as it might conceivably be economically and emotionally, is always considered incomplete sociologically. This is true of every society, whether patrilineal or matrilineal (that is, whether the father or the maternal uncle has the main control over the children). Even in matrilineal societies, where both mother and offspring belong to the mother's clan, the dyad of mother and child is not considered a complete social unit and has no well-defined social identity. The *Principle of Legitimacy* holds that every child shall have a father, and one father only.

If the institution of social fatherhood makes the nuclear family necessary as a social fact, the nuclear family, in turn, makes society possible by defining social relationships through a patterned exchange of sexual partners. This reciprocity, which is the essence of social life, is assured by the incest taboo. Whatever the type of social control of sexual activities—whether premarital sexual intercourse is permitted, preferred, prescribed, or proscribed (Merton, 1957a)—all societies control pregnancy and birth; and no matter how much license there is for some kind of uncontrolled sexual behavior, generally a rule of exogamy requires that sexual partners be obtained from outside one's own family. Since emotional ties between members of the family must necessarily be strong, sexual partners from within the same family group would naturally be the preferred choice; ties between individuals would become ever more limited to fewer and fewer persons and social atomization would replace social process (Middleton, 1964). Despite the fact that "one's own" may be defined differently in various cultures—it may or may not include cousins, for example—the incest taboo makes possible a patterned form of exchange *between* instead of *within* families; it thereby assures the occurrence of social exchange and provides the basis for the operation of the *Principle of Reciprocity.*

. . . The rule of exogamy implies, of course, that feelings of mutual attraction as a motivation for marital selection not be left to chance alone. Cravings for sexual intimacy and mutual attraction through love are not permitted to reign supreme, since they might threaten patterned social alliances. Furthermore, there is a reluctance in every social milieu to let pairs of lovers withdraw completely from the social scene; if this were possible, society could no longer exist. Indeed, marriage itself, with all its social obligations, is a means of control of love relationships.

If exogamy serves to bring into relation with one another otherwise separated groups—that is, if marriage is to be seen as a form of social alliance—there must exist different degrees of preference for different types of alliance. Alliance with strangers entails the risk of disparity of values, of differences in financial or social status, or other discrepancies between the potential allies that are socially defined as undesirable. Since the families and friends of the future partners are concerned about the alliance that comes from marriage, the nuclear families and other social groups may express socially prescribed interest in the marital choices of children and peers. Such controls may be exercised by subtle encouragement, as is often the case in our society, or by enforcement of prescribed behavior, as among the traditional Chinese gentry; but in no society is marriage determined by spontaneous sentiments alone. If sentiments are allowed, or expected, to give the immediate impetus to marital choice, they are to be "appropriate"—that is, they are to be so channeled as not to clash with the value system. In modern Western society, where marital choices are made by the young persons themselves, and where there is strong emphasis on love as a basis of that choice, many subtle and informal pressures may be exercises by parents, as well as, in smaller measure, by peers, to prevent choices that are socially considered "wrong." The freedom in marital selection upon which so much emphasis is placed in our culture does not imply lack of restriction or an unlimited range of choice, but simply freedom of choice within a socially prescribed framework.

. . . The degree of individual choice permitted in marital selection depends on the distribution of duties and obligations within the family system. When the father has absolute power and authority over the members, as among the traditional Chinese gentry, love between the young couple would weaken his son's allegiance and interfere with expected filial piety. Hence, in a society that rests on such rigid and continuous paternal authority, love will have little importance as a basis of marital partnership. In contrast, in a society in which a young man is expected to relinquish his dependency on his parents, it is permissible, even desirable, for the young couple to be bound by strong feelings, since they will have to depend primarily on each other and move to a new abode of their own. The way in which marital choice takes place is related to the internal family organization.

Similarly, the subsequent division of roles within the family is related to the broader social context. Role differentiation would seem to be more sharply marked in societies where the family performs many functions than in those where it is limited mainly to reproduction and socialization. For example, if the family is also a productive unit, division of labor may be expected; if the family is also the center of religious activities, different rituals are assigned to various members. Consequently, the need for role differentiation in the family would seem to be strongly reduced in industrial society, since most of the functions that have traditionally been served by the family are now being performed by outside institutions. Even a large part of the socialization of the young has been taken over by the educational system, and the family as a unit of production has almost disappeared. (Farm families which, by definition, do not actively participate in modern industrial societies, are one notable exception.)

Another reason for a less rigid role differentiation in the modern family is the fact that in industrial societies, where status is based on achievement rather than on ascribed characteristics, sexual differences are correspondingly less emphasized as a basis for role distribution. It is not simply the woman's participation in the economic process that determines her equality in the family, however, for there would be no contradiction between the need for a woman's hard work and her subordination to her husband if "hard work" were seen only as an obligation. But when high *moral* value is given to personal achievement through hard work, the importance of that quality as a criterion for evaluating individuals tends to eliminate from consideration other possible sources of inequality. It is the equalitarian value system of which this moral emphasis on individual achievement is a part rather than a woman's *de facto* participation in economic life, which frees her from her husband's authority and defines her as a companion.

Under this same value system the children are supposed to be treated equally, irrespective of sex, their age a basis for discrimination only insofar as it makes for differences in maturity and ability. Indeed, since the family prepares its young members for their future roles in society at large, it must emphasize those human qualities that are highly valued in the culture. This is why achievement is so highly valued within the middle-class family. The "equalitarian" family, especially cherished in the American middle class, is part of a value system that serves the economic process of an industrial society (Goode, 1963).

However, role differentiation in the family may vary in different subgroups. For example, in the modern middle class the family tends to reveal greater equality between husband and wife, and between siblings in all respects other than age, than in the working class. Similarly, the emphasis on achievement and mobility is not similar in all socio-economic strata. The upper and middle classes tend to be the representatives of the dominant values; in addition, these values are less pervasive as moral guides in strata where there is little opportunity for their attainment. In periods of social revolution in which a formerly underprivileged class is promised a share of the rewards hitherto unattainable, the lower class may well share these values. During the Russian and the Chinese revolutions, for example, the idea of social equality for the working class gave it the necessary impetus for achievement and hard work, and the idea of equality for women freed them from traditional family ties; thus the equalitarian ideology of a social movement became an important force in the rapid industrialization of both these countries.

To be sure, even in modern society, sex remains to some extent a basis for status assignment. In the American occupational sphere women are frequently paid a lower salary than men in the same position, and many types of positions are given to women rarely and reluctantly. Similarly, in the middle-class family full equality between husband and wife is seldom achieved, and there is often a contradiction between the avowed equality of siblings and the actual differential treatment of boys and girls (Komarovsky, 1964). The measure of equality in the family actually attained in industrial societies may differ within as between such societies, and the contradictions stemming from ascribed and achieved status are not always resolved without tension.

The emphasis on personal achievement also bears consequences in regard to relationships within the extended family. Where individuals gain their status through their own achievement rather than through ascription, the family into which they are born, *the family of orientation,* loses importance for the young married couple. Many parents give aid to married children, but such aid is considered temporary and is thought of as "influencing the children's status position" (Sussman, 1953) until *they can be on their own* and *prove themselves.* In spite of the fact that emotional ties between members of an extended family continue to exist (Litwak, 1960), the modern family is more removed, morally if not also physically, from their relatives; it is to the family formed by marriage, *the family of procreation,* that family members owe their primary allegiance. The United States furnishes an excellent example of the modern family in an industrial society. The nuclear family is independent of other consanguine relationships, husband, wife, and children constituting an independent family unit. Compared with preindustrial society, there is a marked decrease in allegiance, in duties as well as rights, between the members of the family of procreation and the members of the family of orientation.

The fact that individuals are placed in the occupational order by ability rather than inherited characteristics, and must therefore be free to accept changes in location and status, also necessitates the severing of close relations with neighborhood friends. Strong emotional attachments to neighbors would impede the geographical movement and status mobility that provide the flexibility needed in modern industry and bureaucratic organization. Studies of army life have shown that privates often refused promotion because they do not want to leave their "buddies." If such nonrational attachments to friends or relatives were typical of the general way of life, it would be difficult for persons to be assigned a place in a rationally organized enterprise on the basis of their abilities.

A binding allegiance to "buddies," whether or not they are family members, may impede individual achievement. The British discovered in India, for example, that as long as people felt bound to their families and were not permitted to enjoy the fruits of their own achievement because they had to support their families, they had no incentive to "better themselves." In 1930 they passed the Gains of Learning Act, which stipulated that "all gains of learning are a man's exclusive and separate property." A consequence of the "administration of the Hindu law by British courts was the disintegration of the joint-family organization (although it was retained in name)" (Kapadia, 1955). Severing family relations made it possible to fill positions in the Civil Service and in industry.

The strength of the ties between the individual and his family of orientation is clearly related to the authority relations within the family. Because of the reduction of the number of the family's social functions and the emphasis on individual achievement, authority in the modern family has been restricted. In an increasingly differentiated social order, controls over individuals emanate from a variety of institutions; authority is segmented just as activities are segmented, not only in social life generally, but to some extent in the family as well. For example, the child, though he is under his parents' authority in general life style, is controlled by teachers in the educational sphere. And the young man who sets out to work is not, as in some rural societies, subjected to his father's management (and often exploitation), but is away from home, under the control of managers or supervisors expressly hired for the job. If achievement is the measure of man, and if it takes place mainly outside of the family, parental figures can judge the performance of their children in only a limited way—especially in a rapidly changing society, where knowledge acquired in the past soon becomes obsolete or supplemented. When children bring new knowledge home from school, for example, parents have to admit that "they don't know everything." Thus authority in the modern family is neither based on the ability of the parent to exploit his children economically, nor on his unequivocal superior knowledge in all spheres.

But the modern family need not be regarded as "breaking down," as some writers on the subject have assumed (except where the limitation of authority within the family is not integrated with the value system, as is sometimes the case, for example, with some immigrant families) (Thomas and Znaniecki, 1958). The loosening of control over the children does not necessarily mean that the family is "weak." Where relaxation of authority by the parents is not accompanied by relinquishment of responsibilities, it is used as a deliberate means to teach the children "to be on their own." In families that value the acquisition of knowledge and individual independence, the child who brings home knowledge responds to the parent's ideal of the "independent achiever;" the father does not so much feel his authority threatened as he delights in his child's conformity to his own values. He may be less secure than the traditional parental figure whose control was unequivocally based on his power position, but he does derive prestige and self-esteem if he acts as an effective representative of the values of the larger social order.

Authority based on this criterion, however, is necessarily limited in a rapidly changing society, and as a result, parental figures lose much of their prestige with advanced age, since they cannot remain models of achievement for younger persons with whom they cannot compete. It would seem that the loss of authority and prestige of the aged is inherent in advanced industrialization.

. . . The high degree of differentiation between social institutions in modern society—in which not only economic, religious, and political activities take place outside of the family, but much of the social control of children is taken over by an outside educational system—might lead one to question whether the family is needed at all. Some utopias have been envisaged where one of the hitherto main functions of parental figures, the upbringing of children, would be taken over completely by outside agencies, such as nurseries and boarding schools, and attempts have actually been made to institute such agencies—notably in the Kibbutzim in Israel. Recent research, however, . . . has shown the importance of the nuclear family for the healthy development of a child's potential and for his adaptation to the broader social order. In the Kibbutzim the earlier de-emphasis on family relationships is gradually being reversed.

Parental figures, be they biological parents or surrogates, are always the important points of reference for the young child. Constituting a group small enough on which the child can focus his yet poorly developed perceptive and discriminating capacities, they serve as the "significant others" whose demands and wishes the child learns to "take into himself." Since they are the persons whose opinions matter and whose responses reveal to the child, explicitly or implicitly, the meaning of his own actions, it is through close contact with them that he learns to internalize social norms, that is, to make social precepts emanate from his own free will. The family is the most effective agent of this process of socialization, "by which people selectively acquire the values and attitudes, the interests, skills and knowledge—in short, the culture—current in the groups of which they are, or seek to become, a member" (Merton, 1957b). Through its power to direct this process by which children are fitted into the social framework and learn to play their roles in it, the family is the main link between the child and the society. But the restricting injunctions that it imposes on the biological organism do not necessarily thwart individual potential. On the contrary, a certain amount of repression is the very condition of human growth, and such injunctions, by setting the boundaries that separate the individual from his environment, help him to become aware of himself as a "self." It is partly through the reaction to and attitude toward his behavior on the part of "significant others" that the child fashions his own picture of himself, including his moral orientations (Mead, 1937).

The affective attachment that is the basis of the child's relationship with the "significant others" is the result of a very early attachment to the mother, which psychoanalysts have called "attachment to a love object." Without such a love object, the infant has no opportunity to develop emotions; even the most primitive biological drives seem to develop only through repeated interaction with a person to whom he has such an attachment.

The two isolated children, Anna and Isabel, about whom Kingsley Davis has written, afford an instructive comparison in this respect. Both had been reared in isolation, secluded from a human environment, until they were found at the age of approximately six. Their physical deterioration was almost complete, and neither could communicate with another person. The child who had been isolated with her deaf-mute mother was able to make a "strange, croaking sound" and react emotionally; in contrast, the other child whose mother was not otherwise exceptional but had deprived her daughter of even her physical presence, was completely apathetic and suspected of being deaf and blind. It would seem that innate biological faculties have to exercise themselves upon an object if they are to function: eyes do not learn to see if there is nothing to be seen, ears do not learn to hear if there is nothing to be heard, and even primitive feelings of rage do not develop without human contact—that is, without a love object.

The family as a small and stable group led by two adults, in addition to fulfilling the infant's basic need for human contact that the mother or mother surrogate must supply, provides a constant pattern of more complex interaction through which the child learns to distinguish between significant adult partners. He learns to articulate his own role in relation to each of them, and thus to understand and judge reality the way adults understand it. The parental figures mediate his exploration of reality and furnish him with the "generalized" values and norms necessary for perceiving and judging as a member of the community. For to understand reality means to understand social reality; every individual sees and responds to other people, and even to objects, in terms of their social definitions and meanings, and relates to them in socially approved ways.

This is not to say that a generalized picture of reality is necessarily homogeneous throughout the community. In a highly differentiated society, where complex division of labor separates the population into many occupational groups and economic strata with varied ethnic and religious affiliations, important value differences among various subgroups often crisscross each other and are reflected in how parents enforce their views upon their children.

The family also serves as a buffer between the potentially harsh impact of social conditions in the larger society and the individual who is attempting to cope with them. As children grow up and enlarge their social vision through exposure to other significant persons, such as teachers and peers, they are introduced to conflicts of value that they must learn to integrate; rapid technological changes may also create expectations contradicting with existing values. At the same time, however, as the family offers such protection, it necessarily reproduces the conflict between the old generation, representatives of tradition and protectors of social continuity, and the young, who are preparing to meet the demands that they will face as new adults.

. . . It should be clear by now that variations in family structure occur, not because of random cultural relativity, but in terms of the differing functional interrelations between the family and the ongoing social order. In some societies, a limited authority over the young within the family may serve to facilitate the direct control by representatives of the political order; in Germany, for example, the Nazis required young people to leave the parental home at prescribed times, to participate in the activities of the Hitler Youth, to go on hikes, or to spend long periods in camps. It put the youngsters for an important part of their lives under the direct control and influence of representatives of the political regime.

But mechanisms of social control may also operate indirectly, through a shared value system. In modern America, for example, adolescents, especially in the middle class, are strongly encouraged to spend four years at college, away from home, though for the purposes of formal education there is no reason why they should not live at home while attending classes. In a society as highly competitive as ours, four years of transition during which young men and women do not yet have to shoulder responsibilities of their own, though they are given the opportunity to be weaned from the sheltered home atmosphere, helps them to develop individual self-reliance and an independent spirit well suited for a competitive way of life. Attending colleges that are geographically removed from home, or living in college dormitories even if the parents' abode is nearby, is believed to be "good for them," since it "encourages independence."

The degree to which the social controls over family patterns are enforced explicitly or operate implicitly may differ, but the range of choice open to an individual family is always limited by two factors. First, definite concepts of public welfare, which are the concern of the state or government, are enforced deliberately either through coercion, as in the case of compulsory education even in democratic societies, or through incentives, as in the case of financial support for *familles nombreuses* in contemporary France. Second, every family must adapt its choices to the economic and political order in which it exists; the choices available may be many or few, but the possible alternatives for such adaptation are always somewhat limited. Since social arrangements within the family may affect development, they must be subject to some control; rigidity of specific controls in different societies seem to depend on the flexibility of the other social institutions and the degree to which innovation is tolerated in these other spheres.

The differential effects of various types of land inheritance illustrate the consequences that family arrangements may have on society. Some investigators believe that in the nineteenth century primogeniture fostered the development of industry in urban centers, while division of property among the sons favored development of local industries in homes or small workshops. In regions where landed property was passed on to a single heir, younger brothers either had to stay as celibates with the main household or, if opportunities were available elsewhere—as they have been since the beginning of industrialization—move to emerging industrial centers. In contrast, where land was divided among all the sons, each set up his own household on his separate piece of land, which frequently proved insufficient for subsistence. As a result, the landholder often had to look for work elsewhere to supplement his income. But rather than leave his land in search of better opportunities, he would try to find work in some local small enterprise (Habakkuk, 1955; Homans, 1941).

Similarly, in our own century, where landholdings in eastern Czechoslovakia were divided among all the sons, and small households were established, many men went to find work in other districts, in order to supplement their income from their small holdings; eventually, however, they returned home. In contrast, in the western part of the country, where land was passed on to a single heir, the younger sons found permanent employment in newly developing industries in the cities. Frequently these sons entered the universities, since their well-to-do fathers could afford to pay for their education, and from these parts of the country rather than from regions where land was fragmented the liberal professions could expect significant recruitment.

One of the most obvious effects of differences of family patterns are differences in fertility rates. They derive from the regulation of sexual life in the family in terms of values which may be those dominant in the society or those peculiar to subgroups. Values, however, are not static. They may undergo changes as a consequence of other changes in the society, or they may be ignored in actual behavior because individuals may be subjected to incompatible values and thus forced to make compromises. Individuals may derive some of their values from the past at the same time as they have some other beliefs in common with various groups that they identify with in the present—the religious, the occupational, the residential, for example, to mention only a few. They may hence be forced to select one set of values rather than another to guide some and not others of their activities. In contemporary America, for example, the difference between Catholics and non-Catholics in their approval and disapproval of birth control is striking indeed; yet, as Freedman *et al.* have pointed out, the similarities between the religious groups are more startling than the differences: compared to the large discrepancies in value concerning birth control, the difference in the actual birth rate, as well as the difference in what is estimated to be a desirable number of children, is quite small (Freedman, 1961). While ideals about desirable sexual behavior tend to be highly differentiated in religious groups, actual fertility tends toward a common norm.

Thus all aspects of family life—marital alliance, the family's role structure, its patterns of socialization—are functionally interrelated with the social order. Inasmuch as the family is society's basic institution, the sociological study of the family must be a study of society.

This must, indeed, be evident if it is understood that the family is the mediator between society and its new members. Even as it is the main agency for transmission of the cultural heritage, passing on the shared values and norms to the next generation, at the same time it cushions its members, especially the more vulnerable young, against the overly harsh impact of society's taboos and prescriptions, thereby helping to transform social constraint into social interdependence and communality. At the same time as it acts as an agency of social control, it provides the ground for individual growth and social reciprocity.

Furthermore, a brief review of the main functions of the family—the institutionalization of social fatherhood, the establishment through marriage of alliances outside of blood relations, the imposition of social norms on the biological organism, and the bestowing of social identity on its members—leads one to the conclusion that in all these tasks the family ensures the victory of the social over the biological. In this basic sense perhaps more than in any other the family can be said to represent the essence of social, that is human, life.

SELECTED REFERENCES

Davis, Kingsley. 1949. *Human Society.* New York: Macmillan.

Freedman, Ronald. 1961. Socio-economic factors in religious differentials in fertility. *American Sociological Review* XXVI:609–14.

Fromm, Erich. 1936. Sozialpsychologischer Teil. In *Autoritaet and Familie,* edited by M. Horkheimer. Paris: Librairie Alcan.

Goode, William J. 1963. *World Revolution and Family Patterns.* New York: Free Press of Glencoe.

Habakkuk, H. J. 1955. Family structure and economic change in nineteenth-century Europe. *Journal of Economic History,* XV:1–12.

Homans, George C. 1941. *English Villagers of the Thirteenth Century.* Cambridge: Harvard University Press.

Kapadia, K. M. 1955. *Marriage and Family in India.* Bombay: John Brown, Oxford University Press.

Komarovsky, Mirra. 1964. Functional Analysis of Sex Roles. In *The Family: Its Structure and Functions,* edited by R. L. Coser. New York: St. Martin's Press.

Levi-Strauss, Claude. 1960. The Family. In *Man, Culture and Society,* edited by H. L. Shapiro. New York: Oxford University Press.

Litwak, Eugene. 1960. Occupational mobility and extended family cohesion. *American Sociological Review* 25:9–21.

Mead, George Herbert. 1937. *Mind, Self and Society.* Chicago: University of Chicago Press.

Merton, Robert K. 1957a. *Social Theory and Social Structure.* Glencoe, IL: Free Press.

Merton, Robert K. 1957b. Socialization: A Terminological Note. In *The Student Physician,* edited by R. Merton, et al. Cambridge: Harvard University Press.

Middleton, Russell. 1964. Brother-Sister Marriage in Ancient Egypt. In *The Family: Its Structure and Functions,* edited by R. L. Coser. New York: St. Martin's Press.

Sussman, Marvin B. 1953. The help pattern in the middle-class family. *American Sociological Review* 18:22–28.

Thomas, William I., and Florian Znaniecki. 1958. *The Polish Peasant in Europe and America.* New York: Dover Publications.

2

The Standard North American Family as an Ideological Code

DOROTHY SMITH
1993

This article proposes that there are "ideological codes" that order and organize texts across discursive sites, concerting discourse focused on divergent topics and sites, often having divergent audiences, and variously hooked into policy or political practice. This ordering and organizing of texts is integral to the coordination and concerting of the complex of evolving discourses.

I am using the term *ideological code* as an analogy to *genetic code.* Genetic codes are orderings of the chemical constituents of DNA molecules that transmit genetic information to cells, reproducing in the cells the original ordering. By analogy, an ideological code is a schema that replicates its organization in multiple and various sites. I want to make clear that an ideological code in this sense is not a determinate concept or idea, although it can be expressed as such. Nor is it a formula or a definite form of words. Rather, it is a constant generator of *procedures for selecting syntax, categories, and vocabulary in the writing of texts and the production of talk and for interpreting sentences,* written or spoken, ordered by it. An ideological code can generate the same order in widely different settings of talk or writing—legislative, social scientific, popular writing, administrative, television advertising, and whatever.

The standard North American family (hereafter SNAF) is an ideological code in this sense. It is a conception of the family as a legally married couple sharing a household. The adult male is in paid employment; his earnings provide the economic basis of the family-household. The adult female may also earn an income, but her primary responsibility is to the care of husband, household, and children. Adult male and female may be parents (in whatever legal sense) of children also resident in the household. Note the language of typification—*man, woman,* and the use of the a temporal present. This universalizing of the schema locates its function as ideological code. It is not identifiable with any particular family; it applies to any. A classic enunciation of the code can be found in George Murdock's use of files containing summaries of ethnographic data accumulated by anthropologists from different parts of the world to establish the universality of the nuclear family (Murdock, 1949). The nuclear family is a theorized version of SNAF. Characteristically, Murdock was able to generate its distinctive form even when ethnographic descriptions contradicted it. Even when the nuclear

From "The Standard North American Family," by Dorothy Smith, from *Journal of Family Issues*, Vol. 14 (1), pp. 51-63.

family is not the "prevailing form," it is "the basic unit from which more complex familial forms are compounded." It is "always recognizable" (Murdock, 1949, p. 2).

Sociobiology deploys SNAF in the similar way:

> The building block of nearly all human societies is the nuclear family. The populace of an American industrial city, no less than a band of hunter-gatherers in the Australian desert, is organized around this unit. In both cases the family moves between regional communities, maintaining complex ties with primary kin by means of visits (or telephone calls and letters) and the exchange of gifts. During the day the women and children remain in the residential area while the men forage for game or its symbolic equivalent in the form of money. (E. O. Wilson, 1978, p. 553).

In this passage, SNAF-ordered terms, sentences, and sequences of sentences synthesize accounts of hunting and gathering forms of society with the contemporary United States. The ideological code generates a common ordering into which descriptive elements from very different societies can be inserted.

We can contrast SNAF-ordered discourse with other representations of family. Rayna Rapp (1978) wrote a sharp critique of it from a feminist standpoint. Carol Stack's (1974) ethnography of the kin relations in a Black neighborhood describes kin relations that cross household boundaries and connect people in ways that cannot be described in SNAF-generated descriptions. For example, women may have children with more than one father, all of whom maintain their relationship to their children; kin terms are extended to those connected in networks of reciprocal support and exchange; and "daddies" and "mommas" may be identified by the care they have provided a child and not by legal status. Jaber Gubrium and James Holstein (1990) have sought a total dissolution of SNAF, exploring the many and various uses of the notion of family in varying contexts and for diverse purposes, individual and collective—talk among people identifying themselves about themselves and their relationships with others (not exclusively "blood" kin), among people in health care or judicial settings, and more. Their book defies the reader to find a determinate unit corresponding to the concept of the family. Although such initiatives may in the long run prevail, they are at present marginal to the ubiquity of SNAF. Indeed the latter is often preserved in the identification of deviant instances, such as "female-headed families." SNAF-defined nonintact families appear to be female-headed families only. The single male-headed family does not appear to be defined as deviant by SNAF. It seems likely that this is because the SNAF code will not generate a male-headed family. It simply falls outside what SNAF "knows" how to "order."

I have stressed above the potential ubiquity of ideological codes. They operate to coordinate multiple sites of representation. Two examples of the operation of SNAF as an ideological code are examined here. Many could be found. They exhibit different ways in which the ideological code operates. The first examines the structuring effect of SNAF in organizing research done by Alison Griffith and myself on the work that mothers do in relation to their children's schooling; the second, its effect in William Julius Wilson's (1987) *The Truly Disadvantaged: The Inner City, the Underclass, and Public Policy.*

RESEARCHING WOMEN'S WORK AS MOTHERS

When Alison Griffith and I started research concerned with the work that mothers do in relation to their children's schooling, we began by talking about our experiences as single parents in relation to our children's schools. Indeed, over a period of 2 or 3 years before we decided to undertake the research, we had shared confidences, complaints, miseries, and guilt arising from this relation. Our more systematic discussion resulted in our formulating our problem as that of being, vis à vis the school, "defective families," families that somehow did not match up to the parental roles defined as proper by the professional ideology of the school system. We decided that to understand how our kind of family was defective, we needed to know more about how the "normative" or "intact" family, to use W. J. Wilson's (1987) term, worked in relation to the school. We wanted to learn more about the relation between the "normative family" and the school, and to focus on the work that women in such families were doing in relation to their children's schooling and perhaps also for the school.

We began as feminism recommends from our own experience and worked with some of the notions I have put forward about the everyday world as problematic (Smith, 1987). Hence the relevances of our study were not structured by sociological or educational theory. At the same time, as our work progressed, we became increasingly aware of the ordinary ways in which our thinking and our telling of experience was structured by text-mediated discourse. Telling experience "naturally" makes use of the speech genres (Bakhtin, 1986) of the setting in which experience is being told. Our interest in schools, our own experience, and our feminist concerns with women's unwaged work were embedded in text-mediated discourses. These discourses are SNAF infected through and through. So, therefore, was our research, as gradually became visible to us at a number of points—one being, of course, that our selection of families to interview was SNAF governed. We confounded the ideological code with the working realities of families in seeking a sample of intact families as if we could find the logic of the code in the socially organized relations of intact families to the school or treat them as manifestations of the normative form. This was to operate with the same ordering procedure as George Murdock (1949) used in his SNAF-governed extraction of nuclear families from the Yale Human Relations Area Files.

We learned painfully and sometimes too late to rectify in our "data collection" that our thinking and research plans were deeply structured by what we have come to call *the mothering discourse.* Mothering discourse in North America developed historically during the first two decades of the 20th century. It was and is actively fed by research and thinking produced by psychologists and specialists in child development and is popularly disseminated in women's magazines, television programs, and other popular media. An important aspect of it is directed toward "managing" women's relation to their children's schooling, enlisting their work and thought in support of the public educational system. Women, particularly middle-class women, were much involved in the early development of this discourse. As housewives and mothers, they continued to participate in the school-mother T-discourse, reading women's magazines and books on child development and child rearing, and deriving guidance, ideas, standards, and schemata for interpreting their children, the relationship between what they may be doing and the child's success in school, and interpreting themselves (Griffith & Smith, 1987). Although what Alison and I call mothering discourse is not reducible to SNAF, it is through and through SNAF ordered, and indeed may have been, in the course of its historical development, among the carriers that generalized SNAF throughout English-speaking North America.

The extent of SNAF invasion of our study is disturbing, given that Alison's thesis had studied the ideology of the single-parent family in the educational context (Griffith, 1984). She had explored the creation in child development and social/psychological discourse of effects of the single-parent (female-headed) families on children, showing how they provide for teachers' procedures for reading back from what can be observed in the classroom to the defective family. She had also shown how the concept operates in educational administration, in the professional formulations of teachers, and in the news media. Yet, when we embarked on our study, we failed to register the extent to which our thinking and research design were organized by the mothering discourse and by conceptions of SNAF. This is one of the reasons for introducing the notion of an ideological code. What I am calling ideological codes are not necessarily committed to a determinate set of categories or conceptual system, or to a determinate content. The code reproduces its *organization* in the discursive texts (including, of course, our interviews) without necessarily being manifested in the use of specific terms. For example, we never thought of, may even have consciously avoided, concepts such as the nuclear family. The discourse on mothering that has flourished in North America since the 1920s is SNAF ordered; so also is the complementary discourse on the family of professional educators; sociological research in education such as James Coleman's (Coleman et al., 1966), is also through and through SNAF structured; and so was our research.

The school-mother discourse lays on the family the primary responsibility for the individual child's school achievement and even his or her success as an adult. SNAF enables interpretation in practical settings to translate *family* into *the mother.* The intact family means that the child's mother is available to do the work for the school that is done invisibly in the home. As Griffith (1984) showed for the ideology of the single-parent family, school-mother discourse provides for schools and teachers a "documentary" method of reading back from a child's behavior in school to its cause in the family, and from a knowledge of family problems to an interpretation of the child's behavior in school. The documentary method of interpretation (Mannheim, 1971) is a circular process. What we see and hear is interpreted in relation to an underlying pattern or schema; the underlying pattern or schema selects and orders the way we attend to things and hence what we see and hear. The school-mother discourse worked both ways. Both Alison and I (and presumably other mothers) and the teachers, counselors, and principals of our children's schools knew how to operate this interpretive circle. We were viewed at school as defective families; defective families produce defective children; any problem our children might have at school indexed the defective family as its underlying interpreter; we were always guilty. SNAF coordinated our relation to school, the school's relation to us, and the school's relation to our children. And children learn SNAF from children's readers. As my small son said one day, arriving home from school, "There's something awfully wrong with our family."

In retrospect, the SNAF structuring of our own experience in being single parents for our children's schools was a powerful effect in our research. Our interest in how SNAF-conforming families worked with the school originated in our experience of being, vis à vis the school our children attended, defective families. This is a classic SNAF procedure of reading. Here of course we return to where Alison and I started our research, now recognizing what we did not recognize in designing our interview topics: namely, the operation of the school-mother discourse in the interview design. Topics such as the typicality of the school day, how mothers kept track of their children's homework, what part they played in school activities, and so forth—these were generated by our ordinary competence in the school mother discourse. Our own participation in this discourse and how it had organized our inventory of interview topics became visible to us rather late in the game and accidentally (Griffith & Smith, 1987).

SNAF was perhaps most obviously at work when our interviews addressed women's employment. We and most of the women we talked to took for granted, at least for the purposes of the interview, that employment outside the home was not "standard" for women with small children. Respondents who did work outside the home were careful to describe how they provided for child care or the exceptional economic situation that required them to do this. The interviewers (there were three, Alison, myself, and a graduate student) were sometimes very tentative in asking the woman about her employment, taking care, as in the following instance, to normalize working outside the home:

> Interviewer: O.K. so the first thing that we want to start talking about is . . . because

> what we are finding is that some mothers work outside the home part-time, some full-time and some don't work outside the home, and so we can just sort of start with a background and then move on. So have you ever worked outside the home?

SNAF is very much at work here. An account of women working outside the home that normalizes it in terms of what other women do, recognizes that paid employment for women with small children *is* a deviation from the normative intact family. The introduction to the question, redefining employment outside the home as normal, responds implicitly to the SNAF ordering that defines it as deviant. SNAF was at work even here.

The SNAF effect was not confined to the design of our study. Many, if not most, of the women we interviewed also operated with the interpretive schema of the school-mother discourse. They intended the very interpretive schemata that we used in writing our questionnaire and in the sense their answers made to us. Our questions and our responses to their answers would demonstrate to them our interpretive competence in the (documentary) method they themselves used and we would find in their answers the sense that could be made as we found them as indices of it. They were practitioners in the discourse of mothering; they were competent. They knew how to address and interpret their lives using the schemata of the school-mother discourse. Our use of open-ended interviews focused on their work practices means that we have accounts of their work that are not wholly constrained or reduced by SNAF. And, of course, we were committed to explicating the actual ways in which mothers' work vis à vis the way the school is organized so that our presuppositions were subject to challenge from the actuality. Nonetheless, it has meant that our analysis of our interview data (both questions and responses) has had to take up how they are ordered by the school-mother discourse and the SNAF code operating within it.

SNAF AS A READING OF THE BLACK FAMILY

In his study of the poverty and problems of Black people in the inner cities of the United States, W. J. Wilson (1987) distinguishes between intact families and those that are not intact. SNAF-governed descriptions of families represent families lacking an adult male head as deviant from the normative intact family with a male head. Wilson's nonintact families are headed by females.

Wilson argues, in opposition to earlier sociological work on the Black family, that the nonintact family supposed to characterize the Black population in the United States is not the result of slavery, or a survival of a traditional Black family form, or an effect of welfare policies. Rather, it is the result of poverty caused by Black male unemployment. "In the early twentieth century, the vast majority of both Black and White low-income families were intact" (W. J. Wilson, 1987, p. 90). The disproportionate concentration of female-headed, nonintact, families in the Black population is an effect of the relatively greater difficulty that Black men have had in finding employment. "Given their disproportionate concentration among the poor in America, black families were more strongly affected by these conditions [poverty, high rates of mortality among black men, need of men to travel in search of jobs] and therefore were more likely than white families to be female-headed" (W. J. Wilson, 1987, p. 64). The "good" intact family is damaged by "bad" conditions. The SNAF code is the key textual structuring device. The movement from a majority of intact families to a majority of female-headed families in the Black population in the United States is interpreted as a movement from a normal to a deviant or damaged form. A theory of male dominance in family and economy is imported into the analysis without being enunciated. SNAF does its quiet work behind the scenes, generating concepts such as the intact family with the female-headed family as its deviant form.

The code operates to suppress alternative structuring devices that might, for example, treat men and women in parallel ways, say, in the context of discussions of unemployment and poverty. When SNAF rules, its constitutive gender differentiation is operative in the text. Wilson's section on the 20th-century emergence of increasing proportions of defective (female-headed) families in the Black population addresses women exclusively. The SNAF code, operating as a set of conventions selecting vocabulary, as well as relations posited syntactically, selects women as protagonists of the decline of the intact and the increase in the nonintact, female-headed family. Men who separate from their wives disappear from view. We do not know, as Stack's (1974) data might suggest, whether they continue to have an ongoing relationship with their children. Even when it would be equally appropriate to insert men into the sentences, women are preferred. "Whereas white women are far more likely to be divorced than separated, black women are more likely to be separated than divorced. Indeed, a startling 22 percent of all married black women are separated from their husbands" (W. J. Wilson, 1987, p. 68). The sentence could just as well have been written thus: "Whereas White men are far more likely to be divorced than separated, Black men are more likely to be separated than divorced," and so on. SNAF ordering procedures govern the selection of women rather than men as the protagonists. By contrast, the section on the contribution made by unemployment to the erosion of the intact family among Black people addresses men almost exclusively. The contribution that Black women's problems of unemployment and low-paid work makes to Black poverty is not a focus.

In consolidating the SNAF-ordered picture of "the" Black family that Wilson draws, he relies on statistical data produced by agencies such as the U.S. Bureau of Census and the National Center for Health Statistics. Although these data do not replicate the normative substance of SNAF, they preserve its order in skeletal form. Mary Romero (1992) did an ethnographic study evaluating census data of undocumented immigrants in San Francisco. She compared census data already collected on people living in a census block of some 105 units in an ethnically mixed, although predominantly Hispanic neighborhood, with what she learned in a house-to-house study. Census categories and procedures significantly misrepresented the actual household composition of people living in these units. The typical distortions she describes can be understood as resulting from the application of SNAF-ordered categories, forms, and procedures applied to situations that were not SNAF ordered.

The historical effects of SNAF are visible in the requirement that the skeletal equivalent of a head of household be designated. At one time, a head of household would be designated; preference was given to men over women. Currently, one individual is designated, identified as that person in whose name a unit is owned or rented. The status of other household members must be defined relative to this person. He or she may not, in fact, be particularly central to the household organization. Because members of the household have to be identified in terms of their relationship to the person accorded this position, extended families sharing residence may be misrepresented. The only category available on the census form to which family members who are neither spouse, child, parent, or sibling of the designated head of household is that of "boarder."

Romero found that in some instances, several families shared one unit. Yet only one individual can be accorded the central position, and all others must be related to her or him. Further, the numbers of categories available on the form may not be enough to record all those in residence. And because multifamily residences were in breach of municipal housing regulations (also SNAF governed), landlords, fearing that the census data might somehow leak out to the municipal housing authorities, might intervene in filling out census forms to ensure that each unit was represented as inhabited by only one family. Such SNAF-designed data collection devices generate a representation of the population that reproduces

SNAF. I do not mean by this that nonintact families, for example, are represented as intact. Rather I mean that the data is ordered according to SNAF. Where data that can be ordered according to SNAF, it will be so ordered; where it can not, SNAF-ordering generates deviant cases. It is obvious, for example, that the kind of kin relations described by Carol Stack (1974) would wholly escape the census categories.

The empirical ground that W. J. Wilson (1987) relies on is not ethnographic. It is largely census data or data from other government agencies. The studies that he draws on and cites are similarly grounded. The empirical grounding of his argument is thus already SNAF ordered. Consider, by contrast, a radical alternative to his SNAF-governed narrative here put forward by a feminist anthropologist, Henrietta Moore. Female-headed families are not, she suggests, to be seen as the result of poverty (and hence by implication, as damaged in some way). Drawing on her African experience, she writes, "Some women are choosing not to marry . . . and . . . significant numbers of married women are choosing to live separately from their husbands" (Moore, 1988, p. 64). Such choices may also be made by women in North America. Of course there may be a relation between poverty and such choices. Some of the patterns described by Carol Stack (1974) suggest a good practical economy in women's choices to have several children by different men; apparently the child's father has an ongoing relationship with his child. Hence a household consisting of a woman with several children broadens its range of kin contacts and consequently its range of potential economic support. Patterns of familial and kin relations that maximize connections among households and among kin make good sense in such conditions. Women's ability to provide for their children and themselves autonomously may be enhanced by avoiding dependence on a particular man. It is only when such patterns of domestic economy are represented in SNAF-ordered terms that they (rather than poverty) appear as a defect. The very language that Wilson uses betrays the normative effect of SNAF.

A SNAF-ordered representation of the patterns described by Stack (1974) might well appear in text as an intergenerational reproduction of defective forms of family. W. J. Wilson (1987) draws on such studies to sustain his argument. For example,

> Hogan and Kitagawa's analysis of the influences of family background personal characteristics, and social milieu on the probability of premarital pregnancy among black teenagers in Chicago indicates that those from nonintact families, lower social class, and poor and highly segregated neighborhoods have significantly higher fertility rates. Hogan and Kitagawa estimated that 57 percent of the teenage girls from high-risk social environments (lower class, poor inner-city neighborhood residence, *female-headed family, five or more siblings, a sister who is a teenage mother* [itals added], and loose parental supervision of dating) will become pregnant by age eighteen compared to only 9 percent of the girls from low-risk social backgrounds. (p. 75)

Wilson writes of female-headed families and out-of-wedlock births as forms of dislocation (W. J. Wilson, 1987, p. 72). Both of these are SNAF-generated deviant forms. Conceptions of low- and high-risk in the passage above follow this conceptual track. A high-risk social environment, that is, a social environment in which a young woman is at high risk of bearing a child out-of-wedlock is one that includes a female-headed family. But, in the end, we might find social organization maximizing the range of reciprocal economic and other benefits under tough conditions, rather than dislocation. My emphases in the above passage locate items that could have originated in a non-SNAF-ordered social organization, perhaps more like that described by Stack (1974). Note, too, that SNAF equates family and household; so do W. J. Wilson, and Hogan and Kitagawa. Hence interhousehold connections are invisible.

We can begin to see how the operation of SNAF in distinct discursive and administrative sites provides for the in-text "reality" of W. J. Wilson's (1987) account. Wilson's SNAF-governed theorizing and that of studies he draws on rely for the most part on SNAF-generated government statistics and are validated by analyses for which they are a resource. Although they are independent, the ordering procedure generating both is the same. The government statistics have not been created for Wilson or dovetailed to his study or the studies he cites. They establish a common empirical ground for the substantial range of studies on which Wilson can call. The operation of SNAF as an ideological code creates a conceptual isomorphy between the data base, the studies Wilson cites, and Wilson's own argument. The empirical "solidity" of Wilson's text depends on this. No wonder, then, that Wilson is able to conclude (for this section of his study) that "[a]vailable evidence supports the argument that among blacks, increasing male joblessness is related to the rising proportions of families headed by women" (W. J. Wilson, 1987, p. 83). Policy implications follow: "The available evidence justifies renewed scholarly and public policy attention to the connection between the disintegration of poor families and black male prospects for stable employment" (W. J. Wilson, 1987, p. 90).

DISCUSSION

. . . The coordination of SNAF-governed texts with SNAF-governed representations of the society through statistics collected by and vested in state agencies provides for the in-text production of simulated realities such as Wilson's. I stress here the discursive concerting of independent sites. Ideological codes do more than order discursive productions. They also provide for the kind of interchange relying on conceptual isomorphy. Where the ordering procedures generating terms and sentences are the same, different content can be lined up as Edward Wilson (1978) lined up (his version of) the family organization of hunting and gathering societies with (his version of) the family organization of contemporary America in the sociobiological quotation above. In William Julius Wilson's (1987) study his SNAF-ordered thesis about Black families is directly coordinated with the SNAF-generated data of the U.S. Bureau of Census. The statistics realize and express the thesis; the thesis interprets the statistics.

Such observations of ideological codes rely both on analytic strategies such as those I have used above and on experiences—of doing research or of reading and writing sociology—that analysis explicates. They rely not only on attention to texts as the inert bearers of statements but also on experience of and in courses of action in discourse, reading texts, such as William Julius Wilson's (1987), going back and forth between text and research practice, as Alison and I did in our study of mother's work in relation to their children's schooling. I have not stressed the experience of reading Wilson's text in which the analysis above was grounded, but it was a distinctive experience of reading in which I found myself resisting as a feminist the argument put forward and yet not finding in the text the ground for such reservation. This difficulty was, of course, compounded by my being White and hence feeling that I had no right to contend with a Black man's authoritative account of Black people's lives. When I came to see that Wilson's simulated reality was organized by the same schema that invaded Alison's and my research, I could begin to isolate the operation of the same ideological code, SNAF, in both sites.

Chatting with Mary Romero one day while hiking up a mountain in Oregon contributed observations of the operation of the code in the routine production of government statistical data. SNAF-infected texts are all around us. They give discursive body and substance to a version of the family that masks the actualities of people's lives or at best inserts an implicit evaluation into accounts of ways of living together in households or forming economically and emotionally supportive relationships outside that do not accord with SNAF. The conceptual containment of alternatives may be very important politically. I suspect that speaking and writing may often be governed by ideological codes even when we dispute their specifics. Until I talked to Mary, I had not realized how a skeletal SNAF still governed the census data format. SNAF may also govern powerfully the formulation of welfare policies. Thus ideological codes may have a peculiar and important political force, carrying forward modes of representing the world even among those who overtly resist the representations they generate. Like Alison and me, we may, without noticing, be trapped by our own competencies as readers and writers of social science, as participants in text-mediated discourse, into regenerating a discursive politics to which we are opposed.

SELECTED REFERENCES

Bakhtin, M. M. 1986. The Problem of Speech Genres. In *Speech Genres and Other Late Essays,* edited by M. M. Bakhtin, V. W. McGee. Austin: University of Texas Press.

Coleman, J. S., with E. Q. Campbell, C. J. Hobson, J. Mc-Partland, A. M. Mood, F. D. Weinfeld, and R. L. York. 1966. *Equality of Educational Opportunity.* Washington, DC: Department of Health, Education and Welfare.

Griffith, A. (1984). Ideology, education and single parent families: The Normative Ordering of Families Through Schooling. Ph.D. diss., University of Toronto.

Griffith, A. I., and D. E. Smith. 1987. Constructing Cultural Knowledge: Mothering as Discourse. In *Women and Education: A Canadian Perspective,* edited by J. Gaskell and A. McLaren. Calgary: Detselig.

Gubrium, J. R., and J. A. Holstein. 1990. *What Is Family?* Mountain View, CA: Mayfield.

Mannheim, K. 1971. On the Interpretation of *Weltanschauung.* In *From Karl Mannheim,* edited by K. Wolff. New York: Oxford University Press.

Moore, H. L. 1988. *Feminism and Anthropology.* Minneapolis: University of Minnesota Press.

Murdock, G. P. *Social Structure.* New York: Macmillan.

Rapp, R. 1978. Family and class in contemporary America: Notes towards an understanding of ideology. *Science and Society* 42:278–300.

Romero, M. 1992. *Ethnographical Evaluation of Behavioral Causes of Census Undercounts of Undocumented Immigrants and Salvadorans in the Mission District of San Francisco, CA* (Report submitted to the Center for Survey Methods Research). Washington, DC: U.S. Bureau of the Census.

Smith, D. E. 1987. *The Everyday World as Problematic: A Feminist Sociology.* Boston: Northeastern University Press.

Stack, C. B. 1974. *All Our Kin.* New York: Harper & Row.

Wilson, E. O. 1978. *On Human Nature.* Cambridge, MA: Harvard University Press.

Wilson, W. J. 1987. *The Truly Disadvantaged: The Inner City, the Underclass, and Public Policy.* Chicago: University of Chicago Press.

3

The Simpsons

Atomistic Politics and the Nuclear Family

PAUL CANTOR
1999

When Senator Charles Schumer (D-N.Y.) visited a high school in upstate New York in May 1999, he received an unexpected civics lesson from an unexpected source. Speaking on the timely subject of school violence, Senator Schumer praised the Brady Bill, which he helped sponsor, for its role in preventing crime. Rising to question the effectiveness of this effort at gun control, a student named Kevin Davis cited an example no doubt familiar to his classmates but unknown to the senator from New York:

> It reminds me of a *Simpsons* episode. Homer wanted to get a gun but he had been in jail twice and in a mental institution. They label him as "potentially dangerous." So Homer asks what that means and the gun dealer says: "It just means you need an extra week before you can get the gun." (Henry, 1999).

Without going into the pros and cons of gun control legislation, one can recognize in this incident how the Fox Network's cartoon series *The Simpsons* shapes the way Americans think, particularly the younger generation. It may therefore be worthwhile taking a look at the television program to see what sort of political lessons it is teaching. *The Simpsons* may seem like mindless entertainment to many, but in fact, it offers some of the most sophisticated comedy and satire ever to appear on American television. Over the years, the show has taken on many serious issues: nuclear power safety, environmentalism, immigration, gay rights, women in the military, and so on. Paradoxically, it is the farcical nature of the show that allows it to be serious in ways that many other television shows are not.

. . . Setting aside the surface issue of political partisanship, I am interested in the deep politics of *The Simpsons,* what the show most fundamentally suggests about political life in the United States. The show broaches the question of politics through the question of the family, and this in itself is a political statement. By dealing centrally with the family, *The Simpsons* takes up real human issues everybody can recognize and thus ends up in many respects less "cartoonish" than other television programs. Its cartoon characters are more human, more fully rounded, than the supposedly real human beings in many situation comedies. Above all, the show has created a believable human community: Springfield, USA. *The Simpsons* shows the family as part of a larger community and in effect affirms the kind of community that can sustain the family. That is at one and the same time the secret of the show's popularity with the American public and the most interesting political statement it has to make.

The Simpsons indeed offers one of the most important images of the family in contemporary American culture and, in particular, an image of the nuclear family. With the names taken from creator Matt Groening's own childhood home, *The Simpsons* portrays the average American family: father (Homer), mother (Marge), and 2.2 children (Bart, Lisa, and little Maggie). Many commentators have lamented the fact that *The Simpsons* now serves as one of the representative images of American family life, claiming that the show provides horrible role models for parents and children. The popularity of the show is often cited as evidence of the decline of family values in the United States. But critics of *The Simpsons* need to take a closer look at the show and view it in the context of television history. For all its slapstick nature and its mocking of certain aspects of family life, *The Simpsons* has an affirmative side and ends up celebrating the nuclear family as an institution. For television, this is no minor achievement. For decades, American television has tended to downplay the importance of the nuclear family and offer various one-parent families or other nontraditional arrangements as alternatives to it. The one-parent situation comedy actually dates back almost to the beginning of network television, at least as early as *My Little Margie* (1952–1955). But the classic one-parent situation comedies, like *The Andy Griffith Show* (1960–1968) or *My Three Sons* (1960–1972), generally found ways to reconstitute the nuclear family in one form or another (often through the presence of an aunt or uncle) and thus still presented it as the norm (sometimes the story line actually moved in the direction of the widower getting remarried, as happened to Steve Douglas, the Fred MacMurray character, in *My Three Sons*).

But starting with shows in the 1970s like *Alice* (1976–1985), American television genuinely began to move away from the nuclear family as the norm and suggest that other patterns of child rearing might be equally valid or perhaps even superior. Television in the 1980s and 1990s experimented with all sorts of permutations on the theme of the nonnuclear family, in shows such as *Love, Sidney* (1981–83), *Punky Brewster* (1984–1986), and *My Two Dads* (1987–1990). This development partly resulted from the standard Hollywood procedure of generating new series by simply varying successful formulas. But the trend toward nonnuclear families also expressed the ideological bent of Hollywood and its impulse to call traditional family values into question. Above all, though television shows usually traced the absence of one or more parents to deaths in the family, the trend away from the nuclear family obviously reflected the reality of divorce in American life (and especially in Hollywood). Wanting to be progressive, television producers set out to endorse contemporary social trends away from the stable, traditional, nuclear family. With the typical momentum of the entertainment industry, Hollywood eventually took this development to its logical conclusion: the no-parent family. Another popular Fox program, *Party of Five* (1994–), . . . shows a family of children gallantly raising themselves after both their parents were killed in an automobile accident.

Party of Five cleverly conveys a message some television producers evidently think their contemporary audience wants to hear—that children can do quite well without one parent and preferably without both. The children in the audience want to hear this message because it flatters their sense of independence. The parents want to hear this message because it soothes their sense of guilt, either about abandoning their children completely (as sometimes happens in cases of divorce) or just not devoting enough "quality time" to them. Absent or negligent parents can console themselves with the thought that their children really are better off without them, "just like those cool—and incredibly good-looking—kids on *Party of Five*." In short, for roughly the past two decades, much of American television has been suggesting that the breakdown of the American family does not constitute a social crisis or even a serious problem. In fact, it should be regarded as a form of liberation from an image of the family that may have been good enough for the 1950s but is no longer valid in the 1990s. It is against this historical background that the statement *The Simpsons* has to make about the nuclear family has to be appreciated.

Of course television never completely abandoned the nuclear family, even in the 1980s, as shown by the success of such shows as *All in the Family* (1971–1983), *Family Ties* (1982–1989), and *The Cosby Show* (1984–1992). And when *The Simpsons* debuted as a regular series in 1989, it was by no means unique in its reaffirmation of the value of the nuclear family. Several other shows took the same path in the past decade, reflecting larger social and political trends in society, in particular the reassertion of family values that has by now been adopted as a program by both political parties in the United States. Fox's own *Married with Children* (1987–1998) preceded *The Simpsons* in portraying an amusingly dysfunctional nuclear family. Another interesting portrayal of the nuclear family can be found in ABC's *Home Improvement* (1991–1999), which tries to recuperate traditional family values and even gender roles within a postmodern television context. But *The Simpsons* is in many respects the most interesting example of this return to the nuclear family. Though it strikes many people as trying to subvert the American family or to undermine its authority, in fact, it reminds us that antiauthoritarianism is itself an American tradition and that family authority has always been problematic in democratic America. What makes *The Simpsons* so interesting is the way it combines traditionalism with antitraditionalism. It continually makes fun of the traditional American family. But it continually offers an enduring image of the nuclear family in the very act of satirizing it. Many of the traditional values of the American family survive this satire, above all the value of the nuclear family itself.

As I have suggested, one can understand this point partly in terms of television history. *The Simpsons* is a hip, postmodern, self-aware show. But its self-awareness focuses on the traditional representation of the American family on television. It therefore presents the paradox of an untraditional show that is deeply rooted in television tradition. *The Simpsons* can be traced back to earlier television cartoons that dealt with families, such as *The Flintstones* or *The Jetsons.* But these cartoons must themselves be traced back to the famous nuclear-family sitcoms of the 1950s: *I Love Lucy, The Adventures of Ozzie and Harriet, Father Knows Best,* and *Leave it to Beaver. The Simpsons* is a postmodern re-creation of the first generation of family sitcoms on television. Looking back on those shows, we easily see the transformations and discontinuities *The Simpsons* has brought about. In *The Simpsons,* father emphatically does not know best. And it clearly is more dangerous to leave it to Bart than to Beaver. Obviously, *The Simpsons* does not offer a simple return to the family shows of the 1950s. But even in the act of re-creation and transformation, the show provides elements of continuity that make *The Simpsons* more traditional than may at first appear.

The Simpsons has indeed found its own odd way to defend the nuclear family. In effect, the show says, "Take the worst-case scenario—the Simpsons—and even that family is better than no family." In fact, the Simpson family is not all that bad. Some people are appalled at the idea of young boys imitating Bart, in particular his disrespect for authority and especially for his teachers. These critics of *The Simpsons* forget that Bart's rebelliousness conforms to a venerable American archetype and that this country was founded on disrespect for authority and an act of rebellion. Bart is an American icon, an updated version of Tom Sawyer and Huck Finn rolled into one. For all his troublemaking—precisely because of his troublemaking—Bart behaves just the way a young boy is supposed to in American mythology, from *Dennis the Menace* comics to the *Our Gang* comedies.

As for the mother and daughter in *The Simpsons,* Marge and Lisa are not bad role models at all. Marge Simpson is very much the devoted mother and housekeeper; she also often displays a feminist streak, particularly in the episode in which she goes off on a jaunt à la *Thelma and Louise.* Indeed, she is very modern in her attempts to combine certain feminist impulses with the traditional role of a mother. Lisa is in many ways the ideal child in contemporary terms. She is an overachiever in school, and as a feminist, a vegetarian, and an environmentalist, she is politically correct across the spectrum.

The real issue, then, is Homer. Many people have criticized *The Simpsons* for its portrayal of the father as dumb, uneducated, weak in character, and morally unprincipled. Homer is all those things, but at least he is there. He fulfills the bare minimum of a father: he is present for his wife and above all his children. To be sure, he lacks many of the qualities we would like to see in the ideal father. He is selfish, often putting his own interest above that of his family. As we learn in one of the Halloween episodes, Homer would sell his soul to the devil for a donut (though fortunately it turns out that Marge already owned his soul and therefore it was not Homer's to sell). Homer is undeniably crass, vulgar, and incapable of appreciating the finer things in life. He has a hard time sharing interests with Lisa, except when she develops a remarkable knack for predicting the outcome of pro football games and allows her father to become a big winner in the betting pool at Moe's Tavern. Moreover, Homer gets angry easily and takes his anger out on his children, as his many attempts to strangle Bart attest.

In all these respects, Homer fails as a father. But upon reflection, it is surprising to realize how many decent qualities he has. First and foremost, he is attached to his own—he loves his family because it is his. His motto basically is, "My family, right or wrong." This is hardly a philosophic position, but it may well provide the bedrock of the family as an institution, which is why Plato's *Republic* must subvert the power of the family. Homer Simpson is the opposite of a philosopher-king; he is devoted not to what is best but to what is his own. That position has its problems, but it does help explain how the seemingly dysfunctional Simpson family manages to function.

For example, Homer is willing to work to support his family, even in the dangerous job of nuclear power plant safety supervisor, a job made all the more dangerous by the fact that he is the one doing it. In the episode in which Lisa comes to want a pony desperately, Homer even takes a second job working for Apu Nahasapeemapetilon at the Kwik-E-Mart to earn the money for the pony's upkeep and nearly kills himself in the process. In such actions, Homer manifests his genuine concern for his family, and as he repeatedly proves, he will defend them if necessary, sometimes at great personal risk. Often, Homer is not effective in such actions, but that makes his devotion to his family in some ways all the more touching. Homer is the distillation of pure fatherhood. Take away all the qualities that make for a genuinely good father—wisdom, compassion, even temper, selflessness—and what you have left is Homer Simpson with his pure, mindless, dogged devotion to his family. That is why for all his stupidity, bigotry, and self-centered quality, we cannot hate Homer. He continually fails at being a good father, but he never gives up trying, and in some basic and important sense that makes him a good father.

The most effective defense of the family in the series comes in the episode in which the Simpsons are actually broken up as a unit. This episode pointedly begins with an image of Marge as a good mother, preparing breakfast and school lunches simultaneously for her children. She even gives Bart and Lisa careful instructions about their sandwiches: "Keep the lettuce separate until 11:30." But after this promising parental beginning, a series of mishaps occurs. Homer and Marge go off to the Mingled Waters Health Spa for a well-deserved afternoon of relaxation. In their haste, they leave their house dirty, especially a pile of unwashed dishes in the kitchen sink. Meanwhile, things are unfortunately not going well for the children at school. Bart has accidentally picked up lice from the monkey of his best friend Millhouse, prompting Principal Skinner to ask, "What kind of parents would permit such a lapse in scalpal hygiene?" The evidence against the Simpson parents mounts when Skinner sends for Bart's sister. With her prescription shoes stolen by her classmates and her feet accordingly covered with mud, Lisa looks like some street urchin straight out of Dickens.

Faced with all this evidence of parental neglect, the horrified principal alerts the Child Welfare Board, who are themselves shocked when they take Bart and Lisa home and explore the premises. The officials completely misinterpret the situation. Confronted by a pile of old newspapers, they assume that Marge is a bad housekeeper, when in fact she had assembled the documents to help Lisa with a history project. Jumping to conclusions, the bureaucrats decide that Marge and Homer are unfit parents and lodge specific charges that the Simpson household is a "squalid hellhole and the toilet paper is hung in improper overhand fashion." The authorities determine that the Simpson children must be given to foster parents. Bart, Lisa, and Maggie are accordingly handed over to the family next door, presided over by the patriarchal Ned Flanders. Throughout the series, the Flanders family serves as the doppelgänger of the Simpsons. Flanders and his brood are in fact the perfect family according to old-style morality and religion. In marked contrast to Bart, the Flanders boys, Rod and Todd, are well behaved and obedient. Above all, the Flanders family is pious, devoted to activities like Bible reading, and more zealous than even the local Reverend Lovejoy. When Ned offers to play "bombardment" with Bart and Lisa, what he has in mind is bombardment with questions about the Bible. The Flanders family is shocked to learn that their neighbors do not know of the serpent of Rehoboam, not to mention the Well of Zahassadar or the bridal feast of Beth Chadruharazzeb.

Exploring the question of whether the Simpson family really is dysfunctional, the foster parent episode offers two alternatives to it: on one hand, the old-style moral/religious family; on the other, the therapeutic state, what is often now called the nanny state. Who is best able to raise the Simpson children? The civil authorities intervene, claiming that Homer and Marge are unfit as parents. They must be reeducated and are sent off to a "family skills class" based on the premise that experts know better how to raise children. Child rearing is a matter of a certain kind of expertise, which can be taught. This is the modern answer: the family is inadequate as an institution and hence the state must intervene to make it function. At the same time, the episode offers the old-style moral/religious answer: what children need is God-fearing parents in order to make them God-fearing themselves. Indeed, Ned Flanders does everything he can to get Bart and Lisa to reform and behave with the piety of his own children.

But the answer the show offers is that the Simpson children are better off with their real parents—not because they are more intelligent or learned in child rearing, and not because they are superior in morality or piety, but simply because Homer and Marge are the people most genuinely attached to Bart, Lisa, and Maggie, since the children are their own offspring. The episode works particularly well to show the horror of the supposedly omniscient and omnicompetent state intruding in every aspect of family life. When Homer desperately tries to call up Bart and Lisa, he hears the official message: "The number you have dialed can no longer be reached from this phone, you negligent monster."

At the same time, we see the defects of the old-style religion. The Flanders may be righteous as parents but they are also self-righteous. Mrs. Flanders says, "I don't judge Homer and Marge; that's for a vengeful God to do." Ned's piety is so extreme that he eventually exasperates even Reverend Lovejoy, who at one point asks him, "Have you thought of one of the other major religions? They're all pretty much the same."

In the end, Bart, Lisa, and Maggie are joyously reunited with Homer and Marge. Despite charges of being dysfunctional, the Simpson family functions quite well because the children are attached to their parents and the parents are attached to their children. The premise of those who tried to take the Simpson children away is that there is a principle external to the family by which it can be judged dysfunctional, whether the principle of contemporary child-rearing theories or that of the old-style religion. The foster parent episode suggests the contrary—that the family contains its own principle of legitimacy. The family knows best. This episode thus illustrates the strange combination of traditionalism and antitraditionalism in *The Simpsons.* Even as the show rejects the idea of a simple return to the traditional moral/religious idea of the family, it refuses to accept contemporary statist attempts to subvert the family completely and reasserts the enduring value of the family as an institution.

. . . The treatment of the family in *The Simpsons* links up with its treatment of politics. Although the show focuses on the nuclear family, it relates the family to larger institutions in American life, like the church, the school, and even political institutions themselves, like city government. In all these cases, *The Simpsons* satirizes these institutions, making them look laughable and often even hollow. But at the same time, the show acknowledges their importance and especially their importance for the family. Over the past few decades, television has increasingly tended to isolate the family—to show it largely removed from any larger institutional framework or context. This is another trend to which *The Simpsons* runs counter, partly as a result of its being a postmodern re-creation of 1950s sitcoms. Shows like *Father Knows Best* or *Leave it to Beaver* tended to be set in small-town America, with all the intricate web of institutions into which family life was woven. In re-creating this world, even while mocking it, *The Simpsons* cannot help re-creating its ambience and even at times its ethos.

SELECTED REFERENCES

Henry, Ed. 1999. Heard on the hill. *Roll Call* 13 May, 44, no. 81.

Richmond, Ray, and Antonia Coffman, eds. 1997. *The Simpsons: A Complete Guide to Our Favorite Family.* New York: HarperCollins.

CHAPTER 2

Sex and Romance

The popular media are perennially preoccupied with sex and romance. Whether the medium is television, movies, radio, magazines, or music videos, we are bombarded with erotic images, romantic plots, and love tales with both happy and sad endings. The selections in this chapter focus on how our preoccupation with sex and romance is socially constructed and the product of a unique set of historical and cultural factors. Even though popular stories about love and courtship seem timeless and universal, a sociological analysis shows how specific social forces shape them. And even though most of us want to believe that our romantic ideals and partners are individually chosen, these readings show how love and sex can be analyzed using sociological concepts such as sexual property, interaction ritual, and gendered entitlement.

Because social scientists typically analyze romance and **courtship** as part of larger kinship systems and mate selection processes, they often invoke market metaphors and draw on theoretical understandings about sexual property. As noted in the next chapter, in virtually all societies, marriage establishes the right to sexual intercourse between spouses. To call this arrangement "property" sounds odd, because we are talking about someone else's body rather than about some inanimate object, but there are good reasons in sociological theory for extending the concept of property to erotic rights in marriage.

The opening article in this chapter is from Kingsley Davis, the sociologist who first coined the term **sexual property.** He argues that property is not the thing itself that someone owns, nor one's constant possession of it, but an agreement among people about how they will act toward it. If you own a car, for example, that does not mean there is a bond between you and your car. What it means is that (1) you have the right to use the car; (2) other people do not have the right to use your car; and (3) you can call on the rest of society to enforce your rights. For example, you can call the police if someone takes your car without your permission. Davis suggests that property rights in this sense can be used to understand sexual property in marriage (Coltrane and Collins, 2001; Davis, 1936, 1949). Thus, marriage is socially defined as (1) the husband and wife have the right to have sexual intercourse with each other; (2) other people do not have the right to sexual intercourse with either of them while they are married; and (3) if violations occur, the aggrieved party can go to court and demand damages (as in divorce). As an "owner" of his wife's sexuality, Davis points out, a husband (e.g., some Eskimos) may loan his wife to another man for a night, in much the same way that the owner of an automobile may loan it to another to drive.

As we will see in later chapters, sexual property rights evolve and change over time and sometimes the analogy with physical objects does not work. For example, a husband in ancient Rome had the legal right to kill or sell his wife or children, whereas our legal codes do not allow husbands or wives to destroy or sell their sexual property. The fact that we do not allow this does not infer a lack of property relationship; rather, the state does not allow people to do certain things with their sexual possessions, more or less in the same way that the state does not allow people to do just anything they want with their cars, such as drive them without a license or faster than the speed limit (Coltrane and Collins, 2001, p. 42).

In this first reading, Davis illustrates how the regulation of sexual property and **emotional property** (what he calls love-property) follow predictable rules. He observes that jealousy is linked to the normative and institutional structure of any given society and discusses how competition for lovers, rivalry, and trespass in romantic relationships are institutionally regulated and culturally variable. Adopting a comparative framework, he finds that in a culture with a custom of exclusive possession of an individual's entire love, jealousy will demand that exclusiveness. In contrast, in a culture that is structured for dividing one's affections, jealousy will reinforce those specific divisions. In general, Davis thus finds that jealousy encourages and maintains fundamental relations of property within any given society. He argues that jealousy can only be understood when analyzed with respect to patterns of power, dominance, and subordination in each society. In what ways is Davis' argument similar to the theoretical approach taken by Coser in the first reading of chapter 1?

In a selection from *Consuming the Romantic Utopia: Love and the Cultural Contradictions of Capitalism,* Eva Illouz analyzes courtship as a ritual. Durkheim (1893/1947) and other early social scientists studied formal religious ceremonies to understand how societies promoted a collective identity by mobilizing the emotional energy of their members through ritual practices. Sociologists since Erving Goffman (1967), in contrast, have been more interested in understanding how smaller rituals help construct and reinforce social bonds between people. Randall Collins (1975) defines a ritual (or **interaction ritual**) as a custom that (1) brings people together face to face; (2) focuses their attention on some common object or activity; (3) promotes a shared emotional tone among the participants, which grows as the ritual proceeds; and (4) produces an emotionally charged symbol, which represents a sense of membership in the group. Extending the insights of both Durkheim and Goffman, Collins describes how love can be seen as a high-intensity ritual: (1) Lovers want to be in each other's sight, they stare into each other's eyes, and they often ignore other people around them. (2) Lovers have a common focus of attention: themselves. Passionate love means being obsessed with the person one loves: forgetting everything else, constantly talking about the relationship, touching each other, thinking about each other. (3) Love is a shared emotion and the feelings associated with it are heightened by the kinds of exclusivity, shared focus, and touching in which lovers frequently engage. (4) Finally, love is known for its symbols: hearts, rings, presents, special clothes, songs or places, and so on. As a high-intensity ritual, love attaches its shared emotions to an emblem of the relationship that serves as a symbolic marker between insiders and outsiders. The ritual of love thus creates a little private cult with its own idealized object of veneration—the loved person.

The excerpt from Illouz's book builds on such formulations to describe the symbolic boundaries that help make an interaction romantic. Drawing on interview data from informants who talked about their own romances, she examines how "temporal, emotional, spatial, and artifactual boundaries" create ritualized romantic moments that set the love relationship apart from everyday life. Delving into the stories that people told, she focuses on the ways that romantic activities are shaped by the consumption of products or services. Romantic moments tend to be constructed around going out: dining, drinking, dancing, concerts, movies, museums; staying in hotels or resorts; and taking trips to special places such as the beach, river, woods, mountains, or abroad. These activities are set apart from everyday life, but purchasing something also plays an important part in their accomplishment. Illouz raises important questions about how modern notions of love and romance are mediated by the consumption of products. Even children she interviewed who had not had romantic encounters were able to summon the mental picture of a romantic dinner in an elegant restaurant, presumably predicated on exposure to the ubiquitous media images of romantic consumption that are used to sell us everything from toothpaste to life insurance. The romantic dinner itself becomes a desired ritual, marketed by those who want to make a profit, sought after by those who want to fall in love, and strategically deployed by those who are attempting to recapture the "magic" in their relationships. As you read the Illouz selection, think about your own romantic feelings. Are they tied to rituals or patterns of consumption?

When sociologists identify patterns of mate selection in the courtship process, they usually invoke the metaphor of a market. Prospective romantic partners offer a set of physical characteristics, personality, skills, age, experience, wealth, education, earning potential, and so on, that are symbolically exchanged for those of a partner. In traditional societies, these trades were made by family elders and governed by complex courtship and marriage rules that sometimes resulted in the direct exchange of goods and services (brideprice, brideservice, dowry) for the marital partner. In our society, we tend to assume that such exchanges are governed by individual preference, but the romantic matches that result still follow predictable patterns. In general, people marry similar others—a pattern known as **homogamy.** Although there are exceptions, the typical pattern in dating, sex, and marriage reflects matches of people with common characteristics (social class, education, ethnicity, physical attractiveness, age, etc.). In part, this is because opportunities for social interaction tend to be limited to similar others, and the places where romantic partners typically meet (school, work, neighborhood, church) are usually composed of similar people.

Erich Goode's article on personal ads is not grounded in such naturalistic settings, but it does reveal how people approach trading in the sexual and romantic marketplace. Because personal advertisements usually contain both the characteristics of the advertisers and the qualities they seek in a partner, such ads are a valuable source of information on types of self-presentation, preferred social roles, gender norms, and judgments of attractiveness in both heterosexual and homosexual samples. Studies show how personal advertisements differ according to gender. Ads written by men tend to seek a casual relationship, emphasize the importance of the woman's physical attractiveness and youth, and comment on the man's secure financial status. Ads written by women, in contrast, tend to seek sincerity and more permanent relationships; offer information about their own personalities, physical appearance, and youth; and seek a successful man. Goode's findings support such patterns, but he moves beyond examining the content of the ads to explore who responds to what type of advertisement and compares the relationship seeking styles of men and women. He draws on the concept of **entitlement,** which he defines as "the feeling by someone, or the members of a group or category, that they have a right to certain privileges, that specific rewards should be forthcoming, that the resources they covet are rightfully theirs by virtue of what they have done or who they are." His analysis shows that men are less likely to fear rejection and far more likely than women to display entitlement in their dating choices. Goode's findings suggest that market and equity theories of courtship should focus on the ways that gender influences personal choices and the ways that it shapes prospects for exchange on the romantic market. In your experience, is being a man or a woman connected to patterns of entitlement as Goode suggests? Do you think that internet dating services reflect the same tendencies as the personal ads?

In the final article in this chapter, Pepper Schwartz and Virginia Rutter review recent trends in teenage sexuality and outline the political debates that surround this contentious topic. They discuss several contradictions inherent in the very idea of teenage sexuality. The first reflects cultural ambivalence about adolescent sexuality. On the one hand, the culture offers up seductive clothing and make-up for preteens and sells them with images of flirtatious young girls. On the other hand, the culture promotes ideals about the innocence and asexual nature of young girls. Another contradiction revolves around long-term trends in the timing of adulthood and the onset of menses. Although we are increasingly likely to encourage young adults to go to college and delay the assumption of adult marital roles, young people are reaching puberty sooner than in any previous generation. A third contradictory tendency concerns the tension between the weakening of the **sexual double standard** and new sex education programs promoting a "just say no" message. We used to expect teenage boys and young men to want sex and teenage girls and young women to resist male advances. As other articles in this reader show, this double standard is not gone, but it has certainly weakened. The partial movement toward equalizing gender scripts, especially in a highly sexualized popular culture with delayed onset of adulthood and the availability of birth control, allows young women more opportunities to be sexual on their own terms. These tendencies are contradicted by the successful political push at the national and local levels for **abstinence-only sex education programs.** In the context of these contradictions, Schwartz and Rutter debunk myths about an epidemic of teenage pregnancy and criticize what they see as nostalgic attempts to revitalize dichotomized sexual scripts for adolescent boys and girls. From a sociological perspective, why do you think the sexual double standard has weakened? Do you think that abstinence-only sex education programs will reverse this trend?

The articles in this chapter draw on sociological theories to show how sex and romance, though typically thought of as things that "just happen," are products of larger structural and cultural forces. If more people understood romance as a socially constructed ritual interaction, do you think they would be more or less likely to fall in love? Why?

REFERENCES

Collins, Randall. 1975. *Conflict Sociology: Toward an Explanatory Science.* New York: Academic Press.

Coltrane, Scott, and Randall Collins. 2001. *Sociology of Marriage and the Family: Gender, Love and Property.* 5th ed. Belmont, CA: Wadsworth/ITP.

Davis, Kingsley. 1936. Jealousy and sexual property. *Social Forces* 14:395–405.

Davis, Kingsley. 1949. *Human Society.* New York: Macmillan.

Durkheim, Emile. 1893/1947. *The Division of Labor in Society.* New York: Free Press.

Goffman, Erving. 1967. *Interaction Ritual.* New York: Doubleday.

SUGGESTED READINGS

Bailey, Beth L. 1988. *From Front Porch to Back Seat: Courtship in Twentieth-Century America.* Baltimore: Johns Hopkins University Press.

Bailey, Beth L. 1999. *Sex in the Heartland.* Cambridge, MA: Harvard University Press.

Cancian, Francesca M. 1987. *Love in America: Gender and Self-Development.* Cambridge, NY: Cambridge University Press.

Holland, Dorothy, and Margaret A. Eisenhart. 1990. *Educated in Romance: Achievement, and College Culture.* Chicago: University of Chicago Press.

Laumann, Edward O., John H. Gagnon, Robert T. Michael, and Stuart Michaels. 1994. *The Social Organization of Sexuality: Sexual Practices in the United States.* Chicago: University of Chicago Press.

Odem, Mary E. 1995. *Delinquent Daughters: Protecting and Policing Adolescent Female Sexuality in the United States, 1885–1920.* Chapel Hill: University of North Carolina Press.

Schwartz, Pepper, and Virginia Rutter. 1998. The Gender of Sexuality. Thousand Oaks: Pine Forge Press.

Wexman, Virginia Wright. 1993. *Creating the Couple: Love, Marriage, and Hollywood Performance.* Princeton, NJ: Princeton University Press.

INFOTRAC® COLLEGE EDITION EXERCISES

The exercises that follow allow you to use the InfoTrac® College Edition online database of scholarly articles to explore the sociological implications of the selections in this chapter.

Search Keyword: Marriage market. Search for articles that discuss the concept of the marriage market. How is the marriage market described in the InfoTrac® College Edition articles that you find? What happens in the context of this market? How does the marriage market reflect social notions about gender?

Search Keyword: Homogamy. Find articles that deal with homogamy. How is homogamy defined? How does homogamy work in the marriage market? How does homogamy relate to decisions about marriage partners?

Search Keyword: Romance. Use InfoTrac® College Edition to see how many articles can you find that deal with the subject of romance. What are some romantic rituals discussed in these articles? Of these articles, what proportion invokes consumption of commercial goods to create romantic moments? How are romance and consumerism tied together?

Search Keywords: Courtship and gender. Look for articles that talk about gender differences in rituals of courtship. What are some of those differences? Do stereotypes about gender underlie any of those differences? If so, how?

Search Keyword: Sexual double standard. What is a sexual double standard? Do you think such a double standard exists? Find articles that discuss this notion. Do these authors think the sexual double standard has changed over time?

Search Keyword: Adolescent sexuality. How many articles can you find that discuss the issue of adolescent, or teen, sexuality? What are some of the social influences on adolescent sexuality? What kinds of contradictory messages are being given to teens about their sexuality? Is there a gender difference in attitudes toward teen sexuality?

Search Keyword: Dowry. Using InfoTrac® College Edition, search for articles that discuss the issues of dowry and dowry death. What is dowry? How is it connected to the notion of women as sexual property?

4

Jealousy and Sexual Property

An Illustration

KINGSLEY DAVIS

1949

At first glance jealousy may seem an unlikely topic for illustrating the value of a sociological approach. It is usually regarded as an emotion, an individual or psychological phenomenon having little to do with culture and social organization. This very conception of jealousy, however, makes it a useful illustration. If a sociological mode of analysis can be shown to throw new light on the subject, the value of such analysis will be demonstrated.

Actually, all of the types of sociological analysis . . . seem applicable to jealousy. First of all, it turns out to be a reasonable assumption that jealousy has a function not only with reference to the individual's emotional balance but also with reference to social organization. Secondly, it appears that the manifestations of jealousy are determined by the normative and institutional structure of the given society. This structure defines the situations in which jealousy shows itself and regulates the form of its expression. It follows that unless jealous behavior is observed in different cultures, unless a comparative point of view is adopted, it cannot be intelligently comprehended as a human phenomenon. Thirdly, the situations in which jealousy occurs involve the statuses and roles of various persons with reference to one another. It is through the definition of these statuses and roles that the institutional framework governs the manifestations of jealous emotion. Fourthly, since the various participants in the love entanglement (and it is only sexual jealousy that we are considering) are pursuing ends by various means, an understanding of their behavior requires analysis in terms of means, ends, and conditions. Finally, jealousy involves certain processes of interaction—competition, rivalry, and trespass. These processes, too, must be understood in their social context, for they are institutionally regulated and culturally variable. All in all it would seem clear that a sociological approach holds definite possibilities of contributing to a knowledge of this peculiar kind of emotion.

SEXUAL PROPERTY AS AN INSTITUTION

Descartes defined jealousy as "a kind of fear related to a desire to preserve a possession." If he had in mind what is customarily called jealous behavior, he was eminently correct. In every case it is apparently a fear or rage reaction to a threatened appropriation of one's own or what is desired as one's own property.

. . . Conflicts over property involve four elements: Owner, Object, Public, and Rival or Trespasser. If the conflicts are in the nature of competition rather than trespass, Ego is a

From Kingsley Davis, "Jealousy and Sexual Property: An Illustration," from *Human Society*, pp. 175-193. New York: Macmillan.

would-be owner and his enemy is a rival instead of a trespasser. A popular fallacy has been to conceive the jealous situation as a "triangle." Actually it is a quadrangle because the public, or community, is always an interested element in the situation. The failure to include the public or community element has led to a failure to grasp the social character of jealousy. The relationships between the four elements are culturally regulated. They are current in the given society and constitute the fixed traditional constellation of rights, obligations, and neutralities that may be called sexual property. They are sustained and expressed by the reciprocal attitudes of the interacting parties.

Since property, however, is not always actually in the hands of the owner, ownership must be distinguished from possession, the one being a matter of law and mores, the other a matter of fact. Possession by a person other than the owner may be either licit or illicit. Illicit possession bears witness that the rules of property are susceptible of evasion. Licit possession by one not the owner, as with a borrowed or rented piece of property, emphasizes the strength of these rules.

Acquisition of property proceeds usually according to socially established norms of competition and, in many cases, by stages. In the initial stage the field is generally open to a class of persons, anybody in this class being free to put in a claim. The qualifying rounds of a golf tournament or the sudden entrance of a strange but attractive young woman are cases in point. Gradually a few competitors take the lead. Social order then requires that others recognize the *superiority* of these, quit struggling, and turn their attention elsewhere. Finally, after continued competition among the favored few, one competitor wins. This is the signal for everyone who was initially interested to drop all pretense of a claim and take his defeat in good spirit. The end is no longer a legitimate one for him because competition for this particular piece of property is now, by social edict, either temporarily or permanently over. It is owned by one man, behind whose title stands the authority of the community.

Values, however, do not invariably change hands in any such orderly fashion. The unscrupulous stand always ready to take possession in defiance of the rules, to replace the orderliness of rivalry with the disarrangement of trespass. They may at any stage, under peril of organized retaliation, upset the procedure and seize physical possession of the property.

There are thus two dangers which beset any person with regard to property. The first is that somebody will win out over him in legitimate competition. This is the danger of superior rivalry. The second is that somebody will illegitimately take from him property already acquired. This is the danger of trespass.

Most malignant emotions are concerned with these two dangers; they are directed either at a rival or trespasser or at someone who is helping a rival or trespasser. Such emotions may be either suppressed by the group culture or utilized for maintaining the organized distribution of property. In general fear and hatred of rivals are institutionally suppressed, fear and hatred of trespassers encouraged.

In the initial stages of acquisition fear of rivals is frequently paramount. Such fear is merely the obverse side of strong desire to win. In so far as a society fosters the desire to win and builds up an emotional drive in the individual to that end, it inevitably fosters the fear of losing. By the same token, when defeat actually occurs it implies a frustration of strong desire, hence an inevitable emotion. This emotion occurs frequently, since most competitors cannot win. Yet once the property is in the winner's hands, social organization requires that such emotions be curbed. Society tends necessarily to suppress them and to encourage one-time rivals to be "good sports," "graceful losers."

The successful rival, however, need not suppress these emotions. Once established as owner he is encouraged by the culture to express them toward any trespasser. Free expression of malevolent emotion against a trespasser protects the established distribution of property and maintains the fixed rules for its competitive acquisition.

TYPES OF PROPERTY ATTITUDES

Can the relationship of affection between two persons be conceived as a property relationship? This is a question not to be answered glibly. The affectional relationship is certainly not identical with *economic* property, although sheer sexual gratification (as in prostitution) may be. Affection assumes that the object is desired in and for itself. It therefore cannot be bought and sold; it is not a means to something else, not an economic thing. Yet the affectional relationship has features that are characteristic of property in general. It is, above all, highly institutionalized; it involves some sort of institutionalized exclusiveness, hedged about with rights and obligations. There is competition for possession, a feeling of ownership on the part of the successful competitor, a "hands-off" attitude on the part of the public, and a general resentment against anyone who tries as a trespasser to break up the relationship by "stealing" the object. In view of these considerations it seems possible to apply the term property to the institutionalized possession of affection. There apparently exists no other term that will describe those types of sanctioned possession that are not economic.

. . . Since affection is a phenomenon of will, the question of possession is thereby placed largely in the hands of the object one wishes to possess. Out of this peculiarity grow the other idiosyncrasies of love-property. We find, for instance, that a jealous lover (assuming for the sake of illustration that he is masculine) often attacks the love-object herself, seeking to restrain her from directing her affection elsewhere or to retaliate against her for having already done so. Having control over the vital element in the situation—the goal being sought by the other party—she is in a position to decide the issue. She can bestow affection either on Ego or on his rival, as she chooses. A man might destroy his food in order to keep another from getting it; he might destroy his jewels or other emblems of prestige; he might even renounce and forsake his profession—but unless indulging in an anthropomorphic fit of temper, he would not thus destroy his possession out of resentment against the possession itself. A man who breaks his golf club after a bad shot knows perfectly well that the club had no volition in the matter. Yet in the case of sexual jealousy the resentment may be more against the object of love than against the rival. Everybody is familiar with the various forms of aggression practiced against each other by those in love, varying all the way from outright murder to mental cruelty.

. . . The love-property situation stands out from the others in that the attitude desired is an attitude of the object itself, not of the general public. This has a profound effect on the attitudes involved. Not only does the object become subject to jealous aggression because it is a human object, but there is a mutuality to the relationship that is lacking in the other forms of property. The affectional relationship implies a reciprocal, mutual interchange between owner and object which is not true of the other forms. Indeed, both are owners and both are objects at the same time. The love relationship, unlike other property relations, is an end in itself. The object of affection plays a dynamic role in determining the direction of the conflict situation by his or her ability to determine the character of this relationship. This doubtless explains why conflicts over love generate more emotion than other kinds and exhibit a more dramatic quality. When the object possessed is another person, the universal tendency of the possessor to identify himself with the thing possessed (transmuting "mine" into "I") is given the greatest opportunity to express itself. The lover feels that his love is a part of himself and that his existence would be meaningless without her. Still, simply because the object of love is not inert but willful, this personal identification is probably most tenuous of all in the love situation. The object possessed has it within her power to nullify the possession. This means that the love relationship is at once unusually close and unusually instable or tenuous. As a result it is doubly intense and highly charged with emotion.

. . . A complete sequence in love-property conflict would begin with the rivalry phase. It would depict the changing attitudes of the rivals, and of the object of affection and the public, as some of the competitors are eliminated and one finally wins. The next phase would show the winner in secure possession at some level of ownership such as the "sweetheart," the "fiancée," or the "spouse" level. He is no longer jealous because rivalry is finished and no trespasser is in sight, and the public has an attitude of "don't disturb." The third phase, trespass, would describe the attitude of Ego as he becomes aware of an enemy—his feelings toward the trespasser, the love object, and the public. It would describe also the attitudes of the trespasser, and since the direction of the sequence hinges largely upon the woman (at least in our culture), it would describe her attitudes toward lover, rival, and public. If she favors the trespasser and is willing to risk Ego's and the community's wrath, Ego may lose. On the other hand if she does not favor the trespasser, if he himself is not willing to take the risk, or if Ego or the public uses irresistible force, Ego may win. The multiplicity of attitudes between the four interacting parties grows amazingly complex. Innumerable combinations are possible. To describe them all would be a fascinating adventure into the anatomy of dramatic reality, but it would also require a complete volume. . . .

JEALOUSY AND INTIMACY

Since in love-property the object of possession is the devotion of another person we may expect jealousy to have a direct connection with the sociology of intimacy. . . .

Although jealousy can appear only when there is a presumption of *primary* association in the past, present, or future the fact should be noted that jealousy also indicates at least a partial negation of that rapport between persons which we commonly ascribe to intimacy. It admits that affection has strayed or may possibly stray in the direction of a rival or trespasser. Even when the affection has not actually strayed, jealousy shows on the lover's part a mistrust inimical to the harmony of perfect intimacy.

What, then, is the function of jealousy in intimate association? As a fear reaction in the initial stages of rivalry it is simply the obverse side of the desire to win the object. The desire to win being institutionally cultivated, the fear of losing is unavoidably stimulated also, though its expression is publicly frowned upon. But after ownership has been attained, jealousy is a fear and rage reaction fitted to protect, maintain, and prolong the intimate association of love. It shelters this personal relationship from outside intrusion. This is not to say that it never defeats its own purpose by overshooting the mark. So deeply emotional is jealousy that its appearance in the midst of modern social relationships (which are most profitably manipulated by self-composed shrewdness) is like a bull in a china shop. Nonetheless its intention is protective. It is a denial of the harmony of intimacy only in so far as its presence admits a breach; and is destructive of it only in so far as it muddles its own purpose.

Jealousy stresses two characteristics of the primary relationship: its ultimate and its personal qualities. The relationship is for the jealous person an ultimate end in itself, all other considerations being secondary. This explains the bizarre crimes so frequently connected with jealousy—crimes understandable only upon the assumption that for the criminal the affection of a particular person is the supreme value in life. It also explains the connection between extreme jealousy and romantic love. The "personal" quality of the relationship is manifested by the unwillingness of the jealous person to conceive any substitute for the "one and only." He insists upon the uniqueness of personality. Were the particular person removed, the whole relationship and its accompanying emotion of jealousy would disappear.

An old debate poses the question whether or not affection is divisible. Is it possible to love two people sincerely at the same time? Most authorities on sex relations answer that it is possible and cite cases as proof. Iwan Bloch, for example, asserts that simultaneous passion for several persons happens repeatedly. (Bloch, 1926; Ellis, 1914; Folsom, 1931) He adds that the extensive psychic differentiation between individuals in modern civilization increases the likelihood of such simultaneous love, for it is difficult to find in any single person one's complete complement. Bloch gives numerous examples from history and literature, particularly cases where one aspect of a person's nature (usually the intellectual) is satisfied by one lover, another aspect (usually the sensual) by a different lover.

The conclusion invariably deduced from this is that jealousy is harmful and unjustified. But to end the discussion with this ethical argument is to miss the point. Even though love, like any other distributive value, is divisible, institutions dictate the manner and extent of the division. Where exclusive possession of an individual's entire love is customary, jealousy will demand that exclusiveness. Where love is divided it will be divided according to some scheme, and jealousy will reinforce the division.

RIVALRY, TRESPASS, AND SOCIAL CLASS

While the love-property situation contains a relationship of intimacy and is therefore illuminated by the sociology of primary association, it also contains a diametrically opposite kind of relation—namely, that of power—which concerns the sociology of dominance and subordination. This relationship which obtains between the lover and his rival or trespasser is not a value in itself but a means to an ulterior end; and it connotes an absolute opposition of purpose in the sense that if one succeeds the other fails. The rival or trespasser may be a stranger or a close friend; in either event, so far as the common object is concerned he is an enemy.

Here as elsewhere in the discussion it makes a difference whether the enemy is a trespasser or a rival. Rivalry is most acute in the early stages of acquisition, and jealousy is at this point a fear of not winning the desired object. Toward one's rival one is supposed to show good sport and courtesy, which is to say that society requires the suppression in this context of jealous animosity. Regulated competition constitutes the *sine qua non* of property distribution and hence of stable social organization. But as one person gets ahead and demonstrates a superior claim, his rivals, hiding their feelings of jealous disappointment, must drop away. If any rival persists after the victor has fortified his claim with the proper institutional ritual, he is no longer a rival but a trespasser.

Jealousy toward the trespasser is encouraged rather than suppressed, for it tends to preserve the fundamental institutions of property. Uncles in our society are never jealous of the affection of nephews for their fathers. But uncles in matrilineal societies frequently are, because there a close tie is socially prescribed between uncle and nephew. The nephew's respect is the property of the uncle; if it is given to the father (as sometimes happens because of the close association between father and son) the uncle is jealous. (Malinowski, 1932) Jealousy does not occur in the natural situations—and the "natural situations" are simply those defined in terms of the established institutions. Our malignant emotions, fear, anger, hate, and jealousy, greet any illicit attempt to gain property that we hold. We do not manifest them when a legitimate attempt is made, partly because we do not then have the subjective feeling of "being wronged" and partly because their expression would receive the disapprobation of the community. The social function of jealousy against a trespasser is therefore the extirpation of any obstacle to the smooth running of the institutional machinery.

Discussions of jealousy usually overlook the difference between rivalry and trespass. A case in point is the old problem of whether one can be jealous of a person not one's equal. If the person is a trespasser the answer is that he can be any distance away in the social scale. But if he is a rival he cannot be too far distant. Rivalry implies a certain degree of equality at the start. Each society designates which of its members are eligible to compete for certain properties. While there are some properties for which members of different classes may compete, there are others for which they may not compete. In such cases the thought of competition is inconceivable, the emotions reserved for a rival fail to appear, and the act is regarded not as rivalry but as a detestable thrust at the class structure. Thus it happens that for a given lover some people cannot arouse jealousy as of a rival. If the love-object yields to a member of a distinctly inferior social class, jealousy will turn into moral outrage, no matter if the lover himself has no claim on the love-object. . . .

But jealousy against a trespasser is another matter. Since a trespasser by definition is a breaker of customary rules, the more he breaks, including the rules of class structure, the more of a trespasser he is. A violator of property rights may for this reason occupy any position on the social scale.

The fact that men of native races sometimes prostitute their wives to civilized men without any feeling of jealousy while they are extremely jealous of men of their own race, (Malinowski, 1929) is sometimes pointed out as showing that men are jealous only of their equals. This is true only in so far as jealousy of rivals is meant. The civilized man is not conceived by the natives as a rival, nor as a trespasser. He *may* be conceived as a trespasser—if, for example, he attempts to retain the wife without paying anything. In the case mentioned he is not a trespasser but merely one who has legitimately paid for the temporary use of property. His very payment recognizes the property rights of the husband. The following case is much more illustrative: "A Frenchman of position picks for his mistress a girl who is not his social equal. You can see for yourself that his wife is not jealous. But let him choose a woman of his own social rank—then you'd see the fur fly; . . ." (Lindsey and Evans, 1927) Among some social spheres in France, if we are to believe what we hear, women of different classes customarily exercise proprietary rights in the same man without feeling jealous of each other. But since it is not customary for women of the same rank to share a man, such a condition would be either rivalry or trespass and would arouse intense jealousy.

One may argue that the nearer two people are in every plane, the more intense will be the jealousy of rivalry; while the further apart they are, the greater the jealousy of trespass.

But between the lover and the object of his love the relationship is not one of power. If a woman is regarded simply as a pawn in a game for prestige the pattern is No. 2 in our typology, not No. 4. It is a question of vanity rather than jealousy. In the love situation the jealous person values the affection for itself. It is his fear of losing this intrinsically valuable affection to a rival or a trespasser, rather than his fear of losing prestige in the eyes of his public and his rival, that paralyzes him.

THE SOCIAL FUNCTION OF JEALOUSY

Into every affair of love and into every battle for power steps society. The community has an inherent interest in love not only because future generations depend upon it but also because social cohesion rests upon the peaceful distribution of major values.

A question that all authorities feel compelled to settle concerns the social or anti-social character of jealousy. Forel declares that jealousy "is only the brutal stupidity of an atavistic heritage, or a pathological symptom," (Forel, n.d.) while Havelock Ellis calls it "an anti-social emotion." (Ellis, 1914) The chief arguments are that it is an inheritance from animal ancestors, a hindrance to the emancipation of women, and an obstacle to rational social intercourse.

The hasty readiness to praise or condemn prevents a clear understanding of the relation of jealousy to the social structure. Careful analysis is cut short by the quick conclusion that jealousy is instinctive. The assumption is that certain stimuli call forth a stereotyped, biologically ingrained response. Jealousy is therefore regarded as an animal urge and denounced as anti-social. Such condemnation instead of comprehension illustrates once more the tendency of the moralistic bias to take precedence over the scientific attitude in the handling of social phenomena.

The instinctive view fails to analyze jealous behavior into its different components—to distinguish between the stimulus (a social situation having a meaning only within the culture where it is found) and the physical mechanism involved in the total response. It puts all constituents into the undifferentiated category of instinct. Doubtless the physiological mechanism operating in the jealously aroused person is inherited. But the striking thing about this mechanism is that it is not specific for jealousy, but appears to be exactly the same in other violent emotions such as fear and rage. The sympathetic nervous system seemingly plays the usual role: increased adrenal activity speeding the heart, increasing the sugar content of the blood, toning up the striated muscles and inactivating the smooth muscles.

If we are to differentiate jealousy from the other strong emotions we must speak not in terms of inherited physiology but in terms of the type of situation which provokes it. The conflict situation always contains a particular content, which varies from one culture to the next. The usual mistake in conceiving jealousy is to erect a concrete situation found somewhere (often in the culture of the author) into the universal and inherent stimulus to that emotion. This ignores the fact that each culture distributes the sexual property of the society and defines the conflict situations in its own way, and that therefore the concrete content cannot be regarded as an inherited stimulus to an inherited response.

This mistake is made, I think, by those theorists who seek to explain certain human institutions on the basis of instinctive emotions. In the field of sexual institutions Westermarck is the outstanding theorist who has relied upon this type of explanation. He disproves the hypothesis of primeval promiscuity and proves the primacy of pair marriage largely on the basis of allegedly innate jealousy (Westermarck, 1922). He assumes, indeed, that all types of sexual relationship other than monogamy (as he knows it in his own culture) are native stimuli to instinctive jealous retaliation.

As soon, however, as we admit that other forms of sexual property exist and that they do not arouse but instead are protected by jealousy, the explanation of monogamy breaks down. Whether as the obverse side of the desire to obtain sexual property by legitimate competition or as the anger at having rightful property trespassed upon, jealousy would seem to bolster the institutions where it is found. If these institutions are of an opposite character to monogamy, it bolsters them nonetheless. Whereas Westermarck would say that adultery arouses jealousy and that therefore jealousy causes monogamy, one could maintain that our institution of monogamy causes adultery to be resented and therefore creates jealousy.

Had he confined himself to disproving promiscuity instead of going on to prove monogamy, Westermarck would have remained on surer ground. Promiscuity implies the absence of any sexual property-pattern. Yet sexual affection is, unlike divine grace, a distributive value. To let it go undistributed would introduce anarchy into the group and destroy the social system. Promiscuity can take place only in so far as society has broken down and reached a state of anomie.

The stimulus to jealousy, moreover, is not so much a physical situation as a meaningful one. The same physical act will in one place note ownership, in another place robbery. Westermarck appears to believe that it is the physical act of sexual intercourse between another man and one's wife that instinctively arouses jealousy. But there are cultures where such intercourse merely emphasizes the husband's status as owner, just as lending an automobile presumes and emphasizes one's ownership of it.

We may cite, for example, the whole range of institutions whereby in some manner the wife is given over to a man other than her husband. These run from those highly ritualized single acts in which a priest or relative deflowers the wife (the so-called *jus prima noctis*) to the repeated and more promiscuous acts of sexual hospitality and the more permanent and thoroughgoing agreements of wife-exchange; not to mention the fixed division of sexual function represented by polyandric marriage. In societies where any institution in this range prevails, the behavior implied does not arouse the feeling of jealousy that similar behavior would arouse in our culture. Jealousy does not respond inherently to any particular physical situation; it responds to all those situations, no matter how diverse, which signify a violation of accustomed sexual rights.

THE INTERNAL VERSUS OVERT MANIFESTATION

Possession of a thing of value without any right to it is a prevalent condition in sexual behavior; affection is evidently difficult to govern. The converse—ownership without possession—is equally prevalent. At least in our culture the instances are countless in which there is no overt transgression of convention and yet affection has strayed. Wives and husbands abound who have little or no affection for their mates but who would not actually sully the marriage tie. Their affection is owned by their mates, but not possessed. . . .

CONCLUSION

It should now be clear that a genuine understanding of jealousy requires that it be studied and analyzed from a comparative sociological point of view. A theory of jealousy which derives its empirical fact solely from the manifestations of jealousy in our own culture can hardly be a satisfactory theory, no matter how plausible it may seem to people who live in our culture. When a comparative point of view is adopted it is seen that the situations calling forth a jealous response vary tremendously from one culture to another. Jealousy is an emotion which has a function as a part of an institutional structure. Not only is it normatively controlled but it gives strength to the social norms as well. To understand the social function of jealousy, to see the significance of its variable but inevitable appearance in different societies, one must have a conceptual apparatus at hand. One must, for example, have some conception of the cultural definition of social situations in terms of the statuses and roles of the participants; some notion of means, ends, and conditions; and some idea of processes of interaction such as rivalry, competition, and trespass. In this way one is able to state in fairly brief and definite fashion the major conclusions about jealousy as empirically observed in different cultures. Otherwise one has facts but little else. It is not contended here that the sociological approach to jealousy is the only one that is needed. Jealousy must also be studied from the point of view of the personality as a unit. But it is contended that for a full understanding of jealousy the sociological approach is indispensable because it addresses itself to a real and important aspect of jealous behavior—the social aspect. . . .

SELECTED REFERENCES

Bloch, Iwan. 1926. *Sexual Life of Our Time.* New York: Allied Book.

Ellis, Havelock. 1914. *Studies in the Psychology of Sex,* VI. Philadelphia: F. A. Davis Co.

Folsom, Joseph K. 1931. *Social Psychology.* New York: Harper.

Forel, August, n.d. *The Sexual Question,* trans. by C. F. Marshall. New York: Rebman.

Lindsey, Benjamin B., and Wainwright Evans. 1927. *The Companionate Marriage.* New York: Boni & Liveright.

Malinowski, Bronislaw. 1929. *The Sexual Life of Savages.* New York: Harcourt, Brace.

Malinowski, Bronislaw. 1932. *Crime and Custom in Savage Society.* London: Paul Trench, Trubner.

Westermarck, Edward. 1922. *The History of Human Marriage.* New York: Allerton.

5

An All Consuming Love

EVA ILLOUZ

1997

REENCHANTING THE WORLD

Courtship is commonly referred to as a ritual. The metaphor may have roots in premodern courtship, which followed a highly codified sequence of actions analogous to formal rituals, but because twentieth-century manners have swept away the rigidity of courtship etiquette, the metaphor of ritual would seem obsolete, no longer capturing the symbolic meaning of romance. However, a cultural analysis of modern romantic practices, as reported by my respondents, reveals that the meaning of romance does indeed draw from religious rituals, even more literally than is usually implied by the word "ritual."

For Durkheim, ritual is the category of religious behavior that delineates and demarcates sacred time, space, and feelings from nonsacred, that is, profane. The marking of such boundaries must follow a formal set of rules to make it a collective and binding behavior. Participants in a ritual, says Durkheim, "are so far removed from their ordinary conditions of life, and they are so thoroughly conscious of it, that they must set themselves outside of and above the ordinary morals." (Durkheim, 1973) Sociologists since have expanded the definition of ritual, suggesting that the secular realm of daily life is also susceptible to formalized rules of conduct similar to those of religious rituals. (Goffman, 1967; Wuthnow, 1987; Moore and Myerhoff, 1977; Alexander, 1988) Any symbolic meaning, when it is intensified, can be ritualized within (rather than outside) daily life. Although romance is a secular behavior, it is experienced to its fullest when infused with ritual meaning. More specifically, an interaction becomes "romantic" when four kinds of symbolic boundaries are set: temporal, emotional, spatial, and artifactual. These boundaries carve a symbolic space within which romance is lived in the mode of ritual.

Temporal Boundaries

According to Eviatar Zerubavel, sacred time has boundaries, that is, points at which it starts and ends (Zerubavel, 1981). A person experiences a romantic moment as taking place during a different time from "regular," profane time. Most readers will recognize these words of a woman filmmaker: "You have to set aside time, to set aside time to make things romantic" (interview 2). Such "setting aside" of time is a very conscious and purposeful action, as the following answer given by a working-class woman demonstrates:

> ***How does one work on [a relationship]?***
>
> You have to make the time, or find the time to do special things that are romantic moments with each other. It's much easier after

> a long-term relationship to just follow [*sic*] into a pattern, you do what you have to do and stay home in front of television and watch it and sometimes that's easier than putting yourself out, like getting yourself dressed up a little bit, like finding a place where to go, making an arrangement for your children to be taken care for. (Working-class woman, no profession, interview 16)

"Romantic time" is constructed to mark out "special things," and as such it is explicitly opposed to the "ordinary" time of everyday life. "It [a romantic moment] would be either early in the morning or late at night, when no one else is around" (working-class man, interview 36). Why do these two slices of time, before or after the regular workday, seem equally romantic? Preceding or following work or domestic chores, they are thus located on the margins of the productive and reproductive time of society.

> ***Could you say what was romantic about [a particular moment]?***
>
> Night, I think, is more romantic than the day. And the forest and the natural environment, that I found to be very romantic. (Male lawyer, interview 5)

Here, temporal boundaries are associated with geographical ones; both mark the seclusion of the lovers. The night is more romantic than the day because it facilitates the lovers' symbolic and physical isolation from their ordinary daytime identities. Night is both a typical marker of romantic time and also the private time par excellence, and this suggests that the boundaries of romantic time are often structured by the boundaries dividing private and public life.

Romantic time is often compared to or called "celebration time." This is often seen in women's magazines: "Your intimate relationship is the most precious gift in your life. It deserves to be celebrated every day" (Woman, 1988).

. . . Romantic time feels like holiday time: it is perceived as different and special and is therefore experienced in the mode of celebration. The life of the couple is susceptible to cyclical celebration, similar to the cycles of religious life. For example, the wedding anniversary demarcates the event of marriage and celebrates it regularly. It commemorates periodically the marriage ceremony and reaffirms the bond through the exchange of gifts and a semipublic celebration. Similarly, Valentine's Day is a collective celebration of private feelings of love. Even unmarried couples, feeling the need to celebrate cyclically the romantic bond, often ritually observe the "first day they met" as a substitute for a wedding anniversary. Romantic moments are lived and remembered both as literally "festive" time and analogously to the experience of religious time.

Spatial Boundaries

Related to this reconstruction of time as festive or consecrated is the rearrangement of the space of everyday life. The setting of new spatial boundaries marks the experience of romance, as evidenced by people's frequent need to move away from their regular domestic space. Even when lovers find themselves in the midst of a crowd of people (as, for example, on the street or in a restaurant), they symbolically construct their space as private and isolated from the surrounding people. For example, "When I am at home, it's not special. I have had also romantic evenings at home but if I had to pick my most romantic moments, it seems more special, maybe I am more relaxed then, when I go out" (female university professor, interview 19).

Although isolation and privacy are intrinsic elements of romance, that home is frequently viewed as unromantic suggests that romance is equated with the couple's ability to construct a new private space around them, even in a public place. When the couple reaches such a sense of isolation from the surrounding environment, they can switch to a space properly distinct from that of daily or ordinary life.

> ***What was romantic about this moment [in Mexico]?***
>
> Just like being thousands miles away from home in our little . . . almost kind of fantasy world. Real world seemed so far away, so we

> were just in our little paradise almost. And there was a certain intimacy about being there and being in that type of setting. (Male corporate banker, interview 20)

In this quote, the geographical distance not only marks but also induces the feeling of romance. Because being in a foreign country is a culturally and linguistically isolating experience, it sharpens the construction of romance as an "island of privacy" in the public world.

Artifactual Boundaries

A common way of signaling a romantic moment is to include objects different from those used for daily purposes, ritual objects being more precious or beautiful than ordinary ones. Gifts are one obvious example, but elegant clothes and expensive meals are also associated with romance.

> ***If you wanted to have a romantic moment with someone, what would you do?***
>
> I'd say I had a pretty romantic evening one night recently. I went to a formal dance, so you get dressed up really nice and you go to a pretty place, you do some dancing, and then I think we came back here for the dessert.
>
> ***What was romantic about that evening?***
>
> Getting dressed up is romantic, and going out and doing nice things, and then also spending some private time, and also doing things that are slightly different from sort of the usual. This results from formal dancing, you know waltz and swing, things like that, so that makes it romantic. You are also served champagne after, which is a little bit different from the usual, though I have it pretty often. (Male university professor, interview 1)

This man invoked two elements frequently found in religious rituals: special vestments and beverages. Aesthetic or expensive objects distinguish the romantic interaction from others by making it at once more intense and more formal. Certain objects, like champagne, roses, or candlelight, have fixed attributes of quasi-sacredness that under the appropriate circumstances can generate romantic feelings. Others acquire sacredness by virtue of their association with intense romantic feelings; for example, a scarf or a napkin used on a romantic occasion can become impregnated with sacredness, able to transport the lover back to the "sacred" moment of the first meeting.

Emotional Boundaries

More than other sentiments, the romantic feeling is expected to be different, unique, for if a romantic emotion cannot be unambiguously distinguished from other feelings, the very ability to feel "love" can be jeopardized. The necessity to make romantic love intrinsically different comes from the norm of monogamy, but it also stems from the fact that if boundaries between romantic feelings and other feelings are blurred, as can be the case for example in "amorous friendships," the perceived uniqueness of romantic feelings is put into question. When asked if there is a difference between friendship and love, everyone says with no hesitation that the two are different and *have* to be different, even if they claim at the same time that friendship is a subcomponent of love. In contrast to love, the sentiment of friendship is not jeopardized by having more than one person as its object. Thus, romantic feelings are commonly described as "special," since they are directly related to a person made separate from the rest of the world.

One of my respondents characterized being in love as "a sense that the other person is so wonderful that you don't understand how it can exist, that you look at the person and the rest of the world disappears" (male art gallery owner, interview 43). The same man described his encounter with a woman whom he fell in love with:

> I met someone and I was so attracted by her that I was frozen, I went out with her, I socialized with her, and it was obvious that we were incredibly attracted to each other. There was electricity. It was a tremendous satisfaction that I had these very strong feelings I haven't felt before. There is a sense of

> connection that happens, for a time, an intermingling of the soul, that you become the same person. There seems to be for me, an absence of sexuality, that you become sort of a spirit, more than a male or female, you become a pure essence.

Although this man gave no religious affiliation, his terminology is obviously religious, and his statement evokes motifs familiar to the phenomenology of religion: a strong sense of bonding and deep connection with the God-lover, the replacement of a sense of personal identity by a feeling of merged identities. There is a similarity between the awe and the intensity felt in the religious experience and the romantic sentiment. Both overwhelm the religious person or the lover, and in both the object of worship or love is perceived as unique and overpowering. . . .

The heightening of emotions, as Durkheim noted, is central to the experience of the sacred. In fact, his descriptions of the feeling of the sacred could easily be applied to the feeling of romantic love: "Vital energies are over-excited, passions more active, sensations stronger. . . . A man does not recognize himself" (Durkheim, 1985). The romantic feeling is reminiscent of the exhilaration often felt by the religious person during moments of religious celebration and communion.

In the terminology of phenomenological sociology, a romantic moment stands outside the taken-for-grantedness of the everyday world. This idea was explicitly stated by the respondents who held that romance is and should be different from daily life, for example, in their frequent use of the word "special" to signal the unique character of the relationship. When explaining why romance was so important to his view of marriage, a respondent invoked this idea: "I would like to have that extraspecial, something out of the ordinary" (male lawyer, interview 5). The feeling of "that extraspecial" is so important that when it is not felt, people call into question their romantic attachments. Romance is described as that which prevents one from taking the other "for granted," even when one shares one's daily life with that significant other. It is "that little extra" one adds to a relationship, a small but crucial difference commonly conveyed in such metaphors as a "spice," a "flavor," a "zest." And this difference is more likely to appear when the moment stands outside the realm of day-to-day life.

. . . Romance is opposed to dailiness, routine, and taken-for-grantedness. Romance represents an excursion into another realm of experience in which settings, feelings, and interactions are heightened and out of the ordinary.

. . . The romantic stands above and outside the stream of everyday life conceived of as effortful, practical, routine, and unplayful. The unromantic, by contrast, is practical and requires that one deal with the daily problems of everyday life (lost luggage) and institutions. . . .

Furthermore, the experience of romance exhibits attributes of religious rituals: by isolating the objects of their emotions from all others, dressing in a specific way, eating special foods, rearranging the space, or moving to another space and setting a special time of celebration, people are able to experience a time that has the subjective "texture" of holiday celebrations, in a space physically or symbolically removed from that of everyday life. They reach a domain where their feelings are heightened, their vital energies are regenerated, the bond to their immediate partner is reaffirmed. Romance is lived in the symbolic mode of ritual, but it also displays the properties of the staged dramas of everyday life. The use of artifacts (clothes, music, light, and food), the use of self-contained units of space and time, the exchange of ritual words of love make romance a "staged reality," an act within which public meanings are exchanged and dramatized to intensify the bond.

A CONSUMING ROMANCE

. . . The cultural categories of romance and love are somewhat intermixed. Although respondents could differentiate between romance and long-lasting love, when they were faced with actual representations, the distinction was not always intelligible to them. The emotion of love was intertwined with romance, defined in terms of the props and settings that induce a "special atmosphere." The most significant implication of this amalgam of meanings is that the boundary between romance as an ambience and intimacy as mutual knowledge and trust is fuzzy, even for those who insist that it should remain distinct.

Asked to report how they "usually spend" or "would like to spend" a romantic moment, fifty respondents gave a wide range of answers: having dinner in a restaurant, having dinner at home, drinking champagne by the fireplace, walking in Central Park, canoeing, going to Mexico, walking along the beach, making love, talking, and so forth (The variety of these answers is outlined in table 1.)

Three main categories of activities are apparent: *gastronomic* (e.g., preparing or purchasing food at home or in a restaurant); *cultural* (e.g., going to the movies, the opera, or a sports event), and *touristic* (e.g., going to a vacation spot or a foreign country). The last category includes exploration of the natural landscape, whether in the city in which the respondent lived (e.g., walk in Central Park, walk near the river) or in a foreign country. Tourism in ethnic or cultural centers and travels into nature are lumped together because both imply geographical or symbolic removal from the daily urban landscape. "Making love" is the only romantic activity not included in the table because it stands as a category on its own: it is neither cultural, commercial, nor touristic.

Romantic encounters can be further divided into three categories according to their relation to consumption: The first is *direct,* where the romantic moment is predicated on the purchase of a commodity, either transient or durable (e.g., eating at a restaurant, buying a cruise to the Caribbean, or giving a diamond ring). An example of a direct consumerist gastronomic activity is eating at a French restaurant, while a more indirect consumption of food is having drinks or dinner at home. Directly consumerist cultural activities are going to a movie, a concert, a play, theater, a baseball game, or dancing. A typical "direct consumption-touristic-natural" activity is a weekend trip to New York or to Cape May at the seashore or, even more dramatic, a worldwide cruise. . . .

The second type of consumption is *indirect,* where the romantic moment depends on consumption but is not the direct outcome of the act of purchasing a good. For example, watching television or listening to records presupposes the purchase of the television set or a stereo system,

Table 1. Respondents' Romantic Activities

Activities	Direct consumption	Indirect consumption	No consumption
Gastronomic	Restaurant Drink	Dinner at home	—
Cultural	Movies Theater Dancing: formal/discoteque Opera Museum	Reading a book or newspaper Crossword puzzle Watching TV Listening to music Painting together Looking at photographs	Talking; sharing secrets; talking about art and politics
Touristic	Cruise Skiing Travel abroad Weekend at seashore Waterskiing Canoeing	Fishing Gardening Picnicking	Walking (river or woods; city streets or park)

but the purchase of these products does not circumscribe the romantic moment itself.

Finally, the third type of romantic activity along this dimension takes place *without mediation* of consumption. For example, talking and making love do not in and of themselves imply or presuppose consumption, although in practice they often do (e.g., having a good conversation in a restaurant setting or making love at a hotel on vacation). A nonconsumerist touristic activity consists in walking within an urban or natural landscape (park, street, near a river). But defining this category solely by the absence of purchasing is problematic because many activities that do not demand money are modeled on consumerist ones (e.g., going to a museum with free admission or to a lecture on art) and because these activities often presuppose substantial cultural capital. Hence, I do not include in this category any activity predicated on cultural products funded by outside agencies. A nonconsumerist romantic moment excludes the intervention or mediation of cultural products and artifacts in the romantic interchange, whether the lovers or someone else paid for them, and whether they are purchased during or before the romantic moment. By and large this means that three categories of romantic activities qualify as nonconsumerist: walks (in nature or in the city), domestic interactions (e.g., talking), and making love at home. . . .

A woman, recalling a recent romantic afternoon with her husband, ended her recounting by saying, "and we watched a video and we just stayed home together" (female lawyer, interview 23). . . .

"Just" staying home together or lying around the house was romantic because these activities were informal, comfortable, and therefore intimate. But it is instructive to notice that even in such moments, respondents used modern technologies (video, television) as a focal point of their interaction. Even domestic romantic moments that were removed from the public sphere of consumption were somewhat organized around an external activity (e.g., cooking or watching a movie) or object ("making love" being the most noticeable exception to this rule). . . .

A few conclusions can be drawn from this short review of the variety of romantic activities. Consumerist practices *coexist* with nonconsumerist or moderately consumerist ones and do not completely override them. Although people often mention that "staying home," "talking," and "making love" are romantic, it remains that the range of consumerist activities is much wider and varied than that of the nonconsumerist ones. Finally, technologies and/or commodities are frequently present in romantic moments because through their mediation partners communicate with each other. We can now take this a step further and show that consumerist romantic moments are culturally *more prevalent* than nonconsumerist ones. Although coexisting, consumerist and nonconsumerist romantic practices do not lie on a continuum, nor are they simply opposed to each other; rather, consumerist romantic moments serve as the *standard against which nonconsumerist moments are constructed.* This accounts for the primacy and preeminence of the former. The following analysis of gastronomic and touristic-natural romantic activities will help clarify this.

Dining In, Dining Out

Of all romantic activities taking place within the framework of daily life, the sharing of food is the most frequently cited in my interviews. Undoubtedly, one reason for this is that in most societies, partnership and social bonds are usually marked by the sharing of food. But this observation leaves one of my findings unexplained: dinners at restaurants were viewed as more "typically" romantic than dinners at home. Even if dinners at home and dinners in the restaurants had been cited equally as romantic, the choice of the restaurant differs from that of home in that it corresponds to the "purest" ideal type of romance. Research by Bachen and Illouz clarifies this difference. (Bachen and Illouz, forthcoming).

The research was based on a sample of 180 American children, ranging in age from eight to sixteen, who were interviewed in a major city on the East Coast. Among other questions, they were asked to describe a "romantic dinner." Almost all the children referred to a restaurant dinner and, more specifically, to dinner in an elegant restaurant. . . . Overall, interviews indicated that, although these eight- to sixteen-year-olds had never experienced romance in such settings, they were able to summon the mental picture of a restaurant in highly precise and concrete detail. The authors argue that, in terms of romance, the visual representation of the restaurant dinner precedes and supersedes, cognitively, the representation of the dinner at home. That is, the association of romance with consumption *is acquired earlier and is mentally more salient than the representation of nonconsumerist romantic moments.*

This claim can also be applied to adults, as evidenced by my respondents. One seventeen-year-old interviewee acknowledged that she had no romantic experience, but when asked to describe a romantic moment, she answered, "I would probably go to a restaurant, have wine, just the common thing to do." This young, inexperienced adolescent immediately conjured up the image of a restaurant and commented on it as a standard romantic activity. So too, my adult respondents, who had firsthand experience of romance, frequently referred to the restaurant as "standard" or "cliché" or "routine."

. . . Even if respondents preferred "informal" and "nonfancy" restaurants, they found the "standard" image of a romantic dinner in a luxurious and expensive restaurant ("formal," "glitzy"), rather than at home. Foremost in respondents' minds, the restaurant dinner is the romantic cliché; dinner at home (the private, cozy place) is seen as a deviation from this standard. . . . Although dinner at home is often viewed as romantic, why is the restaurant dinner the standard definition of a romantic dinner, both in the imaginations of ordinary people and in the imagery of the media? This question suggests more general ones. Why are these images pervasive and not others? From where do they get their symbolic power, their ability to impose themselves on one's consciousness and to function as collective representations? Even if we establish that the market is the main purveyor of these images of restaurants, this fact in itself does not explain why people prefer restaurant dinners on dates and why they conceive of restaurants as more stereotypically romantic than dinners at home or picnics. Unless we assume that people's representations and experiences are produced mechanically by symbols and goods provided by the market, we cannot deduce subjective experiences from the market. If dining at a restaurant stands as a standard romantic activity (albeit a scorned one, at least by some of my respondents), it is because this activity accomplishes "things" that are useful and meaningful to people. The problem, then, is to identify what these "things" are. Taking up a previous strand of my argument, I claim that romantic activities involving consumption carry more powerful and resonant symbols than those that do not involve consumption because the former facilitate the ritualization of romance.

Romantic Consumption as Ritual

Restaurants are romantic because they enable people to step out of their daily lives into a setting saturated with ritual meaning. The design of the restaurant reinforces and transcends the temporal, spatial, artifactual, and emotional boundaries discussed earlier. Time in a restaurant is self-contained: entering and leaving set the temporal and spatial boundaries, and furthermore, the self-contained time of the meal is not susceptible to the schedules and constraints of the outside, "profane" world. For example, one respondent mentioned that a restaurant dinner is romantic because it takes place at a much slower pace than dinner at home. If, in the course of everyday life, people are time-conscious and end their meals quickly, or at least under the pressure of outside constraints, then the "slowness" of the restaurant meal is perceived as romantic by contrast, because its slowness implies that the relationship is the center and the goal of the interaction.

The restaurant is usually an alternative space to the home or workplace. For teenagers living with their parents or for people with children, the restaurant provides temporary isolation from the constraints of the family. Within the restaurant, the tables are arranged so as to isolate each group of diners. The typical romantic restaurant is arranged with "niches" and isolated dark corners, in contrast, for example, to the unromantic Irish bar in which many people crowd together. The sense of a spatial boundary was also evoked by respondents, who referred to the feeling of "being separate amidst a crowd": "I guess my idea of romance is an idea of isolation, in a nice setting, in a nice restaurant, being away from everyone else, or in a nice bar, away from everyone else, maybe being able to watch everybody else but being able to be very much in our own world" (female avant-garde filmmaker, interview 2).

The restaurant displays an array of artifactual boundaries that sets it apart from the stream of everyday life. In restaurants, the food is often "different," more "special" or "exotic" than daily foods. . . . The emotional invigoration of the romantic bond that usually accompanies the consumption of food in a restaurant can be traced to the formal, ritualized nature of interactions there and the conventionalized significance of the various articles involved. The clothing, the usually elegant tablecloth and silverware, the service of a waiter, the presentation of the meal, and the meal itself all make the restaurant a more formal experience than dinner at home. As Finkelstein has observed, "dining out is a highly mannered event" (Finkelstein, 1989). People usually dress up to go to restaurants and interact in a more constrained way than usual, and the presentation and reading of the menu delay the immediate satisfaction of hunger. More importantly, the presentation of a written list of food items—a list that often uses unusual or poetic names—brings, at least temporarily, the signifiers of the food to the fore of the diner's consciousness, if not to the fore of the couple's conversation. The menu is a formal cultural device that defuses the taken-for-grantedness of daily consumption of foods and ritualizes the act of eating, turning it into a symbolic event. In a restaurant, food is not only food. It is part of an ordered set of symbols and signifiers that "mark" the restaurant dinner as "romantic," codifying its meaning and making it more cognitively salient than dinner at home. In structuring the interaction between the participants in this way, the restaurant meal intensifies their interaction. . . .

At the same time that the restaurant heightens the private bond between the couple, it takes that bond to the public sphere. In that respect, the restaurant dinner carries contradictory sociological properties: it affirms the "regressive" intimacy of the couple and at the same time makes the meal a public experience. That is, the couple retreat to their "island of privacy" through highly codified and standardized meanings.

Compared to the restaurant script, dinner at home appears on the informal end of a continuum ranging from formality to informality. Domestic nonconsumerist activities in general (staying home, cooking together, gardening together, talking after dinner, etc.) stress comfort, informality, and intimacy, while the restaurant dinner, a more formal and public experience, tends to mobilize higher levels of energy and intensity. . . . The romantic dinner at home uses many of the elements of restaurant rituals: food and dress "nicer" than usual, soft lighting, background music. These elements make the dinner at home a ritual experience analogous to the consumerist version of dinner in the restaurant. This respondent even referred to the moment as a "ritual," thus suggesting that he was aware of the formal character of romance and intentionally constructed it as such. The symbolic meaning of restaurants can be skillfully reconstructed by respondents within their homes through their manipulation of the boundaries implied in ritual. This in turn suggests that the sphere of consumption and the rituals it affords are not straightforwardly opposed to the domestic sphere.

Conversely, going to a restaurant can be often associated with such "at-home" meanings as informality and simplicity. The social groups that Bourdieu viewed as having much cultural capital but little economic capital (e.g., the cultural specialists) prefer informal, simple, unpretentious, "homey" restaurants, in contrast to groups with more money but less cultural capital (the "bourgeoisie"), who are more likely to prefer the classical elegance of a French restaurant (Bourdieu, 1972). However, the *expensive dinner* is still the cultural standard against which other tastes are articulated.

SELECTED REFERENCES

Alexander, J. 1988. *Durkheimian Sociology: Cultural Studies.* Cambridge: Cambridge University Press.

Bachen, C., and E. Illouz. Forthcoming. Visions of Romance: A Cognitive Approach to Media Effects. Paper presented at the International Communication Association, Dublin.

Bourdieu, P. 1972. Marriage Strategies as Strategies of Social Reproduction. In *Family and Society,* edited by Forster and O. Ranum, trans. E. Forster and P. Ranu, 117-44. Baltimore: Johns Hopkins University Press.

Durkheim, E. 1973. *On Morality and Society,* edited and translated by Robert N. Bellah. Chicago: University of Chicago Press.

Durkheim, E. 1985. *Readings from Emile Durkheim,* edited by K. Thompson. London: The Open University.

Finkelstein, J. 1989. *Dining Out: A Sociology of Modern Manners.* New York: New York University Press.

Goffman, E. 1967. *Interaction Ritual.* New York: Pantheon.

Moore, S., and B. Myerhoff, eds. 1977. *Secular Ritual.* Assen: Van Gorcum.

Woman, April 1988, 27.

Wuthnow, R. 1987. *Meaning and Moral Order.* Berkeley: University of California Press.

Zerubavel, E. 1981. *Hidden Rhythms.* Chicago: University of Chicago Press.

6

Gender and Courtship Entitlement

Responses to Personal Ads

ERICH GOODE

1996

At the core of the heterosexual courtship process we find the male and female gender roles: what men and women are expected to do, specifically, to and with one another. The notion of gender roles as complementary, a building-block of the functionalist paradigm (Parsons & Bales, 1955), was abandoned more than a generation ago. Now we realize that we cannot assume that the male and female gender roles are complementary: They often clash and conflict. Given the vastly different expectations that men hold of women's behavior in contrast with those held by women of men's behavior, it should be little short of remarkable that men and women are still capable of meeting, dating, and engaging in courtship at all, not to mention the fact that a sizeable number end up deciding to formalize their relationships by marriage. By now, we are all familiar with the "mixed messages" men and women send to one another (Tavris & Wade, 1984, p. 112). Tannen (1990) has documented the fact that men and women don't "understand" what members of the other sex are saying; Henley and her associates (1985) go so far as to isolate distinctive styles of "womanspeak" and "manspeak"; Goodchilds and Zellman (1984) show us that adolescent boys and girls hold entirely different notions of what sex-linked signals and cues mean, and should mean, to the other party. Male-female relations are fraught with ambiguity, confusion, misunderstanding, and conflict. And, while courtship is not the only game in town, most men continue to seek the romantic company of women, and most women continue to desire men for a romantic, marriage-oriented relationship.

Some version of conflict theory has replaced the functionalist paradigm in the study of gender roles; the touchstone of the conflict approach has been power differentials between men and women. By acting out stereotypical gender role demands, men and women are assumed to "do" gender by exercising power in a distinctively masculine and feminine fashion. Women are expected to be sensitive to the needs and wishes of their role partners and be modest and unassuming about making demands on others. In contrast, men's role expectations include making explicit demands on the world, egocentrism, and attempting to establish dominance, especially over females. It is perhaps in the arena of courtship that these gender-linked roles play out most visibly and sharply. Superficially, it might seem that these role demands are complementary; in the concrete world, they make for a less than perfect system of communication. In fact, in a number of crucial respects, men's and women's expectations differ sharply, specifically with respect to what they expect from members of the other gender. While men are socialized to expect their needs to be satisfied, especially by women, their expectations

From "Gender and Courtship Entitlement: Responses to Personal Ads," by Erich Goode, from *Sex Roles*, vol. 34(3/4): 141-169; 1996.

may be wildly unrealistic. While women are socialized to be nurturant and to satisfy the needs of others, they are forced to be selective as to just whose needs they decide to satisfy. Put another way, one source of gender miscommunication comes about as a consequence of masculine role overreach: Men often expect more than women feel they are entitled to. I would like to argue that a sense of male entitlement in the courtship process is simultaneously institutionalized and out of synchrony with women's expectations. Entitlement is a fixture of the stereotypical masculine role, specifically with respect to what they want from women; even more strongly, in the courtship process, men feel they are entitled to far more than their female role partners feel they deserve.

Entitlement is the feeling by someone, or the members of a group or category, that they have a right to certain privileges, that specific rewards should be forthcoming, that the resources they covet are rightfully theirs by virtue of what they have done or who they are. Some men feel "entitled" to sex on a date after they have paid for an expensive meal; in an earlier era, some women felt "entitled" to an engagement ring after an exclusive dating relationship had lasted for a designated period of time; some dating parties feel "entitled" to date a partner possessing a certain "value" on the romantic marketplace. The concept of entitlement has both a "subjective" dimension—what the actor expects—and an "objective" dimension—what the community feels the actor deserves. These two dimensions may or may not agree with one another. Thus, on the one hand, we must consider the social embeddedness of an audience's evaluation concerning what sorts of actions or utterances qualify as entitlement. While the specific actions or utterances that come to be seen as unreasonable emanate from specific individuals, what qualifies as an unreasonable demand is determined collectively. Unreasonable expectations are exhibited only when individuals step outside the boundaries of what is regarded as fair and reasonable. Unrealistic entitlement exists when and only when an action or utterance is validated as such by specific audiences. Notions of fairness and equity are inextricably bound to a particular cultural system and social and economic structure. Behavior or utterances which are taken to be presumptuous in one context at one time may not be so regarded in another. In early 20th-century Latin America, for instance, female servants were far more vulnerable to the sexual demands of their employers, whereas today, such a demand would be regarded as arrogant, presumptuous, and improper. In 19th-century England, as many historical accounts indicate, it was taken for granted that upper-middle-class men be permitted fairly wide sexual access to lower and working-class women; today, again, such an expectation would be met with self-righteous anger.

PLACING/ANSWERING PERSONAL ADS AS A FORM OF COURTSHIP

Boy has been meeting girl, and girl meeting boy, for millennia, but only during the past generation or so have millions done so by answering a "personals" advertisement. Hundreds, possibly thousands, of periodicals in the United States and Canada, and probably at least as many in other countries around the world, carry a column in which advertisers request replies from dating or romantic partners. Advertisers, and their respondents, run the gamut from heterosexual to homosexual, from single to married, from teenagers to octogenarians, from those advertising for a single lifetime marital partner to those seeking candidates for impersonal, commercial sex, "discreet" afternoon trysts, threesomes, foursomes, or one-night stands. Still, the overwhelming majority of personal ads are written by heterosexual singles between 22 and 65 who at least claim to seek a long-term, marriage-oriented relationship. Most of the publications in which personals appear run them as a side-line. They include respectable, family-oriented daily newspapers; sensationalistic tabloids; alternative, fringe, or "underground" weeklies; intellectual and literary journals;

local advertisement sheets which are distributed free to subscribers; and slick magazines for trendy yuppie consumers. In addition, a sizeable number are newsletters devoted more or less exclusively to personal advertisements. Recently, personals making use of electronic mail have emerged; and last, dating services which show videotapes of prospective dates to clients, likewise, represent a recent technological innovation in "mediated" modes of courtship.

The procedure for using personal ads as a means of meeting dating partners is essentially the same for all periodicals carrying such a column. The publication prints a series of ads, written by advertisers. Typically, the ads describe the characteristics (or putative characteristics) of the advertiser and describe one or more qualities or characteristics sought in a responder, that is, in a potential or ideal dating partner. Sometimes, the nature of the relationship desired will also be spelled out in an ad. Each ad includes a post office box number, which is held by the publication carrying the ad, and/or a telephone code, again, controlled by the ad-bearing publication. Today, voice mail is a more common avenue of responding to personal ads than letters and notes. The responses are forwarded to the subscriber who then makes a decision as to which ones he or she will answer, if any.

Placing and responding to personal advertisements and other "mediated" or "formal" modes of courtship have been studied as a form of dating behavior. Aside from the personals, they include computer matches, videodating, and matchmaking consultants. It is crucial to recognize the fact that there are both similarities and differences between these less orthodox channels and the more traditional ones. There are at least three major points of difference. First, in "mediated" courtship, the initial selection process takes place prior to a face-to-face meeting. Second, participants have a great deal more deliberate control over the information they allow potential dating partners to have access to than is true of more informal, naturalistic avenues of courtship (Woll & Young, 189, p. 483). And third, while courtship through traditional channels is overwhelmingly a product of personal associations (friendship networks, school, place of employment, etc.), mediated courtship transcends this restriction by attempting to establish an intimate relationship with strangers (Laumann et al., 1994, p. 4). . . .

Courtship has been described as an exchange process or a "market system" (W. J. Goode, 1982, p. 52). As a general rule, like marries like with respect to a wide range of characteristics; moreover, homogamy increases as the length and intimacy of the relationship increases. This is especially the case when it comes to marriage. The process of courtship can be looked at in "supply and demand" terms (p. 53); each partner commands a rough "price" on the dating and courtship market, depending on the supply of dating parties who are sought and competed against. An undesirable dating party may dream of a more desirable partner but he or she cannot command sufficient resources to attract such a partner (p. 53). For key characteristics, then, homogamy rules courtship dynamics. Not only will dating parties tend to learn their value or worth on the dating marketplace through routine socialization and interaction, but their friends will also encourage them to seek partners of similar value (p. 53). Of course, any given characteristic will not necessarily possess the same value to and for men and women equally. More specifically, nearly universally, attractiveness tends to be valued more in women by men and their cohorts, whereas economic success tends to be valued more in men by women and their cohorts (p. 53). Once again, a host of additional factors come into play in choosing a romantic partner (personality factors, for one thing). Nonetheless, the literature is unambiguous about this: Our qualities are ranked by actors in the romantic "marketplace," and the partners who gravitate to one another tend to share similar rankings.

. . . Most dating persons not only size up the value of potential partners in terms of their possession of specific qualities, they also determine their own value to their potential partners. That is, if the value of a given potential partner seems excessive, often, a person will not approach him or her for fear of rejection. That is, some persons realistically reach the conclusion that certain persons are so desirable on the dating "marketplace" that they will not be interested in them, so why make the effort to date them? Hence, a key element in the courtship process is not merely the possession by potential dating partners of certain valued characteristics, it is also a question of the possession of a requisite and necessary quantity of those characteristics. This equation is often short-circuited if an extremely desirable person expresses interest in dating the individual in question.

In addition—and here we arrive at the subject of this paper—not all dating persons are intimidated by the extreme desirability of potential partners. In fact, some have a sense of entitlement to partners who, others would feel, are excessively desirable for them. These "deviant cases" make the courtship process all the more interesting. Lastly, the value of the members of a given dating pool to a given person, and his or her value to them, is relative to their numbers and desirability. If, in a given courtship pool, there is an abundance of desirable dating partners, one will evaluate a given person in a more critical light than if there are few, most of which would otherwise be seen as undesirable. The factor of availability always enters into the equation, and will mitigate evaluations of what specific actions constitute instances of unrealistic and inappropriate entitlement.

COURTSHIP IN THE PERSONALS: FOUR GENERALIZATIONS

Four generalizations emerge from the literature on the personal ads (or other mediated, formal, or simulated dating methods similar to personal ads) as a system of courtship and exchange. The first is the central role of physical attractiveness. The second is the process of "matching." The third is the fact that dating parties exchange "packages" of traits. And the fourth is the gender-linked valuation of traits, characteristics, and dimensions.

The Role of Attractiveness

In a "computer dance" experiment in which undergraduate subjects were matched with dates, a team of social psychologists found that the only factor that determined whether subjects liked their dates, wanted to go out with that partner again, and actually asked the partner out, was how attractive the partner was (Walster et al., 1966). A later study found that, in a series of five encounters, rather than diminishing in importance, physical attractiveness had a "continuing and undiminishing effect on liking" (Mathes, 1975). Numerous studies have been conducted on the role of physical attractiveness in the courtship process; the evidence documenting its strength is abundant and convincing, even overwhelming (Berscheid & Walster, 1974; Patzer, 1985, pp. 81–96; Hatfield & Sprecher, 1986, pp.105ff. 139ff.; Woll, 1986).

The Matching Process

The physical attractiveness phenomenon, taken by itself, would predict the "idealistic" or "maximization" strategy among courtship participants. That is, other things being equal, every party in the dating enterprise would seek partners who are regarded as highly attractive, regardless of his or her own level of attractiveness. However, here, a mitigating factor intervenes: the likelihood of rejection. Thus, "ideally, people would prefer to date very attractive others, but, because rejection is costly, they end up choosing someone of about their same level of attractiveness" (Hatfield & Sprecher, 1986, p. 114). The risk of rejection moderates dating aspirations (p. 114). A "4" might wish to date a "10," but why would a "10" wish to date a "4"? After all, if the attractiveness phenomenon really is as strong as researchers insist, a "10" will wish to date only a "10"; it is a bargain that will be not only desired, but it will be attainable as well. More attractive partners in dating situations expect their dates to be highly attractive (Berscheid et al., 1971, p. 186). Each partner not only desires something in a potential relationship, he or she brings something to that relationship as well. In short, partners exchange characteristics. Rejection, and the prospect of not being able to date at all, result in courtship parties drifting toward persons similar to themselves in levels of attractiveness. As a consequence of potential rejection, rather than adopting the "idealistic" or "maximization" strategy—attempting to date the most attractive partners possible—partners are forced to accept a more "realistic" or "matching" strategy in dating. The explanation for the dynamic that rules this strategy is referred to as the "equity" hypothesis (Berscheid et al., 1971).

Dating Partners as "Packages" of Traits

While physical attractiveness is a—perhaps the—crucial factor determining the sifting, weighing, and matching process in courtship, other factors are simultaneously at work. Partners exchange a number of traits and characteristics, along a number of dimensions. Partners exchange a "profile" of traits simultaneously. On one side of the courtship process, each party is offering "a collection of diverse resources"; on the other side, "another collection of diverse resources is being sought" (Hirschman, 1987, pp. 98–99). Some of the elements in this "collection" include love, physical attractiveness or beauty, educational attainment, intellectual status, occupational status, money, and so on. With every courtship exchange, each partner or party seeks and offers a "package" of traits, rather than just one. A "profile" of each potential partner is assembled; a given party may rank high in several dimensions but low in others. Each party seeks the maximization of the overall value of all of the relevant dimensions relative to his or her overall value on these same dimensions. A fair exchange would be represented by equity in this exchange process—trading one total "package" that is roughly equal in all the relevant dimensions to another on the courtship marketplace. Thus, a young woman with average looks who comes from an upper-middle family might be willing to date an ambitious, attractive man who stems from a working-class background. Demanding to date partners whose total value is vastly greater than one's own would not only represent a serious lack of equity; it is also highly likely to elicit rejection.

Gender-Linked Valuation of Traits

While, for all men and women in a dating context, the attractiveness of their partner is a valuable commodity, the precise value of physical attractiveness varies by gender. The physical attractiveness of their partners is valued more strongly by men than by women. In contrast, women weigh occupational and economic success more heavily in their valuation of their potential partners. In personal ads, men are significantly more likely to request attractiveness in a partner and offer financial and occupational success; women are more likely to request financial and occupational success and offer attractiveness (Harrison & Saeed, 1977; Cameron, Oskamp, & Sparks, 1977; Deaux & Hanna, 1984; Koestner & Wheeler, 1988). The system of requests and offers expressed in personal ads seems to match the stereotypical roles and expectations that men and women exhibit in concrete courting and dating situations generally.

. . . To determine the relative importance and role of two hierarchical qualities—physical attractiveness and occupational/financial success—for men and women involved in the personal advertisements-generated courtship process, I placed four ads; each one appeared in four different newspapers or magazines which carry personal advertisements. (All four journals are published in or near large East Coast cities; in two, personal ads are a side-line of the journal and in two, they are its main function.) Thus, a total of 16 personal ads appeared. The wording in the four ads was almost identical except as it pertained to the two crucial variables. One ad referred to a beautiful waitress; she was intended to offer a potential date who ranked high on physical attractiveness and low on financial success. Another ad referred to a financially successful female lawyer who was "average" in appearance, that is, a woman who ranks considerably lower in attractiveness but considerably higher in economic success than the waitress. A third ad referred to a handsome taxicab driver, who ranked high in attractiveness and relatively low in success. And the fourth ad referred to an average-looking male lawyer, who ranked lower in looks and higher in success than the taxicab driver.

As every social scientist knows, a proper experiment would entail manipulating all the relevant variables; in such an experiment, we would have, in addition to the ads I placed, a handsome male lawyer, a beautiful female lawyer, an average-looking cabdriver, and an average-looking waitress. I feared that doubling the number of ads in a single issue of the newspaper or magazine which carried them would arouse some readers' suspicions. Instead of conforming to the dictates of the classic experiment, therefore, I chose to conduct a less-ideal "quasi-experiment." In addition, I anguished over the equivalency of the occupational categories; a taxicab driver is not exactly equivalent, as a man's occupation, to a waitress for a woman. The problem is, in the cities in which the journals that carried these ads appeared, a male waiter is not equivalent to a waitress, either; many waiters in fashionable restaurants earn incomes that would place them in the top quartile of the earning of all occupations, whereas this is very rarely true of waitresses. Again, I chose what I saw as a rough approximation. In addition, in this study, I focused exclusively on heterosexual courtship; the dynamics of homosexual dating are sufficiently different as to demand a separate discussion. Several researchers (for instance, Laner, 1979; Deaux & Hanna, 1984) have explored heterosexual-homosexual differences in personal ad courtship behavior in detail.

Below, the four ads I placed, with alternative wording, are reproduced. Each possible version was chosen at random. The very slight differences in wording were published so that readers' and responders' suspicion that an experiment was being conducted would not be aroused. (No such suspicion was expressed in any of the responses, even though many respondents, as we'll see, did answer two of the ads.) I requested written responses instead of voice mail because they tend to be easier to handle and consult.

> A beautiful [An extremely beautiful/A very beautiful], shapely, slender blonde, 5′7″, waitress, 30. Would like to meet a man for lasting [serious] relationship.
>
> A handsome [An extremely handsome/A very handsome], athletic, SM, broad shoulders, 6′, 32, cab driver. Seeks woman for lasting [serious] relationship.
>
> Successful [Extremely successful/Very successful] woman lawyer, intelligent, financially secure, 30, average appearance. Seeks man for lasting [serious] relationship.
>
> Successful [Extremely successful/Very successful], financially secure, intelligent, SM, attorney, 32, average looking. Wants to meet woman for lasting [serious] relationship.

Table I. Number of Replies to Personal Advertisements, by Type of Ad

Beautiful waitress	668
Successful, average-looking female lawyer	240
Successful, average-looking male lawyer	64
Handsome cabdriver	15
Total	987

Table II. Number of Replies to Personal Advertisements, by Gender

Replies by men to women's ads	908
Replies by women to men's ads	79

The question of which quality is more influential for men and women can be answered at once. Men were overwhelmingly more influenced by the factor of physical attractiveness in a potential date, while women were substantially more influenced by financial success. In fact, so much was this the case that one is led to suspect that, for a substantial number of men, the lack of a successful job is desirable in a female partner. The four "beautiful waitress" ads attracted a total of 668 responses, or a bit more than 160 per ad; the successful, average-looking female lawyer ads garnered 240 replies, or 60 per individual ad. The average-looking male lawyer ads generated 64 replies, or 16 per ad, and the handsome taxicab driver ads generated a total of only 15 responses, or less than two per individual ad (Table I). Another way of looking at the responses is to consider which gender is more likely to respond at all to a personal ad. Again, the answer is immediately forthcoming: The ads placed by women attracted 908 replies from men, while the ads placed by men attracted 79 replies from women, a ratio of over 11 to 1 (Table II). In this study at least, men were vastly more likely to respond to a personal advertisement than women. This ratio is considerably more lopsided than the 11-to-9 ratio Lynn and Shurgot (1984, p. 352) uncovered, but considerably less so than the 30-to-1 figure estimated by Foxman (1982, p. 108).

ENTITLEMENT IN ACTION: RESPONDING TO PERSONAL ADS

In courtship, an inappropriate sense of entitlement entails demanding more than is appropriate, more than one is entitled to, expecting more resources than one can marshal, command, or attract. It is important to stress the fact that, while some features of entitlement-attribution are transsexual, exactly what one demands in a partner is likely to be gender-linked and hence, what demands are likely to be labeled as an expression of entitlement, likewise, are gender-linked. An example of a demand that is likely to result in the application of a label of inappropriate entitlement regardless of the gender of the demander would be a "4" seeking a date with a "10." An example of a gender-linked demand that will be regarded as an expression of inappropriate entitlement is a man insisting on having intercourse with a woman after a 15-minute acquaintance, or a woman insisting on a man paying for a $200 dinner at the Four Seasons on the first date. As we see, the use of the entitlement notion in the courtship process must inevitably take note of the fact that, as in so many other spheres of life, courtship takes place within the context of a sexist society. Our desire for different arrangements should not prevent us from recognizing these still-enduring tendencies. The practitioner of inappropriate entitlement, then, wants to get something for nothing, or at least something of greater value for very little in return, in the dating and courtship process. Some of these practices violate expectations deemed appropriate for the practitioner's gender.

In responding to personal advertisements, inappropriate entitlement may be expressed in a variety of ways.

One form of entitlement expressed by personal ad respondents can be referred to as a minimalist strategy. "Minimalism" comes in when a respondent puts an absolute minimal effort into answering an ad; operationally, it is defined by engaging in at least one of five actions. Minimalism is sending to the advertiser: (1) a sheet of memo paper with nothing more than a name (whether first only or first and last) and a telephone number and/or an address and/or a post office box; (b) a business card; (c) a Xerox copy of a generic, all-purpose letter; (d) a computer-generated generic, all-purpose letter; or (e) a hand-written note containing 25 words or less. If a note or letter of more than 25 words is attached or added to or written on any of (a) through (d), it does not qualify as a minimalist effort. When ad placers receive a response characterized by (a) through (d), they are likely to feel that the responder has not given much time or thought, or made much of an effort, to connect with them. In the context of dating, such behavior is likely to be regarded as arrogant and presumptuous.

The first variety of entitlement inevitably segues into the second: casting one's net extremely widely. Foxman, author of a manual on how to use the personal ads to best advantage, assumes that all minimalist responders are wide net casters. She refers to them as "blitzers" (1982, p. 104); to ward off such a presumption, she advises: "Never send a copy or a form letter" (p. 111). Not only do blitzers answer many ads, inevitably, they do so on a more or less indiscriminate basis rather than a few on a selective, more judgmental basis. Foxman dismisses them as "pests" (p. 104). Echo Beker and Rosenwald, authors of another "how to" manual: "Photocopiers aren't just looking for you—they're looking for the whole world" (1985, p. 53). Block, likewise, chimes in with the following judgment: "bush leaguers send Xerox copies of form letters. . . . It's tacky, it's impersonal, and it looks as if you're shopping wholesale" (1984, p. 113). Inappropriate entitlement seems an apt word to refer to such behavior. Casting one's net extremely widely implies that the qualities or the responses of potential dating partners are of no consequence, that, regardless of their value on the dating marketplace, one has a right to date them.

. . . A third variety of entitlement entails demanding more from a projected dating partner and/or relationship than is deemed conventionally appropriate. That is, it is not only the qualities or characteristics of potential partners that are evaluated; what those parties can do for one may also be seen as holding positive value. Rewards by partners may be dispensed in a variety of ways, but they are likely to do so only after they have been induced to commit a certain measure of investment to the relationship. Expecting certain rewards without having made the requisite investment is likely to court the charge of presumptuousness. Men may indicate that they expect sex immediately, on the first date; women may expect a long-term commitment from a partner they deem desirable and appropriate long before the nature of the relationship has been negotiated or solidified. Certainly such expectations, and actions based on them, are likely to be regarded as instances of inappropriate entitlement. In addition, and related to the arrogance of demanding more than a given relationship can sustain at a given time, there is the arrogance of terms of address: addressing the ad placer in terms of intimacy and endearment before the respondent has earned the right to do so.

A fourth variety of entitlement is seeking a much more desirable partner than oneself, as determined by widely accepted conventional values—a partner, most observers would feel, one does not "deserve" to date. As noted above, much of the literature on dating and courtship generally, and on the personal ads more specifically, has focused on the "market" metaphor in dating and mating (Hirschman, 1987). The validity of the market metaphor rests on dating parties realistically assessing their own courtship "value" and, as a consequence, seeking out a partner of roughly the same value. Seeking out a partner who is considerably more valuable than oneself in a dating and courtship context is likely to attract the charge of arrogance and presumptuousness. While such behavior, as we saw, almost inevitably courts rejection, for the more highly valued party, again, following the marketing metaphor, it also entails wasting valuable resources (time, for instance) that result in no pay-off. Consequently, they are motivated to avoid such encounters (if they can do so).

. . . The first two of these forms of inappropriate entitlement—minimalism and blitzing—can be measured fairly readily and reliably; I will deal with them first. Variety three—demanding an inappropriate relationship—requires a scrutiny of the letters. Variety four, demanding an inappropriately valuable partner, requires a comparison between the characteristics of the person who (presumably) placed the ad and those of the responder.

Entitlement in Action: Men vs. Women

When we compare male versus female responders, we find marked differences in their expression of the more straightforward, objective measures of entitlement. Eleven of the 79 female responses to my male ads (or 14 percent) were "minimalist" in nature, that is, they were other than (and, in fact, considerably less than) personal letters. . . .

In contrast, 79 out of the 240 responders to the female lawyer (33 percent) and 171 of the 668 responses to the waitress (26 percent) sent in minimalist responses—overall, twice the frequency for the men as for the women responders. . . .

Three of the women were "blitzers," that is, operationally, they sent in responses to both the male lawyer and the cabdriver. That means our 79 letters represent 76 women; of these 76, or 4 percent, are blitzers. In contrast, 104 of the men sent in responses to both the waitress and the female lawyer; this represents 804 men, of which 13 percent are blitzers. . . . Not only were men more likely to send in multiple responses to the two female ads; in the issues in all of the journals in which these ads appeared, the total number of "men seeking women" ads were more frequent than the total number of "women seeking men" ads. . . .

Inappropriate Intimacies

My third form of entitlement is presumptuousness of relationship: taking inappropriate liberties or intimacies before a relationship has been allowed to develop. The use of certain forms of address represents one such inappropriate intimacy. The waitress received only seven letters (one percent of all letters) whose authors used inappropriately intimate forms of address, such as "sugar," "honey," "sweetie," and "princess." The female lawyer received only three (again, one percent). Another inappropriate intimacy is represented by explicit sexual references. I coded a letter as explicitly sexual only if two or more such references were used. Thus, if an author claimed to be a "terrific lover" or to give "great massages," and no other references were present, I did not code the letter as explicitly sexual. However, if either was coupled with the statement "and I know how to satisfy a woman," I regarded it as explicitly sexual. With a few, occasionally blatant, exceptions, in general, the letters were remarkably free of explicit sexual references. The waitress received only 19 letters containing explicit sexual references (3 percent) and the female lawyer received six (also 3 percent). Considering the anonymous nature of personal ad interaction, the lack of overt sexuality in the letters is remarkable. The waitress received 10 letters from men who openly stated that they were married and wanted a "discreet" affair; the female lawyer received only two.

In contrast, it is possible that some of the responders to the cabdriver ad envisioned him not as a serious candidate for a romantic relationship but primarily as a sex object or a "boy toy." Two of the 15 responses to the cabdriver enclosed photographs of the responder in very skimpy outfits; one letter stated that the writer desires "a man who is built big and strong"; one began her letter with "Dear sexy" and another went on to explain that she likes "very macho, domineering guys"; another warned "Be discreet—no one knows I'm doing this." One receives the impression that the very act of attempting to date a cabdriver through the personal already represents a daring, almost deviant act for a woman; the sexually explicit nature of such an act seems to be implied by the effort. One-third (5 out of 15) of the responses to the cabdriver included intimacies that might seem inappropriate at a preliminary stage in the courtship process. In contrast, overtly sexual references were almost completely lacking in the responses to the male lawyer ads. One woman did explain that, among other things, she enjoys sex "(yes, I said sex)," she repeated; the rest of the letters were free of such references. Thus, less than 2 percent of the female responses to the male lawyer's ad contained intimacies. . . .

THE PANEL OF JUDGES

I have determined my fourth criterion of inappropriate entitlement, demanding an overly valuable partner, by means of a "panel of judges." . . . All panel members were informed as to the bogus nature of the ad. Each was asked to read the ad placed by the two (fictive) parties of their own sex. They were then asked the following questions: "Try to picture what he/she is like and what sort of woman/man he/she is probably advertising for and wants to go out with. Now, imagine that he/she receives 10 replies. Let's say that one of the replies he/she receives is from this woman/man. Here's her/his photograph and here's her/his letter. Now, what do you think the chances are that the man/woman who placed the ad you read would want to go out with this woman/man. Would he/she get in touch with her/him and try to arrange a date? Why do you think this is the case?" . . . I coded panelists' judgments into three categories: first, the hypothetical ad placer would probably or certainly get in touch with the responder, that is, there was more than a 50/50 chance; second, there was roughly a 50/50 chance of this happening; and third, that the responder's getting a date with the ad placer was not very likely, that is, there was less than a 50/50 chance. Only one-third of the judgments made by my female panelists (10 panelists evaluating 10 letters) assessing whether the hypothetical ad placer would make an effort to date the male respondents who answered their ad said that there was better than a 50/50 chance—32 percent said so for the waitress and 38 percent said so for the lawyer. In contrast, the male panelists were much more optimistic about the female respondents' chances of getting a date with the fictive ad placer: Eight out of 10 of the 20 judgments concerning the respondents' chances of getting a date with the cabdriver were positive (80 percent) and over half (56 percent) of the 80 judgments on the respondents' getting a date with the male lawyer were positive. In short, for only a minority of their judgments did female panelists say that the female ad placers would be likely to make an effort to date the male respondents who answered the ads. In contrast, the female respondents were said to have a much better chance; more than half of the judgments made by my male panelists decided that the male ad placers were likely to make an attempt to date the female respondents. . . .

Clearly, then, men who put themselves forward as dating candidates are strikingly more likely to be deemed by opposite-sex panelists to be unacceptable partners for those whom they approach. To put the matter in the framework I have been discussing, men are far more likely to display inappropriate entitlement in their dating choices: Among men, less desirable partners are far more likely to consider themselves acceptable dating choices for those who are more desirable than is true of women. They are far less likely to be constrained by the fear of rejection, more likely to pursue the "maximization" rather than the "matching" strategy than is true of women. In short, the applicability of "equity" theory is heavily dependent on gender: It works better for the courtship strategy pursued by women than for that of men. Men are far less fearful of rejection—at least in their actions—and far more likely to over-evaluate themselves as acceptable dates for potential partners. I suggest that this is a fixture of the male gender role. . . . Personal advertisements can be looked at as a fragment of the courtship process. Dating and courtship via the personals is not a perfect reflection or slice of romantic processes generally. Still, their differences from more common and conventional modes must be both recognized and exploited as well; they represent, in Robert Merton's fine phrase, a "strategic research site" to study dating interaction. Few other modes of courtship so overtly attempt to transcend the usual constraints of social networks in generating romantic liaisons. Moreover, homogamy in dating almost certainly plays an increasingly influential role the more the relationship advances. This study has attempted to carry this scrutiny a step further than past studies of personals have done, that is, by examining responses to these ads rather than focusing exclusively on the ads themselves. Yet, ideally, we would want data from relationships that were established through personal ads and progress over time. Perhaps, one day, an even more adventurous researcher than I will attempt this study.

Certainly in the process of meeting partners and establishing a sexual and romantic relationship, the question of equity looms large: establishing each partner's relative "worth," and therefore who may date whom and under what circumstances. And yet, although equity certainly operates in who dates whom over a series of encounters, for a sizeable percentage of our respondents, it was not the sole determinant of who approached whom for a date. Clearly, large numbers of the respondents who answered our ads overestimated their worth on the dating marketplace and approached an unrealistically desirable candidate. Moreover, this was far likelier to be true of men than women. For men, in the earliest stages of attempting to establish a sexual relationship, the question of equity is far less pressing than it is for women. In fact, for many men—in spite of the risk of rejection—the maximization strategy reigns. The fact that this is a deviant or minority strategy, even among men, is not the point here; the fact that men are far more likely to employ the maximization strategy is important. The fact that men are far likelier to advance themselves as sexual candidates, in spite of their undesirability, even in this fairly unusual and atypical dating venue, reveals the crucial role of entitlement in heterosexual dating relations.

SELECTED REFERENCES

Beker, G., and C. Rosenwald. 1985. *The Personal: The Safe Way to Mate and Date.* New York: Zebra Books.

Berscheid, E., K. Dion, E. Walster, and G. Walster. 1971. Physical attractiveness and dating choice: A test of the matching hypothesis. *Journal of Experimental Social Psychology* 7:173–89.

Berscheid, E., and E. Walster. 1974. Physical attractiveness. *Advances in Experimental Social Psychology* 7:157–215.

Cameron, C., S. Oskamp, and W. Sparks. 1977. Courtship American style: Newspaper ads. *The Family Coordinator* 26:27–30.

Deaux, K., and R. Hanna. 1984. Courtship in the personals column: The influence of gender and sexual orientation. *Sex Roles* 11:363–75.

Foxman, S. 1982. *Classified Love: A Guide to the Personals.* New York: McGraw-Hill.

Goodchilds, J. D., and G. L. Zellman. 1984. Sexual Signaling and Sexual Aggression in Adolescent Relationships. In *Pornography and Sexual Aggression,* edited by N. M. Malamuth and E. Donnerstein. Orlando, FL: Academic Press.

Goode, W. J. 1982. *The Family.* 2d ed. Englewood Cliffs, NJ: Prentice-Hall.

Harrison, A. A., and L. Saeed. 1977. Let's make a deal: An analysis of revelations and stipulations in lonely hearts advertisements. *Journal of Personality and Social Psychology* 35:257–64.

Hatfield, E., and S. Sprecher. 1986. *Mirror, Mirror . . . The Importance of Looks in Everyday Life.* Albany: State University of New York Press.

Henley, N., M. Hamilton, and B. Thome. 1985. Womanspeak and manspeak: Sex differences in communication, verbal and nonverbal. In *Beyond Sex Roles,* edited by A. G. Sargent. St. Paul, MN: West Publishing.

Hirschman, E. C. 1987. People as products: Analysis of a complex marketing exchange. *Journal of Marketing* 51:98–108.

Koestner, R., and L. Wheeler. 1988. Self-presentation in personal advertisements: The influence of implicit notions of attraction and role expectations. *Journal of Personal Relationships* 5:149–60.

Laner, M. R. 1979. Growing older female: Heterosexual and homosexual. *Journal of Homosexuality* 4:267–75.

Laumann, E. O., J. H. Gagnon, R. T. Michael, and S. Michaels. 1994. *The Social Organization of Sexuality: Sexual Practices in the United States.* Chicago: University of Chicago Press.

Lynn, M., and B. A. Shurgot. 1984. Responses to lonely hearts advertisements: Effects of reported physical attractiveness, physique, and coloration. *Personality and Social Psychology Bulletin* 10:349–57.

Mathes, E. W. 1975. The effects of physical attractiveness and anxiety on heterosexual attraction over a series of five encounters. *Journal of Marriage and the Family* 37:769–73.

Parsons, T., and R. Bales. 1955. *Family Socialization and Interaction Process.* New York: Free Press.

Patzer, G. L. 1985. *The Physical Attractiveness Phenomena.* New York: Plenum Press.

Tannen, D. 1990. *You Just Don't Understand: Women and Men in Conversation.* New York: Ballantine Books.

Tavris, C., and C. Wade. 1984. *The Longest War: Sex Differences in Perspective.* San Diego: Harcourt Brace Jovanovich.

Walster, E., V. Aronson, D. Abrahams, and L. Rottmann. 1966. Importance of physical attractiveness in dating behavior. *Journal of Personality and Social Psychology* 4:508–16.

Woll, S. 1986. So many to choose from: Decision strategies in videodating. *Journal of Social and Personal Relationships* 3:43–52.

Woll, S. B., and P. Young. 1989. Looking for Mr. or Ms. Right: Self-presentation in videodating. *Journal of Marriage and the Family* 51:483–88.

7

Teenage Sexuality

PEPPER SCHWARTZ AND VIRGINIA RUTTER

1998

The publication of the Kinsey reports in the late 1940s stimulated anxiety that they would promote youthful, unconventional, nonmarital sexual experimentation. And, true enough, teens have embarked on sexual careers earlier and with greater freedom in the past century. . . . Kinsey reported that fewer than 6 percent of American women who were teens in the 1920s or earlier had had premarital intercourse. By 1991, 66 percent of high school senior women and 76 percent of high school senior men had had sexual intercourse.

Notice . . . that the shift for women is much greater than the shift for men. Adult commentary has labeled teen sexuality a spontaneous, irresponsible, narcissistic, and ever younger trend that is massively out of control, but the alarm would seem to be associated with the greater increase in women's—not men's—sexual freedom. Teens are demonized as morally wayward—because women have been admitting to engaging in sex at levels that are increasingly similar to men's. The persistent "gender gap" in acceptance of women's and men's sexuality is what we have referred to as the double standard.

The double standard is not the only contradiction attending teen sexuality. Teens may be having sex earlier, but they did not create the seductive adult clothing made for preteens or the ad campaigns with seductive 13-year-olds. Calvin Klein ads for jeans, T-shirts, and fragrance, and a host of other seductive images, present teens as having the sexy bodies that everybody wants. From here, contradictions multiply. On the one hand, prosecutors press statutory rape charges on behalf of girls who look 30 but are in fact 15. On the other, the culture urges those same young girls to be as sexy as they can and tells fashion models they are over the hill at 21. Our vision of sexiness is essentially based on the barely pubescent body, and yet the barely pubescent body is taboo.

ADOLESCENCE AND FAMILIES

If these adult-generated images of sexy teens seem novel, so is the life stage known as adolescence. Interestingly, teenage sexuality surfaced as a public issue just as the phase of life known as adolescence was recognized. Over the past 150 years, the age of puberty for both boys and girls has dropped from around 17 to around 11 or 12 (Rutter 1995), as a consequence of improved health and nutrition among youth. Yet youths don't legally become adults until their late teens. The interval between puberty and adulthood—adolescence—is a stage of life that had not existed biologically or socially until the mid-twentieth century.

A unique feature of this stage of life is youths' sexual maturity without social maturity. The contrast is apparent when one observes the bedroom of a 12-year-old girl who looks 18. She might still treasure stuffed animals or posters of horses or doll houses. And yet this same child is at least partly aware of the allure she holds for boys and men, and she may be eager to test her new powers of enchantment. Not surprisingly, parents are appalled and scared at this turn of biology—especially when it comes with rebellion against the restrictions they deem appropriate for a 12-year-old. Some psychologists have observed that the presence of sexually mature youth living at home has itself produced negativity of parents toward their teens, and sometimes even parental depression (Steinberg 1994). In other periods of history, a child old enough to be sexually active would be married or betrothed and out of the household. But today, youth tend to remain at home well past puberty, often until their 20s, when they finally complete their education.

In contemporary society, because puberty comes earlier but marriage comes later, a youth may experience sexual and romantic longing with no "socially acceptable" outlet for quite a long time. As teens mature and begin to have sexual relationships, they tend to resent parental interference or constraint. Parental judgments and controls might not be too insufferable for the children (and the rebellion too difficult for the parents) if they lasted for only a short time before the launch into true independence. But many families experience severe challenges to parental authority. At the very least, parents tend to feel that when someone lives in their home, they need to follow house rules, both moral and practical (such as when the kids are due home at night). As guardians of their children, they are concerned for their teenagers' safety and well-being. And they also may find it threatening to lose control over their children. In response, adolescents tend to feel that these rules infringe on their right as human beings to live according to their own judgments and desires.

Some teenagers respect and honor their parents' wishes. They may agree that it is in their best interest to curtail sexual desire and wait to have sex until they are older or married. Some parents may respond to teens' emerging sexual curiosity with support and the information they need to remain safe and self-confident. But other families fight about sexuality with everything they can muster. Not surprisingly, many households with teenagers in them also harbor conflict and psychological distress.

In some families, competition between parent and child may arise. Adult insecurities can emerge as aging adults worry about being displaced as sexual beings by their younger, more sexy offspring and grieve over the loss of their own youth. Some research has shown that parents of a same-sex adolescent (fathers who have a son or mothers who have a daughter) experience a decline in psychological well-being and even temporarily lose sexual interest in their spouse (Steinberg 1994).

For a variety of reasons, then, many parents become ambivalent about teen sexuality. And the same ambivalence is present at the societal level. Teens are perceived as having great sexual opportunity and even feared or resented because of it. Ironically, however, teen sex tends not to be nearly so active as adults imagine nor nearly as pleasing as sex between more experienced adults.

THE STATE OF TEEN SEXUAL BEHAVIOR

The early onset of adolescence and the increased autonomy of young people have left teens with little guidance regarding sexuality and relationships. The consequences are serious: Teenage childbearing undermines the futures of the mothers and the children; sexually transmitted diseases (STDs) undermine hope for a healthy adulthood for many sexually active teens. In addition, a compelling body of research indicates that teen sexuality is often experienced in negative ways, undermining teens' self-esteem, and that it frequently involves coercion or harassment.

The ability to address teenagers' needs has been undermined by adults' obsessive attention to teen sex and neglect of most other aspects of teen life. Ironically, contrary to many popular media stories, teens aren't having a great deal of sex. Often they will try intercourse once and then not try it again until they are older (Rubin 1990). Even teens who remain sexually active are generally not as active as adults, and they tend to be serially monogamous—one exclusive sexual relationship at a time—just like adults.

But sufficient numbers of teenagers get involved in sex and get pregnant to make their sexual behavior a social concern. Luker (1996) reports that 11 percent of women experience pregnancy before the age of 19. Although teens often believe that they can't get pregnant the first time they have sexual intercourse, they can—and they do. To state the obvious, pregnancy happens to women, not men, and thus generates a focus on changing teenage women's behavior rather than intervening with teenage men as well.

Much of the concern about teen sexuality has to do with the observation that more women are having children out of wedlock. In 1990, 31 percent of unwed mothers were teenagers (Luker 1996:199). Conservatives decry the perceived epidemic of teen motherhood, but the facts are more complicated. In the United States and western Europe, there has been what demographers call a major fertility transition or a substantial change in fertility rates. The result has been a steady decline in the rate of reproduction to replacement or below-replacement levels. In North America, fertility has declined steadily since the early nineteenth century, with the exception of the post-World War II baby boom era. People have fewer children, causing the population of some groups—particularly better-educated, richer, and mostly white groups—actually to shrink. Teen pregnancy has declined, just like pregnancy rates among older women, but it simply hasn't declined as much. Therefore, teens represent a larger proportion of the total population of reproducers. In other words, adolescents are not having more babies, adults are having fewer. Furthermore, over half of teen mothers in 1990 were 18 or 19. Twelve thousand women 15 or under had a baby in 1990; over 330,000 women 18 or 19 gave birth (Luker 1996). In other words, the image of "children having children" applies to a small proportion of teenage women having babies.

Plenty of the people who are raising the alarm about teen pregnancy know that, in sheer numbers, there is no such thing as an epidemic of teenage pregnancy. But arousing an outcry over rising teen pregnancy rates is of political value. Mention teen mothers and numerous U.S. voters start paying attention. Politicians spark citizens' outrage and use it to get votes by associating teen motherhood with welfare and casting these women as producing unwanted children whom they cannot support economically.

Nevertheless, the rising proportion of teenage mothers who are not married may be a legitimate cause for concern. Parenthood out of wedlock used to be so stigmatized that every pressure possible was applied by families on both sides to make sure a woman married the father of her child. The old phrase "shotgun wedding" was real enough in some parts of the country. But the ability to compel teenagers to marry evaporated as love triumphed over more pragmatic standards for choosing a mate, such as being "a good provider" and "a solid member of society." The sexual revolution and the women's movement were both preaching independence and individual decision making for women. Parents who wanted to force marriage on pregnant teens received little help from law enforcement agencies. Of course, forcing marriages does not promote the well-being of teen mothers or their offspring. Authorities are not particularly good at either helping establish paternity or making sure child support gets paid. Furthermore, poverty and unemployment make it difficult for young fathers to provide the support the government might like to see them deliver. The welfare system evolved to help support some of these young families to ensure that children did not suffer unduly for their parents' lack of preparation. But in the past three decades, the proportion of teenage mothers who weren't married changed from 25 percent to 75 percent of all teenage mothers (Luker 1996). In some communities, particularly those inhabited by the chronically poor, single motherhood became not simply common but normative. Thus, many began associating teen parenthood, young women involved in sex, single parenthood, and poor and minority women leaning on the state to assist with their progeny. Despite political rhetoric to the contrary, unwed, teen motherhood has not grown as a result of welfare to these needy young people. In fact, welfare has reduced in real value over the period when rates of unwed teen motherhood grew.

Although poverty is a common correlate for teenage motherhood, it isn't the only one. An influential 1994 book by Mary Pipher, *Reviving Ophelia,* shows how teenage women who become mothers are undermined early on by basement-level self-esteem. Although hopelessness and poverty are often experienced together, Pipher found that a future-orientation was not always associated with economic advantages. Regardless of their economic status, women who can see a future and who think well of themselves, profiled in several chapters of *Going All the Way* (Thompson 1996), are often able to abstain from sex or to protect themselves from unwanted pregnancy. But most teenage mothers, bereft of parental support or overwhelmed by the pleasure of romance in a life that has few other pleasures, look to their child for love and community.

And what about the fathers? Historically, and even today, young unmarried men have been given license to have sex. At the beginning of the sexual revolution, teenage men's rate of sexual experience was substantially higher than women's. . . . But although it has increased over the past several decades, it has not increased as much as women's. Perhaps in middle- or upper-middle-class homes, where parents fight to protect their child's future career and social status, a son's sexuality is seen as just as dangerous to his future prospects as a daughter's. But even in families where the future holds promise because of their social class, daughters, not sons, seem to receive the more dire warnings about sex outside of marriage. Regardless of class, the consequences of teen sexuality continue to be more costly for women than for men.

This imbalance may be changing somewhat—but not because of a sudden recognition that young men are and should be equal partners in sexual responsibility. Instead, conservative political forces have seen new possibilities for spreading moral blame and sharing social costs. An odd coalition of feminists and the "new right" have created an image of the unwed young father as a predatory young male, knowledgeable and malevolent, who is really to blame for all this teenage pregnancy and should be severely punished. He should be made to support his children. However, many of these young men, like the women they have paired with, have limited prospects in the job market (Lerman and Ooms 1993).

Another problem with this new focus on men's responsibilities is that it still emphasizes the helplessness of young women. Undoubtedly, sexual predators are out there. . . . Only 71 percent of women claimed to have wanted their first sexual experiences; 25 percent said it was unwanted and 4 percent said it was coerced. Nevertheless, many young women participate in sex with confidence and gusto. Some of these women are canny planners of their own lives—even if one doesn't agree with their values or goals—and not all teenage women are victims or potential victims of predatory young men. Like women in other age groups, some teenage women are attracted to older men; many find the interest of an older guy flattering. Some women know the young men with whom they have sex because they grew up together in the neighborhood and are friends. It is also important to remember that some of these older men are still teenagers or are barely in their 20s themselves. Finally, some young women reject seduction or receive the guidance necessary to contracept effectively.

The idea of raising social costs for the sexual activity of young men to induce them to abstain or use birth control is sensible. Everyone needs to take responsibility for the possibility of a child. But seeing men as predators is a simplification and in many cases an injustice. Feminist columnist Katha

Pollitt (1996) made the point clearly: "Construing teen sex as all victimization seems more compassionate than construing it, like Newt Gingrich [Speaker of the House of Representatives], as all sluttishness. But do we really want to say that a 15-year-old girl is always and invariably incapable of giving consent to sex with her 18-year-old boyfriend?" (p. 9).

We can't assume that all young men somehow know more about what they are doing than young women do. Certainly, some men organize their lives to take advantage of women. Men ought to be accountable for sexual activity, but they ought not to be automatically vilified for being sexual any more than women should be vilified for the same behavior.

Our view is that a very real social concern—teen sexuality—has been used to advance the sex-negative and sex-punitive agendas of various religious and political groups. Of course, young men and women should not produce babies before they are ready to be parents, and young people should have a shot at an unfettered youth and productive adulthood. But the vision that teenagers should be nonsexual, especially in a hypersexualized society, is simply unrealistic. Abstinence may be a reasonable suggestion to the younger teenagers, particularly in conservative neighborhoods or in communities that are unified in their support of them, but in most U.S. and European urban areas, this stance is simply unrealistic.

SEXUALITY EDUCATION

Societies like Sweden, where sex education starts early with full public support, have much lower rates of teen pregnancy than we do. Admittedly, Sweden is different in other important ways: It has fewer people in poverty, it is more homogenous ethnically, and it has a vast social safety net for all citizens. Sweden has a record of pro-family social policy along with a record of pro-sex education policies that emphasize that parenthood should be voluntary (Gauthier 1996). But in our country, people still act as if sex is the exclusive privilege of married people, even though everyone knows that unmarried people have sex. Sex education for youth would surely improve the sexual quagmire teens seem to be in today. Sex education for adults would help, too, so that citizens might understand that sex is neither dangerous nor shameful.

Remarkably, at the same time that teens are being blamed for unrestrained sexuality and an excess of illegitimate births, little in the way of comprehensive sex education or services for teens exists across the United States. For example, virginity for its own sake is still prized in many communities, especially for daughters. This moral code, somewhat more flexible for adults, is highly charged when it concerns teenage sons and daughters. Even though research (Kirby et al. 1994) indicates that the "just say no" and abstinence campaigns tend to increase, rather than decease, sexual activity, powerful lobbies have installed many of these programs in churches and school systems. These programs tend also to be conservatively gendered, reconnecting female virginity with marriageability and presenting definitions of propriety that are different for men than for women.

In fairness to many conservative communities, abstinence movements are not all premised on the double standard. Many religious groups preach abstinence for both sexes and apply sanctions equally. But, in general, social conservatives still see women as more vulnerable than men, see them as incapable of making sexual decisions, and believe that purity is the natural preference of and ideal for young women. Thousands of years of controlling of women's sexuality does not quietly fade away.

Even if conservatives did not take issue with sex education as a means for reducing pregnancy and delaying sexual initiation, they would have another argument against it. Conservatives believe that the family should be controlled by parents, not by government. Therefore, parents should decide what information a child should have about sexuality. Liberals, on the other hand, believe that all children deserve adequate information about sexual health, whether or not their parents elect to provide their children with it. This conflict is highlighted in debates regarding the media and censorship; conservative parent activist groups seek to influence broadcasting on the premise that sex and violence on television contributes to the corruption of children.

In contrast, sex education, from the public health point of view, ensures that young people who engage in sex will do so more safely. Public health professionals who believe in "healthy" sexuality emphasize the benefits of sex education to young people. They will have thought more about their own motives and goals and their responsibility to a partner and thus be prepared to act responsibly. Furthermore, sexuality education seems to reduce sexual abuses, including sexual harassment and coercion among teenagers. Groups taking this position—such as the Sexuality Information and Education Council of the United States (SIECUS); Planned Parenthood; and American Association of Sex Educators, Counselors, and Therapists (AASECT)—recognize that teens will continue to experiment with sexuality but believe education may delay some sexual experimentation. For many professionals, however, the main benefit of sex education is to teach young people how to communicate about sex. The goal is to help people treat one another humanely and to reduce the amount of shame, guilt, and manipulation in sexual relationships.

The comprehensive sex education side seems to be losing this debate. Although the number of sex education programs has increased since the 1970s, the goals have become more diverse, and program efficacy is undermined by a lack of consensus and, in some cases, a lack of quality materials or well-trained instructors (Yarber 1994). Policymakers have generally seen the growth in sexual activity among teenagers as a reason to limit services for sexually active young people. Although sex education is not the culprit for young women's increased sexual activity, nevertheless the fear persists that sex education legitimizes teen sex. Vocal bands of parents have succeeded in locking sex education courses out of many school districts. In some, they have substituted "chastity" programs, which are often programs based on religious ethics and traditional gender roles.

The sexual double standard has perhaps not surprisingly persisted for youth. For teenage men, having sex for the first time still counts as a coming-of-age ritual. Young men are often expected to have sex before late adolescence, and many families are relieved when they see some sign that a son has heterosexual capabilities and desires. These same accolades do not exist for young women: Not only do parents avoid direct or subtle approval of their daughters' sexual initiation, but the consequences of heterosexual sex for young women are more evident. Sexually active women are more vulnerable than men to STDs as well as to negative social reactions. Even liberal parents are more conservative for their daughters than their sons.

Ironically, teenage women, whom conservatives feel have more to lose sexually, still get little special instruction on how to negotiate the crosscurrents of their sexual environment. Some women in impoverished or other high-risk environments, where teenage women are more likely to get pregnant, may get special programs directed at controlling their fertility. But coping skills in such matters as negotiating with men about sex or avoiding health risks are rarely provided. When young women receive guidance, most programs assume that they and not their partners are and should be the sexual gatekeepers. Planned Parenthood is an exception: Local chapters often provide education for teenage men and women and their families. But Planned Parenthood reaches only a small percentage of at-risk young people, and its clinics and group meetings are disproportionately attended by young women and mothers.

TEEN HEALTH

The public debate on teen sexuality has focused squarely on teen pregnancy—as if sex were all that teenagers were doing with their spare time (Dreyfoos 1990). Politicians and opinion leaders moralize about youthful promiscuity and the tax burden it entails for supporting the babies of unwed teen mothers. Worried parents are galvanized by office holders and political fund-raisers who tell them that each and every family is vulnerable to teen pregnancy unless their platform of moral renewal is adopted. Of course, the policymakers emphasize their compassion for teens, but it is instructive to note just how little attention they pay to other health costs of teenage sexuality besides pregnancy.

All the concern about teen sexuality has caused health institutions to provide reproductively focused health care to the exclusion of other kinds. Teens in the United States are enormously underserved (Dreyfoos 1990). Paradoxically, the fear of teen sexuality has reduced teen health services rather than sharpening them.

Even sex-related health problems are inadequately addressed by services to youth. STDs are growing among teens (Holmes et al. 1990). Untreated STDs, especially pelvic inflammatory disease (PID) and chlamydia, threaten women's fertility. PID can cause horrific pain as well as major complications, including damage to the urinary tract and the reproductive organs. Chlamydia may be asymptomatic, but it can end a woman's reproductive future. Although infertility rates in the population in general are declining (U.S. Office of Technology Assessment 1988), they are increasing among poor people and the youngest portion of the U.S. population (Scritchfield 1995). Perhaps it is cynical to think that more attention would be paid to PID and chlamydia if they affected men as much as they affect women.

Although the history of medicine can be said, without much distortion, to be the treatment of male problems (Ehrenreich and English 1978; Lorber 1997), an enormous amount of consciousness raising has changed a total blindness to the specific problems of adult women—if not to teens. A women's health movement grew up in the 1960s and 1970s, marked in part by the publication in 1973 of the first edition of *Our Bodies, Ourselves* by the Boston Women's Health Book Collective. (A revised edition was published in 1996, as well as an edition dedicated to women's health and aging.) Growing attention to women's health concerns such as breast cancer prevention, detection, and treatment continues presently.

Turning teenage sexuality into a political issue may be an inevitable consequence of the relatively new, extended stage of adolescence. Ambivalent and punitive attitudes toward youthful sexuality are useful to many politicians who stir up parental fears and then offer comforting but unrealistic promises that voting for them will turn back the clock. But rarely does any of this politicization translate into health and other services for young people. More often political solutions involve rolling back sexual trends and revitalizing dichotomized, gendered sexual scripts for teenagers.

SELECTED REFERENCES

Dreyfoos, J. 1990. *Adolescents at Risk: Prevalence and Prevention.* New York: Oxford University Press.

Ehrenreich, B., and D. English. 1978. *For Her Own Good: 150 Years of the Experts' Advice to Women.* Garden City, NY: Anchor/Doubleday.

Gauthier, A. H. 1996. *The State and the Family: A Comparative Analysis of Family Policies in Industrialized Countries.* Oxford: Clarendon.

Holmes, K. K., P.-A. Mardh, P. F. Sparling, and P. J. Wiesner. 1990. *Sexually Transmitted Diseases.* 2d ed. New York: McGraw-Hill.

Kirby, D., L. Short, J. Collins, D. Rugg, L. Kolbe, M. Howard, B. Miller, F. Sonenstein, and L. S. Zabin. 1994. School-based programs to reduce sexual risk behaviors: A review of effectiveness. *Public Health Reports* 109:339–60.

Lerman, R. I., and T. J. Ooms. 1993. *Young Unwed Fathers: Changing Roles and Emerging Policies.* Philadelphia: Temple University Press.

Lorber, J. 1997. *Gender and the Social Construction of Illness.* Thousand Oaks, CA: Sage.

Luker, K. 1996. *Dubious Conceptions: The Politics of Teenage Pregnancy.* Cambridge, MA: Harvard University Press.

Pipher, M. 1994. *Reviving Ophelia: Saving the Selves of Adolescent Girls.* New York: Putnam.

Pollitt, K. 1996. Motherhood and morality. *Nation.* 27 May, 9.

Rubin, L. B. 1990. *Erotic Wars: What Happened to the Sexual Revolution?* New York: HarperCollins.

Scritchfield, S. 1995. The Social Construction of Infertility. In *Images of Issues,* edited by J. Best. New York: Aldine de Gruyter.

Steinberg, L. 1994. *Crossing Paths: How Your Child's Adolescence Triggers Your Own Crisis.* New York: Simon & Schuster.

Thompson, S. 1996. *Going All the Way: Teenage Girls' Tales of Sex, Romance, and Pregnancy.* New York: Hill & Wang.

Yarber, W. L. 1994. Past, Present, and Future Perspectives on Sexuality Education. In *The Sexuality Education Challenge: Promoting Healthy Sexuality in Young People,* edited by J. Drolet and K. Clark. New York: SIECUS.

U.S. Office of Technology Assessment. 1988. Infertility: Medical and Social Choices. Publication No. OTA BA 358. Washington, DC: U.S. Congress.

CHAPTER 3

Marriage and Cohabitation

The readings in this chapter focus on **cohabitation** (living together) and the institution of marriage. As with families, most of us know a marriage when we see one, but defining marriage is more complicated than it first appears. How is being married different from being in a long-term committed relationship or co-habiting with a lover? Some argue that we should classify couples as married when they love each other, have sex with each other, live together, and acknowledge the relationship publicly, whereas others argue that a marriage is legitimate only if a couple has a church wedding or obtains an official marriage license from a government agency. And what if governmental jurisdictions hold differing views on marriage, as happened in 1996 when a Hawaii court ruled that the state could not deny marriage licenses to same-sex couples (*Baehr v. Miike,* Haw. 91-1394, 1996) or in 2000 when the state of Vermont approved a new "civil union" law, granting some marriage-like benefits to same sex couples? In response, some other states passed laws forbidding gay marriages and Congress enacted the "Defense of Marriage Act" defining the marital union as between a man and a woman and stipulating that no state could be forced to recognize another state's same-sex marriages (104th Congress, HR 3396). Taking a broader historical and cross-cultural view, however, we find that societies have occasionally approved same-sex marriage, and a fair number have condoned and even promoted marriage between a man and multiple women (known as **polygyny**) or between a woman and multiple husbands (known as **polyandry**).

Although no universal set of marriage rules applies to all people in all cultures, a basic definition of marriage can help us understand how it works. In general, scholars have recognized marriage as a *social* arrangement rather than as an individual agreement between a husband and a wife. In other words, marriage is best understood as a social institution. In *The Sociological Imagination,* C. Wright Mills defined an **institution** as "a set of roles graded in authority that have been embodied in consistent patterns of actions that have been legitimated and sanctioned by society or segments of that society; whose purpose is to carry out certain activities or prescribed needs of that society or segments of that society" (1959, p. 30). As you will see in the selections that follow, marriage has typically been accepted as a social institution, but debates about whose interests marriage serves have been waged for centuries.

Marriage is considered an institution because it follows social norms, links the family networks of husbands and wives, and has major consequences for the larger community. All societies have specific rules governing the process of mate selection, and kinship groups typically control elaborate rituals governing who can marry whom, what goods get exchanged at marriage, where newly married couples should live, and how husbands and wives should act once they are married. The marriage itself is usually a highly ritualized ceremonial event, signaling to the larger community that the couple's union is officially approved. Although marriage is always a public matter involving **kinship groups,** in Western nations it has tended to be regulated by the state through marriage laws as well as through local customs (Goode, 1963).

Although most of us do not think of marriage as a legal or financial contract, laws and legal rulings have long defined what marriage is and what obligations husbands and wives should have to each other and to their children (if any). Many of us become aware of these laws only if we get divorced (see chapter 9), and we take for granted that the institution of marriage is influenced by historical tradition and normative constraints. The selections in this chapter focus on some of the links between the normative and legal aspects of marriage, especially as they reflect the rights of women to own property and exercise self-determination. Western law, based on ancient Roman statutes, has defined the wife as the property of the husband who serves as the head of the household. Although this sounds rather sexist, the laws of most Western societies (as well as their Eastern counterparts) have traditionally defined women as inferior beings who must be protected by a man. For example, the ceremonial custom of a father "giving" his daughter away in marriage reflects the legal inferiority of women. In most cases, women have lost legal rights when they marry, and women's obligations in marriage have always been rather severe. As late as 1850, almost all states in the United States had laws that recognized a husband's right to beat his wife if she did not fulfill her wifely duties (see chapter 11).

Traditional marriage law merged the identities of husband and wife. According to the feudal doctrine of **coverture,** the husband and wife became a unity at the time of marriage, and that unity was the husband. Symbolic loss of identity is still evident today in the custom of a married woman legally adopting her husband's name when she marries: Miss Jane Smith marries Mr. John Jones and becomes Jane Jones or even Mrs. John Jones, whereas the man's legal identity remains the same as it was before marriage. Some alternative customs are becoming more common, such as a married woman keeping her own name, or joining her "maiden" name with that of her husband using a hyphen (Smith-Jones). Some husbands also adopt hyphenated names upon marriage, but the practice is relatively rare, providing evidence that we still have a predominantly **patrilineal** marriage system. In the past, marriage laws also required a wife to take her husband's legal domicile (place of residence) and obligated the wife to provide various domestic and sexual services to her husband. Even as late as the 1970s, laws and court rulings in some states treated the earnings of wives as the property of their husbands (but not the reverse). Today, most states have marriage laws that treat husbands and wives similarly, though in many states women still lose specific legal rights to control property, enter into legal contracts, or charge their husbands with crimes when they get married (Coltrane and Collins, 2001, pp. 374–75).

The first selection in this chapter is from a nineteenth-century English visitor to the United States known as "the first woman sociologist" (Rossi, 1973, p. 118). Harriet Martineau (1802–1876), like Alexis de Tocqueville before her, made observations about the young nation focusing on how America's basic moral values framed its institutional structure. An **abolitionist** before the movement gained widespread support, Martineau compared the position of women in England and America to that of American slaves. She decried the social pressure on women to marry and become mothers, commenting that rapid aging and shallow lives following matrimonial wedlock usually replaced the healthy vigor of American women before they married. In the classic selection presented here from *Society in America,* Martineau offers one of the first pictures of the divergent roles of American men and women. Although we tend to think of American institutions as being more democratic and equal than other nations' institutions, European visitors such as Martineau and de Tocqueville commented that early nineteenth-century Americans were compulsively preoccupied with separating the roles of men and women: "In America, more than anywhere else in the world, care has been taken to constantly trace clearly the distinct spheres of action of the two sexes and both are required to keep in step, but along paths that are never the same" (Tocqueville, 1832/1969, p. 601). Martineau observed that even though Americans enjoyed relative material wealth, they tended to reenact old-world "mercenary marriages" in which men waited to marry then took young brides "for the sake of securing a certain style of living." Martineau criticized the "mournful and injurious" health consequences women faced because of American men's "'chivalrous' taste and temper" that lead them to restrict women's access to gainful employment. She also showed sympathy for the plight of less advantaged American women: "the condition of the female working classes is such that if its sufferings were but made known, the emotions of horror and shame would tremble through the whole society." Martineau thus prefigures later **feminist** criticism of marriage as a property relationship and criticizes the unequal roles of men and women within it. Ultimately, she argues for more equal property laws, easier access to divorce, and better pay for women as the only way to ensure women's equality with men. Although current political debates sometimes assume that women's claims for equality arose in the 1970s, this classic reading shows deeper historical roots. Martineau's essay also shows how sociological analyses linking marriage, property, employment, and housework were evident (if not common) even before the Civil War.

The second selection in this chapter is a review of research on **cohabitation** in the United States. Demographer Pamela Smock notes that the majority of marriages and remarriages now begin as cohabiting relationships, that most younger men and women in America will cohabit at some point in their lives, and that about half of previously married cohabitors and over a third of never-married cohabitors have children in their households. These represent dramatic departures from patterns common in earlier decades of the twentieth century, and social scientists are just beginning to understand their causes and consequences. Most explanations for the rapid rise in cohabitation place it in the context of other long-term historical trends including declining fertility, increasing age at marriage, rising rates of marital disruption, and a growing proportion of children being born outside of marriage, all of which are associated with increased emphasis on individual fulfillment, women's increasing economic independence, and the "sexual revolution" (i.e., acceptance that unmarried people could have sex). Smock discusses why, contrary to some expectations, cohabitation is associated with a greater likelihood of relationship disruption. She also compares cohabitors to married couples, reviews research about children of cohabiting couples, and makes predictions about future research on cohabitation. Finally, she notes that although research on cohabitation is still in its infancy, cohabitation itself is changing, with different types of cohabitation likely to have different causes and consequences. This selection, unlike most of the others in the book, is excerpted from a review article written for academic researchers. As you read this demographic review, think about how writing style and citation format differ between articles. Do you find scientific technical writing to be more or less convincing than other styles?

In the third selection in this chapter, Stephanie Coontz notes that marriage is now an option rather than a necessity for women and examines various impacts of what she calls "the deinstitutionalization" of marriage in modern American society. Whereas marriage once held a monopoly over the organization of people's major life transitions, prescribed narrow work roles according to gender, and regulated the distribution of wealth, Coontz now sees a multitude of social forces regulating people's lives. She criticizes recent "family values" campaigns' attempts to reinstate traditional marriage as misguided and doomed to failure, stating, "trying to reverse a historical trend by asking individuals to make personal decisions opposing that trend is usually futile." Coontz characterizes as misguided the idea that welfare payments to mothers "cause" more nonmarital birth (especially for African Americans) and documents lower rates of illegitimacy in states with higher average welfare benefits. Finally, she provides a sociological and historical understanding of divorce, unwed pregnancy, cohabitation, and other recent trends that moves beyond blaming individuals' supposed lack of belief in marriage for society's most crucial social problems. Do you agree with Coontz that there is a "hidden agenda" behind the new marriage movement.

The final selection in this chapter uses social comparison to explore the essential features of marriage. Tackling the question of whether same-sex marriage should be allowed, Angela Bolte examines various legal, sociological, and philosophical premises of marriage definitions. Does marriage always regulate sexuality, include the provision of economic support, and include the expectation that the couple will have children? How important is a ceremonial wedding event to the public recognition—and social significance—of a marriage? In the end, Bolte argues that rather than weakening marriage as critics contend, extending rights of marriage to same-sex couples would strengthen the institution of marriage.

All four of the articles in this chapter analyze marriage as a social institution and suggest reasons why the laws and norms regulating it can change. In particular, they analyze how economics, gender, and sexuality have been associated with marriage. Do you think that cohabitation and domestic partnerships undermine or support the institution of marriage?

REFERENCES

Coltrane, Scott, and Randall Collins. 2001. *Sociology of Marriage and the Family: Gender, Love, and Property.* 5th ed. Belmont, CA: Wadsworth/ITP.

Goode, William J. 1963. *World Revolution and Family Patterns.* New York: Free Press.

Mills, C. Wright. 1959. *The Sociological Imagination.* New York: Oxford University Press.

Rossi, Alice S., ed. 1973. *The Feminist Papers: From Adams to de Beauvoir.* New York: Columbia University Press.

Tocqueville, Alexis de. 1832/1969. *Democracy in America.* New York: Anchor.

SUGGESTED READINGS

Cott, Nancy F. 2000. *Public Vows: A History of Marriage and the Nation.* Cambridge, MA: Harvard University Press.

Dizard, Jan, and Howard Gadlin. 1990. *The Minimal Family.* Amherst: University of Massachusetts Press.

Geller, Jaclyn. 2001. *Here Comes the Bride: Women, Weddings, and the Marriage Mystique.* New York: Four Walls Eight Windows.

Graff, E. J. 1999. *What Is Marriage For?* Boston, MA: Beacon.

Ingraham, Chrys. 1999. *White Weddings: Romancing Heterosexuality in Popular Culture.* New York: Routledge.

Maushart, Susan. 2001. *Wifework: What Marriage Really Means for Women.* New York: Bloomsbury.

Paul, Pamela. 2002. *The Starter Marriage and the Future of Matrimony.* New York: Villard Books.

Yalom, Marilyn. 2001. *A History of the Wife.* New York: HarperCollins.

Yalom, Marilyn, Laura L. Carstensen, eds., Estelle Freedman, and Barbara Gelpi, consulting eds. 2002. *Inside the American Couple: New Thinking/New Challenges.* Berkeley: University of California Press.

INFOTRAC® COLLEGE EDITION EXERCISES

The exercises that follow allow you to use the InfoTrac® College Edition online database of scholarly articles to explore the sociological implications of the selections in this chapter.

Search Keyword: Exogamy. Search for articles that deal with exogamy, or marriage outside of the group. What are some of the different forms that exogamy can take (for instance, religious exogamy)? How does exogamy affect the way families relate to society?

Search Keyword: Institution of marriage. Use InfoTrac® College Edition to search for articles that talk about the institution of marriage. How do these articles see marriage as an institution? How does C. Wright Mills's definition of *institution* (given in the introduction to this chapter) fit the way marriage is discussed in these articles?

Search Keyword: Traditional family values. Find articles that deal with traditional family values. How is this notion defined? According to these articles, what do families with these values look like? What do families that don't adhere to traditional family values look like? How has this notion been used as a tool of politics?

Search Keyword: Marriage rituals or wedding rituals. Using InfoTrac® College Edition, search for articles that discuss rituals of various marriage and wedding ceremonies. What are some of these rituals? How do they differ by social group? What do marriage and wedding rituals symbolize? How do these rituals reflect norms about marriage?

Search Keyword: Same-sex marriage. How many articles can you find that favor same-sex marriage? How many articles can you turn up that oppose it? What are the arguments proposed for each position? How do these arguments reflect social norms about marriage in general?

Search Keyword: Polygamy. Find articles in InfoTrac® College Edition that deal with the topic polygamy, or marriage to multiples spouses. What attitudes are expressed about polygamy? How are social norms about the institution of marriage reflected in attitudes about polygamy?

Search Keyword: Domestic partnership. Look for articles that discuss the issue of domestic partnership. What does it mean for a couple to be involved in a domestic partnership? What legal formalities are involved? Who can be involved in a domestic partnership arrangement? What are the benefits, and what are the disadvantages, of domestic partnerships for the couple and for society? How similar to marriage is a domestic partnership arrangement?

8

Marriage

HARRIET MARTINEAU,

1837

If there is any country on earth where the course of true love may be expected to run smooth, it is America. It is a country where all can marry early, where there need be no anxiety about a worldly provision, and where the troubles arising from conventional considerations of rank and connexion ought to be entirely absent. It is difficult for a stranger to imagine beforehand why all should not love and marry naturally and freely, to the prevention of vice out of the marriage state, and of the common causes of unhappiness within it. The anticipations of the stranger are not, however, fulfilled: and they never can be while the one sex overbears the other. Marriage is in America more nearly universal, more safe, more tranquil, more fortunate than in England: but it is still subject to the troubles which arise from the inequality of the parties in mind and in occupation. It is more nearly universal, from the entire prosperity of the country: it is safer, from the greater freedom of divorce, and consequent discouragement of swindling, and other vicious marriages: it is more tranquil and fortunate from the marriage vows being made absolutely reciprocal; from the arrangements about property being generally far more favorable to the wife than in England; and from her not being made, as in England, to all intents and purposes the property of her husband. The outward requisites to happiness are nearly complete, and the institution is purified from the grossest of the scandals which degrade it in the Old World: but it is still the imperfect institution which it must remain while women continue to be ill-educated, passive, and subservient: or well-educated, vigorous, and free only upon sufferance.

The institution presents a different aspect in the various parts of the country. I have spoken of the early marriages of silly children in the south and west, where, owing to the disproportion of numbers, every woman is married before she well knows how serious a matter human life is. She has an advantage which very few women elsewhere are allowed: she has her own property to manage. It would be a rare sight elsewhere to see a woman of twenty-one in her second widowhood, managing her own farm or plantation; and managing it well, because it had been in her own hands during her marriage. In Louisiana, and also in Missouri, (and probably in other States,) a woman not only has half her husband's property by right at his death, but may always be considered as possessed of half his gains during his life; having at all times power to bequeath that amount. The husband interferes much less with his wife's property in the south, even through her voluntary relinquishment of it, than is at all usual where the cases of women having property during their marriage are rare. In the southern newspapers, advertisements may at any time be seen, running thus:—"Mrs. A, wife of Mr. A, will dispose of &c. &c." When Madame

From Harriet Martineau. "Marriage" from *Society in America,* Vol. III, pp. 107–151. London: Sanders and Otley. Copyright © 1837.

Lalaurie was mobbed in New Orleans, no one meddled with her husband or his possessions; as he was no more responsible for her management of her human property than anybody else. On the whole, the practice seems to be that the weakest and most ignorant women give up their property to their husbands; the husbands of such women being precisely the men most disposed to accept it: and that the strongest-minded and most conscientious women keep their property, and use their rights; the husbands of such women being precisely those who would refuse to deprive their wives of their social duties and privileges. . . .

I have mentioned that divorce is more easily obtained in the United States than in England. In no country, I believe, are the marriage laws so iniquitous as in England, and the conjugal relation, in consequence, so impaired. Whatever may be thought of the principles which are to enter into laws of divorce, whether it be held that pleas for divorce should be one, (as narrow interpreters of the New Testament would have it;) or two, (as the law of England has it;) or several, (as the Continental and United States' laws in many instances allow,) nobody, I believe, defends the arrangement by which, in England, divorce is obtainable only by the very rich. The barbarism of granting that as a privilege to the extremely wealthy, to which money bears no relation whatever, and in which all married persons whatever have an equal interest, needs no exposure beyond the mere statement of the fact. It will be seen at a glance how such an arrangement tends to vitiate marriage: how it offers impunity to adventurers, and encouragement to every kind of mercenary marriages: how absolute is its oppression of the injured party: and how, by vitiating marriage, it originates and aggravates licentiousness to an incalculable extent. To England alone belongs the disgrace of such a method of legislation. . . .

Of the American States, I believe New York approaches nearest to England in its laws of divorce. It is less rigid, in as far as that more is comprehended under the term "cruelty." The husband is supposed to be liable to cruelty from the wife, as well as the wife from the husband. There is no practical distinction made between rich and poor by the process being rendered expensive: and the cause is more easily resumable after a reconciliation of the parties. In Massachusetts, the term "cruelty" is made so comprehensive, and the mode of sustaining the plea is so considerately devised, that divorces are obtainable with peculiar ease. The natural consequence follows: such a thing is never heard of. A long-established and very eminent lawyer of Boston told me that he had known of only one in all his experience. Thus it is wherever the law is relaxed, and, *caeteris paribus,* in proportion to its relaxation: for the obvious reason, that the protection offered by law, to the injured party causes marriages to be entered into with fewer risks, and the conjugal relation carried on with more equality. Retribution is known to impend over violations of conjugal duty. When I was in North Carolina, the wife of a gamester there obtained a divorce without the slightest difficulty. When she had brought evidence of the danger to herself and her children,—danger pecuniary and moral,—from her husband's gambling habits, the bill passed both Houses without a dissenting voice.

It is clear that the sole business which legislation has with marriage is with the arrangement of property; to guard the reciprocal rights of the children of the marriage and the community. There is no further pretence for the interference of the law, in any way. An advance towards the recognition of the true principle of legislative interference in marriage has been made in England, in the new law in which the agreement of marriage is made a civil contract, leaving the religious obligation to the conscience and taste of the parties. It will be probably next perceived that if the civil obligation is fulfilled, if the children of the marriage are legally and satisfactorily provided for by the parties, without the assistance of the legislature, the legislature has, in principle, nothing more to do with the matter. . . .

It is assumed in America, particularly in New England, that the morals of society there are peculiarly pure. I am grieved to doubt the fact: but I do doubt it. Nothing like a comparison between one country and another in different circumstances can be instituted: nor would any one desire to enter upon such a comparison. The bottomless vice, the all-pervading corruption of European society cannot, by possibility, be yet paralleled in America: but neither is it true that any outward prosperity, any arrangement of circumstances, can keep a society pure while there is corruption in its social methods, and among its principles of individual action. Even in America, where every young man may, if he chooses, marry at twenty-one, and appropriate all the best comforts of domestic life,—even here there is vice. Men do not choose to marry early, because they have learned to think other things of more importance than the best comforts of domestic life. A gentleman of Massachusetts, who knows life and the value of most things in it, spoke to me with deep concern of the alteration in manners which is going on: of the increase of bachelors, and of mercenary marriages; and of the fearful consequences. It is too soon for America to be following the old world in its ways. In the old world, the necessity of thinking of a maintenance before thinking of a wife has led to requiring a certain style of living before taking a wife; and then, alas! to taking a wife for the sake of securing a certain style of living. That this species of corruption is already spreading in the new world is beyond a doubt;—in the cities, where the people who live for wealth and for opinion congregate.

I was struck with the great number of New England women whom I saw married to men old enough to be their fathers. One instance which perplexed me exceedingly, on my entrance into the country, was explained very little to my satisfaction. The girl had been engaged to a young man whom she was attached to: her mother broke off the engagement, and married her to a rich old man. This story was a real shock to me; so persuaded had I been that in America, at least, one might escape from the disgusting spectacle of mercenary marriages. But I saw only too many instances afterwards. The practice was ascribed to the often-mentioned fact of the young men migrating westwards in large numbers, leaving those who should be their wives to marry widowers of double their age. The Auld Robin Gray story is a frequently enacted tragedy here, and one of the worst symptoms that struck me was, that there was usually a demand upon my sympathy in such cases. I have no sympathy for those who, under any pressure of circumstances, sacrifice their heart's-love for legal prostitution, and no environment of beauty or sentiment can deprive the fact of its coarseness: and least of all could I sympathise with women who set the example of marrying for an establishment in a new country, where, if anywhere, the conjugal relation should be found in its purity.

The unavoidable consequence of such a mode of marrying is, that the sanctity of marriage is impaired, and that vice succeeds. Any one must see at a glance that if men and women marry those whom they do not love, they must love those whom they do not marry. There are sad tales in country villages, here and there, which attest to this; and yet more in towns, in a rank of society where such things are seldom or never heard of in England. I rather think that married life is immeasurably purer in America than in England: but that there is not otherwise much superiority to boast of. I can only say, that I unavoidably knew of more cases of lapse in highly respectable families in one State than ever came to my knowledge at home; and that they were got over with a disgrace far more temporary and superficial than they could have been visited with in England. I am aware that in Europe the victims are chosen, with deliberate selfishness, from classes which cannot make known their perils and their injuries; while in America, happily, no such class exists. I am aware that this destroys all possibility of a comparison: but the fact remains, that the morals of American society are less pure than they assume to be. If the common boast be meant to apply to the rural population, at least let it not be made, either in pious gratitude, or patriotic conceit, by the aristocratic city classes, who, by introducing the practice of mercenary marriages, have rendered themselves responsible for whatever dreadful consequences may ensue.

The ultimate and very strong impression on the mind of a stranger, pondering the morals of society in America, is that human nature is much the same everywhere, whatever may be its environment of riches or poverty; and that it is justice to the human nature, and not improvement in fortunes, which must be looked to as the promise of a better time. Laws and customs may be creative of vice; and should be therefore perpetually under process of observation and correction: but laws and customs cannot be creative of virtue: they may encourage and help to preserve it; but they cannot originate it.

. . . One consequence, mournful and injurious, of the "chivalrous" taste and temper of a country with regard to its women is that it is difficult, where it is not impossible, for women to earn their bread. Where it is a boast that women do not labour, the encouragement and rewards of labour are not provided. It is so in America. In some parts, there are now so many women dependent on their own exertions for a maintenance, that the evil will give way before the force of circumstances. In the meantime, the lot of poor women is sad. Before the opening of the factories, there were but three resources; teaching, needle-work, and keeping boarding-houses or hotels. Now, there are the mills; and women are employed in printing-offices; as compositors, as well as folders and stitchers.

I dare not trust myself to do more than touch on this topic. There would be little use in dwelling upon it; for the mischief lies in the system by which women are depressed, so as to have the greater number of objects of pursuit placed beyond their reach, more than in any minor arrangements which might be rectified by an exposure of particular evils. I would only ask of philanthropists of all countries to inquire of physicians what is the state of health of sempstresses; and to judge thence whether it is not inconsistent with common humanity that women should depend for bread upon such employment. Let them inquire what is the recompense of this kind of labour, and then wonder if they can that the pleasures of the licentious are chiefly supplied from that class. Let them reverence the strength of such as keep their virtue, when the toil which they know is slowly and surely destroying them will barely afford them bread, while the wages of sin are luxury and idleness. During the present interval between the feudal age and the coming time, when life and its occupations will be freely thrown open to women as to men, the condition of the female working classes is such that if its sufferings were but made known, emotions of horror and shame would tremble through the whole of society.

For women who shrink from the lot of the needle-woman,—almost equally dreadful, from the fashionable milliner down to the humble stocking-darner,—for those who shrink through pride, or fear of sickness, poverty, or temptation, there is little resource but pretension to teach. What office is there which involves more responsibility, which requires more qualifications, and which ought, therefore, to be more honourable, than that of teaching? What work is there for which a decided bent, not to say a genius, is more requisite? Yet are governesses furnished, in America as elsewhere, from among those who teach because they want bread; and who certainly would not teach for any other reason. Teaching and training children is, to a few, a very few, a delightful employment, notwithstanding all its toils and cares. Except to these few it is irksome; and, when accompanied with poverty and mortification, intolerable. Let philanthropists inquire into the proportion of governesses among the inmates of lunatic asylums. The answer to this question will be found to involve a world of rebuke and instruction. What can be the condition of the sex when such an occupation is overcrowded with candidates, qualified and unqualified? What is to be hoped from the generation of children confided to the cares of a class, conscientious perhaps beyond most, but reluctant, harassed, and depressed?

The most accomplished governesses in the United States may obtain 600 dollars a-year in the families of southern planters; provided they will promise to teach everything. In the north they are paid less; and in neither case, is there a possibility of making provision for sickness and old age. Ladies who fully deserve the confidence of society may realise an independence in a few years by school-keeping in the north: but, on the whole, the scanty reward of female labour in America remains the reproach to the country which its philanthropists have for some years proclaimed it to be. I hope they will persevere in their proclamation, though special methods of charity will not avail to cure the evil. It lies deep; it lies in the subordination of the sex: and upon this the exposures and remonstrances of philanthropists may ultimately succeed in fixing the attention of society; particularly of women. The progression or emancipation of any class usually, if not always, takes place through the efforts of individuals of that class: and so it must be here. All women should inform themselves of the condition of their sex, and of their own position. It must necessarily follow that the noblest of them will, sooner or later, put forth a moral power which shall prostrate cant, and burst asunder the bonds, (silken to some, but cold iron to others,) of feudal prejudices and usages. In the meantime, is it to be understood that the principles of the Declaration of Independence bear no relation to half of the human race? If so, what is the ground of the limitation? If not so, how is the restricted and dependent state of women to be reconciled with the proclamation that "all are endowed by their Creator with certain inalienable rights; that among these are life, liberty, and the pursuit of happiness?"

9

Cohabitation in the United States

PAMELA SMOCK,
2000

Unmarried heterosexual cohabitation has increased sharply in recent years in the United States. It has in fact become so prevalent that the majority of marriages and remarriages now begin as cohabiting relationships, and most younger men and women cohabit at some point in their lives. It has become quite clear that understanding and incorporating cohabitation into sociological analyses and thinking is crucial for evaluating family patterns, the life course of individuals, children's well-being, and social change more broadly. . . .

The most widely cited fact about cohabitation, a fact replicated with several different data sources, is that it has increased dramatically over the last two decades or so (Casper & Cohen 2000). It has gone from being a relatively uncommon experience to a commonplace one and has achieved this prominence quite quickly. A few sets of numbers convey both the change and its rapidity. First, the percentage of marriages preceded by cohabitation rose from about 10% for those marrying between 1965 and 1974 to over 50% for those marrying between 1990 and 1994 (Bumpass & Lu 1999, Bumpass & Sweet 1989); the percentage is even higher for remarriages. Second, the percentage of women in their late 30s who report having cohabited at least once rose from 30% in 1987 to 48% in 1995. Given a mere eight-year time window, this is a striking increase. Finally, the proportion of all first unions (including both marriages and cohabitations) that begin as cohabitations rose from 46% for unions formed between 1980 and 1984 to almost 60% for those formed between 1990 and 1994 (Bumpass & Lu 1999).

A second widely cited fact is that, for most couples, cohabitation is a rather short-lived experience with most ending it either by terminating the relationship or by marrying within a few years. The most recent estimates suggest that about 55% of cohabiting couples marry and 40% end the relationship within five years of the beginning of the cohabitation. . . . Only about one sixth of cohabitations last at least three years and only a tenth last five years or more (Bumpass & Lu 1999).

Finally, contrary to popular image, cohabitation is not a childless state. About one half of previously married cohabitors and 35% of never-married cohabitors have children in the household. In most cases (70%), these are the children of only one partner, making the arrangement somewhat akin to step-families, and the rest of the children involved are the biological offspring of the couple (Bumpass et al.

From "Annual Review of Sociology," Vol. 26, 2000, pp. 1–20, by Pamela Smock. With permission from the *Annual Review of Sociology,* Vol. 26,

1991). And contrary to much of the discourse on single motherhood, a very substantial proportion of births conventionally labeled as "nonmarital" are actually occurring in cohabiting families—almost 40% overall, and roughly 50% among white and latino women and a quarter among black women (Bumpass & Lu 1999). Thus, a large share of children born to supposedly "single" mothers today are born into two-parent households. Moreover, the widely cited increase over recent years in nonmarital childbearing is largely due to cohabitation and not to births to women living without a partner (Bumpass & Lu 1999).

DIFFERENTIALS

Although researchers have found statistically significant differences between cohabitors and others on a host of traits ranging from ideal fertility to the use of leisure time (Clarkberg et al 1995, Landale & Fennelly 1992, Nock 1995, Rindfuss & VandenHeuvel 1990), there are two overarching factors that consistently emerge as a basis of differentiation. First, cohabitation tends to be selective of people of slightly lower socioeconomic status, usually measured in terms of educational attainment or income (Bumpass & Lu 1999...). For example, recent data show that the percentage of 19- to 44-year-old women who have cohabited at some point is almost 60% among high school dropouts versus 37% among college graduates (Bumpass & Lu 1999). The other factor can generally be understood in terms of a "traditional" versus "liberal" distinction. Cohabitation tends to be selective of people who are slightly more liberal, less religious, and more supportive of egalitarian gender roles and nontraditional family roles (Clarkberg et al 1995, Lye & Waldron 1997, Thornton et al 1992).

Notably, there are few apparent race-ethnic differences in the likelihood of cohabitation, at least among the groups for which there is adequate representation in surveys. Recent data show that 45% of white and black and 40% of latino women ages 19–44 have cohabited (Bumpass & Lu 1999). This is contrary to the case for marriage; blacks are less likely to marry than whites, and a fairly large sociological literature has emerged examining the possible causes of this disparity (e.g., Lichter et al 1992, Mare & Winship 1991, Raley 1996).

All in all, cohabitation is common in all subgroups, making it important to underscore that any existing differentials are only tendencies. In fact, one could make the case that we ought to invert the framing of past research on group differentials; instead of asking "who cohabits?" we might ask "who does not cohabit?"

WHY HAS COHABITATION BECOME SO COMMON?

In general, the same explanations that have been posed to understand changes in family patterns overall are also used to explain the trend in cohabitation; cohabitation is taken to be just one component, albeit a recent one, of a constellation of longer-term changes occurring in the United States and in Europe (Cherlin & Furstenberg 1988; Kiernan 1988, 1999). Declining fertility levels, increasing age at marriage, rising marital disruption rates, and a growing proportion of children being born outside of marriage are other manifestations of this broader shift. While some trends, including cohabitation in the United States, began in earnest in the 1960s or 1970s, divorce has been gradually rising for over a century, and there is wide consensus that even the most recent trends have quite long-term historical roots (Bumpass 1990...).

Scholars emphasize various aspects of long-term social change to explain cohabitation's rise in both the United States and other Western industrialized countries. Some aspects may be labeled cultural. Rising individualism and secularism figure prominently in this category (Lesthaeghe 1983, Lesthaeghe & Surkyn 1988, Rindfuss &

VandenHeuvel 1990, Thornton 1989). The former refers to the increasing importance of individual goal attainment over the past few centuries, and the latter to the decline in religious adherence and involvement. A second set of factors is generally labeled economic. This set ranges from broad conceptualizations of the massive social changes wrought by industrialization (Goode 1963) to narrower ones focusing on women's changing roles in the labor market and concomitant shifts in values and attitudes about gender roles (Cherlin & Furstenberg 1988). More proximate and direct sources of cohabitation's rise are also recognized, an important one being the "sexual revolution." As Bumpass (1990) notes, this revolution eroded the main grounds for earlier disapproval of cohabitation (i.e., that unmarried persons were having sexual relations). Once this stigma was removed, cohabitation was free to escalate.

There has also been some speculation about contemporary causal *processes.* One idea is that "feedback loops" are particularly important for understanding recent trends in family patterns (e.g., Bumpass 1990, Rindfuss & VandenHeuvel 1990). The idea is straightforward: The various trends are mutually reinforcing, with changes in one domain of family life maintaining and perhaps accelerating those in other domains. As one example, high aggregate levels of marital disruption can increase the likelihood that people will cohabit as they learn either through observation or experience that marriage may not be permanent. . . .

MAJOR RESEARCH QUESTIONS ABOUT COHABITATION

Research on cohabitation has gone well beyond basic documentation, as important as the latter has been and will continue to be to track evolving patterns. Most of the remaining research can be organized around the following three questions: "How does cohabitation affect marital stability?" "Where does cohabitation fit in the U.S. family system?" And "How does cohabitation affect children?" The first and the third questions are more straightforward than the second; I argue below that three types of studies offer important, albeit indirect, clues about the second question.

How Does Cohabitation Affect Marital Stability?

Common sense suggests that premarital cohabitation should provide an opportunity for couples to learn about each other, strengthen their bonds, and increase their chances for a successful marriage. In fact, this notion is echoed in the sentiments of cohabitors themselves: Data from the NSFH indicate that over 50% of cohabitors view cohabitation as a way for couples to ensure they are compatible (Bumpass et al. 1991). Thus, one would predict that those cohabiting prior to marriage ought to have higher-quality and more stable marriages.

The evidence, however, suggests just the opposite. Premarital cohabitation tends to be associated with lower marital quality and to increase the risk of divorce, even after taking account of variables known to be associated with divorce (e.g., education, age at marriage). Given wide variation in data, samples, measures of marital instability, and independent variables, the degree of consensus about this central finding is impressive (Axinn & Thornton 1992, Schoen 1992, Teachman et al 1991, Thomson & Colella 1992). . . .

Most of the research on this issue has been aimed at explanation and not simply documentation. . . . Two main explanations have been posed to explain the association, and both have received empirical support. The first is what is termed the selection explanation. This refers to the idea that people who cohabit before marriage differ in important ways from those who do not, and these ways increase the likelihood of marital instability. In other words, the characteristics that select people into cohabitation in the first place, such as nontraditional values and attitudes or poor relationship skills, are also those that increase the risk of marital instability. The second explanation is that there is something about cohabitation itself, i.e., the *experience* of cohabitation, that increases the likelihood of marital disruption above and beyond one's characteristics at the start of the cohabitation. Through cohabitation people learn about and come to accept the temporary nature of relationships, and in particular that there are alternatives to marriage. Note that the two explanations are not mutually exclusive, the first focusing on the characteristics that select people initially into cohabitation and the second positing that the experience of cohabitation alters these characteristics to make people even more divorce-prone.

There is a reasonable amount of empirical support for the selection argument . . . and at least two studies have suggested that selectivity entirely explains the association between premarital cohabitation and marital instability. . . .

The second explanation has received less attention: that the experience of cohabiting further increases the risk of marital instability by changing people's characteristics. This is probably because data requirements are somewhat steep. One needs comparable data on attitudes and other factors both prior to and following cohabitation. One exception is the work of Axinn & Thornton (1992), who examine whether the experience of cohabitation between the ages of 18 and 23 significantly alters young men's and women's attitudes toward marriage and divorce. They find that it does, with cohabitation changing people's attitudes in ways that make them more prone to divorce (see also Axinn & Barber 1997, Clarkberg 1999b).

Where Does Cohabitation Fit Into the U.S. Family System?

There has been an effort to determine "where" cohabitation fits into the family system in the United States. What is the *meaning* or significance of cohabitation in the United States? This is a complex and rather ambiguous question; cohabitation researchers have thus attempted to frame it in more tractable terms. Two main possibilities have been posed: Cohabitation is either a stage in the marriage process (i.e., a form of engagement that culminates in marriage) or a substitute for marriage. According to the first view, marriage as an institution is not threatened by cohabitation. As part of the process leading to marriage, cohabitation plays much the same role as engagement. The large proportions of cohabitors that subsequently marry or have plans to marry generally support this notion (Brown & Booth 1996, Bumpass 1990). The second view—that cohabitation is an alternative to marriage—implies that marriage as an institution is threatened and losing its centrality in the United States. A third and less common view, advanced by Rindfuss & VandenHeuvel (1990), is that cohabitation is more appropriately viewed as an alternative to singlehood than to marriage. The authors argue that cohabitation represents an extension of dating and sexual relationships and that its ideology does not include permanence.

Three types of studies are relevant to this issue, all of them being comparisons of one kind or another. The first are those that compare various characteristics of individuals or couples in different statuses (i.e., married, cohabiting, or single). In this case, research is often explicitly motivated by the question of the meaning of cohabitation. The second type includes studies of mate selection. These studies evaluate partner similarity in marriages and cohabitations. Assuming that partner homogamy tells us something about the nature of marriage, the reasoning goes, then one can learn about the meaning of cohabitation by comparing patterns across union type. The third type includes studies on childbearing and how it varies among cohabiting, single, and married women. While this research has often been primarily motivated by a demographic interest in fertility, many authors use findings to speculate about the meaning of cohabitation.

General Comparisons. There appear to be differences on a range of characteristics between cohabitors and both married and single people.

Rindfuss & VandenHeuvel (1990) compared childbearing intentions, schooling, home ownership, employment, and other characteristics between the three groups and found that cohabitors are more similar to single than married people in virtually all of the comparisons. For example, 33% of single and cohabiting males were homeowners compared to 80% for married men, the bivariate relationship remaining statistically significant when a number of background factors were controlled (p. 716). These comparisons lead the authors to conclude that cohabitation is not an alternative to marriage, but an alternative to conventional singlehood.

Other studies of this type emphasize *relationships* rather than individual traits (Brines & Joyner 1999, Brown & Booth 1996, Nock 1995). Nock (1995), for example, argues that cohabitation and marriage differ not necessarily because of the type of people drawn into each, but because cohabitation is less institutionalized than marriage. As Nock states, "Cohabitation is an incomplete institution. No matter how widespread the practice, nonmarital unions are not yet governed by strong consensual norms or formal laws" (p. 74). The weak institutionalization of cohabitation, Nock argues, has several implications. For example, there are fewer obstacles to ending a cohabiting relationship than a marriage, cohabitors are less likely to be integrated into important social support networks, and there is much more ambiguity about what it means to be a cohabiting partner than to be a spouse. Consistent with this conceptualization, Nock finds that cohabitors report lower levels of commitment and lower levels of relationship happiness than do married people (see also Thomson & Colella 1992). Brines & Joyner (1999) examine the factors that promote stability in the two types of couples, finding that egalitarian gender roles (based on similarity in employment and earnings of the two partners) reduce the risk of break-up among cohabiting but not married couples. Their findings suggest that the two types of relationships may operate on different principles, with cohabiting unions operating on a principle of equality.

At the same time, a study by Brown & Booth (1996) suggests instead that there may essentially be two types of cohabiting couples: those who have plans to marry and those who do not. They show that the former are quite similar to married couples in terms of several dimensions of relationship quality (e.g., happiness, conflict management); it is only cohabiting couples without plans to marry who report significantly lower-quality relationships. These findings lead the authors to speculate that cohabitation is similar to marriage for the majority of cohabitors; about three quarters of cohabitors in their NSFH sample report plans to marry their partners. . . .

Mate Selection Studies. Assortative mating, or the propensity of people to marry those like themselves, is a well-established area of sociological research (Kalmijn 1998, Mare 1991). Studies most commonly investigate similarity between spouses in terms of race-ethnicity, educational attainment, religion, and age. Recently a few researchers expanded the scope of this literature by examining similarity between cohabiting partners and comparing it to that of spouses (Blackwell & Lichter forthcoming, Qian 1998, Qian & Preston 1993, Schoen & Weinick 1993). The underlying idea is that difference (or similarity) in mate selection patterns between the two kinds of couples ought to tell us something about whether cohabitation resembles marriage. Schoen & Weinick (1993) clearly state this reasoning: "Patterns of partner choice can provide insight into how cohabitations are similar to, or different from, marriages" (p. 412).

Overall, married couples appear to be somewhat more homogamous in age, religion, and race-ethnicity, although findings are mixed regarding education (Blackwell & Lichter forthcoming, Schoen & Weinick 1993). . . .

Childbearing Studies. If we assume that a main purpose of marriage is, or at least has been, reproduction, then examining the fertility behavior of cohabitors and comparing it to that of married or single women can offer clues about the meaning of cohabitation. That is, if cohabitation is increasingly the arena for reproduction, then one might conclude that cohabitation is not merely a step in the process leading to marriage but perhaps an alternative to it.

An example of a relevant fertility study is Manning (1993). She evaluates the likelihood that an unmarried, pregnant woman will marry before the birth of her child, a topic traditionally called "legitimation" in fertility research. Her main empirical question is: Do cohabiting and noncohabiting single women have *equal* tendencies to marry before childbirth? While oversimplifying here, the answer varies for black and white women, even after, controlling for socioeconomic status, and this variation suggests possible differences in the meaning of cohabitation. Essentially, she finds that for white women in their twenties cohabitation (relative to living alone) increases the likelihood of marriage before childbirth. Presumably, if the cohabiting women considered cohabitation an acceptable context for childbearing, there would be no differential and cohabiting women would remain cohabiting in response to a pregnancy. Manning thus concludes that, for white women, cohabitation is a stage in the marriage process.

The main conclusion emerging from studies of this type is that there are race-ethnic differences in the relationship between pregnancy, cohabitation, and marriage and thus possibly race-ethnic differences in the meaning of cohabitation. The overall interpretation of most of the authors is that cohabitation is more an alternative to marriage (at least in terms of childbearing) among black and mainland Puerto Rican women and more a precursor to marriage among non-latino white women (Landale & Fennelly 1992, Landale & Forste 1991, Loomis & Landale 1994, Manning 1993, 1995, 1999, Manning & Landale 1996, Manning & Smock 1995, Oropesa 1996, Raley 1999).

At the same time, it is important to keep these findings and interpretations of them in perspective. The majority of women in the United States overall do not at this time conceive or give birth during cohabitation. Only about 11% of all children are born into cohabiting households, and about 40% of nonmarital births occur in cohabiting unions (Bumpass & Lu 1999). The key point is that patterns vary by race-ethnicity. Astone et al.'s (1999) study of a cohort of black men in Baltimore attests to the fact that a good deal of fatherhood among black men is occurring in the context of cohabitation. Assigning these men the label of "unwed fathers" based on marital status obscures that they are, in fact, co-resident parents.

How Does Cohabitation Affect Children?

Simply because a child isn't born to cohabiting parents does not mean he or she will not experience a parent's cohabitation at some point during childhood. This latter happens when a child has been living with one parent, typically the mother, and that parent enters a cohabiting relationship.

Just how pervasive is parental cohabitation in children's lives? Cross-sectional statistics indicate that only a small proportion of children live in cohabiting households at any one point in time. Data from the 1990 Public Use Microsample of the Census show that about 13% of children in single-parent families were actually living with cohabiting parents, which translates into just 3.5% of all children (Manning & Lichter 1996). However, the proportion of children who will *ever* live in a cohabiting household during childhood is estimated to be a substantial 40%, underscoring the importance of understanding the effects of parental cohabitation on children (Bumpass & Lu 1999). As Bumpass & Lu write, ". . . now that about two-fifths of all children spend some time living with their mother and a cohabiting partner . . . we simply cannot address the changing family experiences of children while ignoring cohabitation" (1999, p. 21).

Past studies have identified two important issues regarding children's experience of parental cohabitation. The first is that children already disadvantaged in terms of parental income and education are relatively more likely to experience this family form (Bumpass & Lu 1999, Graefe &

Lichter 1999). This finding is consistent with other research showing that, on average, cohabiting households tend to be less well-off financially than married-couple households (e.g., Manning & Lichter 1996) and that good economic circumstances increase the likelihood of marriage among both cohabiting and noncohabiting individuals (e.g., Clarkberg 1999a, Lichter et al 1992, Smock & Manning 1997).

Second, children experiencing parental cohabitation are also more likely to undergo further transitions in family structure. Graefe & Lichter (1999) estimate that most children who are born or ever live in a cohabiting family will experience a change in family structure within a few years. These findings do not bode well for children's well-being. A large body of literature has established that family structure has important effects on children, with deleterious ones for children who grow up without both biological parents (McLanahan & Sandefur 1994, Seltzer 1994). While some of this effect is due to income and other factors, there is evidence that the *number* of changes in family structure is particularly important. The fewer the changes, the better for children (Wu & Martinson 1993, Wu 1996). While children in cohabiting households may in fact be living with two biological parents, although more typically one parent and a "step-parent," they are quite likely to experience future family transitions. . . .

ASSESSMENT OF PAST RESEARCH AND FUTURE DIRECTIONS

A first observation is simply that an enormous amount has been accomplished in a very short time. Just 15 years ago very little was known about cohabitation. Since then family sociologists and demographers have rapidly created a solid base of generalizable knowledge about cohabitation in the United States. We know a good deal not only about overall trends, differentials, and patterns but also about the effect of cohabitation on marital stability, the nature of differences between cohabitation and marriage, and the role of cohabitation in nonmarital childbearing. Often in the guise of the latter two issues, researchers are also developing an understanding of the significance or meaning of cohabitation. . . .

A second observation concerns comparisons between cohabitation and marriage, and in particular, the attempt to gauge the meaning of cohabitation by comparing it to marriage (i.e., some comparison studies and mate selection studies). What is sometimes omitted from studies of this ilk is full acknowledgment that the meaning of *marriage* is dynamic and undergoing radical change. In the last few decades, for example, women have been playing increasingly important roles as income providers (Bianchi 1995). The underlying issue is whether we can gauge what cohabitation means if we are using a standard that is also changing. To say that cohabitation is like or unlike marriage is useful only to the extent that we have adequate knowledge of what marriage is indeed "like."

A related issue is the emphasis in some past research on differences between those who cohabit and those who do not. It is probably time for most research on cohabitation to begin from a premise that the majority of men and women will cohabit, prior to marriage or afterwards, and that the cohabiting couples studied today are the married couples of tomorrow. Certainly, many researchers are already well aware of this issue (e.g., Blackwell & Lichter forthcoming, Clarkberg et al 1995, Nock 1995, 1998).

A third issue is that there are strong indications that *cohabitation* is changing in significant ways, even over the last few years. Consider these examples. First, there is evidence that the inverse relationship between premarital cohabitation and marital stability is diminishing. The effect, if any, is trivial for recent birth cohorts (Schoen 1992). Second, between 1987 and 1995 there has been significant change in cohabiting couples' trajectories: lower proportions are marrying and more are breaking up (Bumpass & Lu 1999). Third, Raley

(1999) shows that there has been a significant shift in the relationship between fertility and cohabitation between the early 1980s and 1990s. Pregnant women are increasingly likely to cohabit or remain cohabiting rather than to marry in response to a premarital pregnancy, suggesting perhaps that cohabitation is becoming more a substitute for marriage as time goes on. Finally, the proportion of cohabitations with children present increased from 40% to 50% between 1987 and 1995 (Bumpass & Lu 1999).

Whatever the substantive implications, this assortment of facts illustrates the point that cohabitation is changing in substantial ways over very short spans of time. A continuing task for researchers will thus be simply to keep pace with developments; as new data become available new descriptive studies will be needed.

Future work on cohabitation is likely to elaborate on at least two existing themes in the literature. The first is the effect of cohabitation on children's well-being. This question follows a long line of research on the effects of family structure on children and engages an ongoing concern of policymakers and funding agencies. Given that currently available data are inadequate for discerning the effects of cohabitation versus other family structures (Bumpass & Lu 1999), there will likely be new data collection efforts as well as new content on existing surveys.

It is also probable that future research will pay relatively more attention to diversity than in the past, consistent with the fact that cohabitation has just about become a majority phenomenon; this will complement research that focuses on central tendencies. In particular, there will be a continuing effort to understand whether and how cohabitation's "meaning" may vary across subgroups given that past studies imply that there are at least race-ethnic differences in this meaning (e.g., Manning 1993). Raley's (1999) findings suggest the importance of ongoing temporal change in the meaning of cohabitation as well.

Finally, virtually all sociological knowledge about cohabitation in the United States has been based on quantitative analysis of survey or census data. Certainly, this approach has taught us a great deal about cohabitation in a very short time. Yet the broader question of the meaning of cohabitation remains difficult to address. To begin to answer it and similar questions, it would be useful additionally to be able to draw on qualitative data that ask people what cohabitation means to them. This approach could increase our understanding of possible diversity in the meaning of cohabitation by gender, social class, or race-ethnicity. More generally, it would provide the sort of in-depth data important for nuanced assessments of recent and possible future changes in family patterns. A practical payoff is that it would provide a basis for new survey content.

SELECTED REFERENCES

Astone NM, Schoen R, Ensminger M, Rothert K. 1999. *The family life course of African American men.* Presented at Annu. Meet. Pop. Assoc. Am., New York

Axinn WG, Barber JS. 1997. Living arrangements and family formation attitudes in early adulthood. *J. Marr. Fam.* 59:595–611

Axinn WG, Thornton A. 1992. The relationship between cohabitation and divorce: selectivity or causal influence? *Demography* 29:357–74

Bianchi S. 1995. The changing economic roles of women and men. In *State of the Union: American in the 1990s,* ed. R. Farley, 1:107–54. New York: Russell Sage. 375 pp.

Blackwell DL, Lichter DT 2000. Mate selection among married and cohabiting couples. *J. Fam. Issues.*

Brines J, Joyner K. 1999. The ties that bind: commitment and stability in the modern union. *Am. Sociol. Rev.* 64:333–56

Brown SL, Booth A. 1996. Cohabitation versus marriage: a comparison of relationship quality. *I. Marr. Fam.* 58:668–78

Bumpass LL. 1990. What's happening to the family? Interaction between demographic and institutional change. *Demography* 27:483–98

Bumpass LL, Lu H. 1999. *Trends in cohabitation and implications for children's family contexts in the U.S. CDE Work Pap. No. 98-15.* Cent. Demography Ecol., Univ. Wisc.-Madison

Bumpass LL, Sweet JA. 1989. National estimates of cohabitation. *Demography* 26:615–25

Bumpass LL, Sweet JA, Cherlin A. 1991. The role of cohabitation in declining rates of marriage. *Demography* 53:913–27

Casper LM, Cohen PN. 2000. How does POSSLQ measure up? Historical estimates of cohabitation. *Demography.* In press.

Cherlin A, Furstenberg FF. 1988. The changing European family: lessons for the American reader. *J. Fam. Issues* 9:291–7

Clarkberg ME. 1999a. The price of partnering: the role of economic well-being in young adults' first union experiences. *Soc. Forces* 77:945–68

Clarkberg ME. 1999b. *The cohabitation experience and changing values: the effects of premarital cohabitation on the orientation towards marriage, career and community. BLCC Work. Pap. No. 99-15.* Cornell Employment and Fam. Careers Inst., Cornell

Clarkberg ME, Stolzenberg RM, Waite LJ. 1995. Attitudes, values, and entrance into cohabitational versus marital unions. *Soc. Forces* 74:609–34

Goode WJ. 1963. *World Revolution and Family Patterns.* New York: Free

Graefe DR, Lichter DT. 1999. Life course transitions of American children: parental cohabitation, marriage and single motherhood. *Demography* 36:205–17

Kalmijn M. 1998. Intermarriage and homogamy: causes, patterns, trends. *Annu. Rev. Sociol.* 24:395–421

Kiernan K. 1988. The British family: contemporary trends and issues. *J. Marr. Fam.* 298–316

Kiernan K. 1999. Cohabitation in Western Europe. *Pop. Trends* 96:25–32

Landale NS, Fennelly K. 1992. Informal unions among mainland Puerto Ricans: cohabitation or an alternative to legal marriage? *J. Marr. Fam.* 54:269–80

Landale NS, Forste R. 1991. Patterns of entry into cohabitation and marriage among mainland Puerto Rican women. *Demography* 28:587–607

Lesthaeghe R. 1983. A century of demographic and cultural change in Western Europe: an exploration of underlying dimensions. *Pop. Dev. Rev.* 9:411–35

Lesthaeghe R, Surkyn J. 1988. Cultural dynamics and economic theories of fertility change. *Pop. Dev. Rev.* 14:1–45

Lichter DT, McLaughlin D, LeClere F, Kephart G, Landry D. 1992. Race and the retreat from marriage: a shortage of marriageable men? *Am. Sociol. Rev.* 57:781–99

Loomis LS, Landale NS. 1994. Nonmarital cohabitation and childbearing among black and white American women. *J. Marr. Fam.* 56:949–62

Lye D, Waldron I. 1997. Attitudes toward cohabitation, family, and gender roles: relationships to values and political ideology. *Social. Perspect.* 40:199–25

Manning WD. 1993. Marriage and cohabitation following premarital conception. *J. Marr. Fam.* 55:839–50

Manning WD. 1995. Cohabitation, marriage, and entry into motherhood. *J. Marr. Fam.* 57:191–200

Manning WD. 1999. *Childbearing in cohabiting unions: racial and ethnic differences.* Presented at Annu. Meet. Pop. Assoc. Am., New York

Manning WD, Landale NS. 1996. Racial and ethnic differences in the role of cohabitation in premarital childbearing. *J. Marr. Fam.* 58:63–77

Manning WD, Lichter DT. 1996. Parental cohabitation and children's economic well-being. *J. Marr. Fam.* 58:998–1010

Manning WD, Smock PJ. 1995. Why marry? Race and the transition to marriage among cohabitors. *Demography* 32:509–20

Mare RD. 1991. Five decades of educational assortative mating. *Am. Sociol. Rev.* 56:15–32

Mare RD, Winship C. 1991. Socioeconomic change and the decline of marriage for blacks and whites. In *The Urban Underclass,* ed. C Jencks, PE Peterson, pp. 175–202. Washington DC: Urban. 490 pp.

McLanahan SS, Sandefur G. 1994. *Growing Up With a Single Parent: What Helps, What Hurts.* Cambridge, MA: Harvard Univ. Press. 196 pp.

Nock SL. 1995. A comparison of marriages and cohabiting relationships. *J. Fam. Issues* 16:53–76

Nock SL. 1998. *Marriage in Men's Lives.* Oxford, UK: Oxford Univ. Press, 165 pp.

Oropesa RS. 1996. Normative beliefs about marriage and cohabitation: a comparison of non-Latino whites, Mexican Americans, and Puerto Ricans. *J. Marr. Fam.* 58:49–62

Qian Z. 1998. Changes in assortative mating: the impact of age and education, 1970–1990. *Demography* 35:279–92

Qian Z. Preston SH. 1993. Changes in American marriage, 1972–1987. *Am. Sociol. Rev.* 58:482–95

Raley RK. 1996. A shortage of marriageable men? A note on the role of cohabitation in black-white differences in marriage rates. *Am. Sociol. Rev.* 61:973–83

Raley RK. 1999. *Then comes marriage? Recent changes in women's response to a nonmarital pregnancy.* Presented at Annu. Meet. Pop. Assoc. Am., NY

Rindfuss RR, VandenHeuvel A. 1990. Cohabitation: a precursor to marriage or an alternative to being single? *Pop. Dev. Rev.* 16:703–26

Schoen R. 1992. First unions and the stability of first marriages. *J. Marr. Fam.* 54:281–84

Schoen R, Weinick RM. 1993. Partner choice in marriages and cohabitations. *J. Marr. Fam.* 55:408–14

Seltzer JA. 1994. Consequences of marital dissolution for children. *Annu. Rev. Sociol.* 20:235–66

Smock PJ, Manning WD. 1997. Cohabiting partners' economic circumstances and marriage. *Demography* 34:331–41

Teachman JD, Thomas J, Paasch K. 1991. Legal status and stability of coresidential unions. *Demography* 28:571–86

Thomson E, Colella U. 1992. Cohabitation and marital stability: quality or commitment? *J. Marr. Fam.* 54:259–67

Thornton A. 1989. Changing attitudes toward family issues in the United States. *J. Marr. Fam.* 51:873–93

Thornton A, Axinn WG, Teachman JD. 1995. The influence of school enrollment and accumulation on cohabitation and marriage in early adulthood. *Am. Sociol. Rev.* 60:762–74

Wu LL. 1996. Effects of family instability, income, and income instability on the risk of premarital birth. *Am. Sociol. Rev.* 386–406

Wu LL, Martinson BC. 1993. Family structure and the risk of a premarital birth. *Am. Sociol. Rev.* 58:210–32

10

The Future of Marriage

STEPHANIE COONTZ,
1997

Most Americans support the emergence of alternative ways of organizing parenthood and marriage. They don't want to reestablish the supremacy of the male breadwinner model or to define masculine and feminine roles in any monolithic way. Many people worry, however, about the growth of alternatives to marriage itself. They fear that in some of today's new families parents may not be devoting enough time and resources to their children. The rise of divorce and unwed motherhood is particularly worrisome, because people correctly recognize that children need more than one adult involved in their lives.

As a result, many people who object to the "modified male breadwinner" program of the "new consensus" crusaders are still willing to sign on to the other general goals of that movement: "to increase the proportion of children who grow up with two married parents," to "reclaim the ideal of marital permanence," to keep men "involved in family life," and to establish the principle "that every child deserves a father."[1]

Who could disagree? When we appear on panels together, leaders of "traditional values" groups often ask me if I accept the notion that, on the whole, two parents are better than one. If they would add an adjective such as two *good* parents, or even two *adequate* ones, I'd certainly agree. And of course it's better to try to make a marriage work than to walk away at the first sign of trouble.

As a historian, however, I've learned that when truisms are touted as stunning new research, when aphorisms everyone agrees with are presented as a courageous political program, and when exceptions or complications are ignored for the sake of establishing the basic principles, it's worth taking a close look for a hidden agenda behind the cliches. And, in fact, the new consensus crowd's program for supporting the two-parent family turns out to be far more radical than the feel-good slogans might lead you to believe.

Members of groups such as the Council on Families in America claim they are simply expressing a new consensus when they talk about "reinstitutionalizing enduring marriage," but in the very next breath they declare that it "is time to raise the stakes." They want nothing less than to make lifelong marriage the "primary institutional expression of commitment and obligation to others," the main mechanism for regulating sexuality, male-female relations, economic redistribution, and child rearing. Charles Murray says that the goal is "restoration of marriage as an utterly distinct, legal relationship." Since marriage must be "privileged," other family forms or child-rearing arrangements should not receive tax breaks, insurance benefits, or access to public housing and federal programs. Any reform that would make it easier for divorced parents, singles, unmarried partners, or stepfamilies to function is suspect because it removes "incentives" for people to get and stay married. Thus, these groups argue, adoption and foster care policies should "reinforce marriage as the child-rearing norm." Married couples, and only married couples, should be given special tax relief to raise their children. Some leaders of the Institute for American Values propose that we encourage both private parties and government bodies "to distinguish between married and unmarried *couples* in housing, credit, zoning, and other areas." Divorce and illegitimacy should be stigmatized.[2]

We've come quite a way from the original innocuous statements about the value of two-parent families and the importance of fathers to children. Now we find out that we must make marriage the only socially sanctioned method for organizing male-female roles and fulfilling adult obligations to the young. "There is no realistic alternative to the one we propose," claims the Council on Families in America. To assess this claim, we need to take a close look at what the consensus crusaders mean when they talk about the need to reverse the "deinstitutionalizing" of marriage.[3]

Normally, social scientists have something very specific in mind when they say that a custom or behavior is "institutionalized." They mean it comes "with a well-understood set of obligations and rights," all of which are backed up by law, customs, rituals, and social expectations. In this sense, marriage is still one of America's most important and valued institutions.[4]

But it is true that marriage has lost its former monopoly over the organization of people's major life transitions. Alongside a continuing commitment to marriage, other arrangements for regulating sexual behavior, channeling relations between men and women, and raising children now exist. Marriage was once the primary way of organizing work along lines of age and sex. It determined the roles that men and women played at home and in public. It was the main vehicle for redistributing resources to old and young, and it served as the most important marker of adulthood and respectable status.

All this is no longer the case. Marriage has become an option rather than a necessity for men and women, even during the child-raising years. Today only half of American children live in nuclear families with both biological parents present. One child in five lives in a stepfamily and one in four lives in a "single-parent" home. The number of single parents increased from 3.8 million in 1970 to 6.9 million in 1980, a rate that averages out to a truly unprecedented 6 percent increase each year. In the 1980s, the rate of increase slowed and from 1990 to 1995 it leveled off, but the total numbers have continued to mount, reaching 12.2 million by 1996.[5]

These figures understate how many children actually have two parents in the home, because they confuse marital status with living arrangements. Approximately a quarter of all births to unmarried mothers occur in households where the father is present, so those children have two parents at home in fact if not in law. Focusing solely on the marriage license distorts our understanding of trends in children's living arrangements. For example, the rise in cohabitation between 1970 and 1984 led to more children being classified as living in single-parent families. But when researchers counted unmarried couples living together as two-parent families, they found that children were spending *more* time, not less, with both parents in 1984 than in 1970. Still, this simply confirms the fact that formal marriage no longer organizes as many life decisions and transitions as it did in the past.[6]

Divorce, cohabitation, remarriage, and single motherhood are not the only factors responsible for the eclipse of marriage as the primary institution for organizing sex roles and interpersonal obligations in America today. More people are living on their own before marriage, so that more young adults live outside a family environment than in earlier times. And the dramatic extension of life spans means that more people live alone after the death of a spouse.[7]

The growing number of people living on their own ensures that there are proportionately fewer families of *any* kind than there used to be. The

Census Bureau defines families as residences with more than one householder related by blood, marriage, or adoption. In 1940, under this definition, families accounted for 90 percent of all households in the country. By 1970, they represented just 81 percent of all households, and by 1990 they represented 71 percent. The relative weight of marriage in society has decreased. Social institutions and values have adapted to the needs, buying decisions, and lifestyle choices of singles. Arrangements other than nuclear family transactions have developed to meet people's economic and interpersonal needs. Elders, for example, increasingly depend on Social Security and private pension plans, rather than the family, for their care.[8]

Part of the deinstitutionalization of marriage, then, comes from factors that few people would want to change even if they could. Who wants to shorten the life spans of the elderly, even though that means many more people are living outside the institution of marriage than formerly? Should we lower the age of marriage, even though marrying young makes people more likely to divorce?[9] Or should young people be forced to live at home until they do marry? Do we really want to try to make marriage, once again, the only path for living a productive and fulfilling adult life?

WORKING WOMEN, SINGLEHOOD, AND DIVORCE

If the family values crusaders believe they are the only people interested in preserving marriages, especially where children's well-being is involved, self-righteousness has blinded them to reality. I've watched people of every political persuasion struggle to keep their families together, and I've met very few divorced parents who hadn't tried to make their marriages work. Even the most ardent proponents of reinstitutionalizing marriage recognize that they cannot and should not force everyone to get and stay married. They do not propose outlawing divorce, and they take pains to say that single parents who were not at fault should not be blamed. Yet they still claim that moral exhortations to take marriage more seriously will reduce divorce enough to "revive a culture of enduring marriage."

This is where the radical right wing of the family values movement is far more realistic than most moderates: So long as women continue to make long-term commitments to the workforce, marriage is unlikely to again become the lifelong norm for the vast majority of individuals unless draconian measures are adopted to make people get and stay married. Paid work gives women the option to leave an unsatisfactory marriage. In certain instances, much as liberals may hate to admit it, wives' employment increases dissatisfaction with marriage, sometimes on the part of women, sometimes on the part of their husbands. When a wife spends long hours at work or holds a nontraditional job, the chance of divorce increases.[10]

I'm not saying we can't slow down the divorce rate, lessen emotional and economic disincentives for marriage, and foster longer-lasting commitments. But there is clearly a limit to how many people can be convinced to marry and how many marriages can be made to last when women have the option to be economically self-supporting. In this sense, the right-wing suspicion that women's work destabilizes marriage has a certain logic.

There is, however, a big problem with the conclusion that the radical right draws from this observation. To say that women's employment has *allowed* divorce and singlehood to rise in society as a whole does not mean that women's work *causes* divorce and singlehood at the level of the individual family, or that convincing women to reduce their work hours and career aspirations would reestablish more stable marriages.[11]

Trying to reverse a historical trend by asking individuals to make personal decisions opposing that trend is usually futile. When individuals try to conduct their personal lives as if broader social forces were not in play, they often end up worse off than if they adapted to the changing times. What traditional values spokesman in his right mind would counsel his own daughter not to prepare herself for higher-paid, nontraditional jobs because these might lead to marital instability down the road?

After all, even if a woman *prefers* to have a male breadwinner provide for her, the fact that people can now readily buy substitutes for what used to require a housewife's labor changes marriage dynamics in decisive ways. Most individuals and families can now survive quite easily without a full-time domestic worker. If this frees women to work outside the home, it also frees men from the necessity of supporting a full-time homemaker.

Before the advent of washing machines, frozen foods, wrinkle-resistant fabrics, and 24-hour one-stop shopping, Barbara Ehrenreich has remarked, "the single life was far too strenuous for the average male." Today, though, a man does not really need a woman to take care of cooking, cleaning, decorating, and making life comfortable. Many men still choose marriage for love and companionship. But as Ehrenreich notes, short of outlawing TV dinners and drip-dry shirts, it's hard to see how we can make marriage as indispensable for men as it used to be. And short of reversing laws against job discrimination, there's no way we can force women into more dependence on marriage. Neither men nor women need marriage as much as they used to. Asking people to behave as if they do just sets them up for trouble.[12]

Wives who don't work outside the home, for instance, are at much higher economic and emotional risk if they *do* get deserted or divorced than women who have maintained jobs. They are far more likely to be impoverished by divorce, even if they are awarded child support, and they eventually recover a far lower proportion of the family income that they had during marriage than women who had been working prior to the divorce.[13]

Women who refrain from working during marriage, quit work to raise their children, or keep their career aspirations low to demonstrate that "family comes first" are taking a big gamble, because there are many factors other than female independence that produce divorce. Some of them are associated with women's decisions or expectations *not* to work or *not* to aspire to higher education. For example, couples who marry in their teens are twice as likely to divorce as those who marry in their twenties. Women who marry for the first time at age 30 or more have exceptionally low divorce rates, despite their higher likelihood of commitment to paid work. Women who don't complete high school have higher divorce rates than women who do, and high school graduates have higher divorce rates than women who go on to college. With further higher education, divorce rates go up again, but should we advise a woman to abandon any aspirations she may have developed in college because her statistical chance of staying married will rise if she quits her education now? Besides, the statistics may confuse the impact of higher education with what is called the "glick effect"—people who have started toward *but failed to complete* a particular degree or diploma, whatever its level, are more likely to divorce than those who secure a precise diploma or degree. A more useful piece of advice might be that if you *are* going to do graduate work, hang in there until you've finished.[14]

Highly educated and high-paid women may have a greater chance of divorcing, but women with low earnings and education have lower prospects of getting married in the first place. Men increasingly choose to marry women who have good jobs and strong educational backgrounds. In the 1980s, reversing the pattern of the 1920s, "women with the most economic resources were the most likely to marry." But of course these are also the women most able to leave a bad marriage.[15]

THE ISSUE OF NO-FAULT DIVORCE

Some people believe we could stabilize families, and protect homemakers who sacrifice economic independence, by making divorce harder to get, especially in families with children. This sounds reasonable at first hearing. Divorce tends to disadvantage women economically, and to set children back in several ways. It is hardest of all on women who committed themselves and their children to the bargain implied by the 1950s marriage ideal—forgoing personal economic and educational advancement in order to raise a family, and expecting lifetime financial support from a husband in return. A 47-year-old divorced mother describes what happened to her: "Instead of starting a career for myself, I helped my husband get his business started. I had four children. I made the beds. I cooked the meals. I cleaned the house. I kept my marriage vows. Now I find myself divorced in midlife with no career. My husband makes $100,000 a year, and we're struggling to get by on a quarter of that."[16]

We could avoid such inequities, say the family values crusaders, if today's marriage contract was not "considerably less binding than, for example, a contract to sell a car or a cow." "The first step is to end unilateral divorce," says Maggie Gallagher of the Institute for American Values. She advocates imposing a five- to seven-year waiting period for contested divorces.[17]

The argument that "we ought to enforce marriage just like any other contract" sounds reasonable until you take a historical and sociological perspective on the evolution of divorce law. Then a number of problems become clear. First, requiring people to stay together has nothing to do with enforcing contract law. When someone breaks a contract, the courts don't normally force the violator to go back and provide the services; they merely assign payment of money damages. If an entertainer refuses to perform a concert, for example, no matter how irresponsibly, the promoter cannot call the police to haul the performer into the theater and stand over him while he sings, or even impose a cooling-off period so he can rethink whether he wants to honor the contract. Instead, the promoter is awarded damages. The contract analogy may make a case for seeking damages, but it has no relevance to the issue of making divorce harder to get.

A woman who has sacrificed economic opportunities to do the bulk of child raising ought to get compensation for that when a marriage breaks up, no matter whose "fault" the failure is. Even full-time working wives often give up higher-paying jobs or education in order to take the lion's share of responsibility for family life. The tendency of many courts during the 1970s to reduce alimony and maintenance allowances for wives was based on the mistaken assumption that because more women were working, male–female equality had already been achieved. Maintenance awards need to be rethought, as well as separated from child support payments, a process now occurring in many states. But improving maintenance provisions for the spouse who did family caregiving is a separate question from forcing someone to remain in a marriage against his or her will—or to do *without* support for a protracted period of time while the courts sort through who was "at fault."[18]

Second, making divorces harder to get would often exacerbate the bitterness and conflict that are associated with the *worst* outcomes of divorce for kids. One of the hot new concepts of the consensus school is that we need to preserve the good-enough marriage—where there is not abuse or neglect but merely an "acceptable" amount of adult unhappiness or discontent in comparison to the benefits for children in keeping the family together. But what government agency or private morals committee will decide if a marriage is "good enough"?

One author suggests we might require parents with children under 18 to "demonstrate that the *family* was better off broken than intact. Unhappiness with one's spouse would not then be a compelling argument for divorce. Domestic violence would be." Yet what would prevent the person who wanted out of the marriage from upping the ante—for example, from threatening domestic violence to get his or her way?[19]

Furthermore, it is *women* more than men who have historically needed the protection of divorce. And yes, I mean protection. Because of men's greater economic and personal power, one divorce historian points out, husbands traditionally handled marital dissatisfaction by intimidating or coercing their wives into doing what the men wanted, such as accepting an extramarital affair or living with abuse. Alternatively, the man simply walked away, taking no legal action. Women, with less social and domestic power, "turned to an external agency, the law, for assistance." Women were the majority of petitioners for divorce and legal separation in English and American history long before the emergence of a feminist movement. Access to divorce remains a critical option for women.[20]

Finally, sociological research finds little evidence that no-fault laws have been the main cause of rising divorce rates, or that, on average, women do worse with no-fault than they did in the days when spouses had to hire detectives or perjurers to prove fault. While more than half the states enacted some form of no-fault divorce legislation in the 1970s, the rates of marital dissolution in most of these states were no higher in the 1970s than would be expected from trends in states that did *not* change their laws. Researcher Larry Bumpass suggests that, in some instances, no-fault has speeded up "cases that were already coming down the pipeline," but the rise in divorce seems to be independent of any particular legal or social policy. And, of course, making divorce harder to get does nothing to prevent separation or desertion.[21]

Work on preventing unnecessary divorces, separations, and desertions needs to happen *before* a marriage gets to the point of rupture. We can experiment with numerous ways to do that, from marriage education to moral persuasion to parenting classes to counseling. We can educate young people about the dangers of our society's throwaway mentality, pointing out that it creates emotional as well as material waste. We should certainly warn people that divorce is never easy for the partners or for their children. Still, history suggests that no amount of classes, counseling, or crusades will reinstitutionalize marriage in the sense that the family values crusaders desire.

Divorce rates are the product of long-term social and economic changes, not of a breakdown in values. Individual belief systems are a comparatively minor factor in predicting divorce. Despite the Catholic Church's strong opposition to divorce, for example, practicing Catholics are as likely to divorce as non-Catholics. Studies have shown that prior disapproval of divorce has little bearing on a person's later chance of divorce, although people who do divorce are likely to modify their previous disapproval. Once again, we have a situation where many intricately related factors are involved. Neither legal compulsions nor moral exhortations are likely to wipe out historical transformations that have been building for so long.[22]

UNWED MOTHERHOOD

Many people I talk with think the place we should draw the line on family diversity is at the question of unwed motherhood. If we could just take care of this problem, they say, society could survive even a fairly large number of divorces.

The growth of unwed motherhood is an even more complicated story than that of divorce, because some of it stems from expanding options for women and some from worsening constraints, especially for low-income or poorly educated women. Often, it's a messy mix, with increased sexual freedom interacting in explosive ways with decreased economic opportunities for both sexes and continuing inequities between men and women.

For example, many people consider the rise in pregnancy among very young teens a symptom of what's wrong with today's "liberated" sexuality. But men over age 20 father five times more births among junior high school girls, and two and a half times more births among senior high school girls, than do the girls' male peers. When the mother is 12 years old or younger, the father's average age is 22. These girls may be responding to new sexual options, but they are doing so within a very old pattern of unequal, exploitative power relations. Three-fourths of girls who have sex before age 14 say they were coerced, and research suggests that a majority of teens who give birth have been physically or sexually abused at some point in their past.[23]

Contrary to what you might guess, the most rapid increase in the rate of nonmarital childbearing occurred between 1940 and 1958, when the rate *tripled* from 7.1 births per 1,000 unmarried women to 21.2. The rate of unwed childbearing rose very slowly from 1958 to 1971, then declined until 1976. Then it rose steadily again until 1992, not quite doubling to reach 45.2 unwed births per 1,000 unmarried women in 1992.[24]

This timing obviously calls into question the popular perception that 1960s permissiveness started the rise in unwed childbearing. Indeed, the increase in unwed motherhood between 1960 and 1975 was *not* due to a significant increase in birth rates among unmarried women but to a combination of two other factors. One was the fact that married women's fertility fell by more than 40 percent during the period, which increased the proportion of unwed births among the total number of births. The second was that the absolute number of single women was rising, putting more women "at risk" for an out-of-wedlock birth. Only after 1975 did the *proportion* of unmarried women who gave birth begin to rise—not just their absolute numbers or their contribution to total births in comparison to married women.[25]

So it's wrong to blame the rise of unwed births on the sexual revolution of the 1960s and 1970s. While it's true that the decision to raise a child born out of wedlock is related to changing cultural values, it appears that changing economic relations between men and women are more important than is usually realized. The renewed increase in rates of unmarried childbearing after 1975 *followed* the fall in real wages for young men that began in 1973. Men with low or irregular wages are far less likely to marry a woman they impregnate than men who have a steady job and earn a family wage. Women are less likely to see such men as desirable marriage partners.

The interactions between economic, social, and cultural factors are too complicated to support any single generalization about the cause of unwed motherhood. In a small though highly publicized number of cases, women's new economic independence has allowed them to choose to bear children out of wedlock. Contrary to conventional wisdom, this choice is more common for women in their twenties and thirties than for teens, and the proportion of such cases remains small. Unwed motherhood is seldom actively pursued by women. National statistics reveal that more than 80 percent of pregnancies among unmarried women are unplanned.

Although the consensus crowd may call for a campaign of stigmatization to prevent women from keeping an unplanned child, it is unlikely that women "of independent means" will give up their hard-earned freedom of choice. And stigmatization would not necessarily prevent unwed motherhood among impoverished women. Historically, out-of-wedlock births have often soared during times of economic stress, and the main result of stigmatization or punitive social policy has been child abandonment or even murder.[26]

Economic stress remains a critical factor in many out-of-wedlock births. The overwhelming majority of single mothers have only a high school education or less. Only one in seven unmarried mothers has a family income above $25,000 the year her child is born; 40 percent have incomes of less than $10,000.[27]

I understand why people worry about women who seem not to plan their childbearing wisely. But I also know, both from sociological research and from participating in parenting workshops across the country, that there are hundreds of paths to becoming a single mother. Some of the tales I hear would seem irresponsible to even the most ardent proponent of women's right to choose; others stem from complicated accidents or miscalculations that even the most radical right winger would probably forgive; still others flow from carefully thought-out choices. But I fail to see how we can draw a hard and fast line between the decisions of unmarried and married women. As one unwed mother said to me, "I made some stupid choices to get where I am. But my reasons for having a kid were no worse than my sister's, who had a baby she didn't want in order to save her marriage. Why should I be the one they stand up and preach against on TV?"

Take the three never-married mothers I met at a workshop in Georgia. One was a young girl who had gotten pregnant "by accident/on purpose" in hopes that her unfaithful boyfriend would marry her. It hadn't worked, and in the ensuing year she sometimes left her baby unattended while she went out on new dates. Shocked into self-examination by being reported for neglect, she concluded that going back to school and developing some job skills might be more help in escaping her own neglectful parents than chasing after a new boyfriend. Another, 25 years old at the time we met, had been 15 when she got pregnant. She considered abortion, but she would have had to drive 300 miles, by herself, to the nearest provider, and she didn't even have money for a hotel room.

Both these young mothers had gone onto welfare. The first had gotten off in three years, and never expected to need it again. The second had stayed on for only a year the first time, then married and found her first job. Shortly afterward, though, her husband left her and she went back on welfare because her work didn't cover child care or medical benefits. Within a year she had found another job and left welfare again. But the company relocated. Twice more she had cycled on and off Aid to Families with Dependent Children because of layoffs or ill health.

At one low point in her welfare days, she told me, she tried prostitution, planning to accumulate some cash she could hide from the welfare agency so she'd have a small financial cushion once she landed a new job. Unfortunately, she said wryly, she hadn't anticipated the job-related expenses in that occupation. For a while she spent a lot of her extra earnings on drugs and alcohol, "to get the dirty taste out of my mouth." She was straight and sober now, and she had a job, but she wasn't willing to stand up and make any testimonials, as her minister had urged her to do during the workshop at which we met. "It's going to be one day at a time for a long time."

These two stories are quite typical. Three-fourths of welfare recipients leave welfare within two years (half get out within a year), but the scarcity and instability of the jobs they find often drive them back to welfare in a few years. Even if they keep their jobs, they seldom earn enough to rise above the poverty line. In light of the hardships facing poor and near-poor mothers, including lack of jobs, child care, and work training, the fact that such a tiny proportion of women stay on welfare continuously for five years (about 15 percent) is testimony to a work ethic and determination that many of us, more fortunately placed, have never tested so severely.[28]

The third woman, by contrast, was an older, well-paid professional who superficially fit the stereotype of the selfish, liberated woman "mocking the importance of fatherhood" by choosing to have a child on her own. But this woman had had three wedding plans fall through before she decided to go ahead on her own. She attributed her failures with men to having been caught between two value systems. Old-fashioned enough in her romantic fantasies to be attracted only to powerful, take-charge men, she was modern enough in her accomplishments that such men generally found her threatening once the early excitement of a relationship died down. At age 35, her last fiancé had walked away, leaving her pregnant. With her biological clock ticking "now or never" in her ear, she decided to have the child.

I have also met men who tell angry stories about having been tricked by a woman into thinking it was "safe" to have sex. "Why should I have to pay child support?" demanded one. "Doesn't that just encourage women to have babies outside of marriage?" It is, of course, totally unethical for a woman to assure a man that sex is "safe" when it isn't. But what is the alternative? If a man could get off the hook by claiming "she told me it was safe," no unmarried father would pay child support.

Charles Murray of the American Enterprise Institute thinks that would be just fine. He advocates denying child support to any woman who bears a baby out of wedlock: Girls, he declares, need to grow up knowing that if they want any legal claims whatsoever on the father of their child, "they must marry." Answering objections that this gives men free reign to engage in irresponsible sex, Murray offers a response straight out of a Dickens novel. A man who gets a woman pregnant; he observes, "has approximately the same causal responsibility" for her condition "as a slice of chocolate cake has in determining whether a woman gains weight." It is her responsibility, not the cake's, to resist temptation.[29]

The analogy works nicely if you think of sexual behavior in terms of greedy women cruising an erotic dessert buffet, trying to decide which sperm-filled slice of temptation to sample. But it misses something important about the dynamics of courting. For centuries, parents have had to teach their daughters that it was ultimately their responsibility to avoid pregnancy, whatever promises a boy might make. Now we will have to teach our sons the same thing, and strict child support just might drive the lesson home.

Meanwhile, though, unwed motherhood, like divorce, is rising around the world. In places as diverse as Southern Africa and Northern Europe, more than 20 percent of all births are to unwed mothers. Part of the reason is the rise in women's economic independence. Part is the decline in society's coercive controls over personal life. And part is simply that with falling marital fertility, even comparatively low rates of unmarried childbearing create higher proportions of children born to unwed mothers.

Unwed motherhood creates tremendous hardships in some situations, while it is far less damaging in societies that pay women decent wages and make a commitment to providing parental leave and quality child care. Half of all births in Sweden and a quarter of all births in France are to unmarried women, for example, but the children do not face the same poverty and limited life options as their counterparts in the United States. We will not abolish unwed motherhood; but we can affect its outcome.[30]

UNWED MOTHERHOOD AND THE NEW WELFARE EXPERIMENT

It may also be possible to reduce unmarried, or at least unpartnered, childbearing through educational campaigns, stiffened child support, and restructured relations between men and women. We have particular reason to think we can reduce unwed births among teenagers, by teaching young people of both sexes to take more personal responsibility for sexual behavior, making birth control available, and improving education or job opportunities in poverty-stricken areas.[31]

In order to accomplish this end, though, we need to reject the incredibly widespread double-barreled myth that unwed motherhood has been both cause and effect of out-of-control welfare spending. In 1996 a new welfare bill passed Congress, largely because of this myth. There was good reason to dislike the way that the welfare system had developed over the past few decades, but it is quite astonishing that the politicians who decided to "end welfare as we know it" knew so little about what they voted to end.

When many people think of unwed motherhood among impoverished women, the image that springs to mind is of a black teenager popping out babies to get the average $65 a month hike this entitles her to in states that have not frozen benefits. But the idea that welfare benefits cause unwed motherhood, *especially* among African Americans, is one of the most ill-founded notions in contemporary political discourse. Recently a group of seventy-six leading researchers on welfare and out-of-wedlock childbearing felt compelled to issue a public statement correcting the assumptions that both Democratic and Republican politicians bring to this question. They pointed out that few studies have found any correlation at all between welfare and unwed motherhood, and those few have found at most only a small link. Interestingly, when such a link has been discerned, it has been found only for *whites,* not for African Americans.[32]

In the absence of factual correlations between welfare and unwed motherhood, politicians and pundits have relied on a few personal anecdotes and some "common sense." Conservative columnist George Will says there's "surely" a relationship between welfare and unwed motherhood, and reporters can always find some youngster ignorant and deprived enough to think welfare gives her the option to have a baby without working. Yet, researchers who have studied the issue estimate that at most the welfare system might have been responsible for 15 percent of the increase in unwed motherhood. Sociologist Mark Rank reports that the typical woman on welfare has 1.9 children, fewer than her counterpart who is not on welfare, and the longer she stays on welfare the *less* likely she is to have more children.[33]

Out-of-wedlock birth rates were stationary or falling from 1960 to the early 1970s, the period when welfare benefits were rising. Increases in rates of unmarried motherhood occurred in the late 1970s and the 1980s, a period when welfare benefits were dropping sharply. In fact, illegitimacy rates in America historically have been lowest in the states with the *highest* welfare benefits, and mothers living in states with higher AFDC payments have returned to work *more* quickly than those in states with less generous payments.[34]

A recent study of New York City, challenging both barrels of the myth at once, found that only one in four poor households in the city is headed by a single parent today, and that welfare spending in the city, adjusted for inflation, had declined by 30 percent between 1970 and 1992. Most poverty spending in the city went to pay Medicaid bills for the elderly and disabled. Medical costs of welfare have indeed been rising around the country. Yet while such increases may benefit hospitals and clinics, patient care remains the same or even deteriorates when the price tag goes up.[35]

This is not an issue of "a welfare culture" based on "rampant immorality." International comparisons show that Americans are more likely to be religious, to disapprove of premarital sex, and to limit welfare benefits than people in any of the European nations or Canada. Yet our rates of unwed teen parenthood are *far* higher than theirs. The two factors most likely to produce high percentages of unwed teenage births in a country are high absolute rates of poverty and a big gap between rich and poor. America has become a leader of the industrial world in both.[36]

The new welfare bill is unlikely to improve our dismal record, despite some improvements in child support collection. It cuts approximately $55 billion, spread over the next six years, from programs serving low-income Americans. The largest "savings" are achieved by reducing spending for food stamps and assistance to disabled and elderly poor people, as well as declaring individuals who legally immigrated to America ineligible for Medicaid, food stamps, and other aid programs.[37]

AFDC has been replaced with a program called Temporary Assistance for Needy Families. This program ends the commitment the federal government made during the Great Depression to respond when poverty and economic hardship rise.

. . . Whatever one's views on the new welfare rules, the idea that such changes will help restore a culture of "enduring marriage," far less a culture that cares about children, is absurd. The historic increase in divorce and unwed motherhood has developed independently of the welfare system. It will not disappear no matter how punitive we get with the poor.

. . . The family values crusade may sound appealing in the abstract. But it offers families no constructive way to resolve the new dilemmas of family life. Forbidding unmarried women access to sperm banks, for instance, is hardly going to put the package of child rearing and marriage back together. It would take a lot more repression than that to reinstitutionalize lifelong marriage in today's society.

As Katha Pollitt argues, "we'd have to bring back the whole nineteenth century: Restore the cult of virginity and the double standard, ban birth control, restrict divorce, kick women out of decent jobs, force unwed pregnant women to put their babies up for adoption on pain of social death, make out-of-wedlock children legal nonpersons. That's not going to happen."[38] If it did happen, American families would be worse off, not better, than they are right now.

NOTES

1. "Marriage in America: A Report to the Nation," Council on Families in America, New York, N.Y., March 1995, pp. 10–11,13.
2. David Popenoe, "Modern Marriage: Revising the Cultural Script," in David Popenoe, Jean Bethke Elshtain, and David Blankenhorn, eds., *Promises to Keep: De-dine and Renewal of Marriage in America.* (Lanham, Md.: Rowman and Littlefield, 1996), p. 254; "Marriage in America," p. 4; David Popenoe, *Life Without Father: Compelling New Evidence That Fatherhood and Marriage Are Indispensable for the Good of Children and Society* (New York: The Free Press, 1996), p. 222; David Blankenhorn, *Fatherless America: Confronting Our Most Urgent Social Problem* (New York: Basic Books, 1995), p. 229; Charles Murray, "Keep It in the Family," *Times of London,* November 14, 1993; Maggie Gallagher, *The Abolition of Marriage: How We Destroy Lasting Love* (Washington, D.C.: Regnery Publishing, 1996), pp. 250–257; Barbara Dafoe Whitehead, "Dan Quayle Was Right," *Atlantic Monthly* 271 (April 1993), p. 49.
3. "Marriage in America," p. 4. The "deinstitutionalizing" phrase comes from Blankenhorn, *Fatherless America,* p. 224.
4. William Goode, *World Changes in Divorce Patterns* (New Haven, Conn.: Yale University Press, 1993), p. 330.
5. On the leveling off of family change, see Peter Kilborn, "Shifts in Families Reach a Plateau," *New York Times,* November 27, 1996. Other information in this and the following three paragraphs, unless otherwise noted, come from Steven Rawlings and Arlene Saluter, *Household and Family Characteristics: March* 1994, Current Population Reports Series P2O-483 (Washington, D.C.: Bureau of the Census, U.S. Department of Commerce, September 1995), pp. xviii-ix; Michael Haines, "Long-term Marriage Patterns in the United States from Colonial Times to the Present," *History of the Family 1* (1996); Arthur Norton and Louisa Miller, *Marriage, Divorce, and Remarriage in the 1990s,* Current Population Reports Series P23-i8o (Washington, D.C.: Bureau of the Census, October 1992); Richard Gelles, *Contemporary Families: A Sociological View* (Thousand Oaks, Calif.: Sage, 1995), pp. 116–120, 176; Shirley Zimmerman, "Family Trends: What Implications for Family Policy?" *Family Relations* 41 (1992), p. 424; Margaret Usdansky, "Single Motherhood: Stereotypes vs. Statistics," *New York Times,* February 11, 1996, p. 4; *New York Times,* August 30, 1994, p. A9; *New York Times,* March 10, 1996, p. A11, and March 17, 1996, p. A8; U.S. Bureau of the Census, *Statistical Abstracts of the United States* (Washington, D.C., 1992); McLanahan and Casper, "Growing Diversity and Inequality in the American Family," in Reynolds Farley, ed., *State of the Union: America in the 1990s,* vol. 1 (New York: Russell Sage, 1995).
6. Larry Bumpass, "Patterns, Causes, and Consequences of Out-of-Wedlock Childbearing: What Can Government Do?" *Focus,* 17 (University of Wisconsin-Madison Institute for Research on Poverty, 1995), p. 42; Larry Bumpass and R. Kelly Raley, "Redefining Single-Parent Families: Cohabitation and Changing Family Reality," *Demography* 32 (1995), p. 98. See also note 33.
7. *Olympian.* February 26, 1996, p. D6.
8. Susan Watkins, Jane Menken, and John Bongaarts, "Demographic Foundations of Family Change," *American Sociological Review* 52(1987), pp. 346–358.
9. Barbara Wilson and Sally Clarke, "Remarriages: A Demographic Profile," *Journal of Family Issues* 13 (1992).
10. Gelles, *Contemporary Families,* pp. 344–345; Alan Booth, David Johnson, Lynn White, and John Edwards, "Women, Outside Employment, and Marital Instability," *American Journal of Sociology* 90 (1989), pp. 567–583.
11. Saul Hoffman and Greg Duncan, "The Effect of Incomes, Wages, and AFDC Benefits on Marital Disruption," *Journal of Human Resources* 30 (1993), pp. 1–41.
12. Barbara Ehrenreich, "On the Family," *Z Magazine,* November 1995, p. 10; Ailsa Burns and Cath Scott, *Mother-Headed Families and Why They Have Increased* (Hillsdale, N.J.: Lawrence Erlbaum, 1994), p. 183.
13. Terry Arendell, "Women and the Economics of Divorce in the Contemporary United States," *Signs* 13 (1987), p. 125.
14. Norton and Miller, *Marriage, Divorce, and Remarriage in the 1990s,* pp. 6–7; Gelles, *Contemporary Families,* pp. 344, 398; Lynn White, "Determinants of Divorce:

A Review of Research in the Eighties," *Journal of Marriage and the Family* 52 (1990), pp. 904–912.

15. Valerie Oppenheimer and Vivian Lew, "American Marriage Formation in the Eighties: How Important Was Women's Economic Independence?" in K. O. Mason and A. Jensen, eds., *Gender and Family Change in Industrialized Countries* (Oxford: Oxford University Press, 1994); Aimee Dechter and Pamela Smock, "The Fading Breadwinner Role and the Economic Implications for Young Couples," Institute for Research on Poverty, Discussion Paper 1051–94, December 1994, p. 2; Marian Wright Edelman, *Families in Peril: An Agenda for Social Change* (Cambridge, Mass.: Harvard University Press, 1987), p. 55; Lawrence Lynn and Michael McGeary, eds., *Inner-City Poverty in the United States* (Washington, D.C.: National Academy Press, 1990), pp. 163–167; University of Michigan researcher Greg Duncan, Testimony before the House Select Committee on Children, Youth and Families, February 19, 1992; *New York Times,* September 4, 1992, p. A1; Daniel Lichter, Diane McLaughlin, George Kephart, and David Landry, "Race and the Retreat from Marriage: A Shortage of Marriageable Men?" *American Sociological Review* 57 (1992), p. 797; Kristin Luker, "Dubious Conceptions—The Controversy Over Teen Pregnancy," *American Prospect,* Spring 1991.
16. Dirk Johnson, "Attacking No-Fault Notion, Conservatives Try to Put Blame Back in Divorce," *New York Times,* February 12, 1996, p. A8.
17. Maggie Gallagher, "Why Make Divorce Easy?" *New York Times,* February 20, 1996; "Welfare Reform and Tax Incentives Can Reverse the Anti-Marriage Tilt," *Insight,* April 15, 1996, p. 24; Suzanne Fields, "The Fault-Lines of Today's Divorce Polities," *Washington Times,* April 22, 1996. See also Maggie Gallagher, *The Abolition of Marriage: How We Destroy Lasting Love* (Washington, D.C.: Regnery Publishing, 1996).
18. Stephen Sugarman and Herma Hill Kay, eds., *Divorce Reform at the Crossroads* (New Haven, Conn.: Yale University Press, 1990); Cynthia Stearns, "Divorce and the Displaced Homemaker: A Discourse on Playing with Dolls, Partnership Buyouts and Dissociation Under No-Fault," *University of Chicago Law Review* 60 (1993), pp. 128–139; Ann Luquer Estin, "Maintenance, Alimony, and the Rehabilitation of Family Care," *North Carolina Law Review* 71 (1993). For my understanding of recent legal trends, I am greatly indebted to conversations with Olympia Attorney Christina Meserve.
19. Maggie Gallagher, "Recreating Marriage," in Popenoe, Elshtain, and Blankenhorn, eds., *Promises to Keep,* p. 237; Debra Friedman, *Towards a Structure of Indifference: The Social Origins of Maternal Custody* (New York: Aldine de Gruyter, 1995), p. 134; Bumpass is quoted in Johnson, "Attacking No-Fault Notion," *New York Times,* February 12, 1996.
20. Roderick Phillips, *Untying the Knot: A Short History of Divorce* (Cambridge: Cambridge University Press, 1991), p. 232. See Burns and Scott, *Mother-Headed Families and Why They Have Increased,* p. 182.
21. Andrew Cherlin, *Marriage, Divorce, Remarriage* (Cambridge, Mass.: Harvard University Press, 1981), p. 49; Johnson, "Attacking No-Fault Notion," *New York Times,* February 12, 1996; William Goode, *World Changes in Divorce Patterns* (New Haven, Conn.: Yale University Press, 1993), p. 318; Shirley Zimmerman, "The Welfare State and Family Breakup: The Mythical Connection," *Family Relations* 40 (1991)5 p. 141.
22. Larry Bumpass, "What's Happening to the Family? Interactions Between Demographic and Institutional Change," *Demography* 27 (1990), p. 485.
23. Mike Males, "Poverty, Rape, Adult/Teen Sex: Why Pregnancy Prevention Programs Don't Work," *Phi Delta Kappan,* January 1994, p. 409; Ellen Goodman, "Return to Statutory Rape Laws," *Olympian,* Feb. 22, 1995, p. A9; Mike Males, *The Scapegoat Generation: America's War on Adolescents* (Monroe, Me.: Common Courage Press, 1996), pp. 17–18; Debra Boyer and David Fine, "Sexual Abuse as a Factor in Adolescent Pregnancy and Child Maltreatment," *Family Planning Perspectives* 24 (1992).
24. *Vital and Health Statistics: Births to Unmarried Mothers,* series 21, no. 53 (Hyattsville, Md.: National Center for Health Statistics, Department of Health and Human Services, 1995), table 1, p. 27.
25. Sara McLanahan and Lynne Casper, "Growing Diversity and Inequality in the American Family," in Reynolds Farley, ed., *State of the Union: America in the 1990s,* vol. 2 (New York: Russell Sage, 1995), pp. 10–11; Stephanie Ventura, *Vital and Health Statistics: Births to Unmarried Mothers: United States, 1980–92,* series 21, Data on Natality, Marriage, and Divorce, no. 53 (Hyattsville, Md.: Department of Health and Human Services), no. PH5 95–1931, table 1, p. 27.
26. "More 'Murphy Brown' Moms," *Olympian,* July 14, 1993; *New York Times,* July 14, 1993, p. A1; Rachel Fuchs, *Poor and Pregnant in Paris: Strategies for Survival in the Nineteenth Century* (New Brunswick, N.J.: Rutgers University Press, 1992); Elizabeth Kuznesof, "Household Composition and Headship as Related to Changes in Mode of Production: Sao Paulo 1715 to 1836," *Society for Comparative Study of Society and History* 41 (1980), p. 100.

27. Usdansky, "Single Motherhood."

28. Katherine Edin, *Welfare Myths: Fact or Fiction? Exploring the Truth About Welfare* (New York: Center on Social Welfare Policy and Law, 1996); Joel Handler, " 'Ending Welfare As We Know It': Another Exercise in Symbolic Politics," University of Wisconsin-Madison Institute for Research on Poverty, Discussion Paper 1053-95, January 1995, pp. 7–9.

29. Charles Murray, "Keep It in the Family," *London Times,* November 14, 1993.

30. Judith Bruce, Cynthia Lloyd, and Ann Leonard, with Patrice Engle and Niev Duffy, *Families in Focus: New Perspectives on Mothers, Fathers, and Children* (New York: The Population Council, 1995), p. 19; Tamar Lewin, "Decay of Families Is Global," *New York Times,* May 30, 1995, p. A5; Leon Eisenberg, "Is the Family Obsolete?" *Key Reporter 60* (1995), pp. 1–5; Usdansky, "Single Motherhood"; Bumpass, "Patterns, Causes, and Consequences of Out-of-Wedlock Childbearing," p. 42.

31. John Billy and David Moore, "A Multilevel Analysis of Marital and Non-marital Fertility in the U.S.," *Social Forces* 70 (1992), pp. 977–1011; Elaine McCrate, "Expectation of Adult Wages and Teenage Childbearing," *International Review of Applied Economics 6* (1992); Lawrence Lynn and Michael McGeary, eds., *Inner-City Poverty in the United States* (Washington, D.C.: National Academy Press, 1990), pp. 163–167. The best recent book on the causes and possible responses to teen pregnancy is Kristin Luker, *Dubious Conceptions: The Politics of Teenage Pregnancy* (Cambridge, Mass.: Harvard University Press, 1996).

32. Robert Moffitt, "Welfare Reform: An Economist's Perspective," *Yale Policy Review* 11 (1993); Sharon Parrott and Robert Greenstein, "Welfare, Out-of-Wedlock Childbearing, and Poverty: What Is the Connection?." (Washington, D.C.: Center on Budget and Policy Priorities, 1995), pp. 15–17.

33. Rebecca Blank, "What Are the Trends in Non-marital Births?" in R. Kent Weaver and William T. Dickens, eds., *Looking Before We Leap: Social Science and Welfare Reform* (Washington, D.C.: Brookings Institution, 1995); Mark Rank, *Living on the Edge: The Realities of Welfare in America* (New York: Columbia University Press, 1994).

34. Gregory Acs, "Do Welfare Benefits Promote Out-of-Wedlock Childbearing?" in Isabel Sawhill, ed., *Welfare Reform: An Analysis of the Issues* (Washington, D.C.: The Urban Institute, 1995); Usdansky, "Single Motherhood"; McLanahan and Casper, "Growing Diversity and Inequality in the American Family," pp. 10–11; Sara McLanahan and Irwin Garfinkel, "Welfare Is No Incentive," *New York Times,* July 29, 1994, p. A13; "Work and Child Care Choices," *Urban Institute Policy and Research Report,* Summer 1994, p. 17.

35. *New York Times,* August 30, 1994, p. A14.

36. Elaine McCrate, "Expectations of Adult Wages and Teenage Childbearing," *International Review of Applied Economics 6* (1992); Ruth Conniff, "The Culture of Cruelty," *The Progressive,* September 1992, p. 16; Barbara Vobejda, "Gauging Welfare's Role in Motherhood: Sociologists Question Whether 'Family Caps' Are a Legitimate Solution," *Washington Post,* June 2, 1994; Spencer Rich, "Generous Welfare May Not Quash Upward Mobility," *Washington Post National Weekly Edition,* August 30-September 5, 1993, p. 28; "Statement on Key Welfare Reform Issues: The Empirical Evidence," Poverty and Nutrition Policy (Medford, Mass.: Tufts University Center on Hunger, 1995), pp. 7,11,15,19; Barbara Crosselde, "U.N. Survey Finds World Rich-Poor Gap Widening," *New York Times,* July 15, 1996, p. A3; Mike Males, "In Defense of Teenage Mothers," *The Progressive,* August 1994, p. 22; Lynn and McGeary, eds., *Inner-City Poverty in the United States,* pp. 163–167; University of Michigan researcher Greg Duncan, Testimony before the House Select Committee on Children, Youth and Families, February 19, 1992; *New York Times,* September 4, 1992, p. A1. On America's higher degree of religious moralism and sexual conservatism, compared to countries with lower rates of teen pregnancy, see James Morone, "The Corrosive Politics of Virtue," *American Prospect* 26 (May–June 1996), pp. 31–32.

37. For sources on the following figures, consult the nonpartisan Congressional Budget Office. See also David Super et al., *The New Welfare Law* (Washington D.C.: Center on Budget and Policy Priorities, 1996); Alison Mitchell, "Two Clinton Aides Resign to Protest New Welfare Law," *New York Times,* September 12, 1996, pp. A1, A14; Robert Pear, "State Welfare Chiefs Ask for More U.S. Guidance," *New York Times,* September 10, 1996.

38. Katha Pollitt, "Bothered and Bewildered," *New York Times,* July 22, 1993.

11

Do Wedding Dresses Come in Lavender?

The Prospects and Implications of Same-Sex Marriage

ANGELA BOLTE,
1998

On 3 December 1996, a Hawaiian state court upheld the legalization of same-sex marriage, although the next day the presiding judge set the ruling aside pending the outcome of an appeal by the state. This ruling rekindled fears that exist regarding the potential legalization of same-sex marriage throughout the United States. To answer concerns regarding same-sex marriage, I will examine several issues. First, the concept of marriage must be defined. Opponents often argue that same-sex marriage violates the definition of marriage and, thus, same-sex marriage should not be legalized. Second, given the difficulty of defining the concept of marriage, I will undertake an examination of the features of marriage so as to illustrate the compatibility of same-sex marriage with those features. Third, I will examine traditional legal arguments against same-sex marriage for their validity. Fourth, I will evaluate Claudia Card's arguments against same-sex marriage and offer a response to these arguments. Finally, I will explore domestic partnerships as a potential alternative to same-sex marriage. Once these steps are completed, it will be seen that same-sex marriage would be beneficial, rather than harmful, to the institution of marriage.

1.

One of the most important issues surrounding the same-sex marriage debate is whether or not same-sex marriages can even exist. All too often, same-sex marriages are discounted as impossible because marriage is viewed as existing only between a man and a woman. Gays and lesbians cannot meet the definition of marriage and, thus, "same-sex marriage" is an oxymoron.

One method of providing evidence for the possibility of same-sex marriages involves recognizing what is meant by the definition of marriage. Richard Mohr illustrates the difficulty of this task by pointing out the following:

> Most commonly, dictionaries define marriage in terms of spouses, spouses in terms of husband and wife, and husband and wife in terms of marriage. In consequence, the various definitions do no work in explaining what marriage is and so simply end up assuming or stipulating that marriage must be between people of different sexes. (Mohr, 1995).

It does seem that Mohr, for the most part, is correct. In most dictionaries marriage is defined

From "Do Wedding Dresses Come in Lavender?: The prospects and implications of same sex marriage," by Angela Bolte, from *Social Theory and Practice*, Vol. 24(1), Spring 1998, pp. 111–118 and 122–128.

in these terms, although at times any reference to "spouses" is dropped, but the point remains. Dictionary definitions usually focus on the concepts of husband and wife and spend little time explaining anything about the actual nature of marriage.

The legal definition of marriage fares no better. Although the marriage laws in most states refer generically to spouses, *Black's Law Dictionary* relies on the 1974 case *Singer v. Hara* to express the legal definition of marriage as "the legal union of one man and woman as man and wife." Since the legal definition of marriage specifically excludes same-sex marriage, the courts have successfully argued that same-sex marriages cannot be allowed because these unions would destroy the very notion of marriage.

Some courts have tried to place an additional stipulation on the definition of marriage; namely, that marriage is a vehicle for the creation and raising of children. This traditional argument against same-sex marriage is also inadequate. With the elimination of the fertility clause in the marriage laws, the courts have removed the raising of children from the core of marriage law and the ability of spouses to create children together is no longer considered a requirement of marriage. Thus, there is no legitimate legal basis from which to deny the right of marriage to gays and lesbians because they cannot create children together. Moreover, if this were legitimate, withholding marriage licenses from elderly or infertile couples would also be legitimate. In each of these cases, the couple cannot create children together and, to be consistent, denying the right of marriage to all of them would be necessary on these grounds. Due to the revised marriage laws, such a position would have little, if any, legal support.

It is also unclear how the stipulation that marriage is for the creation and raising of children truly excludes gays and lesbians. Gays and lesbians are increasingly becoming parents through new avenues such as adoption or artificial insemination. Many gays and lesbians also have children from previous marriages. This occurrence is presently so widespread that some mainstream and gay media have labeled this phenomenon a "gay baby boom."

It is estimated that there are three to four million gay and lesbian parents raising between six and fourteen million children (Henson, 1994). This number is especially significant if it is argued that one of marriage's goals is the protection and care of the children that exist within it. There is an expectation that children are to be protected and cared for by the spouses while a marriage exists and this is reflected in the legal process that occurs when a marriage ends. The divorce laws are devised to help protect children by ensuring that child support is paid if necessary for the welfare of a particular child. Moreover, when a spouse dies, custody of the children is designed to pass to the living spouse, thus ensuring that the children are not removed from a familiar environment.

While traditional arguments claim that same-sex marriages should be banned because the children within those families will be subject to harm both through ridicule and confusion over sexual roles, it is rather the case that children are directly harmed through the banning of same-sex marriages. Currently, when a same-sex relationship ends, no institutions are in place to ensure the protection of the children, as there are when traditional marriages dissolve. A partner who leaves a same-sex relationship is under no obligation to provide financial support for children that he or she may have cared for and supported for years. In a case where the biological parent dies, the children could be left without either parent, if the living partner had not adopted them in a second-parent adoption. Moreover, in most states the courts do not allow second-parent adoptions by gays and lesbians. Given the vast numbers of children with same-sex parents, by not allowing same-sex marriage, many children are adversely affected.

As for the traditional arguments against same-sex marriage, Fredrick Elliston states that "in the case of . . . homosexual marriage, the source of the harm to children is social prejudice" against gays and lesbians (Elliston, 1984). Elliston argues that if same-sex marriages were legalized, this social prejudice would be diminished. As for the second argument, that the children will suffer from confusion over sex roles, this is not necessarily a problem. According to feminist thinking, traditional sex roles are not desirable, and Elliston similarly argues that same-sex marriage "may help to combat this evil [of traditional sex roles]." (Elliston, 1984).

Given the failure of legal definitions of marriage, perhaps sociology or anthropology would be a better source for a definition of marriage. Definitions from these areas are more informative, but do not provide a complete picture of marriage. For example, marriage is described in the following manner:

> [M]arriage has been defined as a culturally approved relationship of one man and one woman (*monogamy*): or of one man and two or more women (*polygamy*), in which sexual intercourse is usually endorsed between the opposite sex partners, and there is generally an expectation that children will be born of the union and enjoy the full birth status rights of their society. These conditions of sexual intercourse between spouses and reproduction of legitimate and socially recognized offspring are not, however, always fulfilled.

The most important aspect of the above definition is the suggestion that a culture can define what is considered a marriage. In fact, a society could define and redefine marriage as often as it chooses.

Given that marriage is a perpetually evolving notion, same-sex marriages would not necessarily have a "negative" impact on the institution of marriage. The fear of such a negative impact is seen in a traditional argument that claims that the recognition of same-sex marriage would lead to the recognition of multiple forms of marriage, such as polygamy or group marriage, and traditional Western marriages would eventually be eradicated. If same-sex marriages lead to the eradication of traditional Western marriages, there is no reason to believe that this would somehow be negative. A move away from traditional Western marriages could be positive and stabilizing for the community because all citizens would be accepted, no matter what form their marriage takes.

Another traditional argument worries: "What if everybody did that [i.e., entered a same-sex marriage]:?" (Elliston, 1984). Although there eventually could be heterosexuals who choose to marry someone of the same sex, this would not necessarily be "negative." In fact, such marriages could be used to help provide basic needs and protections for those who are unable to support themselves. Unfortunately, in our society numerous citizens do not have equal access to basic rights and protections such as adequate housing or health insurance, because access to these rights and protections is limited by financial resources. If an individual lacks the financial means necessary to attain these rights and protections, he or she is usually forced to do without or to accept a substandard replacement.

Moreover, same-sex marriages would grant to gays and lesbians the rights attached to marriage that are presently denied to them. Unlike married opposite-sex couples, gays and lesbians with children are unable to have custody automatically passed to their partner at their death. Similarly, without a will, gays and lesbians cannot ensure that their estates will pass to their partner. Without being married, gays and lesbians cannot even file joint tax returns. Finally, the right of gays and lesbians to live in the community of their choice is limited, if the community specifies by law that only married couples may purchase a house within it. The denial of such rights to these citizens, while presently legal, is heterosexist and unjust.

Same-sex marriage could be used to provide access to these basic rights and protections. For example, a same-sex couple could pool their resources and attain adequate health insurance, better housing, or simply provide themselves financial security. Presently, many gays and lesbians enter traditional marriages with a heterosexual or a friend who is gay or lesbian to obtain these basic rights and protections. Heterosexuals could join a same-sex marriage for the same reason and this would not "make" them homosexual; just as gays or lesbians who are presently in traditional marriages for similar reasons are not "made" heterosexual.

Although some heterosexuals could choose to enter same-sex marriages, the percentage of heterosexuals and homosexuals would not change greatly as some opponents of same-sex marriage fear. Most estimates of the actual number of gays and lesbians in the United States range between one and ten percent. This number would most likely stay the same, although even if this number were to change, there is no reason to believe that this would be "negative." If a person had not previously been attracted to someone of the same sex, the ability of gays and lesbians to marry would not alter the fundamental sexual behavior of this person. For example, with the advent of the gay and lesbian rights movement, gays and lesbians have become increasingly visible, but the percentage of gays and lesbians has not radically changed. What has happened is that gays and lesbians who had felt compelled to live a "straight" life have become more visible. If the legalization of same-sex marriage were to occur, the same phenomenon would be likely, with some gays and lesbians leaving, or not choosing, heterosexual marriages because of the more acceptable alternative.

Some might worry that by legalizing same-sex marriage, the number of children raised in such households would increase and those children would be more likely to be gay or lesbian. Given studies on the sexual orientation of the children of gays and lesbians, these children are no more likely to be gay or lesbian than other children. Moreover, it is unclear why an increase in the number of gays and lesbians would be a "negative." If those who construe the increase of gays and lesbians as negative are actually concerned about the extinction of the human race, it is unclear why this would necessarily happen. The increasing number of gays and lesbians with children illustrates that a reproductive drive exists within gays and lesbians and there is no reason to think that humans would disappear if everyone was gay or lesbian.

2.

Although society can redefine what is meant by the concept of marriage, it is possible that same-sex marriage can be accepted while retaining the key features of marriage as it is presently known within Western society. Although describing marriage is difficult, the following list of features can be obtained.

> [Marriage] is usually a temporally extended relationship between or among two or more individuals; this usually involves (1) a sexual relationship; (2) the expectation of procreation; (3) certain expectations or even agreements to provide economic, physical, or psychological support for one another; and (4) a ceremonial event recognizing the condition of marriage. (Palmer, 1984).

Palmer points out, however, that "none of these is a *necessary* condition, and if they are logically *sufficient* conditions when taken jointly, it is probably because of the inclusion of feature number four." (Palmer, 1984).

Richard Mohr characterizes marriage similarly. He believes that marriage is "intimacy given substance in the medium of everyday life, the day-to-day. Marriage is the fused intersection of love's sanctity and necessity's demand" (Mohr, 1995). This characterization appears adequate to explain an institution infused with vagueness. Taken together, Palmer's and Mohr's conditions for marriage cover what the partners in a contemporary Western marriage generally expect. Moreover, their conditions for marriage dovetail with the definition from sociology and anthropology, but the following reduction is possible: an adequate level of commitment between partners; the joint raising of children, if the partners want them; and love. Although this reduction can be made, it is uncertain what each of these conditions entails for same-sex marriages.

The issue of commitment in marriage is interesting, yet controversial, and upon its discussion, questions concerning both the required amount of commitment and how best to define commitment immediately arise. These questions can be reduced to the following: either a committed marriage must be monogamous or it may be non-monogamous. In the traditional Western view of marriage, monogamy in the form of sexual exclusivity is an essential ingredient in all marriages. Since gays and lesbians are often thought, under this traditional Western view, to be incapable of being sexually exclusive, it is claimed that they should not be allowed to marry, because they cannot meet a "necessary condition" of marriage (Hoaglond, 1988).

Although it is unclear that gays and lesbians are any less sexually exclusive or monogamous than heterosexuals, no marriage must be totally monogamous. If the partners in a marriage choose to have an "open" marriage, this does not mean that their marriage is somehow voided. The marriage is not voided, because monogamy is not a requirement of marriage; in fact, many types of marriage are non-monogamous by definition, such as polygamy or group marriage. Moreover, with the advent of no-fault divorce laws, the lack of monogamy within a marriage is no longer even legal ground for divorce. This shift in the law may be due to a recognition of the fact that the majority of society no longer considers monogamy a necessary condition of marriage. Therefore, if neither of the partners is coerced into giving assent to an open marriage, this decision should be respected.

Simultaneously, a decrease in monogamy need not amount to a decrease in commitment. While it is the case that a partner who is "cheating" on his or her partner most likely is guilty of a lack of commitment, if the relationship is open, a decrease in commitment is not necessary. While judging the degree of commitment in an open marriage is almost impossible, one indicator of commitment could simply be the continuation of the marriage.

In fact, Richard Mohr believes that if same-sex marriages were legalized, the marriages of gay men would help to improve mainstream views about monogamy. Mohr believes that monogamy is not essential for love and commitment in marriage, as evidenced by many long-term gay male relationships that incorporate non-monogamy (Mohr 1995). Instead, Mohr believes that traditional couples should look at the relationships of gay men to rethink the traditional Western model of the family. Mohr's basic position on monogamy and commitment seems correct. A lack of monogamy should not undermine commitment if the partners have agreed not to be monogamous.

Turning to the question of children, as argued above, no legitimate reason exists for preventing gays and lesbians from marrying simply because they are seen as "incapable" of having children. Moreover, nothing prevents gays and lesbians from having children; as previously mentioned, many gays and lesbians do have children. Finally, there is the condition of love. Although love is closely tied to the issue of commitment, love and monogamy are separate issues. Viewed on its own, love is probably the one issue where some agreement can be reached, because gays and lesbians can easily meet this feature of marriage.

With these partial characterizations of marriage, it can be seen that gays and lesbians can meet its features. This means that the courts should not dismiss cases regarding same-sex marriage because these marriages are viewed as impossible by virtue of being between same-sex couples. Instead, a rigorous examination of the legal arguments against same-sex marriage must be undertaken to see if legitimate legal and moral reasons exist for not allowing same-sex marriage.

3.

During the 1970s, several court cases were aimed directly at allowing same-sex marriage, but none succeeded. Because of these failures, and the United States Supreme Court's majority opinion in *Bowers v. Hardwick,* no other major cases regarding same-sex marriage were filed until the early 1990s. In 1993, *Baehr v. Lewin* was brought before the Hawaiian Supreme Court and the court sent the issue back to the lower courts, ruling that the state must demonstrate a compelling state interest to ban same-sex marriage. In December of 1996, a Hawaiian state court ruled that compelling state interest was not illustrated.

With the state court's ruling, there has been enormous speculation as to its potential effect on the rest of the United States. In particular, it has been questioned whether the other forty-nine states will be forced to recognize same-sex marriages performed in Hawaii. A related concern is the impact this ruling will have on opening other states' marriage laws to similar challenges. Many legal scholars believe that most, if not all, states' marriage laws will be challenged and many of those states will eventually be forced to recognize same-sex marriages....

A ... factor that would limit the demonstration of a substantial public policy against same-sex marriage by some states, but not all, is the presence of laws that protect gays and lesbians. Most relevant are the eight states that have laws forbidding discrimination against gays and lesbians, but there are several other important laws (Menson, 1994). The courts in eleven states have rejected presumption against gays and lesbians in custody cases (Swart, 1994). Eight states and the District of Columbia allow custody to gays and lesbians in second-parent adoptions (Swart, 1994). Finally, cities in ten states and the District of Columbia have domestic partnerships for gays and lesbians (Swart, 1994). In each of these individual cases, the state would have a hard time proving a legitimate substantial public policy against gays and lesbians....

While some states may affirm same-sex marriages, this will not necessarily occur in all states. It could be that all states will be able both to eliminate laws that protect gays and lesbians and to put the necessary legislation in place to display a substantial public policy against same-sex marriage. Another scenario could be that the courts would allow any state to bar same-sex marriage due to the heterosexist tradition that exists regarding marriage within the United States. Only real challenges to the marriage laws of numerous states can demonstrate what will actually occur.

4.

Many within the gay and lesbian community also have articulated concerns regarding the acceptance of same-sex marriage. Claudia Card voices many of those concerns in her recent article "Against Marriage and Motherhood." Card believes that while it is wrong for the state to ban same-sex marriages, gays and lesbians should not be eager to enter marriage, because same-sex marriages will create multiple problems for the gay and lesbian community. Card rejects same-sex marriage because of the possible intrusiveness of the state into same-sex relationships, and she raises four specific problems that she believes should concern advocates of same-sex marriage (Card, 1996).

First, gays and lesbians might be pressured into marriage to receive benefits such as health insurance. Second, once gays and lesbians enter these relationships, they could find the negative consequences of divorce too high because they could lose some of their economic resources. Third, marriage, due to its monogamous nature, will be too limiting and will distort the actual nature of many gay and lesbian relationships. Finally, Card claims that the legal access granted to spouses by marriage could open the door to all types of same-sex partner abuse. According to Card, instead of embracing flawed traditions, gays and lesbians should create their own traditions.

The problems that Card points out are significant but not overwhelming. As to her first two concerns, these problems are not specifically gay or lesbian in nature. Many of those working for same-sex marriage realize that basic benefits such as health care should be available to everyone. The expansion of such benefits to everyone, not just to those who are married, is a major goal of these activists. Although the lack of such benefits for everyone within society should be a concern, it need not be tied only to the movement for same-sex marriage.

The basic problem underlying Card's second objection is one that is not new to the gay and lesbian community. Presently, palimony lawsuits often take place when a same-sex partnership dissolves. Card briefly mentions this fact and states that palimony is problematic specifically because it both prevents a partner from easily leaving a same-sex relationship and applies "the idea of 'common law' marriage to same-sex couples" (Card, 1996). Card suggests that instead of palimony or marriage, couples who want a "contractual relationship" should engage in developing a specific legal contract that defines their relationship.

Although Card may find the restrictions placed on relationships both by divorce and palimony too limiting, these legal structures serve to protect both partners in ways that individualized "relationship contracts" might not. First, for such a move to be equitable to both partners, each partner would require a lawyer to ensure that the contract was fair. While this might be a simple requirement for some, many would find the lawyers' fees to be a burden and would reject a move toward a contract for that reason. Presumably, under Card's position, if the partners do not have a relationship contract, there would be no system in place to ensure a fair division of assets following a separation. Second, some might attempt to negotiate their own contracts, and such a move could easily lead to one partner unwittingly accepting an unfair contract. Third, even with a contract negotiated by a lawyer, there is no guarantee that one partner will not be taken advantage of by the other. Fraud and deception often arise in other contractual situations, and there is no reason to believe that relationship contracts would be immune to such problems. Finally, circumstances surrounding a relationship can easily change, and there is no guarantee that the legal contract that encompasses the relationship is flexible enough to span the changes or, if it is not, that it will be altered in response to the changes in the relationship. While a few relationships would not greatly change over time, it cannot be assumed that this would be true for all relationships. Relationship contracts require additional expense, have the potential for unfairness and fraud, and can be made irrelevant by changing circumstances. Most important, by making relationship contracts the sole option for same-sex couples, those who are unable or unwilling to have such a contract will lose their current protections. Thus, while divorce and palimony may have some negative consequences, eliminating these institutions would cause even greater harm.

Card's third problem regarding the limiting nature of marriage is a common concern of some within the gay and lesbian rights movement (Ettelbrick, 1992). Many of those in the gay and lesbian rights movement who stand opposed to same-sex marriage worry that marriage would become even more entrenched as the only acceptable type of relationship. They are concerned that those who are at the fringes of the gay and lesbian rights movement would be excluded because they choose not to live the more "acceptable" life of marriage.

I believe that this fear of accelerated exclusion due to the legalization of same-sex marriage is overstated. The fringe elements of the gay and lesbian rights movement are excluded currently by both the traditional and the gay and lesbian mainstreams with the ban on same-sex marriage. Since these elements are currently ostracized, legalizing same-sex marriage could not do a great deal more to harm those who are excluded.

It could be that those gays and lesbians who choose not to be married will be discriminated against, at times. Single and especially young heterosexuals are often thought not to be as stable as their married peers, and are, at times, at a disadvantage because they are not married. The same would likely hold true for gays and lesbians. Nevertheless, it does seem that, overall, the legalization of same-sex marriage would have a liberalizing influence rather than cause a move toward increased conservatism.

The legalization of same-sex marriage would bring what had once been determined to be "other," that is, what had been determined to be separate and inferior, into the mainstream. In other words, legalizing same-sex marriage would allow one form of difference to be included in what is deemed acceptable. By broadening the definition of what is considered acceptable, other forms of difference could become more accepted. For example, the gay and lesbian rights movement illustrates this occurrence. As gays and lesbians have become more visible and accepted, many issues surrounding their community have moved from unthinkable to potentially realizable. In fact, same-sex marriage has moved from unthinkable thirty years ago to potentially being legalized. In this manner, the legalization of same-sex marriage would lead to more rather than less acceptance of difference, and should therefore be supported by the gay and lesbian community.

Finally, Card's fourth concern claims that same-sex marriage could allow an abusive partner to abuse his or her same-sex spouse easily. While same-sex marriages could potentially increase the control of an abusive partner, it is not clear that this would outweigh the potential benefit that marriage could bring to the issue of same-sex partner battering. Presently, gays and lesbians have difficulty finding help when they are in a battering relationship, for many reasons. Perhaps the main reason is the combination of homophobia and heterosexism. The gay and lesbian community does not wish to recognize the problem of same-sex partner battering, and for this reason "it is absolutely vital to work to eliminate homophobia and heterosexism in the shelter environment" (Geraci, 1986). Although much of the theoretical groundwork on battering was done by lesbians, in conjunction with their work in developing many original shelters, these shelters are not always welcoming to those in the gay and lesbian community who need help (Cecere, 1986). Along with the fear and hatred of gays and lesbians that exists within the shelters, there is often disbelief that a woman could batter another woman or that a man could not defend himself against another man.

If same-sex marriages became widespread, there could be a profound effect on opinions regarding gays and lesbians, which would be beneficial for those gays and lesbians who are in battering relationships. Same-sex marriages could help to eliminate heterosexism and homophobia by elevating homosexuality to the level of acceptability. Through the legalization of same-sex marriage, gay and lesbian relationships would be acknowledged as legitimate. Moreover, gays and lesbians could become more visible due to the protections accorded to married couples. This visibility would further increase acceptance of gays and lesbians throughout society. Attributing increased acceptance of gays and lesbians to their growing visibility does seem possible. The gay and lesbian rights movement itself provides similar evidence, because as gays and lesbians have left the "closet," there has been a corresponding increase in their acceptance by society. In addition, when at least some people are aware that someone they know is gay or lesbian, they can be more accepting of gays and lesbians in general.

Same-sex marriage could also revolutionize the institution of marriage with regard to gender roles that support heterosexism and homophobia. Nan Hunter theorizes that same-sex marriage will cause a "subversion of gender" (Hunter, 1995). Hunter believes that statements made by opponents of same-sex marriage—such as "Who would be the husband?"—illustrate the fear that exists regarding the revolutionary power of same-sex marriages. (Hunter, 1995). While such statements are meant to ridicule the very idea of same-sex marriages, they also illustrate the speaker's fear that he or she will no longer be able to depend on the power granted by a social category such as "husband." Hunter believes that the subversion of gender would revolutionize marriage as it is known today and would begin the process of moving marriage away from its oppressive roots.

If gender roles were eroded, heterosexism and homophobia would be reduced, which would directly benefit those who are in abusive same-sex relationships by making shelter workers more willing to recognize and help them. Thus, it is likely that legalizing same-sex marriage would have a positive impact on abused gays and lesbians, regardless of their marital status.

5.

Although Card is concerned with any type of state regulation of same-sex relationships, she views domestic partnerships as less problematic, in some ways, than same-sex marriage (Card, 1996). In this respect Card agrees with many of those in the gay and lesbian community who have turned away from marriage and have embraced domestic partnerships (Findlen, 1995). While Card is not truly supportive of domestic partnerships, she is mistaken to consider them as a possible alternative to marriage.

Domestic partnerships may appear to be an attractive alternative to same-sex marriage, but the benefits of these partnerships are usually limited in scope. Municipalities offering the option of a domestic partnership usually offer the same benefits to both partners and spouses of employees. Non-employees, on the other hand, generally only have access to family memberships at city-owned attractions or the right to hospital or jail visitations. Private sector businesses offering domestic partnership policies usually restrict benefits to health insurance, although sometimes they are restricted further.

Although domestic partnerships are presently limited in their scope, advocates are working to expand their coverage. Their goal is to have the benefits attached to domestic partnerships equivalent to the benefits attached to traditional Western marriages. If domestic partnerships are implemented and expanded to become marriages by a different name, no fundamental difference between marriages and domestic partnerships would exist, because, presumably, domestic partnerships would be just as difficult to leave as marriages. If there were no difference between the two practices, there would be no real method by which to distinguish them. It appears that these advocates are engaged in self-deception, in that they want the rights and benefits of marriage, but not the label.

Supporters of domestic partnerships often claim that the problems surrounding domestic partnerships should be overlooked because these partnerships could be used to generate some significant benefits. One argument claims that domestic partnerships would be beneficial because they could be used to help educate the public in an effort to "pave the way" for same-sex marriages. A second argument is based on the idea that there is a much more realistic chance that domestic partnerships will become widespread than there is that same-sex marriage will be implemented soon. By having same-sex domestic partnerships in place, gays and lesbians can enjoy some actual benefits within their lifetimes, a guarantee that cannot be made regarding same-sex marriage.

While supporters of domestic partnerships try to illustrate their potential benefits, these relationships remain extremely problematic and could potentially contribute to the unjust treatment of gays and lesbians. If domestic partnerships are expanded, these partnerships could be viewed as "separate, but equal" to marriages between opposite-sex couples. As the civil rights movement illustrates, such situations are very rarely equal. Gays and lesbians must not be taken in by possible benefits, but must examine all the possibilities. There is a very real chance that domestic partnerships could be used to sidestep justice issues regarding the treatment of gays and lesbians. If domestic partnerships were granted, any request for the legalization of same-sex marriage could be seen as unnecessary and selfish. Moreover, while there might be an attempt to increase the legal rights of domestic partnerships, such as adding adoption or tax rights, access to these rights would have to occur at a state or national level through the legislative process. The prospects for state or nation-wide same-sex domestic partnerships at the legislative level are no greater than the legalization of same-sex marriage. Thus, any promise of faster access to these rights is unlikely to be realized. These problems illustrate that domestic partnerships are not a legitimate alternative to same-sex marriage and it is only through marriage that gays and lesbians will achieve the rights they deserve as citizens.

6.

While it has been shown that major changes will occur in marriage when same-sex marriage is legalized, I do not feel that this would lead to the destruction of the institution itself. While some practices that marriage supports will be affected, marriage itself will continue. The political right will argue that any change in marriage will serve to undermine the institution, but this argument is flawed. Change cannot simply be equated with undermining. If this were the case, then marriage has already been undermined. Marriage is significantly different from its original incarnation. Even in the past twenty years there have been many changes in marriage. From the elimination of fertility laws to the advent of no-fault divorce, marriage has changed, but it has not faded from existence. There is no legitimate reason to believe that by allowing gays and lesbians to marry, the institution of marriage will disappear. In fact, it could be that allowing everyone to marry the mate of his or her choice will strengthen marriage by furthering the natural evolution of this diverse and widespread institution.

SELECTED REFERENCES

Card, Claudia. 1996. "Against Marriage and Motherhood." *Hypatia* 11: 1–23.

Cecere, Donna J. 1986. The Second Closet: Battered Lesbians." Pp. 21–31 in *Naming the Violence: Speaking Out About Lesbian Battering,* edited by K. Lobel. Seattle, WA: Seal Press.

Elliston, Fredrick. 1984. "Gay Marriage." Pp. 146–166 in *Philosophy and Sex,* 2nd ed., edited by R. Baker and F. Ellison. New York: Prometheus Books.

Ettelbrick, Paula L. 1992. "Since When is Marriage a Path to Liberation?" Pp. 20–26 in *Lesbian and Gay Marriage: Private Commitments, Public Ceremonies,* edited by S. Sherman. Philadelphia, PA: Temple University Press.

Findlen, Barbara. May/June, 1995. "Is Marriage the Answer?" *Ms.*:86–91.

Geraci, Linda. 1986. "Making Shelters Safe for Lesbians." Pp. 77–79 in *Naming the Violence: Speaking Out About Lesbian Battering,* edited by K. Lobel. Seattle, WA: Seal Press.

Henson, Deborah M. 1994. "Will Same-Sex Marriages Be Recognized in Sister States?: Full Faith and Credit and Due Process Limitations on States' Choice of Law Regarding the Status and Incidents of Homosexual Marriages Following Hawaii's *Baehr v. Lewin." University of Louisville Journal of Family Law* 32:551–600.

Hoagland, Sarah Lucia. 1988. *Lesbian Ethics: Toward New Values.* Palo Alfo: Institute of Lesbian Studies.

Hunter, Nan D. 1995. "Marriage, Law and Gender: A Feminist Inquiry." Pp. 221–233 in *Radical Philosophy of Law: Contemporary Challenges to Mainstream Legal Theory and Practice,* edited by D. S. Caudill and S. J. Gold. Englewood Cliffs, NJ: Humanities Press.

Mohr, Richard D. 1995. "The Case for Same-Sex Marriage." *Notre Dame Journal of Law, Ethics & Public Policy* 95:215–239.

Palmer, David. 1984. "The Consolation of the Wedded." Pp. 119–129 in *Philosophy and Sex,* 2nd ed., edited by R. Baker and F. Ellison. New York: Prometheus Books.

Swart, Jeffery J. 1994. "The Wedding Luau—Who is Invited?: Hawaii, Same-Sex Marriage, and Emerging Realities." *Emory Law Journal* 43:1577–1616.

CHAPTER 4

Reproduction, Birth, and Babies

Throughout history, human civilizations have regulated childbearing to ensure survival. Societies have responded to the population holding capacities of their natural environments in both intentional and unintentional ways. For example, **infant mortality,** warfare, malnutrition, and disease have typically combined to limit population growth, thereby allowing societies to avoid overpopulation. In the modern era, advances in subsistence techniques, nutrition, medicine, and contraception have altered the balance between population and environment and opened the door for more intentional control over decisions about whether or when to have children. Although historical demographers have been able to document trends in population growth as they relate to subsistence practices and technological advances, they have been less successful in identifying the specific mechanisms through which societies regulate their birthrates and sustain themselves. Social scientists are only just beginning to understand how birth and parenting practices are related to institutional structures across time and how these structures are related to men's and women's views about sex and making babies.

We do know that regulation of fertility is always part of larger marriage, kinship, and gender systems. Marriage systems and "legitimacy" standards have confined most births to married couples, and nutritional factors have influenced the duration of nursing and the timing of subsequent births. Social standards and practices, however, have been the most important proximate ingredients in regulating

birthrates. As noted above, high infant mortality rates from disease or malnutrition have typically limited population growth, but because **contraception** was historically ineffective, infanticide (the killing of infants) was the most common intentional form of birth control. For example, in ancient times Plato and Aristotle advocated **infanticide** as a means of population and disease control, and in eighteenth-century England and America, infanticide was common. Abandoning children or putting them in foundling homes, where most infants died, was also typical in eighteenth-century Europe. These practices occurred in societies we consider the most "civilized" in the world, suggesting that modern ideals about the sacredness of infant life are neither longstanding nor universal. Studies drawing on cross-cultural and historical evidence also demonstrate that when mothers enjoy access to material resources, they regulate the timing and frequency of birth, generally choosing to have only a few children (Coltrane and Collins, 2001).

The articles in this chapter focus on birth and the social forces that encourage women to bear and raise children. The first, entitled "Social Devices for Impelling Women to Bear and Rear Children" by Leta Hollingworth, was published in the *American Journal of Sociology* in 1916. In the decade before this article appeared, the American reformer Margaret Sanger coined the term **birth control** and launched a campaign to bring information about it to the public. Contraceptive methods of various types had been used for centuries, but it was not until the invention of vulcanized rubber in 1844 that condoms and diaphragms became effective in preventing conception. They remained illegal in most states, however, and if American women traveled to Europe and brought them back during the early twentieth century, U.S. Customs agents at the dock could confiscate them. Leta Hollingworth focuses less on the debates about contraceptive devices and more on the maternalist ideals of the day that labeled women unnatural and unpatriotic if they wanted to limit the number of children they conceived or bore. She suggests that the idea of a "maternal instinct" propelling all women to equally desire motherhood is both inaccurate and a cultural invention that serves specific social interests. This was the era just before World War I, and Hollingworth was reacting to both social norms and public propaganda that insisted that bearing and rearing children was necessary for the good of the country and its war effort. This was also a time when high levels of immigration and high birthrates among immigrants were used by race-conscious nationalists to call for "improving" the race by increasing the white birthrate and limiting the fertility of immigrants and other "undesirables" through eugenic sterilization. Hollingworth's essay foreshadows later sociological analyses by focusing on how medical, political, legal, educational, and artistic institutions portray a positive view of motherhood, minimizing its "disagreeable features," and labeling women who do not conform to the maternal ideal as deviant. She labels cultural myths about the natural benefits of early and frequent motherhood as "bugaboos" perpetrated by men and argues instead for variability in maternal desires and for a woman's right to participate in decisions about whether or when she will bear and rear children. Some social critics draw parallels between the early twentieth century described by Hollingworth and the recent past, insofar as both eras were characterized by U.S. involvement in global wars, successive waves of immigration, changing gender roles, and calls for sexual abstinence for unmarried women and the idealization of motherhood for married white women.

In a selection from the book *A World of Their Own Making,* John Gillis illustrates how particular historical and cultural circumstances socially construct motherhood and mother love. Arguing that the meaning of motherhood is never stable or transparent, he describes how social customs and practices produce interesting differences in maternity that transcend the biology of conception, pregnancy, and birth. Wide variation in mothering is evident in the various images, symbols, and rituals associated with birth, and Gillis shows how these social factors give motherhood different meanings in different times and places. His key insights derive from the observation that childhood was a more collective enterprise in previous eras and that before the nineteenth century, nurturing was thought of as an acquired talent more than as a sex-specific natural quality (e.g., stemming from a maternal instinct that all women possess). Gillis observes that motherhood was largely subordinated to wifehood from the sixteenth to the nineteenth century, so modern ideals about women fulfilling their true identity by giving birth would have seemed odd to people living then. Rather than the intense focus on mother–infant bonding characteristic of modern American society, earlier eras promoted a ritual separation of mother and child following the birth. The context of birth varied dramatically from our present-day practices, as there was little medical regulation of, or intervention into, pregnancy and birth. At the same time there were relatively high rates of infant and maternal death at childbirth, a fact that might help explain minimal ritual anticipation of the birth itself. This selection reminds us how even the physical and emotional aspects of family transitions such as birth and death are differentially regulated and experienced depending on the religious and cultural beliefs of the time.

In a selection from *Thinking about the Baby,* Susan Walzer describes how new parents in contemporary America talk about the emotions and obligations associated with becoming a mother or father. Her research draws on interviews with parents of one-year-olds to describe how their thoughts, feelings, and self-images were transformed in relation to their children, their partners, and the people around them, as well as in relation to the cultural ideals of what it means to be a mother or a father. One of the concepts she uses is **parental consciousness,** meaning not only mindfulness or awareness ("how babies fill parents' minds"), but also parents' judgments about these thoughts ("how they think they *should* be thinking about their babies"). She finds that parental consciousness is heavily influenced by gender and by cultural ideals about the ways that mothers and fathers should worry about baby care and financial security. Finally, she discusses how gendered processes that treat mothers as "ultimate managers" and men as "helpers" shape divisions of family labor.

The final reading in this chapter focuses on **reproductive technology.** Barbara Katz Rothman addresses legal scholars in this selection, drawing on her research about the regulation of techniques such as artificial insemination, in vitro fertilization, frozen embryos, surrogacy, and ultimately cloning. She points out that the theoretical suppositions of laws designed to control these technologies are fundamentally flawed. Our legal system is founded on social contract theories that ask the question: "How do people, as rational self-interested individuals, come together to create social order?" Katz Rothman, on the other hand, assumes that sociability is natural, emphasizing how we are inevitably embedded in social life and dependent on others. Rather than people being essentially individual separate beings, she sees people as being essentially connected beings that are the product of social order. Invoking the term *patriarchy* ("rule of fathers"), Katz Rothman describes how **patriarchal kinship systems** have focused on the man's "seed" and perpetuated kinship and lineage systems that have controlled women's bodies and their reproductive capacity. She suggests alternative symbolism, in which care and nurturance bring forth babies into the world and ensure their future well-being. Mothers' bodies have always been essential to the birthing process, but even as science creates new reproductive possibilities, Katz Rothman advocates combining recognition of the birthing process with emphasis on the importance of nurturance. At the same time that she champions motherhood, she recognizes that privileged women benefit from "modified patriarchy," insofar as they are able to purchase the reproductive and domestic services of other women—mostly poor women of color. Her final point is that rather than being based solely on contracts and individual rights, theoretical models underpinning our legal system could and should honor nurturing and connection. The natural sociality of humans, exemplified by the dependence of newborn infants on parents, ought to be institutionalized in our legal codes.

The four readings in this chapter focus attention on the ways that women's fertility is regulated by society. Specific social contexts shape ideals about maternity and infant care that serve various interests. When contraception, birth, or infant care practices violate normative expectations, women are labeled deviant and their reproductive capacity is called into question. In the modern context, infertility, nonmarital child birth, and same-sex parenting through artificial insemination challenge patriarchal control of women's reproduction. In the coming decades, the politics of gender and sexuality are likely to exert as much influence over debates about women's reproduction as any scientific breakthroughs in reproductive technologies.

REFERENCES

Coltrane, Scott, and Randall Collins. 2001. *Sociology of Marriage and the Family: Gender, Love, and Property.* 5th ed. Belmont, CA: Wadsworth/ITP.

SUGGESTED READINGS

Andrews, Lori B. 2001. *Future Perfect: Confronting Decisions about Genetics.* New York: Columbia University Press.

Arendell, Terry, ed. 1997. *Contemporary Parenting: Challenges and Issues.* Thousand Oaks, CA: Sage Publications.

Cowan, Carolyn Pape, and Philip A. Cowan. 2000. *When Partners Become Parents: The Big Life Change for Couples.* Mahweh, NJ: Lawrence Erlbaum Associates.

Davis-Floyd, Robbie E., and Carolyn F. Sargent, eds. 1997. *Childbirth and Authoritative Knowledge: Cross-Cultural Perspectives.* Berkeley: University of California Press.

Dolgin, Janet L. 1997. *Defining the Family: Law, Technology, and Reproduction in an Uneasy Age.* New York: New York University Press.

Ginsburg, Faye D., and Rayna Rapp, eds. 1995. *Conceiving the New World Order: The Global Politics of Reproduction.* Berkeley: University of California Press.

Gosden, Roger G. 1999. *Designing Babies: The Brave New World of Reproductive Technology.* New York: W. H. Freeman.

Martin, Emily. 1992. *The Woman in the Body: A Cultural Analysis of Reproduction.* Boston: Beacon Press.

May, Elaine Tyler. 1995. *Barren in the Promised Land: Childless Americans and the Pursuit of Happiness.* New York: Basic Books.

McCann, Carole R. 1994. *Birth Control Politics in the United States, 1916–1945.* Ithaca, NY: Cornell University Press.

Rothman, Barbara Katz. 2001. *The Book of Life: A Personal and Ethical Guide to Race, Normality, and the Implications of the Human Genome Project.* Boston: Beacon Press.

INFOTRAC® COLLEGE EDITION EXERCISES

The exercises that follow allow you to use the InfoTrac® College Edition online database of scholarly articles to explore the sociological implications of the selections in this chapter.

Search Keywords: Birth and rituals. Using InfoTrac® College Edition, find articles dealing with rituals of childbirth. Compare and contrast some of the rituals associated with the process of birth that occur across different cultures.

Search Keyword: Birth control. Search for articles that deal with birth control. What are some of the current debates about types of birth control and dissemination of birth control information? How do these discussions relate to debates on traditional family values? How do these discussions relate to the issue of reproduction as a means of social control?

Search Keyword: Infertility. How many articles can you find that discuss infertility? What is the definition of this term? Are there differences in the ways these articles define or describe infertility? What does this suggest to you about the social construction of the concept of infertility? Why and how is it socially constructed?

Search Keyword: Eugenics. Draw on InfoTrac® College Edition to locate articles that deal with eugenics. What is meant by this term? How does the notion of eugenics relate to reproduction as social control?

Search Keyword: Assisted reproductive technology. Search for articles that talk about types and implications of assisted reproductive technologies (also known as ART). What does ART mean for how we define families? What does it mean for how we define parenthood? How does ART relate to the notion that parenthood is socially constructed?

Search Keyword: Maternal instinct. How many articles can you find in the InfoTrac® College Edition online database that refer to maternal instinct? How is the concept discussed in these articles? Based on what you read, do you think that maternal instinct can be used as a form of social control over women? If so, how?

12

Social Devices for Impelling Women to Bear and Rear Children

LETA HOLLINGWORTH,
1916

. . . Child-bearing is in many respects analogous to the work of soldiers: it is necessary for tribal or national existence; it means great sacrifice of personal advantage; it involves danger and suffering, and, in a certain percentage of cases, the actual loss of life. Thus we should expect that there would be a continuous social effort to insure the group-interest in respect to population, just as there is a continuous social effort to insure the defense of the nation in time of war. It is clear, indeed, that the social devices employed to get children born, and to get soldiers slain, are in many respects similar.

But once the young are brought into the world they still must be reared, if society's ends are to be served, and here again the need for and exercise of social control may be seen. Since the period of helpless infancy is very prolonged in the human species, and since the care of infants is an onerous and exacting labor, it would be natural for all persons not biologically attached to infants to use all possible devices for fastening the whole burden of infant-tending upon those who are so attached. We should expect this to happen, and we shall see, in fact, that there has been consistent social effort to establish as a norm the woman whose vocational proclivities are completely and "naturally" satisfied by child-bearing and child-rearing, with the related domestic activities.

There is, to be sure, a strong and fervid insistence on the "maternal instinct," which is popularly supposed to characterize all women equally, and to furnish them with an all-consuming desire for parenthood, regardless of the personal pain, sacrifice, and disadvantage involved. In the absence of all verifiable data, however, it is only common-sense to guard against accepting as a fact of human nature a doctrine which we might well expect to find in use as a means of social control. Since we possess no scientific data at all on this phase of human psychology, the most reasonable assumption is that if it were possible to obtain a quantitative measurement of maternal instinct, we should find this trait distributed among women, just as we have found all other traits distributed which have yielded to quantitative measurement. It is most reasonable to assume that we should obtain a curve of distribution, varying from an extreme where individuals have a zero or negative interest in caring for infants, through a mode where there is a moderate amount of impulse to such duties, to an extreme where the only vocational or personal interest lies in maternal activities.

From Leta Hollingworth "Social Devices for Impelling Women to Bear and Rear Children" *American Journal of Sociology,* vol. 22(1), pp. 19–29.

The facts, shorn of sentiment, then, are: (1) The bearing and rearing of children is necessary for tribal or national existence and aggrandizement. (2) The bearing and rearing of children is painful, dangerous to life, and involves long years of exacting labor and self-sacrifice. (3) There is no verifiable evidence to show that a maternal instinct exists in women of such all-consuming strength and fervor as to impel them voluntarily to seek the pain, danger, and exacting labor involved in maintaining a high birth rate.

We should expect, therefore, that those in control of society would invent and employ devices for impelling women to maintain a birth rate sufficient to insure enough increase in the population to offset the wastage of war and disease. It is the purpose of this paper to cite specific illustrations to show just how the various social institutions have been brought to bear on women to this end. Ross has classified the means which society takes and has taken to secure order, and insure that individuals will act in such a way as to promote the interests of the group, *as those interests are conceived by those who form "the radiant points of social control."* These means, according to the analysis of Ross, are public opinion, law, belief, social suggestion, education, custom, social religion, personal ideals (the type), art, personality, enlightenment, illusion, and social valuation. Let us see how some of these means have been applied in the control of women.

Personal ideals (the type).—The first means of control to which I wish to call attention in the present connection is that which Ross calls "personal ideals." It is pointed out that "a developed society presents itself as a system of unlike individuals, strenuously pursuing their personal ends." Now, for each person there is a "certain zone of requirement," and since "altruism is quite incompetent to hold each unswervingly to the particular activities and forbearances belonging to his place in the social system," the development of such allegiance must be—

> effected by means of types or patterns, which society induces its members to adopt as their guiding ideals. . . . . To this end are elaborated various patterns of conduct and of character, which may be termed social types. These types may become in the course of time personal ideals, each for that category of persons for which it is intended.

For women, obviously enough, the first and most primitive "zone of requirement" is and has been to produce and rear families large enough to admit of national warfare being carried on, and of colonization.

Thus has been evolved the social type of the "womanly woman," "the normal woman," the chief criterion of normality being a willingness to engage enthusiastically in maternal and allied activities. All those classes and professions which form "the radiant points of social control" unite upon this criterion. Men of science announce it with calm assurance (though failing to say on what kind or amount of scientific data they base their remarks). For instance, McDougall (1998) writes:

> The highest stage is reached by those species in which each female produces at birth but one or two young, and protects them so efficiently that most of the young born reach maturity; the maintenance of the species thus becomes in the main the work of the parental instinct. In such species the protection and cherishing of the young is the constant and all-absorbing occupation of the mother, to which she devotes all her energies, and in the course of which she will at any time undergo privation, pain, and death. The instinct (maternal instinct) becomes more powerful than any other, and can override any other, even fear itself.

Professor Jastrow (1915) writes:

> . . . . *charm* is the technique of the maiden, and *sacrifice* the passion of the mother. One set of feminine interests expresses more distinctly the issues of courtship and attraction; the other of qualities of motherhood and devotion.

The medical profession insistently proclaims desire for numerous children as the criterion of normality for women, scornfully branding those so ill-advised as to deny such desires as "abnormal." As one example among thousands of such attempts at social control let me quote the following, which appeared in a New York newspaper on November 29, 1915:

> Only abnormal women want no babies. Trenchant criticism of modern life was made by Dr. Max G. Schlapp, internationally known as a neurologist. Dr. Schlapp addressed his remarks to the congregation of the Park Avenue M. E. Church. He said, "The birth rate is falling off. Rich people are the ones who have no children, and the poor have the greatest number of offspring. Any woman who does not desire offspring is abnormal. We have a large number, particularly among the women, who do not want children. Our social society is becoming intensely unstable."

And this from the *New York Times,* September 5, 1915:

> Normally woman lives through her children; man lives through his work.

Scores of such implicit attempts to determine and present the type or norm meet us on every hand. This norm has the sanction of authority, being announced by men of greatest prestige in the community. No one wishes to be regarded by her fellow-creatures as "abnormal" or "decayed." The stream of suggestions playing from all points inevitably has its influence, so that it is or was, until recently, well-nigh impossible to find a married woman who would admit any conflicting interests equal or paramount to the interest of caring for children. There is a universal refusal to admit that the maternal instinct, like every other trait of human nature, might be distributed according to the probability curve.

Public opinion.—Let us turn next to public opinion as a means of control over women in relation to the birth rate. In speaking of public opinion Ross says:

> Haman is at the mercy of Mordecai. Rarely can one regard his deed as fair when others find it foul, or count himself a hero when the world deems him a wretch. . . . For the mass of men the blame and the praise of the community are the very lords of life.

If we inquire now what are the organs or media of expression of public opinion we shall see how it is brought to bear on women. The newspapers are perhaps the chief agents, in modern times, in the formation of public opinion, and their columns abound in interviews with the eminent, deploring the decay of the population. Magazines print articles based on statistics of depopulation, appealing to the patriotism of women. In the year just passed fifty-five articles on the birth rate have chanced to come to the notice of the present writer. Fifty-four were written by men, including editors, statesmen, educators, ex-presidents, etc. Only one was written by a woman. The following quotation is illustrative of the trend of all of them:

> M. Emil Reymond has made this melancholy announcement in the Senate: "We are living in an age when women have pronounced upon themselves a judgment that is dangerous in the highest degree to the development of the population. . . . We have the right to do what we will with the life that is in us, say they."

Thus the desire for the development of interests and aptitudes other than the maternal is stigmatized as "dangerous," "melancholy," "degrading," "abnormal," "indicative of decay." On the other hand, excessive maternity receives many cheap but effective rewards. For example, the Jesuit priests hold special meetings to laud maternity. The German Kaiser announces that he will now be godfather to seventh, eighth, and ninth sons, even if daughters intervene. The ex-President has written a letter of congratulation to the mother of nine.

Law.—Since its beginning as a human institution, law has been a powerful instrument for the control of women. The subjection of women was originally an irrational consequence of sex differences in reproductive function. It was not *intended* by either men or women, but simply resulted from the natural physiological handicaps of women, and the attempts of humanity to adapt itself to physiological nature through the crude methods of trial and error. When law was formulated, this subjection was defined, and thus furthered. It would take too long to cite all the legal provisions that contribute, indirectly, to keep women from developing individualistic interests and capacities. Among the most important indirect forces in law which affect women to keep them child-bearers and child-rearers only are those provisions that tend to restrain them from possessing and controlling property. Such provisions have made of women a comparatively possessionless class, and have thus deprived them of the fundamentals of power. While affirming the essential nature of woman to be satisfied with maternity and with maternal duties only, society has always taken every precaution to close the avenues to ways of escape therefrom.

Two legal provisions which bear directly on women to compel them to keep up the birth rate may be mentioned here. The first of these is the provision whereby sterility in the wife may be made a cause of divorce. This would be a powerful inducement to women who loved their husbands to bear children if they could. The second provision is that which forbids the communication of the data of science in the matter of the means of birth control. The American laws are very drastic on this point. Recently in New York City a man was sentenced to prison for violating this law. The more advanced democratic nations have ceased to practice military conscription. They no longer conscript their men to bear arms, depending on the volunteer army. But they conscript their women to bear children by legally prohibiting the publication or communication of the knowledge which would make child-bearing voluntary.

Child-rearing is also legally insured by those provisions which forbid and punish abortion, infanticide, and infant desertion. There could be no better proof of the insufficiency of maternal instinct as a guaranty of population than the drastic laws which we have against birth control, abortion, infanticide, and infant desertion.

Belief.—Belief, "which controls the hidden portions of life," has been used powerfully in the interests of population. Orthodox women, for example, regard family limitation as a sin, punishable in the hereafter. Few explicit exhortations concerning the birth rate are discoverable in the various "Words" of God. The belief that family limitation will be punished in the hereafter seems to have been evolved mainly by priests out of the slender materials of a few quotations from Holy Writ, such as "God said unto them, 'Multiply and replenish the earth,' " and from the scriptural allusion to children as the gifts of God. Being gifts from God, it follows that they may not be refused except at the peril of incurring God's displeasure.

Education.—The education of women has always, until the end of the nineteenth century, been limited to such matters as would become a creature who could and should have no aspirations for a life of her own. We find the proper education for girls outlined in the writings of such educators as Rousseau, Fénelon, St. Jerome, and in Godey's *Lady's Book.* Not only have the "social guardians" used education as a negative means of control, by failing to provide any real enlightenment for women, but education has been made a positive instrument for control. This was accomplished by drilling into the young and unformed mind, while yet it was too immature to reason independently, such facts and notions as would give the girl a conception of herself only as future wife and mother. Rousseau, for instance, demanded freedom and individual liberty of development for everybody except Sophia, who was to be deliberately trained up as a means to an end. In the latter half of the nineteenth century when the hard battle for the real enlightenment of women was being fought, one of the most frequently recurring objections to admitting women to knowledge was that "the population would suffer," "the essential nature of woman would be changed," "the family would decay," and "the birth rate would fall." Those in control of society yielded up the old prescribed education of women only after a stubborn struggle, realizing that with the passing of the old training an important means of social control was slipping out of their hands.

Art.—A very long paper might be written to describe the various uses to which art has been put in holding up the ideal of motherhood. The mother, with children at her breast, is the favorite theme of artists. The galleries of Europe are hung full of Madonnas of every age and degree. Poetry abounds in allusions to the sacredness and charm of motherhood, depicting the yearning of the adult for his mother's knee. Fiction is replete with happy and adoring mothers. Thousands of songs are written and sung concerning the ideal relation which exists between mother and child. In pursuing the mother-child theme through art one would not be led to suspect that society finds it necessary to make laws against contraconception, infanticide, abortion, and infant desertion. Art holds up to view only the compensations of motherhood, leaving the other half of the theme in obscurity, and thus acting as a subtle ally of population.

Illusion.—This is the last of Ross's categories to which I wish to refer. Ross says:

> In the taming of men there must be provided coil after coil to entangle the unruly one. Mankind must use snares as well as leading-strings, will-o-the-wisps as well as lanterns. The truth by all means, if it will promote obedience, but in any case obedience! We shall examine not creeds now, but the films, veils, hidden mirrors, and half lights by which men are duped as to that which lies nearest them, their own experience. This time we shall see men led captive, not by dogmas concerning a world beyond experience, but by artfully fostered misconceptions of the pains, satisfactions, and values lying under their very noses.

One of the most effective ways of creating the desired illusion about any matter is by concealing and tabooing the mention of all the painful and disagreeable circumstances connected with it. Thus there is a very stern social taboo on conversation about the processes of birth. The utmost care is taken to conceal the agonies and risks of child-birth from the young. Announcement is rarely made of the true cause of deaths from child-birth. The statistics of maternal mortality have been neglected by departments of health, and the few compilations which have been made have not achieved any wide publicity or popular discussion. Says Katharine Anthony, in her recent book on *Feminism in Germany and Scandinavia* (1915):

> There is no evidence that the death rate of women from child-birth has caused the governing classes many sleepless nights.

Anthony gives some statistics from Prussia (where the figures have been calculated), showing that

> between 1891 and 1900 11 per cent of the deaths of all women between the ages of twenty-five and forty years occurred in child-birth. . . . . During forty years of peace Germany lost 400,000 mothers' lives, that is, ten times what she lost in soldiers' lives in the campaign of 1870 and 1871.

Such facts would be of wide public interest, especially to women, yet there is no tendency at all to spread them broadcast or to make propaganda of them. Public attention is constantly being called to the statistics of infant mortality, but the statistics of maternal mortality are neglected and suppressed.

The pains, the dangers, and risks of child-bearing are tabooed as subjects of conversation. The drudgery, the monotonous labor, and other disagreeable features of child-rearing are minimized by "the social guardians." On the other hand, the joys and compensations of motherhood are magnified and presented to consciousness on every hand. Thus the tendency is to create an illusion whereby motherhood will appear to consist of compensations only, and thus come to be desired by those for whom the illusion is intended.

There is one further class of devices for controlling women that does not seem to fit any of the categories mentioned by Ross. I refer to threats of evil consequence to those who refrain from child-bearing. This class of social devices I shall call "bugaboos." Medical men have done much to help population (and at the same time to increase obstetrical practice!) by inventing bugaboos. For example, it is frequently stated by medical men, and is quite generally believed by women, that if first child-birth is delayed until the age of thirty years the pains and dangers of the process will be very gravely increased, and that therefore women will find it advantageous to begin bearing children early in life. It is added that the younger the woman begins to bear the less suffering will be experienced. One looks in vain, however, for any objective evidence that such is the case. The statements appear to be founded on no array of facts whatever, and until they are so founded they lie under the suspicion of being merely devices for social control.

One also reads that women who bear children live longer on the average than those who do not, which is taken to mean that child-bearing has a favorable influence on longevity. It may well be that women who bear many children live longer than those who do not, but the only implication probably is that those women who could not endure the strain of repeated births died young, and thus naturally did not have many children. The facts may indeed be as above stated, and yet child-bearing may be distinctly prejudicial to longevity.

A third bugaboo is that if a child is reared alone, without brothers and sisters, he will grow up selfish, egoistic, and an undesirable citizen. Figures are, however, so far lacking to show the disastrous consequences of being an only child.

From these brief instances it seems very clear that "the social guardians" have not really believed that maternal instinct is alone a sufficient guaranty of population. They have made use of all possible social devices to insure not only child-bearing, but child-rearing. Belief, law, public opinion, illusion, education, art, and bugaboos have all been used to re-enforce maternal instinct. We shall never know just how much maternal instinct alone will do for population until all the forces and influences exemplified above have become inoperative. As soon as women become fully conscious of the fact that they have been and are controlled by these devices the latter will become useless, and we shall get a truer measure of maternal feeling.

> One who learns why society is urging him into the straight and narrow way will resist its pressure. One who sees clearly how he is controlled will thence-forth be emancipated. To betray the secrets of ascendancy is to forearm the individual in his struggle with society.

The time is coming, and is indeed almost at hand, when all the most intelligent women of the community, who are the most desirable child-bearers, will become conscious of the methods of social control. The type of normality will be questioned; the laws will be repealed and changed; enlightenment will prevail; belief will be seen to rest upon dogmas; illusion will fade away and give place to clearness of view; the bugaboos will lose their power to frighten. How will "the social guardians" induce women to bear a surplus population when all these cheap, effective methods no longer work?

The natural desire for children may, and probably will, always guarantee a stationary population, even if child-bearing should become a voluntary matter. But if a surplus population is desired for national aggrandizement, it would seem that there will remain but one effective social device whereby this can be secured, namely, *adequate compensation,* either in money or in fame. If it were possible to become rich or famous by bearing numerous fine children, many a woman would no doubt be eager to bring up eight or ten, though if acting at the dictation of maternal instinct only, she would have brought up but one or two. When the cheap devices no longer work, we shall expect expensive devices to replace them, if the same result is still desired by the governors of society.

If these matters could be clearly raised to consciousness, so that this aspect of human life could be managed rationally, instead of irrationally as at present, the social gain would be enormous—assuming always that the increased happiness and usefulness of women would, in general, be regarded as social gain.

SELECTED REFERENCES

Jastrow, J. 1915. *Character and Temperament.* New York: Appleton.

McDougall, W. 1908. *Social Psychology.* London: Methuen & Co.

13

Mothers Giving Birth to Motherhood

JOHN GILLIS,
1996

"There have always been mothers but motherhood was invented."

ANN DALLY, *INVENTING MOTHERHOOD*[1]

Because we assume that the physical act of giving birth naturally produces the desire and ability to nurture, we are stunned when we learn of birth mothers abusing or murdering their children, even though almost two of every three infants who die violently are killed by their own parents. When Susan V. Smith of South Carolina drowned both her sons in 1994, many of her neighbors found themselves searching for the answer to the question: "How could a mother do that to her children?"[2] We simply cannot believe that in giving birth a woman does not also give birth to herself as a mother. Yet many cultures make a distinction between maternity and motherhood, and even in Western society the connection between giving birth and giving nurture is surprisingly recent. It was not until 1875 that English-speaking people began talking about "true motherhood" as if maternity and motherhood were one and the same. Only in our own century have these terms become so completely identified that we have felt compelled to invent a new vocabulary—surrogate mothers, adoptive mothers, foster mothers—to describe those who do not combine maternity and motherhood in the prescribed manner.

The meanings of motherhood and fatherhood are never stable or transparent but forever contested and changing. Whatever may be universal about the biology of conception, pregnancy, and birth, maternity has no predetermined relationship to motherhood, and paternity no fixed relationship to fatherhood; both vary enormously across cultures and over time.[3] The many meanings of motherhood and fatherhood are not only reflected in the various images, symbols, and rituals associated with birth but are shaped by them. Faced with the ultimate mystery of human reproduction, we turn to rituals to provide us with a sense of meaningfulness. Birth has always been marked culturally, but whereas its rites once served to create and sustain a distinction between maternity and motherhood, today they underline the identity between these concepts. When a woman gives birth in the late twentieth century, she does so not once but four times: to the child, to herself as mother, to the man as father, and to the group that in our culture we are most likely to call family.[4]

Our equation of maternity with motherhood is not only relatively recent but historically unprecedented by the standards of the Western world. In earlier centuries, giving birth and giving nurture were often incompatible for demographic and economic reasons as well as cultural ones. Because of the high levels of both fertility and mortality that prevailed in Europe and North America until the nineteenth century, there was simply no way that all women who gave birth could also mother all their children. Maternal mortality never fell below 7 percent until this century. Until about a century ago, infant mortality rates, calculated as the percentage of infants who die before they reach their first birthday, ranged from 15 to 25 percent; only about half of all those born lived to the age of twenty-one. To replace these losses, women's fertility rates remained very high. Children came so quickly that it was often impossible for a woman to nurture all who were born to her, and she was likely to die before all her children left home.[5] The lifelong, intensive involvement with the individual child that has become the standard of motherhood in our own times was simply impossible for many women before the twentieth century. As a consequence, maternity and motherhood were understood as quite separable, not unlike the current understanding of fatherhood, in which the term "to father" means merely to generate and implies none of the nurturing capacities that currently attach to the words "to mother."[6]

Children in earlier periods did not lack for mothering, however. There existed a wide range of alternative sources of nurture. Wet-nursing had always been practiced and seems to have increased in the seventeenth and eighteenth centuries.[7] Placing out infants to women who would suckle them for an extended period was common not only among upper-class women, many of whom considered breast-feeding distasteful and unfashionable, but among working women who had neither the time nor the energy for the task.[8] Until Jean-Jacques Rousseau and his followers managed to convince the literate public that using wet-nurses was a violation of the laws of nature, deciding how to nurse an infant was usually done on the basis of convenience, with no particular symbolic value attached to the maternal breast. It was not until the nineteenth century that wet-nursing went into a precipitous decline, which began when infants were no longer being sent out to nurse and nurses were required to live in under close maternal supervision. Eventually, the very idea of the nonmaternal breast became incompatible with good motherhood, and by 1900 the wet-nurse had become a thing of the past, associated with so-called primitive cultures but having no place in civilized society. Mothers either suckled their own infants or bottle-fed, a method made safe by the milk pasteurization techniques developed late in the nineteenth century.

The convergence of maternity and motherhood proceeded fastest among the middle classes, but even at that social level it was still common in the nineteenth century for older children to be informally adopted by relatives.[9] This practice remained quite widespread among the working classes well into the early decades of this century.[10] In neighborhoods where kin lived nearby, children often took meals and slept apart from their biological parents; in families with many offspring, older siblings were frequently sent to live with more distant relatives—to "claim kin," as it was called in England—a form of intrafamily relief.[11] This was merely an extension of the ancient practice of circulating children for their own good, which in earlier times had been more likely to involve movement among unrelated households. In sixteenth-century England, 60 percent of those between the ages of fifteen and twenty-four were living apart from their parents, mainly as servants.[12] From the early nineteenth century onward, however, the movement of the children of the poor into the households of the better-off began to slow down. Girls continued to be sent away into domestic service, but working-class boys stayed closer to home. Still, it was not until the interwar period of the twentieth century that parents could expect most if not all of their children to be their responsibility until they saw them married, and even then the newlyweds were likely to return to one of the parental homes until they could find a place of their own.

As John Boswell has shown, Christian culture in earlier times never held parents wholly responsible for bringing up all their children. Giving up a child to the church through the medieval institution of oblation was regarded as an act of both piety and good sense. Protestants eliminated this practice but established foundling hospitals and orphanages, which served a similar purpose. They too saw nothing immoral or unnatural about giving up one's children to the "kindness of strangers." In the eighteenth century, one-quarter of all the children born in Toulouse, France, were turned over to the care of others.[13]

In this country as well, the foundling home and the orphanage remained vital institutions until the early twentieth century, housing large numbers of children—mainly for short rather than permanent placement, however. It was only after the Second World War that these institutions were closed down and Western societies turned to adoptive families and foster homes as the exclusive means of caring for displaced children. This change followed the general shift in thinking of parenting as an individual rather than a collective responsibility, one best carried out by one set of parents rather than several. Even now we assume that foster or adoptive mothers are second-best to what we call "real" or "natural" mothers. There has always been a certain suspicion of stepmothers, but the fine lines we draw between different kinds of mothers, always maintaining biological motherhood as the norm, is a distinctly recent phenomenon. Stepmotherhood and grandmotherhood were not sharply defined categories until the nineteenth century, when, as Ann Dally points out, motherhood itself finally emerged "as a concept rather than a mere statement of fact."[14] Until that time, anyone who mothered was called "Mother," regardless of biology. The term was applied to the mistresses of brothels and to the keepers of journeymen's hostels. In colonial New England, all older women, whether they had children or not, were called mothers.[15] In Europe it was common to call midwives "good mothers."[16] Mothering knew no age, race, or gender boundaries. Older sisters who brought up their siblings were referred to as "little mothers"; slave women who nursed white children were called "mammies"; and nurturing qualities were attributed to men as well as to women throughout the medieval and early modern periods.[17] Until the nineteenth century, the term "to father" still retained nurturing as well as generative connotations.[18] Only in this century has maternity come to bear all the weight of the symbolic as well as practical meanings that were once attached to all who mothered rather than to the one particular person who gave birth.

Our contemporary notion that individual mothers are wholly responsible for the physical, spiritual, and emotional well-being of their children had no place in earlier understandings of reproduction. For most of human history, birth has been one of those things, like death, for which no human could claim responsibility, for the organs of reproduction, like the body itself, were seen as part of a larger cosmos that determined the timing and nature of all human events. Some human control over life and death was attributed to magical practices, which may not have been effective but were symbolically important: they gave people a feeling of predictability and security. While hostile to pagan magic, Christianity incorporated many of these practices into its own rites of birth, marriage, and death; until the Reformation, combinations of sacred and secular magic provided the symbolic reassurance lacking in the real world.[19] Magical notions of the cosmos were not overcome easily, and the Protestant project of disenchantment was only partially effective before the nineteenth century. Likewise, older understandings of the body changed only very gradually.

In the ancient world, a variety of mother goddesses provided the sense of security and comfort that real mothers often could not offer their own children. Their relics and images became pilgrimage sites for infertile women as well as orphaned children, for Christians as well as pagans. Mother goddesses continued to hold sway during the first millennium of Christendom, for "the insecurity and real dangers of medieval childhood created powerful, persistent fantasies of protection and rescue by an omnipotent, loving mother."[20] Nor was fantasy directed only toward female figures: Jesus was often portrayed as having maternal features, and monks were known to present themselves in similar terms.[21]

A plethora of maternal figures, male and female, were required to meet these needs, because until the twelfth century Christianity had no central mother figure. In her earliest representations,

Mary was associated more with virginity than maternity, and the mother of Jesus was envisioned by the church more as the queen of heaven than as a mother as such. Only when Mary began to be represented as a mother in the late Middle Ages did she attract the devotion previously attached to pagan goddesses. For the first time it became possible to envision maternal as well as virgin saints. "In the Virgin," writes Clarissa Atkinson, "Christians discovered and made manifest in art and worship the powers of a sacred female common to many of the world's religions."[22]

In Mary, mother of Christ, late medieval Christians found a symbolic mother to live by. Her cult reached its apogee during the fourteenth and fifteenth centuries, when, as we have already seen, the Holy Family also became central to Catholic devotions, offering safe storage for the ideals of an emerging family system in which the nuclear unit was conceived for the first time as a moral core. It was at this time that Jesus acquired a father as well as a mother; for the next three centuries, the father figure was to compete with the mother figure for the right to symbolize nurturance. This was particularly true in Protestant lands, where the Reformation of the sixteenth century brought the Holy Family down to earth, finding new sources of symbolic reassurance within its own communities of godly households. A similar shift was apparent in Catholic countries, where the Holy Family also ceased to have a sacramental value and became a model for real families. The cult of Mary would continue to be a source of comfort to Catholics right up to the present day, but everywhere there was an increased emphasis on the heads of households providing the sense of protection and security that had once been sought elsewhere, at the roadside shrine or in monastic institutions.

Until the nineteenth century, nurturing capacity was thought of as an acquired talent more than as a sex-specific natural quality. Seventeenth- and eighteenth-century books on parenting were directed more to fathers than to mothers.[23] That there was as yet no equation between womanhood and motherliness was also evidenced in the witch persecutions of the sixteenth and seventeenth centuries. At no time before or since have so many women been associated with "the image of the witch as anti-wife and anti-mother—a sexual threat instead of a helpmate, a frightful danger to reproduction and the Christianization of children."[24] It was not accidental that in both Europe and North America many of those who were condemned and killed as witches were older women, many of them midwives and women past their childbearing years, onto whom could be projected the negative feelings of those who felt in some way deprived of nurture and protection.[25] Although patriarchs were the logical targets of such anger, they were too powerful to be confronted with these feelings. Instead, 80 percent of all those tried and executed for witchcraft were older women, most of them widowed, poor, and living on their own.

Midwives were particularly targeted because they were suspected of having knowledge of the old magic that Protestants wished to banish from their theocentric universe. The power to do good, traditionally assigned to these women, was easily stigmatized as the power of frightful evil.[26] Christina Larner's profile of the accused witch of the seventeenth century bears a striking resemblance to negative stereotypes of women in our own time: "She is assertive; she does not require or give love (though she may enchant); she does not nurture men or children, or care for the weak."[27] Accusations were rarely directed against wives and birth mothers as such, or even against wet-nurses or other women involved in child-rearing. Perhaps the anger was amplified by the guilt some fathers and mothers felt about not providing sufficiently for their own children. In any case, the patriarch and his goodwife were much too powerful to be criticized directly. If early modern Europeans and Americans could find nurturance in a much wider community than we can, they were also capable of wreaking vengeance on people they scarcely knew, persons onto whom it was easy to displace violently negative fantasies.

From the sixteenth through the early nineteenth centuries, motherhood was still subordinated to wifehood. As Clarissa Atkinson has described it, Protestantism placed a greater premium on "woman's role as a wife, consort, helpmeet, and lover—like Eve before the Fall, a central figure but secondary and complementary to her husband."[28] In the household economy of the commercial phase of capitalism, women were often partners, though normally junior partners, in farming and proto-industrial enterprises. Even as employment outside the household became closed to them, they were increasingly active in production both for household use and for the expanding market economy. Our notion of "housework," a term invented only in 1841 to describe domestic tasks, is incapable of encompassing all the skills acquired and practiced by women prior to the mid-nineteenth century.[29] The demands on a wife's time and energy were such that many found it difficult, if not impossible, to mother full-time. Until very late, the goodwife took precedence over the good mother.

Premodern rituals of pregnancy and birth reflected the tensions between wifehood and motherhood and reconciled the two by representing maternity as an episode in a woman's life rather than the beginning of an all-consuming career. Contrary to what we have been led to believe, birthing rites in earlier periods were not necessarily more elaborate than those of the modern era. Today women are the subject of intense and highly ritualized attention from conception onward, culminating with hospitalized birth, which Robbie Davis-Floyd has described as an "event more elaborate than any heretofore known in the 'primitive' world."[30] Through this modern rite of female passage a modern woman is left with few doubts about her primary identity. She may be a wife, consort, helpmeet, and lover, but she is above all a mother.

All societies mark birth and give it a meaning consistent with their material conditions and cultures. The birthing rites of the seventeenth and eighteenth centuries acknowledged maternity but reconciled it to women's other roles. They did so by representing pregnancy and maternity as something that happened to a women, as an episode in her life in which she was more the object of natural and supernatural forces than a subject in control of her own body. As Jacques Gelis reminds us: "To the country mind, in times gone by, men had to wait for nature to accomplish her work within the time she herself had set. It could be neither hindered nor precipitated. In a word, nature must go at her own pace, and the child 'come' in its own time."[31] When Protestants substituted the will of God for the whim of nature, they did not grant any additional agency to mothers. Indeed, their completely theocentric universe deprived women of access to the charms and potions that had provided comfort to Catholics. In America as well, the Protestant woman "could only throw herself on the mercy of God, and the midwife dared do nothing that might appear magical."[32] As a result, the prospect of birth became more rather than less terrifying.

Like all rites of passage, birth during this period consisted of three stages—separation, transition, and reincorporation.[33] But unlike our contemporary version of the birthing ritual, which heavily emphasizes the social separation of the pregnant woman, the premodern birth process placed the most symbolic weight on the final phase, the incorporative rites of baptism of the child and the churching of the woman. These are best described as communal rites of progression rather than as individual rites of passage, for their purpose was not to underline the separateness of mother and child but to restore the household and communal relationships disrupted by the arrival of the little stranger.[34]

Traditional rites made little of the preparation for birth and a great deal of its consequences. Prior to the nineteenth century, births were hardly anticipated, for it was thought unlucky to preempt either nature or divine will by preparing for birth in too overt a manner. There was no sure way of confirming pregnancy until the woman felt the movement of the fetus, the so-called quickening, though much effort was put into divining conception by various magical means. According to the contemporary understanding of fertilization, male seed was endowed with the greatest generative powers. The position of intercourse was said to determine the sex of the child, and men believed they could tell at ejaculation whether or not conception had occurred. But there were myriad other ways to divine pregnancy, none of which required medical attention.[35]

Prenatal medical care was in fact quite rare, though there was a great deal of lore about how a woman should conduct herself either to ensure a healthy birth or, if the baby was unwanted, to end the pregnancy. Her thoughts and actions were assumed to affect the child in her womb, though it was also believed that others, especially the father, could also influence it.[36] As for what we would call the fetus, it was thought of as a fully formed child from the seventh month onward, with a will of its own, just biding its time before entering the world. Although the child was thought to be influenced by its mother's behavior, it was assumed to have as much, if not more, control over the woman's body as did the woman herself, lending further credence to the notion that pregnancy was something that happened to her rather than a condition she was entirely responsible for.[37]

There is evidence that husbands monitored the health of their wives very closely, keeping diary records, as did the Reverend Ralph Josselin.[38] But for the most part, pregnancy went unmarked. There were no changes in behavior or dress, no efforts to collect baby clothes or pick out a name; such acts were thought to be presumptuous, even unlucky.[39] Indeed, there was as much to fear as to celebrate since, as mentioned earlier, the maternal death rate never fell below 7 percent and often went higher.[40] There was no way to avoid morning sickness or labor pains, which doctors, midwives, and mothers alike still thought of as either naturally or divinely ordained, brought upon women by Eve's misconduct and therefore something they should accept rather than resist. From the fourteenth century onward, maternal suffering replaced virginal status as a way of demonstrating female holiness. "The definition of a good mother as a suffering mother was firmly lodged in the ideologies of sanctity and of motherhood," notes Atkinson.[41] And this was as true of Protestantism as of Catholicism, though Catholic women still had the sufferings of Mary to give them psychological comfort.

Not that women did nothing to ward off natural or supernatural torments. For centuries pagans had resorted to charms and potions to ease the pains of childbirth, and the Catholic church had offered its own shrines and relics for the same magical purposes. From the late Middle Ages onward, the girdle of Mary and the relics of other mother saints were the object of female pilgrimage. When a Catholic woman could not go herself, she would send a "traveling girl" to bring back some of the magic.[42] Their Protestant sisters were deprived of these means, but many visited holy wells and sacred stones clandestinely. Their primary resource was prayer, often offered as a kind of incantation designed to provide some measure of comfort.[43]

Most women continued their normal routines right up to the moment of labor. In the ordinary language of the seventeenth and eighteenth centuries, pregnancy and birth were described not as a condition but as an activity—"breeding"—not all that different from a housewife's other enterprises.[44] While many women sought a little rest and indulgence during their pregnancies, eighteenth-century medical advice books encouraged them to remain active in everyday tasks, for they were thought to be plethoric, requiring leaner diets and more rather than less exercise.[45] In any case, a pregnant woman hardly stood out in a population in which virtually all married women were bearing children until illness or death prevented it. As Jacques Gelis has noted, "This simultaneous and permanent presence of pregnancy was an essential element of the 'human landscape' in past centuries. The community was perpetually pregnant with itself."[46]

Few women did much in anticipation of birth itself. A few who could afford to leave their busy households seem to have returned to their mothers; in England aristocratic women "went to Town," London being the favorite place to deliver.[47] But most households could not afford to dispense with their female members even for a short period. Thus, most women were at or near home when birth pains began, and often as surprised by the onset of labor as were their husbands and neighbors. The beginning of labor was interpreted as the child's efforts to get out of the womb. The flurry of activity that it precipitated may strike us as chaotic, even careless, but it was consistent with the traditional understanding of the body as subject to natural and supernatural forces beyond human control.

The first step once birth was imminent was to call for the midwife and to separate and isolate the birthing mother.[48] A "lying-in chamber" was designated and closed off, the doors shut and the windows draped, so that none of the normal sounds, smells, or activities of the household could penetrate.[49] Sometimes birth would take place in the warmest place, normally the kitchen. If a bedroom was chosen, it was completely rearranged so as to deconstruct its familiar features, including the bed itself. Few women gave birth in their own beds, for the birthing position of the time was either standing or squatting, and it was normal to substitute a special cot or birthing chair.[50] Suddenly and deliberately removed from her role as wife and helpmeet, the expectant mother was as isolated as if she had been removed to a birthing hut in the African rain forest.[51]

The midwife was joined by a half-dozen "gossips," neighborhood matrons who were there to witness the birth and assist as best they could. They busied themselves preparing special food and drink, usually a caudle of either hot wine or spiced porridge, sharing birth stories, and praying for a safe delivery. Birth was considered a women's affair, and only when the life was threatened was the male doctor called in.[52] Every effort was made to prevent husbands from seeing or hearing what was often a painful and sometimes a lethal process. They awaited news in the company of male friends, drinking the "groaning malt," drowning the anxiety that birth invariably evoked—the fear of losing not only a mother and child but their indispensable helpmeet and companion.[53]

The father's absence from the birthing room might suggest the lack of a concept of paternity, but in reality the opposite was the case. Men were said to feel a pregnancy, to share morning sickness and suffer the so-called husband's toothache, even experiencing labor pains.[54] The rituals of couvade, common to virtually all societies, constituted a parallel rite of progression in which paternity was formally acknowledged by the father and the community. In addition, the law recognized paternity by assigning rights in the child to the father rather than to the mother.[55]

Once the ritual separation had been accomplished, the waiting began. Little effort was made to hasten nature's pace or to substitute human for divine will. Midwives sometimes liked to hurry events for their own convenience, but medical opinion prior to the nineteenth century was against inducing labor. In fact, midwives were no more sensitive or tolerant of the mother's wishes than were male doctors.[56] The common language of the time—"with child," "brought to bed," "lying-in"—reinforced the notion of the mother as an object of forces beyond her control.[57] Birth was the moment of greatest danger to her, to the child, and to everyone attending her. It had once been a moment when every available magical means would be brought to bear, but by the eighteenth century these were largely unavailable to women of the middle and upper classes. In Protestant America it was said that "no midwives can do what angels do," a reference to the fatalism that attended birth during this period.[58]

Once the child was delivered, it was the task of the midwife to cut the umbilical cord, separating the child symbolically as well as physically from the mother. Instead of being brought immediately to the breast, the child was often taken to the hearth, symbolically identifying it with the house rather than with the mother.[59] It was then swaddled and shown to the father, his friends, and the other neighbors gathered at the house at the news of the event. The neglect of the mother in the immediate postpartum period was not as cruel as it may seem. She was regarded as out of danger and needing rest, but, of equal significance, she was not supposed to show too much affection toward the child. The mother love that is so much celebrated in our own day was regarded with great suspicion during the eighteenth century.[60] Any display of emotion suggested that the woman was still under the control of the natural and supernatural forces associated with birth. Both she and the child needed protection from these; both needed a time and a space to gain, or regain, the full measure of their humanity before reincorporation into family and community.[61]

The birthing ritual did not end with the biological event but continued for some days and even weeks until all its most important phase, the rites of incorporation, were complete. For the child this meant a second birth through the rite of baptism. Ralph Josselin wrote at the birth of his first child in 1642, "God wash it from it[s] corruption and sanctify it and make it his owne."[62] Cleansing was one of the traditional functions of baptism, but even before the child was brought to the font it had ordinarily undergone several folk rites of purification, separating it symbolically from the womb and forces of nature that had brought it into the world. Great attention was given to shaping the child's head, as if it had to be remade in a human image.[63] Swaddling served similar symbolic purposes, for it "was these clothes which made the child human, just as the wider ceremony of childbirth of which swaddling was a part made the delivery an act of culture, not merely of nature."[64]

In the traditional narrative of birth, the wife "presented" the husband with the child, who in turn re-presented it to the world with great flare and ceremony. Among the eighteenth-century gentry, birth, especially of a first-born son, was celebrated with bonfires, feasting, and distribution of largesse. Among the middle classes, patriarchal rites were more restrained but followed a similar pattern of celebrating with family and friends.[65]

Until a child was given this second birth, it was deemed very important to keep it isolated from all natural or supernatural influences. Even its name was not made known for fear that external powers would take possession of it.[66] The haste to baptize newborns reflected the same fear. When a child was in danger of dying, midwives had the right to baptize it themselves.[67] Naturally, Protestant theology's rejection of infant baptism posed a problem for those who regarded the rite as vital to the making of a human being; the great mass of the population continued to bring infants to the church font, while educated Protestants opted for private christenings, which avoided the appearance of infant baptism but still represented a symbolic second birth, providing the newborn with a name and a social identity. On all social levels, this was a time for communal feasting and gift-giving. "The birth was a convivial affair, welding the family and the community together round the child."[68]

Mothers, still confined to the lying-in chamber by the strict conventions of the day, were rarely present at the church when their children were baptized.[69] They played little part in the immediate postpartum festivities, which were presided over by the paterfamilias. If the infant was to be wet-nursed, the mother might get only a brief glimpse before it was taken away, not to be seen again for months, sometimes even years. In the meantime, she had entered into her period of "lying in," a ritualized period of up to a month when she was in a transition state, betwixt and between, neither fully a wife nor fully a mother. The physical act of giving birth was not at that time deemed sufficient to endow her with the wholeness and sanctity we now see as naturally conferred by maternity. Maternity was an event, not a cultural category capable of endowing a woman with motherhood as we would understand it. Indeed, seeing birth as something that happened to her allowed a woman to return to her household roles relatively unchanged by the biological experience.

Restoring a new mother to the fullness of womanhood required a period of several weeks, called "her month"; this final stage of the traditional ritual of progression would return her to her role as wife and coworker. For the first week the new mother was supposed to remain immobilized in bed, drinking the special caudle and eating a restricted diet. She would receive a carefully orchestrated series of visitors—women relations first, later female friends—sharing caudle with them.[70] The husband was the first male to enter the lying-in chamber, but it was thought dangerous to have sex during "her month," and even the mother's breast, which has such erotic meaning in modern culture, was then regarded with distaste, even fear.[71] Gradually the lying-in chamber would be opened up and restored to its original order, and the new mother would venture into the other rooms of the house. She would not leave the house, however, until the end of the month, which was normally marked by the religious rite of "churching," the religious ceremony of purification and thanksgiving to which was attached so much significance during this period.[72]

It was popularly assumed that the unchurched mother, sometimes known as the "green woman," was so dangerous that she could kill the grass she stepped on, induce unwanted pregnancies, and bewitch both people and animals.[73] In the sixteenth century, churching involved a public procession from the house to the church, the mother surrounded by her "gossips," sometimes led by the midwife herself. It was not until a woman had received the priest's blessing that "she may now put off her veiling kerchief, and look her husband and neighbors in the face again," wrote Henry Barrow, who, like other reformers, saw too much pagan and papist magic in the churching ritual.[74] Protestant sects substituted the simple rite of thanksgiving prayers for the older purification rites, but these were not always satisfactory to ordinary women who demanded that parsons church them in the old manner; when they refused, some women were known to church themselves.[75]

Even as the literate classes turned increasingly to private thanksgiving, public churching remained extremely popular throughout the eighteenth and nineteenth centuries, kept alive by women who valued it for its power to reconnect them with their households and communities.[76] During "her month," the domestic order was turned upside down, and husbands took on many of the wifely duties. New mothers were even spared their usual sexual duties until they were churched, a rite of incorporation that, as David Cressy writes, "established a ritual closure to this state of affairs, allowing the resumption of sexual relations between husband and wife and the restoration of normal domestic order."[77] According to John Brand's eighteenth-century description, "on the day when such a Woman was Churched, every Family, favoured with a call, were bound to set Meat and Drink before her."[78] In America this event was called the "groaning party." While the womenfolk rejoiced indoors, the men drank and fired off guns in recognition that a moment of danger had passed and their symbolic universe was once again in proper order.[79]

It is not surprising that churching remained popular among women of all classes throughout the eighteenth century. The upper classes differed from the lower only in their preference that it be a private ceremony, performed at home.[80] For all women, however, it signaled a return to their primary identity as wife and helpmeet, and as part of the community of women. Today we think of birth (and especially first birth) as a new beginning that initiates motherhood and starts a family, thereby bringing a woman into the fullness of her femininity. Earlier generations, who did not equate maternity with motherhood or insist that nurturance was the sole responsibility of the individual mother, endowed birth with an entirely different meaning, making of it less a rite of individual passage and more a rite of progression for the entire community.

Rituals like these did not simply reflect behavior, they shaped it. The traditional rites of churching and baptism incorporated mothers and children into the community in a way that underlined, not a woman's own individual motherhood, but her connection to all mothers, and her children's connection to all children. The rites encouraged women to see their offspring as separate from themselves and to see mothering as one task among many, one that could be shared with others, including men. This is not to say that they did not care deeply about their children. It was precisely because parents were so concerned with the well-being of their offspring under conditions of high mortality and economic uncertainty that they were willing to entrust them to the kindness of strangers for both the short and the long term.[81] Taking care of children was central to a house mistress's duties, but in the era of the patriarchal household the role of wife subsumed that of mother. Until the nineteenth century, children looked beyond their own natural families for mothering and fathering. In turn, mothers and fathers looked to children who were not their own to fulfill their duty and desire to be good parents.

NOTES

1. Ann Dally, *Inventing Motherhood: The Consequences of an Ideal* (London: Burnett Books, 1982), p. 17.
2. "Disillusioned Town Reviles Woman Accused of Killings," *New York Times,* November, 1994; see also Susan Chira, "Murdered Children: In Most Cases, a Parent Did It," ibid.
3. Dana Raphael, "Matrescence, Becoming a Mother: An 'Old/New' Rite of Passage," in *Being Female: Reproduction, Power, and Change,* ed. Dana Raphael (The Hague: Mouton, 1975), pp. 65–71.
4. Robbie E. Davis-Floyd, *Birth as an American Rite of Passage* (Berkeley: University of California Press, 1992), pp. 13, 38.
5. Even among the relatively healthy American populations, rates of orphanage were very high; see Richard Wertz and Dorothy Wertz, *Lying-in: A History of Childbirth in America* (New York: Free Press, 1977), p. 3; Peter Laslett, *Family Life and Illicit Love in Earlier Generations* (Cambridge: Cambridge University Press, 1977), chap. 4.
6. Shari Thurer, *Myths of Motherhood: How Culture Reinvents the Good Mother* (Boston: Houghton Mifflin, 1994), p. 213.
7. Ibid., p. 177; Elisabeth Badinter, *Mother Love: Myth and Reality: Motherhood in Modern History* (New York: Macmillan, 1981), p. 48; Valerie Fildes, *Wet-nursing: A History from Antiquity to the Present* (Oxford: Basil Blackwell, 1988), chaps. 6–8.
8. Fildes, *Wetnursing,* chap. 8.
9. Leonore Davidoff and Catherine Hall, *Family Fortunes: Men and Women of the English Middle Class, 1780–1850* (Chicago: University of Chicago Press, 1987), pp. 222–23.
10. Ellen Ross, *Love and Toil: Motherhood in Outcast London, 1870–1918* (New York: Oxford University Press, 1993), pp. 133–37.
11. Carl Chinn, *They Worked All Their Lives: Women of the Urban Poor in England, 1880–1939* (New York: St. Martins Press, 1988), chaps. 2–4; Michael Anderson, *Family Structure in Nineteenth-Century Lancashire* (Cambridge: Cambridge University Press, 1971), pt. 3.
12. Illana Krausman Ben-Amos, *Adolescence and Youth in Early Modern England* (New Haven, Conn.: Yale University Press, 1994), p. 2.
13. John Boswell, *The Kindness of Strangers: The Abandonment of Children in Western Europe, Late Antiquity to the Renaissance* (New York: Pantheon, 1988), p. 11.
14. Dally, *Inventing Motherhood,* p. 17.
15. Laura Thatcher Ulrich, *Good Wives: Image and Reality in the Lives of Women in Northern New England, 1650–1750* (New York: Vintage, 1980), p. 158.
16. Jacques Gelis, *History of Childbirth: Fertility, Pregnancy, and Birth in Early Modern Europe* (Cambridge: Polity, 1991), p. 105.
17. On the phenomenon of "little mothers," see Elizabeth Roberts, *A Woman's Place: An Oral History of Working-Class Women, 1890–1940* (Oxford: Basil Blackwell, 1984), pp. 24–25, 173; Chinn, *They Worked All Their Lives,* pp. 26–36; on the familial relationship of whites and blacks on American slave plantations, see Mechal Sobel, *The World They Made Together: Black and White Values in Eighteenth-Century Virginia* (Princeton, N.J.: Princeton University Press, 1987), chap. 10; on the nurturing qualities attributed to men, see Davidoff and Hall, *Family Fortunes,* pp. 329–35.
18. For a fuller discussion, see Gillis, 1996.
19. Keith Thomas, *Religion and the Decline of Magic* (New York: Scribners, 1971), chaps. 2–3.
20. Clarissa Atkinson, *The Oldest Vocation: Christian Motherhood in the Middle Ages* (Ithaca, N.Y.: Cornell University Press, 1991), p. 137.
21. Caroline Walker Bynum, "Jesus and Mother and Abbot as Mother: Some Themes in Twelfth-Century Cistercian Writing," in her *Jesus as Mother: Studies in the Spirituality of the High Middle Ages* (Berkeley: University of California Press, 1983), pp. 110–59; on the continuation of this trend, see David Leverenz, *The Language of Puritan Feeling: An Exploration in Literature, Psychology, and Social History* (New Brunswick, N.J.: Rutgers University Press, 1980).
22. Atkinson, *The Oldest Vocation,* pp. 115, 143.
23. Steven Ozment, *When Fathers Ruled: Family Life in Reformation Europe* (Cambridge, Mass.: Harvard

University Press, 1983); Thurer, *Myths of Motherhood,* pp. 166–67; Mary Ryan, *The Empire of Mother: American Writing about Domesticity, 1830–1860* (New York: Haworth, 1982), pp. 18–22.

24. Atkinson, *The Oldest Vocation,* p. 232; Thurer, *Myths of Motherhood,* p. 157.
25. Atkinson, *The Oldest Vocation,* p. 232.
26. On the powers of midwives, see Gelis, *History of Childbirth,* pp. 105–10.
27. Christina Larner; *Witchcraft and Religion,* p. 84.
28. Atkinson, *The Oldest Vocation,* p. 220.
29. Thurer, *Myths of Motherhood,* p. 90.
30. Davis-Floyd, *Birth as an American Rite of Passage,* pp. 1–2.
31. Gelis, *History of Childbirth,* p. 65.
32. Wertz and Wertz, *Lying-in,* p. 23.
33. On rites of passage generally, see Arnold van Gennep, *Rites of Passage* (Chicago: University of Chicago Press, 1960).
34. On rites of progression, see David Cheal, "Relationships in Time: Ritual, Social Structure, and the Life Course," *Studies in Symbolic Interaction* 9(1988): 98.
35. Audrey Eccles, *Obstetrics and Gynecology in Tudor and Stuart England* (Kent, Ohio: Kent State University Press, 1982), pp. 24–26, 60; Angus McClaren, *Reproductive Rituals: The Perception of Fertility in England from the Sixteenth to the Nineteenth Century* (London: Methuen, 1984), chaps. 1–2.
36. McLaren, *Reproductive Rituals,* pp. 13–30; Gelis, *History of Childbirth,* pp. 47–56, and chap. 6.
37. Gelis, *History of Childbirth,* p. 58.
38. Alan Macfarlane, *The Family Life of the Reverend Ralph Josselin* (New York: W. W. Norton, 1970), pp. 81–91.
39. Gelis, *History of Childbirth,* pp. 67ff.; American death records in the eighteenth century include many infants who died without names, Sandra Brant and Elissa Cullman, *Small Folk: A Celebration of Childhood in America* (New York: E. P. Dutton, 1980), p. 43.
40. Thurer, *Myths of Motherhood,* p. 171.
41. Atkinson, *The Oldest Vocation,* p. 193.
42. Gelis, *History of Childbirth,* pp. 70–75.
43. Thomas, *Religion and the Decline of Magic,* pp. 508, 516.
44. Ralph Josselin used this language; and it continued among the upper classes until the late eighteenth century; Macfarlane, *The Family Life of the Reverend Ralph Josselin,* pp. 84–85; see also Judith S. Lewis, *In the Family Way: Child-bearing in the English Aristocracy, 1760–1860* (New Brunswick, N.J.: Rutgers University Press, 1986), p. 72; Madeleine Riley, *Brought to Bed* (South Brunswick, N.J.: A. S. Barnes, 1968), p. 4.
45. Eccles, *Obstetrics and Gynecology in Tudor and Stuart England,* pp. 45–47, 60–65; Ann Oakley, *The Captured Womb: A History of the Medical Care of Pregnant Women* (Oxford: Basil Blackwell, 1984), pp. 22–24.
46. Gelis, *History of Childbirth,* p. 45.
47. On aristocratic women, see Lewis, *In the Family Way,* pp. 52–54, 156–58.
48. Van Gennep, *Rites of Passage,* p. 41.
49. Eccles, *Obstetrics and Gynecology in Tudor and Stuart England,* pp. 94–95; Adrian Wilson, "Participant or Patient? Seventeenth-Century Childbirth from the Mother's Point of View," in *Patients and Practitioners: Lay Principles of Medicine in Pre-Industrial Societies,* ed. Roy Porter (Cambridge: Cambridge University Press, 1985), p. 135; Wertz and Wertz, *Lying-in,* chap. 1.
50. Lewis, *In the Family Way,* p. 151; Wilson, "Participant or Patient?" p. 135; Edward Shorter, *The Making of the Modern Family* (New York: Basic Books, 1975), p. 145; Eccles, *Obstetrics and Gynecology in Tudor and Stuart England,* p. 92; Gelis, *History of Childbirth,* pp. 97–98, 130–32.
51. Wilson, "Participant or Patient?" pp. 132–35; Shorter, *The Making of the Modern Family,* pp. 48–56; Ralph Houlbrooke, *English Family Life, 1576–1716: An Anthology from Diaries* (Oxford: Basil Blackwell, 1989), pp. 129–30; Wertz and Wertz, *Lying-in,* pp. 12–14.
52. Shorter, *The Making of the Modern Family,* pp. 293–94.
53. Wilson, "Participant or Patient?" pp. 133–36.
54. Gelis, *History of Childbirth,* pp. 38, 155; Lisa Cody, "The Politics of Body Contact: The Discipline of Reproduction in Britain, 1688–1834" (Ph.D. dissertation, University of California at Berkeley, 1993), conclusion.
55. Nigel Lowe, "The Legal Status of Father: Past and Present," in *The Father Figure,* eds. L. McKee and M. O'Brien (London: Tavistock, 1982), pp. 26–28; Riley, *Brought to Bed,* pp. 68, 105–13.
56. Wertz and Wertz, *Lying-in,* pp. 20–23; Wilson, "Participant or Patient?" pp. 129–30; Shorter, *The Making of the Modern Family,* pp. 38–39; Gelis, *History of Childbirth,* pp. 134–35.
57. Riley, *Brought to Bed,* pp. 3–4; Lewis, *In the Family Way,* p. 72.
58. Quoted in Wertz and Wertz, *Lying-in,* p. 21.
59. Gelis, *History of Childbirth,* p. 163.

60. Ruth Bloch, "American Feminine Ideals in Transition: The Rise of the Moral Mother, 1785–1815," *Feminist Studies* 4, no. 2 (1978): 101–26; Badinter, *Mother Love,* pt. 1.
61. Gelis, *History of Childbirth,* p. 183.
62. Alan Macfarlane, *The Family Life of Ralph Josselin* (New York: W. W. Norton, 1970), p. 88.
63. Eccles, *Obstetrics and Gynecology in Tudor and Stuart England,* p. 83; Joseph Illick, "Childrearing in Seventeenth-Century England and America," in *The History of Childhood,* ed. Lloyd deMause (New York: Psychohistory Press, 1974), p. 307; *Notes and Queries,* 5th series (September 14, 1878): 205, and (September 28, 1878): 255–26.
64. Wilson, "Participant or Patient?" p. 137.
65. Macfarlane, *The Family Life of Ralph Josselin,* pp. 88–89; Houlbrooke, *English Family Life,* p. 131; on folk rites, see John Brand, *Observations on Popular Antiquities* (London: Chatto and Windus, 1877), pp. 340–41.
66. Gelis, *History of Childbirth,* p. 195.
67. Houlbrooke, *English Family Life,* pp. 130–31; B. Midi Berry and Roger Schofield, "Age of Baptism in Preindustrial England," *Population Studies* 25 (1971): 453–63; Thomas, *Religion and the Decline of Magic,* pp. 36–7, 56.
68. Gelis, *History of Childbirth,* p. 188.
69. Wilson, "Participant or Patient?" p. 138.
70. Eccles, *Obstetrics and Gynecology in Tudor and Stuart England,* pp. 95–97; Wilson, "Participant or Patient?" pp. 137–38; Lewis, *In the Family Way,* pp. 194–99; van Gennep, *Rites of Passage,* p. 48; Gelis, *History of Childbirth,* pp. 188–94.
71. Linda Pollock, *Forgotten Children: Parent-Child Relations from 1500 to 1900* (Cambridge: Cambridge University Press, 1983), p. 215; Eccles, *Obstetrics and Gynecology in Tudor and Stuart England,* pp. 14, 98.
72. Wilson, "Participant or Patient?" p. 138; Lewis, *In the Family Way,* pp. 195–97.
73. Thomas, *Religion and the Decline of Magic,* pp. 15, 38–39; David Cressy, "Thanksgiving and the Churching of Women in Post-Reformation England," *Past and Present* 141 (November 1993): 115.
74. Quoted in Thomas, *Religion and the Decline of Magic,* p. 60.
75. Cressy, "Thanksgiving and the Churching of Women," pp. 123–32.
76. Peter Rushton, "Purification or Social Contract? Ideologies of Reproduction and the Churching of Women after Childbirth," in *The Public and Private,* eds. Eva Gamarnikow, et al. (London: Heinemann, 1983), pp. 124–31.
77. Cressy, "Thanksgiving and the Churching of Women," p. 115.
78. Brand, *Observations on Popular Antiquities,* p. 228.
79. Wertz and Wertz, *Lying-in,* pp. 5–10.
80. Lewis, *In the Family Way,* pp. 201–2; Adrian Wilson, "The Ceremony of Childbirth and Its Interpretation," in *Women as Mothers in Pre-Industrial England,* ed. Valerie Fildes (London: Routledge, 1990), p. 92.
81. Pollock, *Forgotten Children,* pp. 111–13; Boswell, *Same-Sex Unions,* pp. 428–34.

14

Parental Consciousness and Gender

SUSAN WALZER,
1998

When I began interviewing new parents, I was on the lookout for variations; I wanted to understand what made some mothers and fathers more gender-differentiated than others. As my research progressed, however, I was struck more by the similarities than by the differences in the new parents I was meeting. While varying in employment status, gender ideology, and divisions of physical labor, new mothers seemed to be having experiences more like each other than not, and distinct from the transitions of their husbands and partners.

What I realized was that the ways in which new mothers and fathers perceived themselves and interacted with their partners were both more complicated and more simple than dichotomies such as employed/unemployed and involved/uninvolved suggested. What was complicated was that generally individuals did not perceive themselves, nor did their partners, as one thing or another. A father might be relatively uninvolved in caregiving tasks, but both he and his wife perceived him as an involved father. A mother might be working outside the home, but both she and her husband framed her employment as something that did not interfere too much with her primary responsibility: mothering.

That was the complicated part. What was simple was that, with some exceptions that I will describe as we go along, there did seem to be one basic dichotomy: mothers and fathers. New mothers were having a different experience than new fathers were, and their interactions with each other appeared to acknowledge and sustain this difference. In some cases the difference was reflected in divisions of caregiving and economic arrangements, but in more cases, it was manifested in nonbehavioral ways. I do not mean to imply here that if you've seen one mother or father you've seen them all. There have been class differences noted among mothers, for example, but these same researchers also point out that there is an overriding similarity in women's approaches to mothering (Hays 1996; McMahon 1995).

Gender differentiation is reflected not only in the concrete divisions of labor of new parents, but in their thoughts and feelings about their babies—what I refer to as their "parental consciousness." I use "consciousness" in two senses of the word. Consciousness refers to mindfulness or awareness, so when I talk about parental consciousness, I am talking about how babies fill parents' minds. I also use it to underscore that the internal experiences that new mothers and fathers report are, in part, a

From *Thinking about the Baby: Gender and Transitions into Parenthood,* by Susan Walzer. Reprinted by permission of Temple University Press.

social product—born out of interaction and shaping their actions (Ritzer 1983). In other words, parents think about their babies, and they also think about these thoughts; they judge these thoughts by how they think they *should* be thinking about their babies.

The process I am referring to is captured in George Herbert Mead's concept of the self as both subject and object. In Mead's view, the self arises out of social experience; we experience our selves by observing ourselves as others see us. New parents' identities emerge in a social process: they observe others' responses to them as mothers and fathers, and this observation shapes their identities as mothers and fathers. Bill illustrates this in talking about what happens when he is out in the world with his daughter: "When I'm with her, people call me sir. I don't know, I look at myself in the mirror and picture myself as seventeen, but I have a lot more responsibilities. I'm not as carefree as I was before." When people call Bill "sir," he thinks about how he views himself. On one hand, he does not see himself as others do, and yet he goes on to claim in himself the characteristics that make him a "sir," a father: he has responsibilities and is less carefree than he was before his daughter was born.

In "doing gender" terms, the parents I interviewed carried particular images of what mothers and fathers were supposed to think about—what their responsibilities and feelings were supposed to be—and they were accountable to these images. The process of claiming the socially defined characteristics of a mother or father was not necessarily always conscious, but there were parents who described thinking that they should be worried about something that they weren't in fact worried about, or thinking that they should be feeling something different from what they were actually feeling. Parents also actually had the thoughts and feelings that they thought mothers and fathers should have, as Bill did, and in the group of parents that I interviewed, mothers tended to have different parental consciousness than fathers did. These differences seemed to emerge, in part, out of the experience of caring for the baby, but they also emerged in the context of other social processes. And they played a role in reproducing gendered divisions of baby care. . . .

Several of the women that I interviewed reported feeling a particular kind of social acceptance upon becoming mothers. The metaphor used was of having become part of a "club" that conferred on them a new status with other women, as Laura described: "I feel like I've joined a whole new dimension of other people. I wasn't in that club before. And we have this person that I have a lot of respect for that I work with and. . .she came up to me, she put her arm around me, and she said, 'You're about to venture into something that you've never experienced before in your entire life.' She said, 'You're about to become a member of this very unique club.' And I was like Wow! That's cool!" New fathers also felt a commonality with other parents, but they did not tend to focus on being accepted. Rather, they described themselves having a new acceptance of other people's behavior:

> I've always seen other people with kids and I've always been real bored with them doting over their kids and all the little hurdles and landmarks they reach, but when you have your own kid, it really is different. (Peter)
>
> I guess you become more forgiving of other adults. You never knew why they were so wacky and it's because they have kids. (Tom)

The ways in which new mothers and fathers talked about losing time, as well as their experiences of their babies' dependence, resulted in different internal responses to the loss of autonomy and mobility that caring for a baby generates.

Virtually all of the parents I interviewed remarked on the change that having a baby made in the amount of time they had for themselves and their partners. Constance said of her son, "My time is his." Even a father who declined to participate in my study said on the telephone, "Maybe you could put that in your paper—there's no time." Yet the stresses that the mothers and fathers I spoke with experienced in relation to time tended to be different. To summarize simply, many mothers experienced stress about time they *didn't* spend with their children, while many fathers were stressed by time they *did* spend with their children in relation to other things they were not getting done. I do not mean to imply that mothers wanted to spend all of their time with their babies or that fathers did not, but that they felt a different accountability about their time, which came through in how they talked about this issue.

Fathers were more apt to address the time that a child requires in negative terms. Phil remarked, "It's almost like you're held hostage to the kid," and Chad said, "I wish I could enjoy it more. For whatever reason at times I can't. And maybe when he gets older and can talk and all that then we will have it easier but right now it's like a one-way street and you got to take care of him. That's just the nature of things."

While it is probable that some mothers shared these sentiments, they did not express them as openly as fathers, who expressed discomfort with the lack of productivity they experienced while spending time with their child:

> I mean if I'm watching him during the day—she's at work—forget about doing anything. It's constant attention. You can't read, you can't study, you can't paint, you can't do anything. You really got to sit there and watch him. (Peter)

> Sitting two hours playing with him, when I first did it was like, this is a waste of my time. I said, "I have more important things to do." And I'm still thinking, "Look at the time I've spent with him. What would I have done otherwise?" (Chad)

One father, Jack, talked about having difficulty connecting with his baby during the first month. When I asked whether he thought there was something about the baby's behavior, he responded, "Yeah, the fact that he keeps laying there and not doing anything." Jack referred to his feelings as normal, but it is unlikely that a mother who did not feel a connection to her baby for a month would consider her response normal. In fact, for many women, the onset of postpartum depression is signaled by guilt brought on by the fact that they do not feel about their newborn babies what they expect to feel (Taylor 1996).

Phil admitted that at times he thought it might have been a mistake to have a child. Unlike Jack, he did not think that his response was normal and described himself as feeling closed in: "Kind of like my life was never going to be the same. Kind of like spring came around, normally do more stuff outside, away from home. It kind of dawned on me that things weren't getting any less demanding dealing with Louise. Things that I would normally do, go out and play tennis or go to the driving range, hit some balls, it just wasn't working out."

Phil's despair about losing certain leisure activities was in contrast to what one mother, Ruth, described as how she deals with her "working mom's syndrome": "I find if I can be home at five, I can have five to eight, but if I stop at the gym and work out, I'm not home till 6:30. I only have an hour with him. So I've pretty much given up that." Another mother, Liza, talked about making the "obvious" choice to give up contact with her friends: "I used to on my days off always go out with my friends and stuff and I don't do that, obviously, now, and it's great. I just love being here with him."

For mothers, the question was not what else they could be doing with their time, but whether they were giving their children enough time:

> There are still times when I'm like, am I not spending enough time with her? But I spend *all* my time with her. (Melissa)

> Sometimes I feel a little guilty that... I have a little bit more that I should be giving him. (Peggy)

> I just think it is very hard both working full-time and trying to keep the important part of family to me, you know, spending time with her. (Harriet)

In contrast to Harriet's notion that the important part of family is spending time with her daughter, some fathers felt that the time they spent with their children was unimportant relative to other things they had to do. Mothers were less apt to allow themselves to prioritize anything beyond the time they had to spend in their workplaces as important enough to trade off for time with their children. These reactions may be related to the different ways in which the men and women that I interviewed responded to their children's dependence on them.

Many of the mothers and fathers that I spoke with described a new consciousness that evolved with having a baby: a sense that there was someone relying on them. The ways that they talked about this dependence, however, often reflected a gendered dichotomy between financial and other kinds of care. One father, Brett, who described himself as preoccupied with financial matters, joked that the meaning of becoming a father for him was "one more deduction" on his taxes. Other fathers were more sober in describing their concern with providing and planning for the future:

> It's more pressure. You want better things for your baby so you apply more pressure to yourself.... Pressure to succeed so that you can give your family and your children more than what you perceived as what you had. (Jake)

> I'm concerned a lot more with the future. I mean, before we had plans, but now you've got to take into account saving money for college. (Gil)

> I think about everything a little bit differently.... Everything from simply when to take a vacation day, and now you think about, what would [child] like to do? Should we take that trip next year because he'll be old enough to enjoy it? Just everything in general. Financially you start thinking about have you been doing the wisest thing with your funds with the price of college and all going up. (Ted)

Financial responsibility was perceived by some of the men I interviewed as a unique way in which babies need their fathers. One father, Jay, talked about other caregiving as something he did if his wife was not available; economic provision was something that his baby needed particularly from him: "Someone needs me. I mean Dylan depends on me, you know? Whether it is feeding him or changing his diapers and playing with him when Gloria is not here or whatever. Or just the financial aspect to work, cause I mean if something happened to me, Gloria would just carry on, I mean she could survive and stuff. But Dylan, you know, he couldn't."

Sometimes this sense of financial responsibility was reinforced by wives who had left jobs or were working part-time:

> Jay is going to be job hunting and stuff and that really consumes him right now. Of course he is worried, you know, he's the breadwinner in the family, so I think it kind of scares him. (Gloria)

> I think it scares him a lot more than he will let on...because he knows that he has to take care of her. I could get a job and I could take care of myself, but he has to take care of her. (Whitney)

> I think he now feels differently, like he has to take care of his family and he has to be the breadwinner or whatever. Even though we probably make just as much money, you know, our salaries are compatible. But I just feel that he has that different dimension. (Laura)

Fathers did not tend to question the need to go to work, nor did they see their work demands as affecting their identities as good parents.... Some experienced loss in relation to time they could not spend with their babies when they were at work, but they described their babies' abilities to make them forget their work days when they came home, and there was pleasure in their descriptions of going to work and then coming home:

> I go to work, I come home, and I feel so refreshed when I'm around him. (Todd)

> You can always come home and you can be guaranteed that he'll be there smiling. It helps you forget about your job. It puts things in perspective; it's just a job. (Chip)

> It's nice coming home and having a bad day at work and seeing Dylan and he'll come running up with his arms up, you know, and it just makes all your worries just melt away. (Jay)

For Bill, the tensions around employment and parenthood came up in his ambivalence about doing jobs besides his primary job:

> I do that for extra money to make sure that we do have things but then I think as I'm doing it, I'm not home with her. I'll get home, I'll be home for a half hour, and she'll go to bed... In one respect I want to have the money. In another respect, is that going to make me all that happy? Having maybe a little bit extra but missing out on, I think I've missed out on, I've been there for everything, but I haven't been there right on the spot. Like the first time she really walked. I mean like I came home, I was there, but I wasn't right there. I don't want to miss out on that stuff.

Fathers like Bill experienced their absence from their babies as a loss for themselves while mothers expressed concern that their absence would result in loss for their babies (and perhaps more unconsciously, for their sense of themselves as mothers). It is clear that these fathers felt a sense of responsibility in relation to their babies. As Elliot said, "I've become somebody's father figure, somebody's role model." Yet as much responsibility as fathers felt to support their children financially and otherwise, the mothers I interviewed tended to describe a more minute-to-minute, pervasive sense of ultimate responsibility, as Mandy illustrated: "The minute they're born you just become this protective thing that takes care of this baby. You become less of yourself and more of something that's there for the baby."

The experience that Mandy describes of becoming less of herself is one that has been identified in other research about parenthood. Motherhood appears to take up more of women's identities than fatherhood does of men's. All invest in parenthood, but fathers tend to hold onto other parts of themselves more than mothers do (Cowan and Cowan 1992).

Motherhood is a state of "being" while fatherhood is something that men "do," according to Diane Ehrensaft (1990: 98), who argues that it is harder for mothers to create boundaries between themselves and their children than it is for fathers: "Mothers are connected while fathers are separate." This sense of connection, McMahon (1995: 268) notes, is one of the greatest rewards of being a mother for the women that she interviewed, but its flip side—feeling responsible—is the worst thing about being a mother. And it is this sense of ultimate responsibility for children that the women McMahon spoke with perceived as one of the great differences between mothers and fathers.

This difference appeared between the mothers and fathers that I interviewed as well. It is not that fathers did not feel responsibility for their babies, but it tended to take a different form than the ultimate responsibility that mothers described. This, McMahon reports, is the biggest downside of being a mother, and many of the women I interviewed also presented this as a source of stress. . . .

While mothers felt more ultimately responsible for babies than fathers, differences in parental consciousness between men and women were also reinforced by their divisions of the care of their babies, of which women tended to do a disproportionate amount (see also Belsky and Volling 1987; Berman and Pedersen 1987; Dickie 1987; Thompson and Walker 1989).

DIVISIONS OF BABY CARE

Approximately one-third of the mothers and fathers that I interviewed reported having a relatively equal division of labor, while two-thirds did not. In this discussion of baby care, I want to emphasize the interactional climate in which women's additional caregiving takes place, as well as forms of more "invisible" or mental labor (see DeVault 1991). I discuss three categories of mental labor involved in taking care of a baby—worrying, processing information, and managing the division of labor—and I suggest that even in cases in which fathers are participating in physical care, mothers tend to be in charge of mental labor. This is consistent with Jay Belsky's finding that even dedicated fathers rarely assume managerial chores such as scheduling doctor's appointments or knowing when it is time to buy new diapers (Belsky and

Kelly 1994). I suggest, however, that the difference in men's and women's participation in mental baby care goes beyond remembering to do things; it is both an outcome of, and a sustaining force in, gender-differentiated parental consciousness.

My use of the term "mental labor" is meant to differentiate this less visible work from physical tasks—to capture the internal and interpersonal work that is part of infant care. I include in this category what has been called "emotion," "thought," and "invisible" work in other sociological analyses (see Hochschild 1983; DeVault 1991); that is, I identify aspects of baby care that involve thinking or feeling, managing thoughts or feelings, and that are not necessarily identified as work by the person doing it.

Worrying

> My mind is always on something, you know, how is he? Or how's he eating? Or how he's this or that, how he's doing in day care. (Sylvia)

> I worry about her getting cavities in teeth that are not even gonna be there for her whole life. Everything is so important to me now. I worry about everything. (Miranda)

> It's like now you have this person and you're always responsible for them, the baby. You can have a sitter and go out, yes, and have a break, but in the back of your mind, you're still responsible for that person. You're always thinking about that person. (Peggy)

Regardless of their employment status, the mothers that I interviewed tended to worry (see also Ehrensaft 1990; Hays 1996). Eileen described thinking about their babies as something that mothers do: "Mothers worry a lot." Worrying was such an expected part of mothering that the absence of it might challenge one's definition as a good mother. Alison said of her first day back at her job after being home with her baby: "I went to work and I basically had to remind myself to call and check on him once, I felt, or I'd be a bad mother."

This is a good example of "doing gender" accountability. Alison wasn't actually worried about her baby, but she felt that she had to behave as though she was or she would be a bad mother. Fathers do not necessarily think about their children while they are at work, nor do they worry that not thinking about their child reflects on them as parents (Ehrensaft 1990).

The tendency for mothers to think and worry about babies appeared to be an important source of differentiation within the couples I interviewed, and it presented a paradox for women. On one hand, worrying was associated with irrationality and unnecessary anxiety, and some fathers suggested that their partners worried too much about their babies. As Stuart said of Laura: "Sometimes I say 'He's fine, he's fine,' but he's not fine enough for her." On the other hand, worrying was perceived as something that good mothers do. A number of fathers made an explicit connection between good mothering and their wives' mental vigilance:

> She's a very good mother. She worries a lot. (Peter)

> She's always concerned about how she's doing or she's always worried about if Carrie's feelings are hurt or did she say something wrong to her. (Gil)

Why is worrying associated with being a mother? I suggest two general reasons, which generate two kinds of worry. The first is that worrying is an integral part of taking care of a baby. It evokes, for example, the scheduling of medical appointments, babyproofing, or a change in the baby's diet. What I refer to as *baby worry* is generated by the question: What does the baby need? And babies need a lot. Baby worry is usually performed by mothers because they tend to be the primary caregivers; however, it can also be carried by fathers in cases in which they take primary responsibility for their babies.

The two fathers that I met who spent more time with their babies than their wives also experienced worry more typical of mothers. One of these fathers, Tom, described a subsuming of himself into the care of his child similar to what some mothers described: "If I'm going to be Hannah's dad, I have trouble with watching her all day and then going to work and trying to see who I am at the same time. So I don't think I'm doing much personal development, learning skills or anything like that. Until she gets old enough that I don't have to watch her all the time."

Other research also suggests that there is a connection between taking responsibility for physical care and baby worry (see Coltrane 1989, 1996; DeVault 1991). It may be, as Sara Ruddick (1983) suggests, that behaving like a mother makes one think like a mother. And although I have only two men as examples, I will speculate that not behaving like a father makes one not think like a father. What I mean is that the two fathers in my sample who spent more time with their babies than their wives did were men who were not particularly identified with their employment; their parental consciousness did not revolve around their planning for the future and being breadwinners.

Arnie described himself as having a checkered work history and was openly ambivalent about a job that he had switched to because the hours were better for his family (a move more typical of mothers). He described guilt about his daughter spending time with a baby-sitter in a way that was also more typical of mothers: "Well I'll tell you that's like ninety nine and nine tenths of the battle to be able to go to work and know she's not going to be propped with a bottle, and even still we got the best of the best with the sitter and it still affects you."

I asked him how it affects him, and he replied, "How in just that you want to know what's going on for your daughter and you want the best for her and you just want a good environment and I don't think there is any place in the world that I could have more comfort, but you still wonder and you still care and you want to know what's going on and you feel guilty."

This leads to the second reason that new mothers tend to worry, which is that social norms make it particularly difficult for mothers to feel that they are doing the right thing. I call this *mother worry,* and it is generated by the question: Am I being a good mother? While there may be psychological explanations for the tendency for mothers to worry, and in shared parenting families, for mothers to worry differently from fathers (see Ehrensaft 1990), what I want to emphasize here is that mother worry is induced by external mechanisms as well. That is, mothers feel connected to their children and see their children as extensions of themselves in a way that prompts worry (Ehrensaft 1990); mothers are also aware that their children are perceived by others as reflecting on them. Some of the mothers that I interviewed expressed worry about how others evaluated them as mothers:

> I think that people don't look at you and say, "oh there's a good mother," but they will look at people and say, "oh there's a bad mother." (Sarah)

> Being a mother I worry about what everyone else is going to think. (Maggie)

Perhaps Maggie worried about what people thought of her as a mother because she shared the view that mothers are ultimately responsible for children: "The behavior of the child reflects the mother's parenting . . . I mean kids, you have all these things with kids shooting people, and I blame it on . . . mothers not being around."

Baby worry and mother worry are different though related forms of worry. It is evident how baby worry can be characterized as mental labor; it is an integral part of the more obvious physical tasks involved in taking care of a baby. Worrying that a baby is cold, for example, leads to clothing the baby warmly. While the productiveness of mother worry may be less apparent, it is connected to mental labor. Worry about whether one is being a good mother reinforces mothers enacting baby worry as well as other forms of mental labor such as seeking information about baby development and illness.

Worrying is therefore not only induced by mothers' desires to be perceived as taking care of their babies correctly, it is also part of how they *do* take care of their babies. Miranda described her "stressing out" as linked to her getting things done on time; Gil's job was to tell her to lighten up: "I'm the one who stresses out more. He is very laid back. He doesn't worry about things. In fact he procrastinates. And I'm the one, run run run run run. . . . But one of us has to get things done on time and the other one has to keep the other one from totally losing it and make them be more relaxed. So it kind of balances us out."

Liza said of her interactions with Peter: "He'll say, 'Whoa it's time to go to bed' and I'll say, 'Well Peter, you know, I've got to make bottles. I've been working all day and you think they're just going to get done by themselves?' " Peter confirmed that Liza's worrying ensured care for their child: "She worries a lot. I'm probably too easygoing, but she makes sure he goes to the doctor, makes sure he has fluoride, makes sure he has all of his immunizations. She's hypervigilant to any time he might be acting sick. She's kind of that way herself. I kid her about being a hypochondriac. She makes sure he gets to bed on time, makes sure he's eating enough, whereas I'm a little more lackadaisical on that."

The two couples described above. . . presented a kind of complementarity or balance between the mother and father—the mother worried, the father didn't—his job, in fact, might be to tell her not to worry. This dynamic reinforced a gendered division of baby worry. Although there was a subtext that the mother's worrying was unnecessary or neurotic, she did not stop. In fact, the suggestion that the mother relax served to reinforce her worrying, because although she did not recognize it as work, she did recognize that worrying got things done for the baby that might not get done if no one worried.

If the father offered to share the worrying rather than telling the mother to stop, the outcome might be quite different. This was suggested to me by my observation that mothers whose husbands spent more time with their babies—and worried more than other fathers did—appeared to worry less than other mothers. But I have too few counterexamples to do anything but speculate about this. . . .

Managing the Division of Labor

I want to expand the concept of managing that has already been applied to infant care in past studies and suggest that it is not only the baby's appointments and supplies that mothers tend to manage, but their babies' fathers as well (see also Ehrensaft 1990). To use the language from *What to Expect,* "enticing" fathers into helping out with their babies is an invisible mental job performed by new mothers. Liza said, for example: "Peter is very good at helping out. If I say, 'Peter, I'm tired, I'm sick, you've got to do this for me, you've got to do that,' that's fine, he's been more than willing to do that."

Embedded in the use of the verb "help" is the notion that parenting is ultimately the mother's responsibility—that fathers are doing a favor when they parent (McMahon 1995). The default position, which is a factor in mothers' parental consciousness, is that the mother is on duty unless she asks for or is offered help. This is a state of affairs that created dissonance for some of the couples I met, and wives especially, who expected their marriages to be partnerships.

Husbands who reported that they did not do as much caregiving as their wives tended to perceive themselves as helpers to their wives, or as Richard said, the "secondary line of defense." Some fathers expressed guilt about not helping out more, admitting that they simply let their partners do more; some, who became primary caregivers while their wives were at work, relinquished the role upon their wives' return, as in the case of Brendan and Eileen. Even in situations in which fathers reported that they and their partners split tasks equally, mothers often played a role in delegating the work (Coltrane [1989] and Ehrensaft [1990] also describe "manager-helper" dynamics in couples who share child care):

> I don't change her [diaper] too often—as much as I can get out of it. (Eddie)
>
> She'll hear him when I won't sometimes or whatever, but she doesn't believe me. (Michael)
>
> Then at night either one of us will give him a bath. She'll always give him a bath, or if she can't, she'll tell me to do it because I won't do it unless she tells me, but if she asks me to do it I'll do it. (Peter)

The commentary from these fathers, who perceived that they split tasks equally with their partners, reflects a division of labor in which their female partners were the ultimate managers. They "shared" tasks with their wives—when their wives told them to.

Diaper changes were a particular area in which the work of enticing was evident:

> I mean diapering, that's hard to say. He won't volunteer, but if I say, "Honey, she needs a diaper change, could you do it?" he does it. (Whitney)

> It took me a little while to get him to change the nasty diapers...but now he changes 'em all. He's a pro. (Miranda)

Mothers also made decisions about when not to delegate:

> I do diapers. Joel can't handle it well. You know, he does diapers too, but not if there's poop in them. (Maggie)

> I'm pretty much in charge of that, which is fine, because it's really not that big of a deal. And she's more, it seems like she's easier for me than she is for him when it comes to diapering cause I just all the time do it, you know? (Nancy)

Nancy illustrates how habitual patterns can become perceived as making sense—doing becomes a kind of knowing (Daniels 1987; DeVault 1991)—just as being the one to read the book makes the mother the expert. Melissa described what is involved in feeding the baby (and her husband Brett):

> I know what has to be done. I know that like when we sit down for dinner, she [child] has to have everything cut up, and then you give it to her, you know, where he sits down and he eats his dinner. Then I have to get everything on the table, get her stuff all done. By the time I'm starting to eat, he's almost finished. Then I have to clean up and I also have to get her cleaned up and I know that like she'll always have to have a bath, and if she has to have a bath and if I need him to give it to her, "Can you do it?" I have to ask . . . because he just wouldn't do it if I didn't ask him. You know, it's just assumed that he doesn't have to do it.

While on one level it appears that women are in charge of the division of labor, the assumption of female responsibility means that, on another level, men are in charge—because it is only with their permission and cooperation that mothers can relinquish their duties. Maggie complained that Joel would leave the house while their child was taking a nap: "It's always the father that can just say, 'Okay, I'm gonna go.' Well I obviously can't leave, he's ready for a nap, you know? It's nap time. Mommy seems to always have to stay. I think that fathers have more freedom."

These kinds of statements go against suggestions that mothers may not want to relinquish control to their male partners because motherhood is a source of power for women. It may be more accurate to speak, as Coltrane (1996: 230) does, of some mothers trying to hold on to control and self-esteem by maintaining their primary responsibility for family work. I also think that the desire of mothers to be perceived as good mothers is quite powerful, and this may be what they feel they are trading off if they are not the primary caregivers. Children validate women's characters, McMahon (1995: 234) argues: "For a woman to be remiss in feeling responsible for her child would implicate her whole moral character." Note that McMahon talks about what the woman feels, not what she does.

In the context of feeling that they were ultimately responsible for their children, the women I interviewed were often satisfied with their partners' willingness to help and appreciated gestures from their husbands. Gloria, for example, talked about Jay bringing a bottle upstairs for their son before leaving for work in the morning. Mandy, another stay-at-home mother, spoke with appreciation about Chip not "holding it against" her when she goes out with friends: "He never says, 'Go, but I'm gonna remember this.' He doesn't ever do that."

When mothers delegated tasks to their husbands, men's compliance with orders was not compulsory (see also DeVault 1991). Fathers who considered themselves equal participants in the division of labor would use the fact that they were willing to do diapers as an example:

> We each will do whatever we have to do. It's not like I won't change diapers. (Chad)
>
> We tried to make it pretty equally divided. I mean I don't have any aversion to doing diapers or any of that kind of stuff. (David)

Mothers did not necessarily see any baby task as optional for them, as Sarah illustrated: "It's kind of give and take. As far as diaper changing, I think I do more . . . It's not one of his favorite tasks."

In the separation of mental and physical care of babies, women were the "bosses" in the sense that they created the organizational plan and delegated tasks to their partners. But they managed without the privileges of paid managers. If their "workers" did not do the job, they blamed themselves:

> I think I myself have a problem with relationships, with trusting, and I'm afraid to trust that he would get the things done that need to be done. But you know, I think he probably would. (Maggie)
>
> I guess I'm not demanding enough. (Ann Marie)
>
> I tend to get his stuff ready for day care and Sean could do it very easily. It's just a pattern. Part of it is my problem that I don't say, "You do it tonight." (Sylvia)

Some women just preferred to absorb the tasks themselves rather than train and compensate their partners. Mothers may not want to pay the price of having fathers help more, Ralph and Maureen LaRossa (1989: 146) suggest: "They may not be comfortable with the deferential stance they are expected to take to offset their husband's gratuities." Ironically, carrying the primary and ultimate responsibility for baby care may disempower women in relation to their husbands—leading to greater, rather than less, dependence and losses in interpersonal and economic power (Waldron and Routh 1981; Blumberg and Coleman 1989).

SELECTED REFERENCES

Belsky, Jay, and John Kelly. 1994. *The Transition to Parenthood: How a First Child Changes a Marriage.* New York: Delacorte Press.

Blumberg, Rae Lesser, and Marion Tolbert Coleman. 1989. A Theoretical Look at the Gender Balance of Power in the American Couple. *Journal of Family Issues* 10:225–250.

Coltrane, Scott. 1989. Household Labor and the Routine Production of Gender. *Social Problems* 36:473–490.

_____. 1996. *Family Man: Fatherhood, Housework, and Gender Equity.* New York: Oxford University Press.

Cowan, Carolyn Pape, and Philip A. Cowan. 1992. *When Partners Become Parents: The Big Life Change for Couples.* New York: BasicBooks.

DeVault, Marjorie L. 1991. *Feeding the Family: The Social Organization of Caring as Gendered Work.* Chicago: University of Chicago Press.

Ehrensaft, Diane. 1990. *Parenting Together: Men and Women Sharing the Care of Their Children.* Urbana: University of Illinois Press.

Hays, Sharon. 1996. *The Cultural Contradictions of Motherhood.* New Haven: Yale University Press.

Hochschild, Arlie Russell. 1983. *The Managed Heart: Commercialization of Human Feeling.* Berkeley: University of California Press.

LaRossa, Ralph, and Maureen Mulligan LaRossa. 1989. Baby Care: Fathers vs. Mothers. In *Gender in Intimate Relationships: A Microstructural Approach,* edited by Barbara J. Risman and Pepper Schwartz. Belmont, CA: Wadsworth Publishing Company.

McMahon, Martha. 1995. *Engendering Motherhood: Identity and Self-Transformation in Women's Lives.* New York: The Guilford Press.

Ritzer, George. 1983, 1988. *Sociological Theory.* New York: Alfred A. Knopf.

Ruddick, Sara. 1983. Maternal Thinking. In *Mothering: Essays in Feminist Theory.* Savage, MD: Rowman & Littlefield Publishers, Inc.

Taylor, Verta. 1996. *Rock-a-by Baby: Feminism, Self-Help, and Postpartum Depression.* New York: Routledge.

Waldron, Holly, and Donald K. Routh. 1981. The Effect of the First Child on the Marital Relationship. *Journal of Marriage and the Family* (Nov.):785–788.

15

Daddy Plants a Seed: Personhood Under Patriarchy

BARBARA KATZ ROTHMAN,
1996

For a sociologist, a day-long law conference is a challenge. It is not that the concepts are unfamiliar; they are the daily concepts of American life. It is that these concepts, these values, these ideologies that underlie American life go unchallenged in this setting. They are, in essence, the rules of the game, the delimiting system within which you must work.

From the perspective of the social scientist or social philosopher, there is an irony in watching the law attempt to address medicine. From where I sit the same social world—with all of its limitations, flaws, biases, and injustices—produces both medicine and the law. The same system that produces the technologies produces the legal structure that presumes to control them. And from where I sit, not only is it not working, but it cannot be expected to work. The rights of those individuals most in need of protection cannot be protected, neither in law nor in medicine, when approached from the perspective of individual rights. That is the built-in tension in, most especially, any feminist critique of the laws surrounding procreation.

Our legal system is founded on social contract theories, and contracts in American life are more than means. They are themselves deeply held moral values, symbols of goodness, fairness, and justice. Throughout the legal discussion of the new reproductive technologies, the language of contracts hums and buzzes. The solution to all problems appears to lie in the protection of individual liberties, and the way to accomplish that is through some form of contract protection. No matter what the problem—frozen embryos fought over or abandoned; babies torn from the breasts of recalcitrant "surrogates"; insurance companies' eugenic/cost-saving pressures to abort for "preexisting conditions" in a fetus—the solution lies in a better contract, a more informed individual, pulling clauses from the fullest array of choices.

But social contract theory is itself a development of a social philosophy developed to answer the founding question: "How is society possible?" The question is absurd. It grows out of an assumption that in the beginning, there are individuals. The argument, with its flaws, was brought most clearly to the fore in the classic contract theories, such as those of Hobbes, Mill, Spencer, and Locke, but remains current in social philosophy. The central question with which these philosophers wrestle is, How do people, as rational, self-interested individuals, come together to create

From "Daddy Plants a Seed: Personhood under Patriarchy," by Barbara Katz Rothman, *Hastings Law Journal,* vol. 47, no. 4, 1996.

social order? Although contract theory has been systematically, creatively, thoughtfully, and thoroughly critiqued, do bear in mind that it is the philosophical underpinning of American society, winding its way throughout our entire legal system, setting the basis for all our legal thinking.

But "How is society possible?" is the wrong question. I remember many years ago in an introduction to philosophy class being taught one of the classic proofs of the existence of God. Since everything is naturally still, God is needed to explain the existence of motion. But who said everything is naturally still? What is "still"? Atomic particles whir around in rocks, and all life is inherently in motion. Even in death, the body is in the motion of decomposition. Stillness is an illusion, an outcome of the limited vision of the human eye.

So it is with the question of where the social comes from. That is not what needs explaining. Like motion, the social is what there is—its absence is an illusion. We look at something like the Los Angeles riots on television, and think we are seeing the absence of social order. Look again, and we see people passing televisions to each other out of store windows; we see gangs, groups, clusters of interacting, social people. This was, of course, the key insight in early "slum" research and other Chicago School urban sociology which found patterns and organization where chaos and anarchy were thought to be. Discovering order in the world, from social life in the slum (reborn "ghetto" or "inner city") to the ecosystem of the jungle (reborn "rain forest") continues to be startling to mainstream, Eurocentric researchers, who presumably thought they had cornered the market on order.

The teasing out, recognition, naming, acknowledgement of the social order is the stock in trade of sociology; it is what we do. Many of us would argue that sociology began in earnest with Durkheim's critique of Spencer's utilitarianism, as Durkheim attempted to explain social solidarity. Durkheim, as a sociologist, argued from social structure—we are placed socially, and from that placement come the feelings that permit solidarity. Durkheim said that the utilitarian argument would not work. We could not come together as individuals to form a social contract, because without a *preceding* trust, there can be no contract. He placed the source of that trust, of our social solidarity, in our ritual coming together—the "collective conscience" or "collective representations" that are social life (Durkheim, 1964).

This argument at the level of the social is what distinguishes sociology. But I am not alone in not quite seeing how this explains why we join together, why we develop these rituals, why feelings emerge as they do in this joining together. No question, ten people joining together in song is quite different from ten people singing in their own showers. In ritual we lose ourselves in something larger than ourselves. We become part of a whole. But where is the need for that closeness grounded? Why do we seek out sociability over and over again?

Are people essentially individuals, separate beings, who must come together to form a social order? Or are people essentially interconnected beings, products of the social order? American society, American law, operates on the basis of the first assumption. I am operating on the basis of the second.

Contract theories are problematic for answering many of life's important questions, for addressing many of our value-laden concerns. But this is a theoretical, philosophical approach that is particularly badly situated for addressing issues of procreation. I will argue that the very question of where the social comes from—and all of the law that flows from that—is itself grounded in a patriarchal ideology.

I. PATRIARCHY

The term "patriarchy" is often loosely used as a synonym for "sexism" or to refer to any social system in which men rule. The term technically means "rule of fathers," though in its current practical usage it more often refers to any system of male superiority and female inferiority. But male dominance and patriarchal rule are not quite the same thing, and the distinction is important.

Patriarchal kinship is the core of what is meant by patriarchy—the idea that paternity is the central social relationship. A very clear statement of patriarchal kinship is found in the book of Genesis, in the "begets." Each man, from Adam onward, is described as having "begotten a son in his likeness, after his image." After the birth of this firstborn son, the men are described as having lived so many years and begotten sons and daughters. The text then turns to that firstborn son, and in turn, his firstborn son after him. Women appear as "the daughters of men who bore them offspring." In a patriarchal kinship system, children are reckoned as being born to men, out of women. Women, in this system, bear the children of men.

The essential concept here is the "seed," the part of man that grows into the child of his likeness within the body of woman. Such a system is inevitably male-dominated, but it is a particular kind of male domination. Men control women as daughters, much as they control their sons. But they also control women as the mothers of men's children. It is a woman's motherhood that men must control to maintain patriarchy. Any reading of the history of family law lays bare these control mechanisms. In a patriarchy, because what is valued is the relationship of a man to his son, women are a vulnerability that men have; to beget these sons, men must pass their seed through the body of a woman.

While all societies appear to be male-dominated to some degree, not all are patriarchal. In some, the line of descent is not from father to son, but along the line of women. In these matrilineal societies, it is a shared mother that makes for a lineage or family group. Men still rule in these groups, but they do not rule as fathers. They rule the women and children who are related to them through their mother's line. Women in such a system are not a vulnerability, but a source of connection. People are not men's children coming through the bodies of women, but the children of women.

Let me put this in everyday language. In Western, patriarchal societies, the classic where-do-babies-come-from tale we tell children is a variation on "Daddy plants a seed in Mommy." Contrast this with a tale Pearl Buck wrote for children. Johnny wants to know where he came from. You were in me, his mother explains. But BEFORE, Johnny wants to know. You were *always* in me, his mother explains. When I was in my mother, you were in me. When she was in her mother, you were in me. You were always in me (Buck, 1954). That is matrilineal thinking.

Modern thinking is no longer classically patriarchal. We have acknowledged that women have seeds too, and have extended to women some of the privileges of patriarchy. Women are recognized as also being connected to their children, through their seed. This modified patriarchy is not at all like matrilineal thinking. It maintains absolutely the primacy of the seed. Children are "half his, half hers," we say—and might as well have grown in the backyard. Or in a "gestational host," an animal host or a machine.

In patriarchal thinking, including our own modified version, when people talk about "blood ties," they are talking about a genetic tie—the only truly bloodless part of procreation, a connection by seed. In a mother-based system, the blood tie is the mingled blood of mothers and their children. Children grow out of the blood of their mothers, of their bodies and being. The maternal tie is based on the growing of children; the patriarchal tie is based on genetics, the seed connection.

Each of these ways of thinking leads to different ideas about what a person is, and ultimately, what society itself is. In a mother-based system, a person is what mothers grow. People are made of the care and nurturance that bring a baby forth into the world and turn that baby into a member of society. In a patriarchal system, a person is what grows out of a seed; originally a man's seed, but now expanded to the sex-neutral language of "gametes." The essence of what a person is, in patriarchal thinking, is there when the seed is planted. Motherhood becomes, in such thinking, a place. Providing the place becomes a service.

Under patriarchy, the place in which the seed grows does not really matter. It can be a wife, a "surrogate," or an artificial womb.

Such a system brings us the dismissal of the significance of nurturance. In a classic patriarchy, men—and particularly men of the upper classes—had the rights to hire whatever services and to use whatever women they needed in order to achieve paternity. In some times and places, these men have used their wives for procreation and kept mistresses; in others, wives assumed more of the role of mistress, while wet nurses, nannies, governesses, and child-tenders took on the nurturance needed to accomplish paternity.

In our own modified patriarchy, women—and particularly women of the upper classes—are acquiring many of the privileges of patriarchy. How will the developing technologies of procreation be used in this context? You have only to look at the poor women of color tending their white affluent charges in the playgrounds of every American city to understand which women will be carrying valued white babies in their bellies as a cheap service.

This expansion of the privileges of patriarchy to women is not the unique contradiction it might appear, but flows directly out of liberal thinking and liberal progress. The social history of America has been an expansion of the category of "individual"—first a truer, and then a broader interpretation of "all men" who are created equal. It is this expansion—from white men of property to all white men to all men to women—that has been the saving grace of the American system. The principle of the contract is upheld; the definition of the individual who may enter into contract is expanded.

When applied to issues of procreation, procreative law, and procreative technology, the contradiction emerges. The liberal theorists in this area assure us that these technologies pose no fundamental threat to family law as we understand it, and they tell us that this technology will expand the "reproductive options" of unpartnered men and women, of gay men and lesbians. They are, oddly, right. The fundamentals—the primacy of the seed and the fungibility of nurturance—are maintained. The definition of the individuals who may use nurturance services to grow their gametes into children of their likeness expands. In this arena as well, the principle of the contract is upheld, and the definition of the individual expanded.

II. BEYOND PATRIARCHY

People do not begin as separate beings, disparate individuals, scattered gametes. We begin as parts of our mothers' bodies. We don't, as the language of patriarchy would have it, "enter the world" or "arrive." From where? Women who give birth, I have often pointed out, don't feel babies arrive. We feel them leave. But the very language we have for understanding the origins of people, for explaining where babies come from, is the language of patriarchy. Nurturance is not a service provided to gamete owners, but the fundamental human condition.

What does liberal, individualist theory, in which our legal, social, and political institutions are grounded, tell us about this connectedness? What are the origin theories that accompany the social contract theory? Consider what Hobbes had to say: It is as if we spring up like mushrooms (Hobbes, 1841). Put as baldly as that, or implicitly stated, the social contract approach is based on the emergence of people as fully formed individuals: rational, self-interested, and ready to come together. And from where do these people emerge? They emerge from the bodies, kitchens, and lives of women: out of the "private" and into the "public" world.

People do not spring forth asocially, like mushrooms. We are conceived inside human bodies. We come forth after months of hearing voices, feeling the rhythm of the body, cradled in the pelvic rock of our mother's walk. We move from inside the body to outside. Right outside. We spend years in intimate physical contact with other bodies. Being cradled, carried, held, suckled. That is who we are and how we got to be who we are. We are not separate beings who must learn to cope with others, but attached beings who learn how to separate. Cradle a child, sit by the side of someone frightened or in pain or dying—we cling to one another. For a few moments, now and again, when everything is going okay, it is possible to hold on to the illusion that we are separate individuals. But the connectedness is the reality, the separation the illusion. We seek the connectedness in which we are grounded. We are social.

So what does this tell us about the reproductive technologies and reproductive laws we develop? Our patriarchal ideology tells us that the essence of humanity lies in our seeds, that what makes a person is fundamentally there in that seed from which the person springs. The person whose seed it is, is the person whose child it is; reproductive technologies focus on getting particular persons' seeds to grow. On the one hand, this brings us such technologies as microinjection of sperm. On the other hand, it brings us the relatively crude application of prenatal diagnosis: you start a pregnancy, you test the seed, and then you keep it or abort it.

But we are developing ever more sophisticated technologies out of this ideology, mechanisms to sort, read, and manipulate seeds. As industrial society took the machine as its model for the body, post-industrial society takes the computer, and sees the seed as containing not the entire object within (the homunculus) but the "program." And programs, we think, can be fixed. New technologies of procreation have moved the seed outside of the body, brought it into the laboratory and made it manipulable. The genetic engineering of human beings, all the way up and down the slippery slope, is coming our way.

Ideologies bring us technologies, and technologies create the world in their own image. That is the context in which I see both the reproductive technologies and the growing body of law surrounding them—one part of patriarchal ideology creating the world in its own image. But that image is one of scattered, separate individuals, an image in which social life is problematic, in which connection requires some explanation. We are moving in the direction of creating a world in which Hobbes will be right; we will spring up like mushrooms. There are already eager scientists actively pursuing the artificial womb, the totally controlled environment in which to grow the "perfect" engineered embryos.

But if Hobbes might turn out to be right, so too will Durkheim. Hobbes' world will not work. Without the trust, the social solidarity, the fabric of connectedness, the social order is not possible.

The answer offered within the American legal system is not sufficient to protect us in this. More protection of individual rights, more contracts, more informed consent—these are all necessary, but far from sufficient to address the far reaching implications if reproductive technology continues to recreate the world in its own image.

SELECTED REFERENCES

Buck, Pearl S. 1954. *Johnny Jack and His Beginnings.* New York: J. Day Co.

Durkheim, Emile. [1933] 1964. *The Division of Labor in Society.* Translated by G. Simpson. New York: Free Press of Glencoe.

Hobbes, Thomas. 1841. "Philosophical Rudiments Concerning Government and Society." In *The English Works of Thomas Hobbes of Malmesbury 1,* edited by W. Molesworth. London, UK: J. Bohn.

CHAPTER 5

Childhood Socialization

How are children socialized to fit in to society? Older models of child development assumed that socialization was a one-way process in which children passed through various age-graded stages with parents making them conform to a fixed set of cultural standards. More recently, social scientists have documented how standards of behavior vary across groups, how parental standards and goals undergo change, how children socialize parents (as well as the reverse), and how children learn different things from their parents than what they are consciously trying to teach them. Most research now recognizes that **socialization** is complex and bidirectional, and that factors such as social class, neighborhood, ethnicity, schools, churches, media, extended family, and peer groups play an important role in children's acquisition of the skills, knowledge, and dispositions necessary to become successful adult members of society. The readings in this chapter call attention to the ways that social institutions shape the development of children and place them into the larger social system.

The history of childhood socialization reveals two contradictory tendencies: at the same time that parenting has become more individualistic, parents seem to exercise less control over their children. Without question, one of the biggest changes in childhood socialization practices in America over the past two centuries is the movement toward more individualized and privatized parenting: that is, individual parents assume more responsibility for the care of their offspring and consequently are also blamed for how their children turn out. In the past, extended kin, neighbors, and other local adults collectively were much more involved in regulating the behavior of young people and also assumed more responsibility for how they behaved as they made the transition to adulthood. Child socialization, in this sense, has become more privatized and is now assumed to be the province of the **nuclear family,** or even more narrowly, of the parent who has responsibility for the child—in most cases the mother.

At the same time, however, mass influences on the growth and development of children have increased. With the modern advent of compulsory public education, children became exposed to more similar types of instruction and training (at least those children whose parents could not afford to send them to private schools). Not surprisingly, the rise in compulsory education paralleled a phenomenal growth in the industrial economy. As manufacturing increased, children were less needed on the farm or in family trades, and they were subject to school routines that taught punctuality and conformity to rules, traits highly valued in the emerging wage labor economy. More recently, children are spending more time watching television than interacting with teachers or family members. Mass media exposes children to standardized styles, values, and stereotypes, thereby conditioning them to be passively entertained mass consumers. The simultaneous promotion of individualized parenting and mass consumption seems contradictory but, from a sociological perspective, reflects both the tensions and preoccupations of our society.

The articles in this chapter describe how parenting practices and cultural ideals socialize children to assume their places in society. In the first selection, anthropologist Margaret Mead draws on research from the 1930s and 1940s to address questions about the fit between a society's gender ideals and the temperament of males and females (Mead 1935/1963, 1949). Her analysis describes how children in every society face the task of understanding gender categories and applying them to the self. Mead's cross-cultural studies focus simultaneously on **sex** (the physiological aspects of being male or female) and **gender** (the social aspects of being masculine or feminine), but her conceptual discussions prefigure later social scientific distinctions between the two. A key to her approach is the recognition that all men and all women have variable degrees of what we now call masculinity and femininity. What matters, according to Mead, is which traits or physical characteristics are valued by the society into which one is born, the place on the gender or sexual-body-type continuum where one falls, and the amount of latitude in gender or sexual display allowed by the particular society. In the selection reproduced here, Mead argues for recognizing variation in both gender and sexual characteristics across cultures and within groups of men and women, with special attention to the ways children conform to those expectations.

Mead refers to her earlier research on three New Guinea tribes to show how gender stereotypes relate to modern social prejudices. One of the tribes she studied—the docile mountain-dwelling Arapesh—had little gender differentiation, with both men and women conforming to Westerners' expectations for feminine behavior. They were a gentle people who cherished children, and both men and women exhibited sensitive and nurturing behaviors. The Mundugumor, a neighboring tribe rumored to be fierce headhunters, in contrast, were very competitive and antagonistic. Both men and women conformed to Westerners' expectations for masculine behavior, insofar as they were independent, aggressive, callous, and rejecting of children. The third group Mead studied in New Guinea, the Tchambuli, seemed to reverse the typical gender expectation of Westerners. Tchambuli women were described as brisk, businesslike, and coolly cooperative, routinely making decisions that affected men's lives. Tchambuli men, in contrast, were characterized as sensitive and flighty, traits that Westerners would describe as feminine (Coltrane, 1996, pp. 178–79). According to Mead, Tchambuli men were preoccupied with trivial gossip, tended to be moody, and were subject to sudden fits of jealousy. Prefiguring a **social constructionist** approach to gender, Mead concluded that "many, if not all, of the personality traits which we have called masculine or feminine are as lightly linked to sex as are clothing, the manners, and the form of head-dress that a society at a given period assigns to either sex" (Mead, 1935, p. 280).

The second article in this chapter explores historical changes in the ideology of childhood. Betty Farrell reminds us that human infants need adult care for a longer time than any other species, but she suggests that this dependence leads to substantial variability in beliefs about children and childhood. She focuses on the fact that children have not always been seen as innocent beings in need of vigilant parental protection. Seventeenth-century Puritans believed that children were naturally depraved and therefore parents felt compelled to mete out strict discipline in an effort to gain their salvation. Farrell discusses how the demographic facts of large households, close living quarters, short birth intervals, economic necessity, and high mortality led to relatively unsentimental relations with children in earlier historical eras. She describes how a middle-class model of the vulnerable and defenseless child who required a mother's full-time care gained cultural dominance during the nineteenth century. She traces the development of the "priceless" child, remarking that an unintended consequence of the sentimentalization of childhood has been the commercialization of the child, both as object and as consumer. Finally, Farrell documents how threats to children have emerged as a powerful symbol of the innocence of childhood at the same time that our social institutions increasingly neglect the needs of real children. As you read this selection, think about the ways that Farrell's argument is similar to, or different from, Illouz' discussion of romance in chapter 2.

In the last selection in this chapter, Michele Adams and I focus on the personal and social implications of parents treating sons and daughters differently. This reading illustrates how gender **stratification** in the larger society is related to socialization processes. We show how families encourage boys to be masculine by rejecting things feminine, which tends to drive them away from families. The corollary is that feminine ideals tend to enmesh girls within families. We document some of the troubles faced by boys as they attempt to become men and highlight problems they encounter when they attempt to start families of their own. The dilemmas men face reconciling their ideals of masculinity with their positions as husbands and fathers are indicative of larger social issues addressed in other selections within this book. This article, like the one from Margaret Mead, focuses on the ways that children's gender socialization is influenced by cultural expectations. And like the Farrell article and the Kohn article in chapter 8, this analysis reveals how parents' child-rearing values are linked to larger social structures and patterns of age, class, and race stratification. As you read these selections, think about whether your own experiences confirm or contradict what these authors are saying. Do you think people can change socialization practices on their own?

REFERENCES

Coltrane, Scott. 1996. *Family Man: Fatherhood, Housework, and Gender Equity.* New York: Oxford University Press.

Mead, Margaret. 1935/1963. *Sex and Temperament in Three Primitive Societies.* New York: William Morrow.

Mead, Margaret. 1949. *Male and Female.* New York: William Morrow.

SUGGESTED READINGS

Adler, Patricia A., and Peter Adler. 1998. *Peer Power: Preadolescent Culture and Identity.* New Brunswick, NJ: Rutgers University Press.

Bottoms, Bette L., Margaret Bull Kovera, and Bradley D. McAuliff, eds. 2002. *Children, Social Science, and the Law.* Cambridge, UK: Cambridge University Press.

Corsaro, William A. 1997. *The Sociology of Childhood*. Thousand Oaks, CA: Pine Forge Press.

Eder, Donna. 1995. *School Talk: Gender and Adolescent Culture*. New Brunswick, NJ: Rutgers University Press.

James, Allison, and Alan Prout, eds. 1997. *Constructing and Reconstructing Childhood: Contemporary Issues in the Sociological Study of Childhood*. 2nd ed. London; Washington, D.C.: Falmer Press.

Maccoby, Eleanor E. 1998. *The Two Sexes: Growing Up Apart, Coming Together*. Cambridge, MA: Belknap Press of Harvard University Press.

Steinberg, Shirley R., and Joe L. Kincheloe, eds. 1997. *Kinderculture: The Corporate Construction of Childhood*. Boulder, CO: Westview Press.

Thorne, Barrie. 1993. *Gender Play: Girls and Boys in School*. New Brunswick, NJ: Rutgers University Press.

INFOTRAC® COLLEGE EDITION EXERCISES

The exercises that follow allow you to use the InfoTrac® College Edition online database of scholarly articles to explore the sociological implications of the selections in this chapter.

Search Keywords: Childhood and innocence. Draw on the InfoTrac® College Edition database to find articles that discuss the relationship of childhood and innocence. Why do we consider childhood to be a time of innocence? How does the ideal of childhood innocence match the reality of most children's lives?

Search Keyword: Child-rearing practices. How do various groups bring up their children in different ways? Search for articles that discuss child rearing among different groups and think about the effects of different child-rearing practices on children's socialization.

Search Keyword: Gender socialization. Find articles that deal with the topic of children's gender socialization. According to these articles, what are some of the ways that children are socialized into existing social attitudes about gender? What are the results or outcomes of these processes?

Search Keywords: Children as consumers. Look for articles in InfoTrac® College Edition that discuss the issue of children as consumers. What types of products are promoted to children? How has childhood been used to promote consumerism over the long term? How does consumerism support continuation of the existing class structure?

Search Keywords: Children and television. How many articles can you find that deal with the impact of television on children? What are some of those impacts? In what ways are these impacts beneficial to children? In what ways are they detrimental?

Search Keyword: Violent children. Use InfoTrac® College Edition to search for articles that discuss violence perpetrated by children. According to these articles, to what extent are children responsible for their violent behaviors? To what extent is society responsible? What are some of the social factors that could contribute to children's violence?

16

Sex and Temperament

MARGARET MEAD,
1949

It is not enough for a child to decide simply and fully that it belongs to its own sex, is anatomically a male or a female, with a given reproductive rôle in the world. For growing children are faced with another problem: "*How male,* how female, am I?" He hears men branded as feminine, women condemned as masculine, others extolled as real men, and as true women. He hears occupations labelled as more or less manly, for a man, or more or less likely to derogate her womanhood, for a woman. He hears types of responsiveness, fastidiousness, sensitivity, guts, stoicism, and endurance voted as belonging to one sex rather than the other. In his world he sees not a single model but many as he measures himself against them; so that he will judge himself, and feel proud and secure, worried and inferior and uncertain, or despairing and ready to give up the task altogether.

In any human group it is possible to arrange men and women on a scale in such a way that between a most masculine group and a most feminine group there will be others who seem to fall in the middle, to display fewer of the pronounced physical features that are more characteristic of one sex than of the other. This is so whether one deals entirely in secondary sex characters, such as arrangement of pubic hair, beard, layers of fat, and so on, or whether one deals with such primary sex characters as breasts, pelvic measurements, hip-torso proportions and so on. These differences are even more conspicuous when one considers such matters as skin sensitivity, depth of voice, modulation of movement. Also, one finds in most groups of any size that there are very few individuals who insist on playing the rôle of the opposite sex in occupation or dress or interpersonal sex activities. Whether full transvestitism will occur seems to be a question of cultural recognition of this possibility. Among many American Indian tribes the *berdache,* the man who dressed and lived as a woman, was a recognized social institution, counterpointed to the excessive emphasis upon bravery and hardiness for men. In other parts of the world, such as the South Pacific, although a large number of ritual reversals of sex on ceremonial occasions may occur, there are many tribes where there is no expectation that any single individual will make the complete shift. Peoples may provide sex-reversal rôles for both sexes—as among the Siberian aborigines, where sex reversal is associated with shamanism; they may permit it to men but deny it to women; or they may not provide any pattern at all. But between the conspicuous transvestitism of the Mohave Indians (Devereux,

1937)—where the transvestite men mimic pregnancy and child-birth, going aside from the camp to be ceremonially delivered of stones—and the Samoans—who recognize no transvestitism, but among whom I found one boy who preferred to sit among the women and weave mats—the contrast is clearly one of social patterning. A society can provide elaborate rôles that will attract many individuals who would never spontaneously seek them. Fear that boys will be feminine in behaviour may drive many boys into taking refuge in explicit femininity. Identification of a little less hairiness on the chin, or a slightly straighter bust-line, as fitting one for membership in the opposite sex may create social deviance. If we are to interpret these experiences, which all children have, we look for some theory of what these differences mean.

We strip away all this superstructure when we have invoked the presence or absence, the recognition and toleration, of transvestite social institutions, or the explicit suppression of homosexual practice, but we still find differences that need explanation. After we have gathered together the insights from detailed case-histories in Western society that show how accidents of upbringing, faulty identifications with the wrong parent, or excessive fear of the parent of the opposite sex may drive both boys and girls into sexual inversion, still we are left with a basic problem. Set end to end, standing in a line, the men of any group will show a range in explicit masculinity of appearance as well as in masculinity of behaviour. The females of any group will show a comparable variety, even more, in fact, if we have X-ray pictures to add to their deceptive pelvic profiles, which do not reveal their feminine reproductive capacities accurately (Greulich and Thoms, 1939). Is this apparent range to be set down to differences in endocrine balance, set against our recognition that each sex depends for full functioning upon both male and female hormones and the interaction between these hormones and the other endocrines? Has every individual a bisexual potential that may be physiologically evoked by hormone deficit or surplus, which may be psychologically evoked by abnormalities in the process of individual maturation, which may be sociologically invoked by rearing boys with women only, or segregating boys away from women entirely, or by prescribing and encouraging various forms of social inversion? When human beings—or rats—are conditioned by social circumstances to respond sexually to members of their own sex as adults and in preference to members of the opposite sex, is this conditioning playing on a real bisexual base in the personality, which varies greatly in its structure as between one member of a group and another?

At first blush, it seems exceedingly likely that we have to advance some such hypothesis. If one looks at a group of little boys, it would seem fairly obvious that it would be easier to condition those who now appear "girlish" to an inverted rôle, and that from a group of little girls, the "boyish" girl would be the easiest to train into identification with the opposite sex. And does not "easiest" here mean the greatest degree of physical bisexuality? Yet the existing data make us pause. The most careful research has failed to tie up endocrine balance with actual homosexual behaviour. Those rare creatures who have both male and female primary sex organs present of course major anomalies and confusions, but so far they have thrown little light on the general problem. The extraordinary lack of correlation between physique that can be regarded as hypermasculine and hyperfeminine and successful reproductivity is marked in every group. The man who shows the most male characteristics may have no children, while some pallid, feminine-looking mouse of a man fathers a large brood. The woman with ample bosom and wide hips may be sterile, or if she bears children she may be incapable of suckling them. Yet we are still continually confronted with what looks like a correlation between the tendency towards sexual inversion of the men and women who deviate most towards the expected physique of the opposite sex. In the primitive tribe that does not recognize inversion, the boy who decides to make mats will look more like the female type for that tribe, the woman who goes out hunting will tend to look more like the male. Does this apparent physical correspondence mean nothing, is it sheerly an accident within a normal range of variation? If the tribe sets hairiness up as a desirable male characteristic, will the less hairy become confused about their sex rôle, while if the tribemen think that hairiness is simply a brutish characteristic, the very hairy may be almost sexually ostracized and the most hairless will not thereby be regarded as less male? This would be the extreme environmental answer, while the invocation of some very subtle, as yet unplumbed structural and functional variation in the biological basis of sex membership would be the extreme genetic answer.

I suggest another hypothesis that seems to me to fit better the behaviour of the seven South Seas peoples whom I have studied. A Balinese male is almost hairless—so hairless that he can pluck his whiskers out one by one with pincers. His breasts are considerably more developed than are a Westerner's. Almost any Balinese male placed in a series of western-European males would look "feminine." A Balinese female, on the other hand, has narrow hips and small high breasts, and almost any Balinese female placed among a group of western-European women would look "boyish." Many of them might be suspected of being unable to suckle children, perhaps accused of having infantile uteruses. But should these facts be interpreted to mean that the Balinese is more bisexual, less sexually differentiated, than the western European, that the men are less masculine, the women less feminine, or simply that the Balinese type of masculinity and femininity is different? The extreme advocates of a varying bisexual balance would claim that in some races the men are less differentiated, are more feminine and so on, than in others, and might also apply the same argument to the women. But on the whole, it would be agreed that at least some of the respects in which a Balinese male would seem feminine are matters that do not really affect his masculinity at all: his height, girth, hairiness, and the like. So it might be fairly readily admitted that as between racial strains that vary as greatly as Balinese and northern Europeans, Andamanese pygmies and Nubian giants, not only would certain of the criteria for masculinity and femininity be inoperative, but also that actual cross-correspondence might occur, as all Andamanese males would fall within the height range for females in some much taller group.

But all human groups of which we have any knowledge show evidence of considerable variation in their biological inheritance. Even among the most inbred and isolated groups, very marked differences in physique and apparent temperament will be found, and despite the high degree of uniformity that characterizes the child-rearing practices of many primitive tribes, each adult will appear as more or less masculine, or more or less feminine, according to the standards of that particular tribe. There will be, furthermore, orders of variation that seem, at least on inspection—for we have no detailed records—to apply from one group to another. Although almost every Balinese would fall within the general configuration that might be classified technically as asthenic, yet the asthenic Balinese continues to contrast with the Balinese who is heavier in bony structure, or shorter and plumper. Within the limits set by the general type, these same differences occur, in both men and women. Not until we have far more delicate methods of measurement, which allow not only for individual constitution but for ancestral strains, will we have any way of knowing whether there is any genuine correspondence, on a behavioural level, among the slender, narrow-bodied of the Arapesh, Tchambuli, Swede, Eskimo, and Hottentot, or whether their behaviour, although possibly in some way constitutionally based, is still in no way referable to something they may be said to have in common. Until such measures are developed, and such studies made, one can only speculate on the basis of careful observation, with no better instrument for comparison than the human eye. But use of this instrument on seven different peoples has suggested to me the hypothesis that within each human group we will find, probably in different proportions and possibly not always in all, representatives of the same constitutional types that we are beginning to distinguish in our own population. And I further suggest that the presence of these contrasting constitutional types is an important condition in children's estimate of the completeness of their sex membership.

If we recognized the presence of comparable ranges of constitutional types in each human society, any single continuum that we now construct from the most masculine to the least masculine can be seen to be misleading, especially to the eye of the growing child. We should instead define a series of continuums, distinguishing between the most masculine and the least sexually differentiated male within each of these several types. The slender little man without beard or muscle who begets a whole brood of children would not then seem such an anomaly, but could be regarded as the masculine version of a human type in which both sexes are slender, small, and relatively hairless. The tall girl whose breasts are scarcely discernible, but who is able to suckle her baby perfectly satisfactorily as her milk seems to spread in an almost even line across her chest, will be seen not as an imperfectly developed female—a diagnosis that is contradicted by the successful way in which she bears and suckles children, and her beautiful carriage in pregnancy—but as the female of a particular constitutional type in which women's breasts are much smaller and less accentuated. The big he-man with hair on his chest, whose masculinity is so often claimed to be pallid and unconvincing, will be seen to be merely a less masculine version of a type in which enormous muscularity and hairiness are the mode. The woman whose low fertility contrasts so strangely with her billowing breasts and hips may be seen as only one of a type of woman with very highly emphasized breasts and hips—her low fertility only conspicuous because most of the women with whom she is compared have smaller bosoms and less full hips. The apparent contradiction between pelvic X-rays and external pelvic measurements might also be resolved if it were considered from this point of view.

And as with physical type, so with other aspects of personality. The fiery, initiating woman would be classified only with fiery, initiating men of her own type, and might be found to look like not a lion, but merely like a lioness in her proper setting. When the meek little Caspar Milquetoast was placed side by side not with a prize-fighter, but with the meekest female version of himself, he might be seen to be much more masculine than she. The plump man with soft breast-tissue, double chin, protruding buttocks, whom one has only to put in a bonnet to make him like a woman, when put beside the equally plump woman will be seen not to have such ambiguous outlines after all; his masculinity is still indubitable when contrasted with the female of his own kind instead of with the male of another kind. And the slender male and female dancers, hipless and breastless, will seem not a feminine male and a boyish female, but male and female of a special type. Just as one would not be able to identify the sex of a male rabbit by comparing its behaviour with that of a lion, a stag, or a peacock as well as by comparing rabbit buck with doe, lion with lioness, stag with doe, and peacock with peahen—so it may well be that if we could disabuse our minds of the habits of lumping all males together and all females together and worrying about the beards of the one and the breasts of the other, and look instead for males and females of different types, we would present to children a much more intelligible problem. . . .

The growing child in any society is confronted then by individuals—adults and adolescents and children—who are classified by his society into two groups, males and females, in terms of their most conspicuous primary sex characters, but who actually show great range and variety both in physique and in behaviour. Because primary sex differences are of such enormous importance, shaping so determinatively the child's experience of the world through its own body and the responses of others to its sex membership, most children take maleness or femaleness as their first identification of themselves. But once this identification is made, the growing child then begins to compare itself not only in physique, but even more importantly in impulse and interest, with those about it. Are all of its interests those of its own sex? "I am a boy," but "I love colour, and colour is something that interests only women." "I am a girl," but "I am fleet of foot and love to run and leap. Running and leaping, and shooting arrows, are for boys, not girls." "I am a boy," but "I love to run soft materials through my fingers; an interest in touch is feminine, and will unsex me." "I am a girl," but "My fingers are clumsy, better at handling an axe-handle than at stringing beads; axe-handles are for men." So the child, experiencing itself, is forced to reject such parts of its particular biological inheritance as conflict sharply with the sex stereotype of its culture.

Moreover, a sex stereotype that decrees the interests and occupations of each sex is usually not completely without a basis. The idea of the male in a given society may conform very closely to the temperament of some one type of male. The idea of the female *may* conform to the female who belongs to the same type, or instead to the female of some other type. For the children who do not belong to these preferred types, only the primary sex characters will be definitive in helping them to classify themselves. Their impulses, their preferences, and later much of their physique will be aberrant. They will be doomed throughout life to sit among the other members of their sex feeling less a man, or less a woman, simply because the cultural ideal is based on a different set of clues, a set of clues no less valid, but different. And the small rabbit man sits sadly, comparing himself with a lionlike male beside whom he is surely not male, and perhaps for that reason alone yearning forever after the lioness woman. Meanwhile the lioness woman, convicted in her inmost soul of lack of femininity when she compares herself with the rabbity little women about her, may in reverse despair decide that she might as well go the whole way and take a rabbity husband. Or the little rabbity man who would have been so gently fierce and definitely masculine if he had been bred in a culture that recognized him as fully male, and quite able to take a mate and fight for her and keep her, may give up altogether and dub himself a female and become a true invert, attaching himself to some male who possesses the magnificent qualities that have been denied him.

Sometimes one has the opportunity to observe two men of comparable physique and behaviour, both artists or musicians, one of whom has placed himself as fully male and with brightly shining hair and gleaming eye can make a roomful of women feel more feminine because he has entered the room. The other has identified himself as a lover of men, and his eye contains no gleam and his step no sureness, but instead an apologetic adaptation when he enters a group of women. And yet, in physical measurement, in tastes, in quality of mind, the two men may be almost interchangeable. One, however, has been presented for example, with a frontier setting, the other with a cosmopolitan European one; one with a world where a man never handles anything except a gun, a hunting-knife, or a riding-whip, the other with a world where men play the most delicate musical instruments. When one studies a pair such as this, it seems much more fruitful to look not at some possible endocrine difference, but rather at the discrepancy, so much more manifest to one than to the other, between his own life preferences and those which his society thinks appropriate for males.

If there are such genuine differences among constitutional types that maleness for one may be so very different from maleness for another, and even appear to have attributes of femaleness—as found in some other type—this has profound implications not only for interpretation of variation within each sex, and for the forms of inversion and sex failure that occur in any society, but also for the pattern of inter-relationships between the sexes. Some simple societies, and some castes within complex societies, seem to have chosen their sex ideals for both sexes from the same constitutional type. The aristocracy, or the cattlemen, or the shopkeeper class may cherish as the ideal the delicate, small-boned, sensitive type for both males and females, or the tall, fiery, infinitely proud, specifically nervously sexed man and woman, or the plump, placid man and woman. But we do not know whether the male and female ideals of a given culture complement each other in this way. When the ideals for the two sexes do seem to be consistently interrelated, it is probable that a more finely meshed, more biologically direct relationship can be established as the ideal marriage, and the marriage forms will have a greater consistency. When those men and women who do not conform to the ideal type try to use the marriage forms—the delicate interwoven ballet, or the fierce proud reserve, or the comfortable post-prandial hot milk, which have become the appropriate and developed forms for that ideal type—they are at least faced with a consistent though alien pattern, which it may be easier to learn.

Let us imagine for instance an aristocracy in which for both men and women the ideal is tall, fiery-tempered, proud, specifically and very sensitively sexed. Into such an aristocratic household is born a boy who is plump, easy-going, fond of eating, diffusely sexed. All through his childhood he will be trained in the behaviour appropriate to a type very different from himself, and this will include accepting as his feminine ideal a girl who is fiery-tempered, reserved, specifically sexed. If he marries such a girl, he will have learned a good part of his proper rôle, which she in turn will have learned to expect of him. If he marries a girl who deviates as much as he from the expected standards, each will nevertheless have learned a consistent rôle, he to treat her as if she were sensitive and proud, she to treat him as if he were sensitive and proud. Their life may have more artificiality in it than that of those who actually approximate the types for which the cultural rôles are designed, but the very clarity of the pattern of male and female rôles may make them rôles that can be played. In every such tightly patterned picture there will be some who will rebel, will commit suicide—if suicide is a culturally recognized way out—will become promiscuous or frigid or withdrawn or insane, or, if they are gifted, will become innovators of some variation in the pattern. But most of them will learn the pattern, alien though it be.

So in each of the societies I have studied it has been possible to distinguish those who deviated most sharply from the expected physique and behaviour, and who made different sorts of adjustment, dependent upon the relationship between own constitutional type and cultural ideal. The boy who will grow up into a tall, proud, restive man whose very pride makes him sensitive and liable to confusion suffers a very different fate in Bali, Samoa, Arapesh, and Manus. In Manus, he takes refuge in the vestiges of rank the Manus retain, takes more interest in ceremonial than in trading, mixes the polemics of acceptable trading invectives with much deeper anger. In Samoa such a man is regarded as too violent to be trusted with the hardship of a family for many, many years; the village waits until his capacity for anger and intense feeling has been worn down by years of erosive soft resistance to his unseemly over-emphases. In Bali, such a man may take more initiative than his fellows only to be thrown back into sulkiness and confusion, unable to carry it through. Among the Maori of New Zealand, it is probable that he would have been the cultural ideal, his capacity for pride matched by the demand for pride, his violence by the demand for violence, and his capacity for fierce gentleness also given perfect expression, since the ideal woman was as proud and fiercely gentle as himself.

But in complex modern societies, there are no such clear expectations, no such perfectly paired expectancies, even for one class or occupational group or rural region. The stereotyped rôles for men and for women do not necessarily correspond, and whatever type of man is the ideal, there is little likelihood that the corresponding female type will also be the ideal. Accidents of migration, of cross-class marriage, of frontier conditions, may take the clues for the female ideal from quite another type from which the male ideal is taken. The stereotype may itself be blurred and confused by several different expectations, and then split again, so that the ideal lover is not the ideal brother or husband. The pattern of inter-relationships between the sexes, of reserve or intimacy, advance or retreat, initiative and response, may be a blend of several biologically congruent types of behaviour instead of clearly related to one. We need much more material on the extent to which this sort of constitutional types may actually be identified and studied before we can answer the next questions about the differential strength and stability and flexibility of cultures in which ideals are a blend, or a composite, or a single lyric theme, ideals that are so inclusive that every male and female finds a rather blurrily defined place within them, or so sharp and narrow that many males and females have to develop counterpointed patterns outside them.

A recognition of these possibilities would change a great deal of our present-day practices of rearing children. We would cease to describe the behaviour of the boy who showed an interest in occupations regarded as female, or a greater sensitivity than his fellows, as "on the female" side, and could ask instead what kind of male he was going to be. We would take instead the primary fact of sex membership as a cross-constitutional classification, just as on a wider scale the fact of sex can be used to classify together male rabbits and male lions and male deer, but would never be permitted to obscure for us their essential rabbit, lion, and deer characteristics. Then the little girl who shows a greater need to take things apart than most of the other little girls need not be classified as a female of a certain kind. In such a world, no child would be forced to deny its sex membership because it was shorter or taller, or thinner or plumper, less hairy or more hairy, than another, nor would any child have to pay with a loss of its sense of its sex membership for the special gifts that made it, though a boy, have a delicate sense of touch, or, though a girl, ride a horse with fierce sureness.

If we are to provide the impetus for surmounting the trials and obstacles of this most difficult period in history, man must be sustained by a vision of a future so rewarding that no sacrifice is too great to continue on the journey towards it. In that picture of the future, the degree to which men and women can feel at home with their own bodies, and at home in their relationships with their own sex and with the opposite sex, is extremely important.

SELECTED REFERENCES

Devereaux, George. 1937. "Institutionalized Homosexuality of the Mohave Indians." *Human Biology* 9:498–527.

Greulich, William Walter and Herbert Thomas, with Ruth Christian Twaddle. 1939. "A Study of Pelvic Type and Its Relationship to Body Build in White Women." *Journal of American Medical Association* 112:485–493.

17

Childhood

BETTY FARRELL,
1999

In the seven months between October 1997 and May 1998, the communities of Pearl, Mississippi; West Paducah, Kentucky; Jonesboro, Arkansas; and Springfield, Oregon, were linked by a common event: They were all sites of a murderous rampage by a child who, armed with guns more befitting a soldier at war than a schoolboy, opened fire on his classmates, teachers, and, in some cases, parents. On June 5, 1998, two other news items appeared side by side on the same page of the *New York Times* recording more tragic events involving children. In one, a father admitted to having used a sledgehammer to kill his two 5-year-old twins when he "just lost it" because they were moving too slowly while getting ready for day care. A blurb in the adjacent column reported that a Dallas jury had returned a guilty verdict in the case of an eleven-year-old boy who, with his seven- and eight-year-old accomplices, had sexually assaulted, beaten, and killed a 3-year-old girl. The seven- and eight-year-olds were too young to face criminal charges; the eleven-year-old, a fourth-grader, could receive a sentence of up to forty years (*New York Times* June 3, 1998, A12; June 5, 1998, A15).

These stories have the power to shock and horrify, even in a society that has become inured to reports of violent crime. They carry greater weight than do other kinds of personal tragedies because they involve children. On the one hand, a child murderer seems profoundly shocking even among people resigned to adult violence because children are presumably innocent by nature. Intense social anguish has followed on the heels of these crimes perpetrated by children, as the survivors and other members of the distraught community, the general public, and social commentators speculate about the causes of the tragedy. An inadequate home life? Inattentive teachers and school officials? Corrupting media influence? Declining moral standards? The availability of guns in a culture of violence? All of these explanations have been offered as to why some children seem to defy the very nature of childhood by engaging in the most violent acts. Because it is commonly assumed that children are inherently good, a murderous child is a deeply disturbing anomaly that must be explained, interpreted, and analyzed in order to be contained as a social threat.

On the other hand, adults, unlike children, are assumed to be dangerous. But any adult—particularly a parent—who harms a child evokes special horror for having violated the implicit trust that the vulnerable young must place in their adult caretakers and protectors. Child victims are often mourned publicly even more than are adult victims, many of whom may be upstanding citizens, community leaders, or key economic providers for their families whose loss has objectively greater social, political, and economic consequences than that of a child. The loss of a child, however, stirs powerful sentiments, symbolizing as it does both the tragedy of unrealized human potential and the culpability of adult negligence.

These two images that are seemingly so contradictory—the dangerous child who is capable of the most horrific acts of violent crime and the vulnerable child who requires but does not always receive vigilant adult protection and nurturance—coexist uneasily in shaping contemporary American perceptions of children and ideas about childhood. Are children more vulnerable and at risk today than ever before? To what extent have changes in the American family been responsible for simultaneously increasing threats to children and producing threatening children? These questions touch a social nerve for all adults—parents and nonparents alike—because they point to social conditions that shape the next generation and hence the future. Murderous children and murdered children are only the most extreme cases, of course. Between these endpoints on the continuum of modern childhood lie a whole set of other social concerns about American children: babies suffering from the inherited effects of AIDS or fetal alcohol syndrome; young children and adolescents affected by the growing public health epidemic of youth violence (Prothrow-Stith 1991); driven, hurried, over-scheduled "children without childhood" (Elkind 1981; Postman 1982; Winn 1983; Medrich 1982); consumer market-targeted and media-saturated youth. All of these concerns seem hallmarks of our time, uniquely modern problems. To help put these concerns about children into historical perspective, I focus on four themes in this chapter; changes in the ideology of childhood; the demographic and social parameters that have shaped children's lives; the shifts in caretaking arrangements that children in the United States have experienced; and the relations between children and the state. Only by looking at the ideologies, social conditions, family structures, and social policies that have shaped American children's lives over time can we begin to make sense of whether children's lives today are more precarious than in the past.

THE IDEOLOGY OF CHILDHOOD

Given the frequency with which children are cited as the country's most precious national resource, the United States would seem to be unambiguously child centered in its commitments and ideology. Even the characterization of children as a national resource should give some pause though, suggesting as it does that children are a commodity, however much a valued one. Yet, at the same time that many adults claim child-centeredness, the troubling statistics about the growing number of children in poverty, being abused and neglected, and among the homeless raise serious questions about the level of national commitment to children's welfare (Children's Defense Fund 1995; U.S. Advisory Board on Child Abuse and Neglect 1995). These contradictions, among others, challenge us to think critically about the rhetoric used in relation to children and how and why it differs from the actual conditions of their lives. How adults think about children—what ideologies about childhood prevail in particular cultural contexts and historical moments—is a critical component in understanding the institutional forces that shape children's lives.

What we do know with certainty about children is that they need adult care for a very long period of time—longer than the young of any other species—if they are to survive to adulthood and to develop with full physical, social, cognitive, and emotional capabilities. Any society that does not provide adequate care for its children can not be sustained over time. Thus, the physical condition of dependency among children and the social importance of providing for their care would seem to imply that it is natural, universal, and inevitable for adults to think of children as vulnerable and innocent vessels on whom the effects of culture and society are imprinted from the moment of birth. But even the universal fact of physical dependency does not account for the wide variation in ideologies about the nature of children and the state of childhood. Childhood is a physiological stage of human development, but it is also a socially defined and culturally constructed stage of life that varies by time and place.

Children in America have not always been seen as innocent beings and potential victims, in need of the most vigilant adult attention. This ideology is relatively new and still considerably ambiguous. For although most twentieth-century American parents consider their own children to be vulnerable and worthy of adult protection and care, they are much less sympathetic toward other people's children. From the lack of support for public education and universal health care to the many indications of a widespread lack of concern about the quality of life for children in poverty, ideas about family privacy and individualism that are widespread in U.S. culture have also infused the way Americans think about children. The origin of the sentimental beliefs about the sanctity and preciousness of childhood that dominate public rhetoric but have come to be applied privately and locally rather than generalized to all children is the starting point for this analysis. In sketching out the history of the idea of childhood as it has developed in American culture, I mean to suggest that Americans should adopt a more critical stance toward the notion of "the best interests of the child." Which interests get defined as "best"? by whom? based on what prevailing ideas about human nature? and with what consequences (Mnookin 1985; Purdy 1992; Fineman 1995)? Ideas about children, as they have been articulated by adults, have had a significant impact on shaping the experiences and life chances of American children.

The relatively distant American past of colonial New England provides the sharpest contrast with current notions of innocent and vulnerable childhood. The seventeenth-century Puritans subscribed to a doctrine of natural human depravity. Salvation, they believed, was a state that had to be achieved through a long and arduous process in which parents actively and vigilantly socialized their children. Some historians have speculated that such seventeenth-century child-rearing customs as putting out, the process of sending one's own children out to be apprenticed to nonrelatives while taking other children into the household in their place, was explained by this ideology of children's natural depravity. Recognizing that the bonds of love and attachment to their own children threatened the kind of strict discipline that would ultimately lead to adult salvation, Puritan parents, some have argued, developed the institutional response of putting out to remove their children from the temptation of overly indulgent parental care (Morgan 1966). Others have suggested alternatively that family crisis or economic need in colonial America was a far more common cause of putting out than the fear of spoiling children (Wall 1990: 97–111, 212, n. 103). Yet, whatever the cause, putting out was a system that affected even very young children and thrust them into the demanding world of work outside a relatively protected environment. A court in seventeenth-century Plymouth Colony, for example, ruled in favor of a master who claimed that his young servant, Joseph Billington, repeatedly left his service to return home. The court ordered Joseph to return to his master's employ and to remain there, and it further ordered his parents to be placed in the stocks if they allowed their son to come home again. At the time of this ruling, Joseph Billington was five years old—a vulnerable, small child by twentieth-century standards, but clearly not so by the measure used in the seventeenth century (Wall 1990: 124).

The psychodynamic relations between parents and young children in colonial New England—in particular, the Puritan theological injunction to break the will of the child for his or her own good, or "better whipt, than damned" in the words of Cotton Mather (Mintz and Kellogg 1988:

15)—also contrast with contemporary perspectives. The demographic facts of large families and short birth intervals must have created the basis for a very different set of family relations than the conditions of the twentieth century have produced (Demos 1970; Greven 1970, 1977; Wells 1982). Close living quarters and many helping hands in the household may have meant constant attention to the needs of infants, but as new siblings took their place in the family order and as parental concerns about the dire consequences of spoiling the child became more pronounced (Greven 1973), children by the age of two may have encountered an environment of harsher discipline than most American adults today would agree is ideal for raising children.

In his landmark book, *Centuries of Childhood* (1962), Philippe Ariès provocatively argued that medieval society lacked the concept of childhood as a separate or unique stage of life. Only in the seventeenth century, he claimed, did themes of childhood emerge in painting and literature through depictions of children's distinct style of dress, play, and speech. Based on evidence drawn from paintings, literary references, and diaries—the kinds of historical sources that have traditionally reflected an upper-class bias—Ariès argued that the "discovery" of childhood helped produce a new set of feelings and attitudes about the family as a more child-centered institution in the seventeenth century. By the eighteenth century, these social relationships intensified as the family became more withdrawn from the larger society, taking on its modern guise as a private fortress in a sea of external social relations.

Ariès's argument about the historically variable nature of childhood has been both contested and refined over the past thirty years. One debate among family historians, building on Ariès's thesis that childhood itself is a relatively new concept, has centered on whether or not parents and children related to each other differently in the past, in particular with parents reserving their emotional investment in infants and young children because of the high rates of infant mortality (Pollock 1987). Although this premise seems questionable, much evidence in the historical record supports the idea that parent-child relationships during the colonial period were more emotionally distant than most twentieth-century Americans would recognize as innate or natural. A ritual of extended mourning did not mark the death of young children in seventeenth- and eighteenth-century America, for example, and parents routinely named their children after older siblings who had died, a custom that may offend contemporary sensibilities about the uniqueness and individualism of even very young children (Smith 1977). Another Massachusetts court case from 1661 provides a glimpse into the complex mix of emotional and pragmatic considerations that characterized parent-child ties in colonial America. In this case, a widow who had apprenticed out her only son moved into the master's home to care for him for three weeks when he became seriously ill. Upon becoming sick herself, she returned to her own home, leaving her son to the care of the master's mother over the next ten weeks. Although the widow had reason to fear that her son was sick enough to die and although she thought the master was overly harsh in his treatment of her child, she clearly expected compensation for any prolonged care of her son (Wall 1990: 124–125). As this case suggests, a fine line between affection and pragmatism characterized the parent-child relationships in colonial America.

There are some indications of regional variations in colonial child-rearing ideas and practices. The Chesapeake gentry families of the eighteenth century, whose distinctive culture was built on the development of a slave-holding plantation economy, were more likely to emphasize the father's paternal influence in family life and to adopt a child-rearing style of affectionate indulgence and tolerance for children's autonomy. Quakers in the Delaware Valley, unlike their Puritan New England neighbors, did subscribe to a belief in childhood innocence, and they sought to create a controlled family environment in which children would be protected from the corruptions of worldly influences (Lewis 1983; Levy 1988). Yet, even with these variations, it is still clear that seventeenth- and eighteenth-century adults saw children through a very different lens than the one most Americans currently use. The socioeconomic context from which family life took its meaning, as well as the larger size of families and households in colonial America, gave children an instrumental role and significance that they lack today. As necessary laborers in the domestic economy and as the future caretakers of aging parents, children were useful additions to the family unit. That quality of usefulness meant that, beyond infancy, children were not understood as being particularly fragile, innocent, or vulnerable, as they have come to be seen since. Instead, the defining characteristics of useful children were sturdiness and an early capacity for responsibility, all the qualities that gave them the appearance of little adults roughly by the age of seven.

By the time of the American Revolution, the distinctive regional variations in family values and practices of the colonial period had largely given way to an emerging national culture in which the affectionate, antipatriarchal family played a central role (Reinier 1996). The shift in beliefs about children and the nature of childhood took root first among the middle class in the early nineteenth century, spreading to encompass working-class children by the early twentieth century. Not surprisingly, middle-class children were redefined as vulnerable innocents in need of full-time care and devoted attention in the same era as their mothers were being redefined as the guardians of virtue and morality, the ideal keepers of the domestic sphere. Nineteenth-century America was the context for the emergence of an ideology of domesticity that still reverberates in family life today, but only a relatively small group of privileged middle-class children were the first recipients of the nurturing attention of newly idealized mothers. As middle-class women were defined by and enclosed within the sanctuary of the private home—an ideological shift that enhanced their domestic power in the short term but resulted in the long-term exclusion from resources and power in the economic and political spheres (Welter 1966; Cott 1977; Ryan 1981)—middle-class children were sentimentalized and redefined as needing their mothers' full-time, loving, but vigilant, attention.

Not all parents could live up to this ideal. Many, by economic or geographic necessity, continued throughout the nineteenth and into the early twentieth century to define their children as little adults, particularly in their capacity as paid and unpaid laborers in the family economy. Those children growing up on the western frontier plains, for example, were always crucial workers in their farming and ranching families (West and Petrik, 1992: 26–37). And class differences, which sharpened as the United States industrialized during the nineteenth century, drew clear distinctions in the life experiences of children. With the decline of the apprenticeship system in the early industrial era, working-class children provided a cheap source of labor in factories, as well as contributing to the family income through a substantial amount of piecework at home. Even very young children could be engaged at home in such tasks as making buttons and artificial flowers, pulling bastings, cutting and gluing boxes, caring for younger siblings, and running errands. Young urban children scavenged on city streets for wood chips, coal, or dropped food items while their older siblings collected rags, nails, and pieces of rope to be sold to junk dealers (Reinier 1996: 138). A Polish immigrant growing up in Chicago in the first decades of the twentieth century recalled in an oral history interview that "no one in my neighborhood ever had to buy any fuel, any oil, or any wood" (Nasaw 1985: 97). The historical evidence suggests that working-class children were active in their scavenging efforts, and most were neither closely watched nor carefully protected, whether at work in factories or fields, on urban streets, or in their own homes.

Despite the enormous variation in children's life experiences in the nineteenth-century United States, it was ultimately the middle-class model of the vulnerable and defenseless child who required a mother's full-time care that gained dominance as the cultural norm, if not the behavioral reality, by the middle of the century. The power of this new ideological construction of childhood was such that it began to spread across the sharp divides of social class and geographic variation in children's worlds, even when it did not fit their actual experience. Between the 1870s and the 1930s, working-class as well as middle-class children began to be sentimentalized and reinterpreted as innocent and vulnerable, a process that involved a significant cultural shift from a belief in the economic usefulness of children to one in which children were understood as economically useless but morally and emotionally priceless (Zelizer 1985).

Sociologist Viviana Zelizer has identified several key indicators of this significant shift in the way Americans thought about children by the early twentieth century. The battle over child labor was waged in terms of competing ideologies about the moral worth of work for "useful children" versus the exploitation of "vulnerable children." Social structural changes certainly contributed to the removal of many children from the industrial workplace in the early twentieth century. The development of industrial capitalism, for example, increased the demand for a more skilled workforce, and, over the course of the nineteenth century, immigrants replaced children in many unskilled and semiskilled jobs. But, along with these structural factors, a growing cultural sense of children as unsuited for most paid work, other than in the exceptional cases of child actors (whose activities were considered more fun than work) and newspaper deliverers (whose tasks were considered wholesome work), separated the young from the productive economy and relegated them to the "domesticated, nonproductive world of lessons, games, and token money" (Zelizer 1985: 11). Child labor aroused the sympathy of Progressive Era reformers, even as part-time work for urban children, wedged between their school and family time, continued to flourish as an acceptable opportunity to earn valued spending money (Nasaw 1985).

A number of paradoxes accompanied this reinterpretation of priceless childhood in the early twentieth century. Child life insurance was introduced to cover the high costs of an elaborate funeral in the case of a child's early death; high court settlements were awarded in cases of accidental death, not so much to offset the loss of a child's labor power as to compensate, at least symbolically, for the parents' emotional loss; and a market developed for the adoption of blond, blue-eyed infants because they were now considered the most priceless children of all. These institutional developments had the paradoxical consequence of redefining the economic value of children's worth, ultimately setting a price on priceless children. The end result, one with which we still live today, is that there has been an unintended commercialization of childhood along with its sentimentalization: Increasingly excluded from the sphere of economic productivity by the early twentieth century, children were nevertheless defined in economic terms by a wide array of institutions and agencies, and today they continue to be enticed into the market as significant consumers at younger and younger ages.

While the new ideology about childhood was spreading in the first three decades of the twentieth century, the Depression and World War II intervened to ensure that many children did continue to experience the role of the "useful" child to midcentury. As Elder (1999) and Clausen (1993) have shown through their longitudinal studies of children who came of age in these decades, many experienced a foreshortened sense of childhood as they contributed to the family income or assumed responsibilities that were defined as unchildlike for earlier and later cohorts. Despite the growing dominance of an ideology promoting the innocence and vulnerability of children in the twentieth century, not all analysts have argued that this earlier exposure to the adult world was harmful. Some have suggested it as one cause of the remarkably familistic values and behaviors of those middle-class Americans who came into adulthood in the immediate postwar era of the 1950s (Cherlin 1992). At least one explanation for this cohort's notable patterns of young age at marriage, high fertility (resulting in the baby boom), and low rate of divorce has been attributed to the early responsibility that they assumed as children of the Depression era.

By many historical accounts now available, the 1950s and early 1960s should be seen as a unique social and economic period in U.S. history, rather than as the benchmark against which to measure current family experiences (May 1988; Skolnick 1991; Coontz 1992). As the war economy of the 1940s was converted into a new growth industry based on durable consumer goods and a service economy in the 1950s and as veterans' benefits fueled massive suburban community development, the social world in which white, middle-class women and children found themselves was increasingly homogeneous and contained. A newly revitalized ideology of domesticity that located women's and children's roles in the home and men's roles in the public world emerged in the postwar decades and became the standard against which nonwhite and non-middle-class mothers and children were judged as well. Even when they did not fit the actual life experience of many people in the United States, the ideologies of proper womanhood, childhood, and middle-class family domesticity had a remarkably powerful capacity to extend beyond their original reach. This era was, not coincidentally, the one in which television grew up and presented its narrowly homogeneous view of the world through the frame of the family sitcom (Taylor 1989). Normal childhood and family life have been measured ever since against the picture seen through this lens, a fictive construction against which contemporary social reality is often judged and found lacking.

The forces of social change that lay just beneath the surface at midcentury and that began to be seen and felt by the mid-1960s had a profound effect on ideas about children and childhood. Whereas previous cohorts of adults had worried about rebellious children (to be dealt with through the legal system), deprived children (to be dealt with through social services agencies), and sick children (to be dealt with through the medical system), a new concern about child victims began to be more pronounced in popular consciousness and national policy debates during the 1970s (Best 1990). Indeed, the concern with the prevalence and extent of child abuse arguably reached the level of a national obsession in the 1980s. Adult fears about child abduction, kidnapping, and molestation—at their height in the 1980s and 1990s—have fueled contemporary concerns about children as an especially threatened and vulnerable group. Every day, newspapers and television reports seem to confirm that ours is an especially difficult and unsafe era for children.

Concerns about child abduction actually had deep historical roots by the time the furor about an epidemic of missing children swept across the United States in the late twentieth century. The first ransom kidnapping of a child to receive widespread national attention occurred in the famous 1874 abduction case of four-year-old Charley Ross, known to a generation of Americans as "the Lost Boy" (Fass 1997: 21–56). In several ways, this highly public and never-resolved case defined the threat of stranger abduction and heightened awareness of childhood vulnerability forevermore in American culture. It showed that neither parents nor the police could ensure children's safety in their homes or communities. It revealed, in the shocking demand for ransom in exchange for Charley's safe return, that the preciousness of childhood now carried a price. In the power of this story to sensationalize and titillate the American public, it "provided the occasion for Americans to discuss an array of social issues: family and parenting, sexuality and gender, policing and law enforcement, criminality and insanity, community norms, and the role of the state" (Fass 1997: 257). This case of child abduction was one of the first times, but by no means the last, that American adults would rally around children and childhood as embodying all their anxieties about the modern world in the throes of social change.

The definition of children as victims crossed another milestone in 1962 with the development of the medical concept of "the battered child syndrome," and the subsequent emergence of a highly successful campaign in the 1970s and 1980s to shape public conceptions of this social problem on a new scale. The initial publicized fear that thousands of children were routinely being subjected to physical and emotional harm by neglectful parents and to abduction, molestation, and ritual abuse by strangers was not borne out by supporting evidence. The subsequent shift from a concern about an epidemic of child victimization to the rhetorical claim that "even one child harmed is too many" suggests how successful the campaign to redefine children as victims (or potential victims) has been. At a cultural juncture when many Americans were experiencing generalized anxiety about the direction and consequences of social change, threats to children emerged as a powerful, yet manageable, symbol of that change. In the place of broader anxiety about the future and the pace of social change, children—as visible symbols of the future—could serve as a specific focus of the concern. When the dangers facing children get defined as those caused by individual deviants—kidnappers who lurk around playgrounds, sadists who poison children's Halloween candy, and the like—structural explanations of and solutions to a broader array of complex social problems affecting children and families are effectively muted.

Although there may be genuine reason for concern about dangers to children in the modern era, it is also clear that the current notion of childhood vulnerability is a powerful idea that lobbyists can manipulate as a political issue. Where the institutionalized neglect of children through inadequate health provisions, underfunded schools, and the decline of affordable housing fails to elicit an immediate call to action, the threats of child abduction and abuse often do. The sentimentalization of childhood that is so much a part of the contemporary ethos—the idea that childhood is the repository of a natural innocence that is constantly threatened by predatory adults—turns out to have great political resonance and the power to mobilize widespread public response. It should not be surprising, then, that threatened children appear on both sides of intensely debated social and political issues: as the "murdered babies" versus the "unwanted children" in the abortion debate, as the "future beneficiaries" versus the "impoverished victims" of current budget cuts and efforts to restructure government priorities and programs. In this context, reports of murderous and dangerous children seem particularly anomalous and horrifying. Having so thoroughly incorporated the idea that children are by nature innocent and harmless and vulnerable to exploitation, Americans are especially susceptible to the shocking discrepancy between the child victim and the child victimizer.

The ideological constructions of childhood that have prevailed in American history matter not only because they have shaped the way most Americans see and understand the world but because they are the foundation for the set of institutions, programs, and policies that adults construct with real consequences for children's lives. The history of the ideology of childhood in America is one that reveals a profound ambivalence: Americans publicly proclaim their child-centeredness but remain more committed to the idea of the innocence and vulnerability that characterize the state of childhood than to improving the lives of actual children.

SELECTED REFERENCES

Ariès, Phillipe. 1962. *Centuries of Childhood: A Social History of Family Life.* Trans. Robert Baldick. New York: Knopf.

Best, Joel. 1990. *Threatened Children: Rhetoric and Concern about Child-Victims.* Chicago: University of Chicago Press.

Cherlin, Andrew J. 1992. *Marriage, Divorce, Remarriage,* revised edition. Cambridge: Harvard University Press.

Children's Defense Fund. 1995. "The State of American Children." Washington, D.C.: Children's Defense Fund.

Clausen, John A. 1993. *American Lives: Looking Back at the Children of the Great Depression.* Berkeley: University of California Press.

Coontz, Stephanie. 1992. *The Way We Never Were: American Families and the Nostalgia Trap.* New York: Basic Books.

Cott, Nancy F. 1977. *The Bonds of Womanhood: Woman's Sphere in New England, 1780–1835.* New Haven, Conn.: Yale University Press.

Demos, John. 1970. *A Little Commonwealth: Family Life in Plymouth Colony.* New York: Oxford University Press.

Elder, Glen H., Jr. 1999. *Children of the Great Depression: Social Change in Life Experience,* 25th anniversary edition. Boulder, Col.: Westview Press.

Elkind, David. 1981. *The Hurried Child: Growing up Too Fast, Too Soon.* Reading, Mass.: Addison-Wesley Publishing.

Fass, Paula S. 1977. *The Damned and the Beautiful: American Youth in the 1920s.* New York: Oxford University Press.

Fineman, Martha Albertson. 1995. *The Neutered Mother, the Sexual Family and Other Twentieth-Century Tragedies.* New York: Routledge.

Greven, Philip J. 1970. *Four Generations: Population, Land, and Family in Colonial Andover, Massachusetts.* Ithaca, N.Y.: Cornell University Press.

———. 1973. *Child-Rearing Concepts, 1628–1861; Historical Sources.* Itasca, III.: F. E. Peacock Publishers.

———. 1977. *The Protestant Temperament: Patterns of Child-Rearing, Religious Experience, and the Self in Early America.* New York: Knopf.

Levy, Barry. 1988. *Quakers and the American Family: A British Settlement in the Delaware Valley.* New York: Oxford University Press.

Lewis, Jan. 1983. *The Pursuit of Happiness: Family and Values in Jefferson's Virginia.* Cambridge: Cambridge University Press.

May, Elaine Tyler. 1988. *Homeward Bound: American Families in the Cold War Era.* New York: Basic Books.

Medrich, Elliott A. 1982. *The Serious Business of Growing Up: A Study of Children's Lives Outside School.* Berkeley: University of California Press.

Mintz, Steven, and Susan Kellogg. 1988. *Domestic Revolutions: A Social History of American Life.* New York: Free Press.

Mnookin, Robert, ed. 1985. *In the Interest of Children: Advocacy, Law Reform and Public Policy.* New York: W. H. Freeman.

Morgan, Edmund S. 1966. *The Puritan Family: Essays on Religion and Domestic Relations in Seventeenth-Century New England.* New York: Harper and Row.

Nasaw, David. 1985. *Children of the City: At Work and at Play.* Garden City, N.Y.: Anchor Press/Doubleday.

Pollock, Linda. 1987. *A Lasting Relationship: Parents and Children over Three Centuries.* Hanover, N.H.: University Press of New England.

Postman, Neil. 1982. *The Disappearance of Childhood.* New York: Delacorte Press.

Prothrow-Stith, Deborah. 1991. *Deadly Consequences.* New York: HarperCollins.

Purdy, Laura M. 1992. *In Their Best Interest?: The Case Against Equal Rights for Children.* Ithaca, N.Y.: Cornell University Press.

Reinier, Jacqueline S. 1996. *From Virtue to Character: American Childhood, 1775–1850.* New York: Twayne Publishers.

Ryan, Mary P. 1981. *Cradle of the Middle Class: The Family in Oneida County, New York, 1790–1865.* Cambridge: Cambridge University Press.

Skolnick, Arlene. 1991. *Embattled Paradise: The American Family in an Age of Uncertainty.* New York: Basic Books.

Smith, Daniel Scott. 1977. "Child Naming Patterns and Family Structure Change: Hingham Mass., 1640–1880," *The Newberry Papers in Family and Community History* Paper 76-5 (January). Chicago: Newberry Library.

Taylor, Ella. 1989. *Prime Time Families: Television Culture in Postwar America.* Berkeley: University of California Press.

U.S. Advisory Board on Child Abuse and Neglect. 1995. "A Nation's Shame: Fatal Child Abuse and Neglect in the United States," 5th report. Washington, D.C.: Department of Health and Human Services, Administration for Children and Families.

Wall, Helena M. 1990. *Fierce Communion: Family and Community in Early America.* Cambridge: Harvard University Press.

Wells, Robert V. 1982. *Revolutions in Americans' Lives: A Demographic Perspective on the History of Americans, Their Families, and Their Society.* Westport, Conn.: Greenwood Press.

Welter, Barbara. 1966. "The Cult of True Womanhood, 1820–60," *American Quarterly* 18 (Summer): 151–174.

West, Elliott, and Paula Petrik, eds. 1992. *Small Worlds: Children and Adolescents in America, 1850–1950.* Lawrence: University of Kansas Press.

Winn, Marie. 1983. *Children Without Childhood.* New York: Penguin.

Zelizer, Viviana A. 1985. *Pricing the Priceless Child: The Changing Social Value of Children.* New York: Basic Books.

18

Boys and Men in Families

MICHELE ADAMS AND SCOTT COLTRANE,
2003

As we raise boys to be masculine men, we often end up getting troubled boys. Snips, snails, and puppy dog tails, little boys are noisier, more active, more competitive, and more aggressive than little girls, according to research and popular cultural stereotypes. They reject (as they are taught) their mothers, their families, and adults, in general. Sometimes they grow up to join gangs, assault young women, attack other young men, or commit suicide. At some (often indeterminate) point, they cross the cultural boundary between boyhood and manhood and become men who are unemotional, withdrawn from their families, aggressive, or violent. The trouble with boys is that they learn the lesson well and assume the cultural mantle of masculinity. "The trouble with boys," according to one British researcher, "is that they must become men" (Phillips 1994:270).

In this chapter, we look at how boys become men within the context of the family, and how, as part of that process, gender inequality is sustained and reproduced. We first examine how the cultural concept of masculinity is based on a proscription against being feminine. Noting how boys and girls are raised differently from the beginning of their lives, we observe how masculine ideals project boys out of and away from the family, while feminine ideals enmesh girls within it. We also point out the troubles faced by boys as they attempt to become men by incorporating ideals of dominant masculinity into their own gender schema. We then follow these boys-turned-men as they confront problems feeling "at home" in family environments. Here we see that the dilemmas men face reconciling their ideals of masculinity with their positions as husbands and fathers are part of a larger set of social problems that stem from separate spheres ideology and structural gender inequality in the society at large. We conclude by suggesting social and individual changes that might help attenuate the alienation that appears to be the plight of men living in today's families.

IDEALS OF MASCULINITY AND FEMININITY

Ideals of masculinity and femininity, passed down from nineteenth-century notions of separate spheres, assume that boys and girls are intrinsically and unalterably different in terms of personality and, therefore, behavior. Men, oriented to the public sphere, are understood to be active, strong,

Michele Adams and Scott Coltrane, "Boys and Men in Families."

independent, powerful, dominant, and aggressive, with masculinity signifying "being in control" (Kaufman 1993). Women, associated with the private sphere, are seen as passive, weak, dependent, powerless, subordinate, and nurturing. While social, economic, demographic, and cultural contexts have changed since the nineteenth century, idealized perceptions of masculinity and femininity have remained remarkably consistent. Even today, the notion of separate spheres and attendant sex differences in temperament are invoked to substantiate gender stratification institutionally (see, for instance, Brush 1999), as well as to privilege male power and interests in the home (Jones 2000; Kimmel 2000). Besides their prescriptive elements, these idealized gender differences in temperament are proscriptive, as well, for "an essential element in becoming masculine is becoming not-feminine" (Maccoby 1998:52). Taken as a whole, the mandate for boys to be not-feminine, unlike (and in direct opposition to) the mandate for girls *to be* feminine, is a mandate that drives them away from family relations, particularly relations with their mothers (Silverstein and Rashbaum 1994). Although assumed to be a baseline requirement for boys' achievement of manhood, this cultural mandate can cause problems for them when they mature into men. As men, they will have little ideological precedent for living harmoniously in a family environment, especially one that is increasingly predicated on ideals of democratic sharing. By continuing to follow the dictates of separate spheres, we may be creating manly men, but we are also crippling men emotionally and creating husbands and fathers who are destined to be outsiders or despots in their own families

SOCIALIZATION: BOYS (AND GIRLS) IN FAMILIES

Society can work only if its members "organize their experience and behavior in terms of shared rules of interpretation and conduct" (Cahill 1986:163). All societies thus socialize children to internalize the shared rules and norms that drive collective behavior, thereby becoming self-regulating participants in society. More formally, socialization is the process through which "we learn the ways of a given society or social group so that we can function within it" (Elkin and Handel 1989:2); while older notions of socialization suggested that the process began and ended in childhood, according to more recent theories, it is a lifelong process that allows us to move in and out of various social groups our entire lives. Part of this process involves gender socialization; that is, learning society's gender rules and regulations (typically dichotomized as either masculine or feminine) and becoming adept at behaving in accordance with the socially accepted gender patterns associated with our sex (male or female). Gender, that is to say, is not the same thing as sex, which generally groups people into categories based on their biologically given reproductive equipment. Gender, on the other hand, is a social construction, emergent, dynamic, variable within and across cultures, and historically situated, but also reflecting certain patterns within a given society (Coltrane 1998). According to sociologists Candace West and Donald Zimmerman (1987), we "do gender" by acting out our culture's perception of those patterns that reflect what it is to be a man or a woman. The family is typically considered the main institution for both production and reproduction of polarized gender values. Although individuals are socialized in many different contexts throughout their life (school, neighborhood, community, peer group, workplace, church, polity), family tends to be the primary initial socialization agent, acting as a microcosm of society, and providing a child's first exposure to interaction with others. It is generally in the family that children first acquire enduring personality characteristics, interpersonal skills, and social values (Maccoby 1992). It is also in the family that children get their first look at what gender means, to them and to others, as they interact in daily life (Coltrane and Adams 1997). Specifically, it is in the family that boys first come to understand their privileged status and the ways in which male privilege equates to power. Finally, it is often in the family that these boys, grown into men, later come to understand the contradictions inherent in that power (Coltrane 1996; Kaufman 1999).

Early Gender Differentiation

Gendered parents transmit gender-laden assumptions and values to their children, starting before the children are born. Procedures such as amniocentesis and sonograms allow parents to find out the sex of their unborn child so that they might plan early for gender-appropriate nurseries and infant wardrobes, as fashion- (and gender-) conscious parents would be loathe, for instance, to bring their newborn son home in a pink or flowered cozie. Knowing the sex of an infant before birth can have other more sinister effects. In some countries, such as India and China, the traditional bias toward males is reflected in a prevalence of sex-selective abortions, as well as female neglect and infanticide after birth (Balakrishnan 1994; Chunkath and Athreya 1997; George, Rajaratnam, and Miller 1992; Weiss 1995). In rural Bangladesh, traditional son preference drives the use of contraceptives by women in their childbearing years (Nosaka 2000). Furthermore, research has shown that more family resources, such as food and medicine, are allocated to sons, whose rate of survival is, thus, higher than daughters (Chen, Huq, and D'Souza 1981; Bhuiya and Streatfield 1991). These gender preference practices, some more extreme than others, are part of patriarchal societies where the notion prevails that sons have more value than daughters. But even in societies such as the United States and Canada, where disappointment over the birth of a girl may be more reserved, technologies allowing for "prenatal discrimination" are becoming more widely used (Bozinoff and Turcotte 1993). In industrialized societies, as well as in less developed ones, notation of difference between boys and girls before birth signals the privilege and power that boys, and later, men, will experience in their lives. Once the baby arrives, new parents advertise the sex of their infant so that no mistake can be made as to its traits or prospects for success: will it be a future president or will it be a wife and mother? Announcements and banners proclaim "It's a boy," or "It's a girl," giving admirers the gender context to remark on the baby's characteristics and potential. Mothers attach cute little pink bows to the bald heads of baby girls to set them apart from the supposedly rough and tumble boy babies (who, it turns out, are not only visually indistinguishable from girl babies but slightly more fragile medically). The baby boy is housed in a nursery painted in bold colors of blue or red and outfitted with sports and adventure paraphernalia; the infant girl is treated to a pink boudoir with plenty of dolls and soft things to cuddle (Pomerleau, Bolduc, Malcuit, and Cossette 1990). If a boy, the newborn is dressed in blue and is given gifts of tiny jeans and bold-colored outfits; if a girl, she is outfitted in pink and receives ruffled, pastel ensembles (Fagot and Leinbach 1993). Moreover, research shows, based on what they are told the newborn's sex is, people (including strangers and especially children) tend to characterize infants based on whether they believe the baby to be a boy or a girl, seeing boys as stronger, bigger, noisier, and (sometimes) smarter than girls, even when the same baby is represented as male to some observers and female to others (Coltrane 1998; Cowan and Hoffman 1986; Stern and Karraker 1989). That is, people draw on a cultural overlay of gender stereotypes to make their first assessment of a baby's personality and potential. Parents also use gender stereotypes when assessing the behavior and characteristics of their newborns (Rubin, Provenzano, and Luria 1974), and interact with them based on these stereotyped preconceptions. For instance, parents (particularly fathers) tend to react to their infant boys by encouraging activity and more whole-body stimulation and tend to enforce gender stereotypes more than mothers, especially in sons. This tendency extends across types of activities, including toy preferences, play styles, chores, discipline, interaction, and personality assessments (Caldera, Huston, and O'Brien 1989; Fagot and Leinbach 1993; Lytton and

Romney 1991). Although both boys and girls receive gender messages from their parents, boys are, nevertheless, encouraged to conform to culturally-valued masculine ideals more than girls are encouraged to conform to lower-status feminine ideals. Boys also receive more rewards for gender conformity (Wood 1994). Because society places greater emphasis on men's gender identity than on women's, there is a tendency for more attention to be paid to boys, reflecting an androcentric cultural bias that values masculine traits over feminine (Bem 1993; Lorber 1994).

Paradoxically, masculine gender identity is also considered to be more fragile than feminine gender identity (Bem 1993; Chodorow 1978; Dinnerstein 1976; Mead 1949), and takes more psychic effort because it requires suppressing human feelings of vulnerability and denying emotional connection (Chodorow 1978; Maccoby and Jacklin 1974). Boys, therefore, are given less gender latitude than girls, and fathers are more intent than mothers on making sure that their sons do not become sissies. Later, as a result, these boys-turned-men will be predisposed to spend considerable amounts of time and energy maintaining gender boundaries and denigrating women and gays (Connell 1995; Kimmel and Messner 1998). Nonetheless, fathers' role in sustaining gender difference is neither fixed nor inevitable. Mothers' relatively lax enforcement of gender stereotypes relates to the amount of time they spend with children. Since they perform most of the childcare, mothers tend to be more pragmatic about the similarities and dissimilarities between children, and their perceptions of an individual child's abilities are somewhat less likely to be influenced by preconceived gender stereotypes. Similarly, when men are single parents or actively co-parent, they behave more like conventional "mothers" than standard "fathers" (Coltrane 1996; Risman 1989). Involved fathers, like most mothers, encourage sons and daughters equally, utilizing similar interaction and play styles for both. They also tend to avoid both rigid gender stereotypes and the single-minded emphasis on rough and tumble play customary among traditional fathers (Coltrane 1989; Parke 1996). As a result, when fathers exhibit close, nurturing, ongoing relationships with children, they develop less stereotyped gender attitudes as teenagers and young adults (Hardesty, Wenk, and Morgan 1995; Williams, Radin, and Allegro 1992).

Different treatment of newborn boys and girls, based on their sex, is a product of the behavior of gendered adults (family members and strangers) and institutionalized expectations about gender derived from society as a whole (Coltrane and Adams 1997). According to psychologist Sandra Bem (1983), gender is not something that is naturally produced in the mind of the child, but instead reflects the gender polarization prevalent in the larger culture. Moreover, gender-differentiated treatment continues as the child grows up; gender-appropriateness is reinforced through toys (trucks, sports equipment, and toy guns for boys; dolls, tea sets, and toy stoves for girls), as well as expectations for behavior that result in praise and reinforcement for "correct" (gender appropriate) behavior and reprimand and punishment for "incorrect" (gender inappropriate) behavior. For instance, taking into account the masculine imperative for emotional distance, studies analyzing a number of northern European countries, as well as the United States, find that parents tend to actively discourage displays of emotion in boys by pressuring them not to cry or otherwise express their feelings (Block 1978, as cited in Maccoby 1998:139). Girls, in contrast, are not only encouraged to express their emotions, but are also taught to pay attention to the feelings of others.

The result of this indoctrination is that, as they become developmentally able, boys and girls incorporate the gendered messages and scripts that parents (and other significant adults) have communicated to them into their own version of an age-appropriate gender schema (Bem 1983). A gender schema is a cognitive way of organizing information, a sort of "network of associations" that "functions as an anticipatory structure" ready to "search for and to assimilate incoming information" in terms of relevant schematized categories

(Bem 1983:603). A kind of perceptual lens, a gender schema predisposes us to see the world in terms of two clearly defined "opposites"—male and female, masculine and feminine. Accordingly, children develop gender schema without even realizing it when the culture in which they live is stereotyped according to gender. Developing networks of associations that guide their perceptions, children come to see the world in gender-polarized ways and to live out the gender-polarization that they have learned to make their own. Children then go about re-creating, according to their own developmental ability, a world in which boys/men and girls/women are not only different, but polar opposites, and where boys/men are generally powerful and privileged. As they grow up, moreover, they come to understand that, while most men are more powerful than most women, not all men are equally powerful, and that some (hegemonic) masculinities entail more privilege than other (subordinated) masculinities (Connell 1995, 1987).

Children's Agency and the Construction of Gendered Behavior

We see evidence of the ways that children create their own gendered worlds in the fact that, from the time they are about three years old, they begin to associate consistently with same-sex playmates, generally without direct provocation or instigation from adult caretakers (Howes and Philipsen 1992; Maccoby 1998; Thorne 1993). In this way, children begin to institute at an early age the gender segregation that traverses adult society. Noting this tendency, sociologist William Corsaro (1997:4) sees children as "active, creative social agents who produce their own unique children's cultures while simultaneously contributing to the production of adult societies." Moreover, forays into cross-gender territory generally herald advances toward a heterosexual romantic culture rather than enduring friendships that cross gender lines (Adler and Adler 1998; Eder with Evans and Parker 1995; Thorne 1993). As these social scientists suggest, romantic "crossings" (Thorne 1993) strengthen traditional gender boundaries and behaviors while reinforcing the gender segregation evident in same-sex friendship groupings. Boys' playgroups and girls' playgroups exhibit distinctive styles of play. One significant difference between them is that boys appear to be more separated from the world of adults (Maccoby 1998), a tradition which begins in the family when boys, between 24 and 36 months old, begin to invite less contact from their mothers (Clarke-Stewart and Hevey 1981; Maccoby 1998; Minton, Kagan, and Levine 1971). What is unclear about this "separation," is exactly how much is initiated by the child, and how much is initiated by the child's mother or parents, who feel that "too much" mothering can be dangerous to a boy's masculinity (Silverstein and Rashbaum 1994). This impulse also conforms to the cultural mythology of "mother-blaming," reminding us (in movies, on television, and in novels) of the over-involved, domineering mother who emasculates her son, makes him into a "sissy," and leaves him unfit and unable to take his place in the patriarchal scheme of oppression (Silverstein and Rashbaum 1994). This separation from the adult world takes the form of increased mischievousness at home, in direct opposition to maternal direction (Minton, Kagan, and Levine 1971) and less sensitivity to teachers (Fagot 1985). Boys also play rougher than girls, with their interaction frequently bordering on aggression, if not outright violence (Maccoby 1998). Boys' rough-and-tumble play appears to be designed to create a dominance hierarchy and to mitigate a presumption of weakness (Jordan and Cowan 1995; Maccoby 1998; Petit, Bakshi, Dodge, and Coie 1990); girls, on the other hand, do select leaders, but draw on leadership qualities other than physical dominance (Charlesworth and Dzur 1987; Maccoby 1998). There is even a difference in styles of discourse, with girls negotiating to keep interaction going, while boys simply command and demand, thus stopping effective interaction (Maccoby 1998:49). Finally, boys' playgroups involve more competition than girls', with boys spending much more time playing competitive games and girls focusing on recreation that entails taking turns (Crombie and Desjardins 1993).

That these tendencies of boys in their same-sex playgroups reflect parentally-encouraged and socially-approved masculine ideals is apparent, as boys display masculinity by withdrawing from adults (mothers, in particular), and by being dominant, competitive, aggressive, and (over)active. Because we take for granted that masculinity is a positive cultural and institutional ideal, we don't tend to view masculinity *per se* as a negative factor that can cause problems for boys as they negotiate their gender performance against a backdrop of broader principles of social order. Most of the time, when boys' behavior runs counter to social norms, we chuckle that "boys will be boys." When that behavior reaches beyond the acceptable, however, we begin to acknowledge that living up to masculine ideals can, indeed, cause trouble.

Boyhood Troubles

The way we raise boys in our society not only reinforces masculine personality ideals, but also encourages behavior that reflects those ideals. We valorize manhood and start, from the beginning of their lives, to transmit that valorization to our children. Children realize, early on, that if they are fortunate enough to be born with the legitimating penis, then they are likely to receive the rewards, rights, privileges, and entitlements that come along with it, although the amount of those rewards is premised on other social factors as well. On the other hand, if they are female, they realize that they are destined to help provide those rewards to their more privileged brothers. That is, children begin to incorporate these ideals into their own perceptions and behaviors and begin to "act out" the gender scripts that they have learned. Moreover, as gendered parents, we expect and encourage boys to pursue our cultural ideals of masculinity. From early in their youth, we teach them (through, for instance, toys and sports) to symbolically correlate competition, violence, power, and domination with masculinity. And, finally, we actively insist on their separation from mothers (in effect, their separation from anything feminine that might sully their budding masculinity). In short, by defining masculinity as "anything not feminine" and by defining femininity in conjunction with the family and domesticity, we are, in effect, defining boys and men away from the family and outside of it. When translated into behavior attenuated by developmental stage, boys often end up in trouble—overactive and inattentive in school (the class clown), competitive and aggressive, even violent. Studies show that elementary school-aged boys are up to four times as likely as girls to be sent to child psychologists, twice as likely to be considered "learning disabled," and much more likely (up to ten times) to be diagnosed with emotional maladies such as attention deficit disorder (Kimmel 2000:160; Pollack 1998). Studies also show that "problem behaviors" of adolescent boys (including school suspension, drinking, use of street drugs, police detainment, sexual activity, number of heterosexual partners, and forcing someone to have sex) are associated with traditional masculine ideology (Christopher and Sprecher 2000; Pleck, Sonenstein, and Ku 1994; Schwartz and Rutter 1998).

Aggression has become a touchstone for American adolescent boys, and violence among them is epidemic. Kaufman (1998) notes that men construct their masculinity amid a triad of violence: men against women, men against men, and men against themselves. This triad of men's violence applies even to adolescent boys, and results, at least in part, from their internalizing the masculine ideal and attempting to live up to its precepts. Young men's violence against women, which Kaufman (1998:4) suggests represents both an individual "acting out" of power relations and an individual's enactment of social power relations (sexism), plays out in instances of rape (acquaintance and stranger) and sexual harassment, perpetrated in all-male enclaves such as fraternities (Lefkowitz 1997; Sanday 1990) and athletic teams (Benedict 1997). Research analyzing rape figures between 1979 and 1987 show that youths 20 years old and younger accounted for 18% of single-offender and 30% of multiple-offender rapes

(Kershner 1996); the FBI reports, moreover, that adolescent males accounted for the greatest increase in arrested rape perpetrators in the United States during the early 1990s (Ingrassia, Annin, Biddie, and Miller 1993; see also Kershner 1996).

Male youth violence against other males is extensive, creating battlefields out of city parks and school playgrounds. Gangs of all racial and ethnic groups flourish in urban areas, as adolescent boys attempt to create "family" with tools honed to incorporate ideals of manhood. In 1997, it was estimated that there were 30,500 youth gangs and 815,896 gang members active in the United States (National Youth Gang Center 1999). Teenaged boys represent, moreover, the majority of victims of violent crimes among youth, as they are also the majority of perpetrators. While pre-teen boys and girls are equally as likely to be homicide victims, once they reach their teen years, boys are significantly more likely than girls to be murdered (Snyder and Sickmund 1999). They are also more murderous than young women, representing 93% of known juvenile homicide offenders between 1980 and 1997. While, during the same time period, fewer than 10 juvenile homicide offenders per year were age 10 or younger, 88% of these offenders were also male (Snyder and Sickmund 1999:53–54).

Nor are other males the only victims of violence committed by boys and young men. In the United States, the school shootings of the 1990s (carried out overwhelmingly by boys, most of whom were from "good" [i.e., unbroken] homes) further attest to the lack of fit between how boys are learning to be men and the men that society wants. Disturbingly, a number of these rampages were orchestrated by boys who were seen by their peers, not as bullies (the masculine ideal), but as bullied (the feminine counterpart), thus highlighting the desperate actions sometimes undertaken by young men to prove their masculinity.

Men's violence against themselves, the final "corner" in Kaufman's triad of male violence, can also manifest itself in adolescence. One of the ways men do violence against themselves is by "stuffing" their emotions, in pursuit of a traditionally masculine ideal that reflects dread of feminine hyper-emotionality. Young men are encouraged to avoid displays of emotion, as are young boys; we even tend to "see" male newborns as less emotional than their female counterparts, reading onto them the expectations of masculine nonemotionality. As boys grow up, "they often fail to learn the language with which they could describe their feelings and without language it is hard for anyone to make sense of what he feels" (Phillips 1994:67). One articulation of this problem is the preponderance of suicide committed by male adolescents. In 1996, for example, 2,119 suicides in the United States involved youth under the age of 19, 80% of whom were male (Snyder and Sickmund 1999:24). Male youth suicide is a trend, moreover, that extends beyond the United States: a Finnish study of adolescent boys, for instance, showed that those young men with no diagnosable psychiatric disorders (that is, the "normal" boys) communicated an intent to commit suicide for the first time shortly before actually taking their own life, suggesting a relative lack of emotional communication to those who might otherwise provide help to them (Marttunen, Henriksson, Isometsa, Heikkinen, Aro, and Lonnqvist 1998).

Boys into Men: Preparation for Family Life

Just as boys are expected to reject their mothers and leave their families (physically and emotionally) in order to achieve manhood, so, too, they are expected to return to family life after a period of time to create and lead families of their own. By the end of adolescence, these young men have been socialized into, and internalized, the norms, values, and entitlements of the masculine ideal on a personal level, largely through interaction with gender-conscious parents and kin, and involvement with same-sex school peer groups. As they leave adolescence, in the interim between being banished from and returning to family life, however, boys-becoming-men are often subjected to a higher level of initiation into manhood involving male bonding and solidification of the collective practice of masculinity; these initiation rites tap into interests that extend, moreover, to corporate, state, and even global, levels (Connell 1998, 1990, 1987) and affect the ways men later interact in families. If athletic, young men join male-only football, basketball, or baseball teams; at college they are encouraged to belong to all-male fraternities; in the army, navy, marines, or air force, they are enlisted in the ranks of, if not an all-male grouping, one that is overwhelmingly so; and in the workplace, they enter sex-stratified occupational organizations. Each of these male-dominated associations has its own rituals that involve strengthening masculine ideals and notions of entitlement, already internalized at a personal level, at an abstract level that makes them appear to be, more than ever, part of the "natural" gender order. Full initiation into such groups usually involves some type of woman- and/or gay-bashing activity that accentuates the boundary between male and not male, masculinity and femininity, heterosexuality and homosexuality. These activities entail a "link between personal experience and power relations" (Connell 1990:507), or, more specifically, *collective* male experiences and power. Through such fratriarchal (Remy, 1990) activity as college fraternity pranks (Lyman 1998), collective condoning of gang or individual rape (Lefkowitz 1997; Sanday 1990), corporate victimization (Szockyj and Fox 1996) and sexual harassment of women and homosexuals (Connell 1992, 1995; Morris 1994), these organizations inaugurate boys into "real" manhood at a social level. With inauguration into the collective production of oppression, men become participants in and supporters of, to a more or lesser extent based on cross-cutting issues of race and class, social institutions of inequality such as sexism, racism, classism, and homosexism.

SELECTED REFERENCES

Adler, Patricia A., and Peter Adler. 1998. *Peer Power: Preadolescent Culture and Identity.* New Brunswick, NJ: Rutgers University Press.

Bem, Sandra Lipsitz. 1983. "Gender Schema Theory and Its Implications for Child Development: Raising Gender-Aschematic Children in a Gender-Schematic Society." *Signs* 8:598–616.

Bem, Sandra Lipsitz. 1993. *The Lenses of Gender: Transforming the Debate on Sexual Inequality.* New Haven, CT: Yale University Press.

Bhuiya, A., and K. Streatfield. 1991 "Mothers' Education and Survival of Female Children in a Rural Area of Bangladesh." *Population Studies* 45:253–264.

Bozinoff, Lorne, and Andre Turcotte. 1993. "Canadians Are Perplexed About Choosing Child's Sex." *Gallup Report:* 1–2.

Brush, Lisa D. 1999. "Gender, Work, Who Cares?!" Pp. 161–189 310 in *Revisioning Gender,* edited by M. M. Ferree, J. Lorder, and B. B. Hess. Thousand Oaks, CA: Sage Publications.

Cahill, Spencer E. 1986. "Childhood Socialization as a Recruitment Process: Some Lessons from the Study of Gender Development." *Sociological Studies of Child Development* 1:163–86.

Caldera, Yvonne M., Aletha C. Huston, and Marion O'Brien. 1989. "Social Interactions and Play Patterns of Parents and Toddlers with Feminine, Masculine, and Neutral Toys." *Child Development* 60:70–76.

Charlesworth, W. R., and C. Dzur. 1987. "Gender Comparisons of Preschoolers' Behavior and Resource Utilization in Group Problem-Solving." *Child Development* 58:191–200.

Chodorow, Nancy J. 1978. *The Reproduction of Mothering.* Berkeley, CA: University of California Press.

Christopher, F. Scott, and Susan Sprecher. 2000. "Sexuality in Marriage, Dating, and Other Relationships: A Decade Review." *Journal of Marriage and the Family* 62(4):999–1017.

Chunkath, Sheela Rani, and V. B. Athreya. 1997. "Female Infanticide in Tamil Nadu: Some Evidence." *Economic and Political Weekly* 32(17):21–28.

Clarke-Stewart, K. Allison, and C. M. Hevey. 1981. "Longitudinal Relations in Repeated Observations of Mother-Child Interaction from 1 to 1-1/2 Years." *Developmental Psychology* 17(2):127–145.

Coltrane, Scott. 1996. *Family Man: Fatherhood, Housework, and Gender Equity.* New York: Oxford University Press.

Coltrane, Scott. 1998. *Gender and Families.* Thousand Oaks, CA: Pine Forge Press.

Coltrane, Scott, and Michele Adams. 1997. "Children and Gender." Pp. 219–253 in *Contemporary Parenting: Challenges and Issues,* edited by T. Arendell. Thousand Oaks, CA: Sage Publications.

Connell, R.W. 1987. *Gender and Power: Society, the Person, and Sexual Politics.* Stanford, CA: Stanford University Press.

Connell, R.W. 1990. "The State, Gender and Sexual Politics." *Theory and Society* 19(5):507–544.

Connell, R.W. 1995. *Masculinities.* Berkeley, CA: University of California Press.

Corsaro, William A. 1997. *The Sociology of Childhood.* Thousand Oaks, CA: Pine Forge Press.

Cowan, G., and C. Hoffman. 1986. "Gender Stereotyping in Young Children: Evidence to Support a Concept-Learning Model." *Sex Roles* 14:211–224.

Eder, Donna, with Catherine Colleen Evans and Stephen Parker. 1995. *School Talk: Gender and Adolescent Culture.* New Brunswick, NJ: Rutgers University Press.

Elkin, Frederick, and Gerald Handel. 1989. *The Child and Society: The Process of Socialization.* New York: McGraw-Hill, Inc.

Fagot, Beverly I. 1985. "Beyond the Reinforcement Principle: Another Step Toward Understanding Sex-Role Development." *Developmental Psychology* 21:1097–1104.

Fagot, Beverly I., and Mary D. Leinbach. 1993. "Gender Role Development in Young Children: From Discrimination to Labeling." *Developmental Review* 13:205–224.

George, Sabu, Abel Rajaratnam, and B. D. Miller. 1992. "Female Infanticide in Rural South India." *Economic and Political Weekly* (May 10):1153–56.

Hardesty, Constance, Deeann Wenk, and Carolyn Stout Morgan. 1995. "Paternal Involvement and the Development of Gender Expectations in Sons and Daughters." *Youth & Society* 26(3):283–297.

Ingrassia, Michele, P. Annin, N. A. Biddle, and S. Miller. 1993. "Life Means Nothing." *Newsweek.* July 19:16–17.

Jones, Ann. 2000. *Next Time She'll Be Dead: Battering and How to Stop It.* Boston, MA: Beacon Press.

Jordan, Ellen, and Angela Cowan. 1995. "Warrior Narratives in the Kindergarten Classroom: Renegotiating the Social Contract?" *Gender and Society* 9(6):727–743.

Kaufman, Michael. 1993. *Cracking the Armour: Power, Pain and the Lives of Men.* Toronto: Viking.

Kaufman, Michael. 1998. "The Construction of Masculinity and the Triad of Men's Violence." Pp. 4–17 in *Men's Lives,* edited by M. S. Kimmel and M. A. Messner. Boston, MA: Allyn and Bacon.

Kaufman, Michael. 1999. "Men, Feminism and Men's Contradictory Experiences of Power." Pp. 75–103 in *Men and Power,* edited by J. A. Kuypers. Amherst, NY: Prometheus Books.

Kershner, Ruth. 1996. "Adolescent Attitudes About Rape." *Adolescence* 31(121):29–33.

Kimmel, Michael S. 2000. *The Gendered Society.* New York: Oxford University Press.

Kimmel, Michael, and Michael Messner, eds. 1998. *Men's Lives,* 4th Edition. Boston, MA: Allyn and Bacon.

Lorber, Judith. 1994. *Paradoxes of Gender.* New Haven, CT: Yale University Press.

Maccoby, Eleanor E. 1992. "The Role of Parents in the Socialization of Children: An Historical Overview." *Developmental Psychology* 28:1006–17.

Maccoby, Eleanor E. 1998. *The Two Sexes: Growing Up Apart, Coming Together.* Cambridge, MA: The Belknap Press of Harvard University Press.

Maccoby, Eleanor E., and Carol Nagy Jacklin. 1974. *The Psychology of Sex Differences.* Stanford, CA: Stanford University Press.

Marttunen, Marui, J., Markus M. Henriksson, Erkki T. Isometsa, Martti E. Heikkinen, Hillevi M. Aro, and Jouko K. Lonnqvist. 1998. "Completed Suicide Among Adolescents With No Diagnosable Psychiatric Disorder." *Adolescence* 33(131):669–681.

National Youth Gang Center. 1999. *1997 National Youth Gang Survey.* Washington, D.C.: Office of Juvenile Justice and Delinquency Prevention.

Nosaka, Akiko. 2000. "Effects of Child Gender Preference on Contraceptive Use in Rural Bangladesh." *Journal of Comparative Family Studies* 31(4):485–501.

Parke, Ross D. 1996. *Fatherhood.* Cambridge, MA: Harvard University Press.

Phillips, Angela. 1994. *The Trouble With Boys: A Wise and Sympathetic Guide to the Risky Business of Raising Sons.* New York: Basic Books.

Pleck, Joseph H., Freya L. Sonenstein, and Leighton C. Ku. 1994. "Problem Behaviors and Masculinity Ideology in Adolescent Males." Pp. 165–186 in *Adolescent Problem Behaviors: Issues and Research,* edited by R. D. Ketterlinus and M. E. Lamb. Hillsdale, NJ: Lawrence Erlbaum Associates, Publishers.

Pollack, William. 1998. *Real Boys: Rescuing Our Sons from the Myths of Boyhood.* New York: Henry Holt and Company.

Pomerleau, Andree, Daniel Bolduc, Gerard Malcuit, and Louise Cossette. 1990. "Pink or Blue:

Environmental Gender Stereotypes in the First Two Years of Life." *Sex Roles: A Journal of Research* 22(5–6):359–367.

Remy, John. 1990. "Patriarchy and Fratriarchy as Forms of Androcracy." Pp. 43–54 in *Men, Masculinities and Social Theory,* edited by J. Hearn and D. H. J. Morgan. London and Cambridge, MA: Unwin Hyman.

Risman, Barbara. 1989. "Can Men 'Mother'? Life as a Single Father." Pp. 155–164 in *Gender in Intimate Relationships: A Microstructural Approach,* edited by B. J. Risman and P. Schwartz. Belmont, CA: Wadsworth Publishing Company.

Rubin, J., R. Provenzano, and Z. Luria. 1974. "The Eye of the Beholder: Parents' Views on Sex of Newborns." *American Journal of Orthopsychiatry* 44:512–519.

Schwartz, Pepper, and Virginia Rutter. 1998. *The Gender of Sexuality.* Thousand Oaks, CA: Pine Forge Press.

Silverstein, Olga, and Beth Rashbaum. 1994. *The Courage to Raise Good Men.* New York: Penguin Books.

Snyder, Howard N., and Melissa Sickmund. 1999. *Juvenile Offenders and Victims: 1999 National Report.* Washington, D.C.: Office of Juvenile Justice and Delinquency Prevention.

Thorne, Barrie. 1993. *Gender Play: Girls and Boys in School.* New Brunswick, N.J.: Rutgers University Press.

Weiss, Gail. 1995. "Sex-Selective Abortion: A Relational Approach." *Hypatia* 10(1):202–217.

West, Candace, and Donald Zimmerman. 1987. "Doing Gender." *Gender and Society* 1:125–151.

Wood, Julia. 1994. *Gendered Lives: Communication, Gender, and Culture.* Belmont, CA: Wadsworth.

CHAPTER 6

Motherhood

Americans harbor unrealistically high expectations for mothers. We not only assume that mothers should and will provide unconditional love and limitless domestic services to children, husbands, and other family members, but also tend to hold them accountable if children or other family members turn out flawed. According to the British sociologist Ann Oakley, modern images of motherhood rest on three myths that are assumed to be eternal truths: (1) all women need to be mothers; (2) all mothers need their children; and (3) all children need their mothers. Although we tend to accept all three without question, Oakley shows how they are not literally true. First, not all women want to be mothers—or can be—as demonstrated by increasing rates of childlessness. Second, not all mothers need children—or very many of them—as seen by observing how women all over the world have attempted to limit the number of children they bear when given the choice. And finally, even though children need regular care, it is not necessary that the biological mother provide all, or even most, of it (Oakley, 1974). The articles in this chapter interrogate such myths by focusing on cultural expectations for maternal "instincts" and exploring variations in motherhood practices according to race, ethnicity, and social class.

As noted in chapter 4, throughout human history, a delicate balance was achieved between ecological carrying capacity and fertility. Because of low levels of protein intake, low ratios of female body fat, and late weaning, women exercised a sort of natural **birth control** and typically spread the births of their children more than four years apart. **Life expectancy** was short, people did not accumulate much wealth, there were relatively few distinctions between groups of people, and men and women cooperated in subsistence activities and some aspects of childcare. Much variation could be seen among the various hunter-gatherer societies around the world, but in general, women were valued for more than just their childbearing and child rearing, and most had some control over their own fertility. In traditional **agrarian societies** (e.g., ancient China, Egypt, medieval Europe) life was harsh for the peasants who made up the majority of the population. Rural landlords took most of the surplus crops and pressured peasant women to have many babies, who, if they survived, became child laborers and field hands. In many of these societies women had little power, so in spite of the difficulty of childbirth under poverty conditions and high maternal death rates, women were treated as breeders—their sexuality was tightly controlled and they were encouraged to bear many children. In societies where women had more power, they took whatever measures they could to keep the birthrate more under their own control. The same is true today in many less developed countries of the Third World. When they acquire more economic resources, women tend to reduce their own birthrates by gaining access to contraceptives or abortions, sometimes in opposition to their husbands (Blumberg, 1984).

Colonial American society had elements of these agrarian practices, as well as a need for more children to help work the land that was usurped from the more nomadic Native Americans who preceded them. As America turned to a commercial economy in the nineteenth century and eventually to urban industrial production in the twentieth century, the moral responsibility for children shifted from fathers to mothers. As men increasingly left the home to go to work for wages, the cult of domesticity glorified motherhood and reassured middle-class women that their natural place was in the home. Motherhood was placed on a pedestal, and the contrast between the outside world and home came to be seen as a contrast between Man and Woman (Coltrane, 1998; Skolnick, 1991). According to the **ideology of separate spheres,** men's impure public activities in business and politics rendered them in need of redemption via the grace and purity of the womanly home. Domestic tasks assumed a spiritual importance as the middle-class home was transformed into a "private" place where women were expected to comfort and civilize both men and children. Homemaking became exclusively the woman's domain. Although working-class women continued to work for pay, middle- and upper-class women transformed homemaking into a profession. The rise of scientific mothering and the home economics movement promoted the idea that the home was a private haven under women's control. The growth of the middle class meant that many more "mistresses" demanded the help of maids and nannies to cook, clean, and raise children. During the early part of the twentieth century, black women were most likely to be servants and laundresses, especially in the South, but in the North as well. In the Southwest, Chicanas were disproportionately concentrated in domestic service, and in the Far West (especially in California and Hawaii), Asian men were most often household servants (Glenn, 1992). As the "hiring class" expanded, middle-class homemakers came to think of themselves as supervisors whose superior knowledge allowed them to manage and oversee the manual labor of the servants they employed (Romero, 1992). Turning the middle-class home into the woman's domain, and treating **homemaking** and motherhood as a profession, gave some wives managerial control over day-to-day domestic activities, but it also subordinated women's needs to those of their husbands and families (Bose, 1987; Skolnick, 1991).

According to the ideology of **separate spheres** and the cult of domesticity (also called the cult of true womanhood), middle-class white women were supposed to realize their "true" nature by marrying, giving birth, and most important, tending children. Motherhood was elevated to a revered status, and wives' **homemaking** came to be seen as a moral calling and a worthy profession. The True Woman was supposed to be inherently unselfish, and her moral purity, nurturant character, and gentle temperament were seen as uniquely qualifying her to rear young children (Coontz, 1992). Arlene Skolnick (1991) notes that the development of the modern private family brought new burdens for middle- and upper-class women, because they were supposed to create a wholesome home life, which, in turn, was seen as the only way to redeem society. Under the previous agricultural economy, the community regulated most family functions and repaired any moral defects of families. In the newer model common to **industrial societies,** families and especially mothers, were supposed to compensate for the moral defects of the larger society. Throughout most of the twentieth century, middle-class women were expected to be consumed and fulfilled by their "natural" wifely and motherly duties. Isolated in suburban houses, many mothers assumed almost sole responsibility for raising children, aided by occasional reference to Dr. Spock or some other child-rearing manual (Hays, 1996).

The idea that all women should be mothers and that they should gain intense satisfaction from it has been characterized as **compulsory motherhood** (Pogrebin, 1983). According to this view, a woman's ultimate purpose is to be a mother and everything else she does is secondary. A woman's well-being is so tied up with mothering that her identity is sometimes assumed to be tenuous and trivial without it. Ironically, researchers have found that mothers of young children have lower levels of well-being than other women, primarily because in the absence of substantial help and support, the day-to-day activities of mothering small children are isolating, tedious, and unrelenting. Compulsory motherhood implies that women should find total fulfillment in having children and taking care of them. Although most women do gain profound satisfaction from caring for children, research shows that women without children have about the same level of personal well-being as women with children. Today, motherhood is rarely the sole basis for a woman's identity, especially since most women are spending less time as mothers of young children than any time in recent history. About one in five American women never do become mothers, and the average American woman now spends only about one-seventh of her adult life either pregnant, nursing, or caring for preschool children (Coltrane, 1998).

Nevertheless, motherhood ideals are so pervasive that we tend to generalize them to all women in all times and places. As potential mothers, women and girls are assumed to be more kind, caring, and nurturant than men and boys. Women *do* provide more care and emotional support to other people than men, but conventional expectations for "intensive mothering" are socially constructed and subject to change (Hays, 1996). History provides plentiful examples of mothers who ignore their children's needs or commit **infanticide** and of fathers who provide nurturing care to their offspring. Biological sex differences are obviously important when we consider men's and women's roles in pregnancy and nursing, but beyond the basic physical facts of bearing children and breast-feeding, parenting and other care-giving behaviors are learned. Women, like men, have to learn how to take care of babies. New mothers report that in spite of their hope for a "maternal instinct" to show them how to parent, their mothering skills actually develop through trial and error (Cowan & Cowan, 1992; Glenn, 1994; Hays, 1996). They read parenting books, talk to other mothers, and figure out what to do by interacting with their children. And in general, women have been more likely to learn how to care for children and other family members than men because personal, cultural, and economic circumstances have encouraged them to do so (Coltrane, 1998).

The first selection in this chapter comes from one of the pioneers of the campaign to gain women the right to vote. Elizabeth Cady Stanton draws on her own nineteenth-century experiences of having a newborn baby to comment on the lack of social preparation for motherhood. Although most people then—as now—assumed that mothering is instinctual and automatic, Elizabeth Cady Stanton's essay shows how good baby care is a combination of knowledge, experience, and reason. Mothers must learn to care for infants, just as all humans learn other important life skills. This theme was reiterated by the sociologist Alice Rossi in 1968, when she described how unprepared most people are when they assume parental duties. Like Stanton, Rossi noted that motherhood has been defined as essential to a woman attaining full adult status in our society, but that unlike most other societies, we isolate mothers in individual households where they rarely receive the help or training that they need to be good mothers. Unlike most roles that adults assume in our society, parents receive little preparation for assuming the tasks of rearing children. Most jobs afford people opportunities to get training or to serve as apprentices before they gradually assume full responsibility for their occupational duties. Parenthood, in contrast, comes all at once and is irrevocable, and its training period—pregnancy—teaches us almost nothing about infant care. We are not only ill prepared for all the details of caring for a newborn baby, but also after the baby is born, have to be "on duty" every day, twenty-four hours a day (Coltrane and Collins, 2001). How do you think our society could do a better job of preparing mothers and fathers to be parents?

The second article in this chapter turns our attention toward issues of race and racial inequality. In "Fictive Kin, Paper Sons, and Compadrazgo," Bonnie Thornton Dill provides an historical account of the ways that women of color in the United States were valued "as workers, breeders, and entertainers of workers, not as family members." Denied the social, legal, and economic supports needed to maintain family life, women of color developed both individual and collective ways to sustain their families, nurture their children, and resist oppression. Thornton Dill first gives a brief review of African American slave families and the enormous difficulties they faced. Not only were families routinely disrupted, but also black women were routinely raped and black men were denied the standard social privileges of manhood. Thornton Dill focuses on the enduring importance of the mother–child tie in African American families and the relatively egalitarian division of labor that evolved between black men and women. She next considers Chinese sojourner families who were denied the right to form families through both law and social custom. Split households, with wives remaining in China and husbands working in the United States, were used as a survival strategy, with families waiting a generation or more before husbands and wives could live together and raise children. The third group Thornton Dill considers is Mexican Americans, who were often forced to live in migratory labor camps or barrios and whose families developed unique survival strategies based on cultural traditions to deal with widespread racism and segregation. As Thornton Dill points out, these three groups of women struggled to maintain family units in the face of many hardships. Their labor in the productive sphere was required to achieve even minimal levels of family subsistence and they suffered high rates of **infant mortality** and disease and had a short life span. Their strong commitment to motherhood cannot be understood without reference to racial oppression they faced and the creative adaptations they devised to maintain their families.

The last selection in chapter 6 reports on the phenomenon of Latina transnational motherhood. Echoing some of the themes in the historical selection from Bonnie Thornton Dill, Pierrette Hondagneu Sotelo and Ernestine Avila focus on the experiences of contemporary Mexican and Central American immigrant women who are compelled to live apart from their children and perform domestic work in the United States. Transnational mothers typically use their own mothers, other relatives, or comadres (co-mothers) to care for their children while they are away, sending earnings back home so that their children can have adequate food, shelter, clothing, and education. In many cases these women have taken jobs and assumed **breadwinning** roles because they can find employment more easily and with more consistency than their husbands who must work as day laborers or migrant field hands in an ethnically stratified labor market. Such choices are difficult and fraught with sadness for most mothers. This article highlights how labor market practices and immigration policies help shape mothering practices, allowing more privileged professional or middle-class women to hire child-rearing and housekeeping services of immigrant women who are faced with difficult choices about caring for their own children back home.

Collectively, the articles in this chapter suggest that motherhood is not biologically predetermined but rather socially constructed within specific historical, cultural, and economic circumstances. The selections also confirm that multiple forms of motherhood coexist in any era. The cult of true womanhood, under which women are confined to the domestic sphere, was (and is) principally a white middle-class phenomenon. If we are to move beyond nostalgic, ethnocentric, and class-bound images of motherhood, we will need to develop more inclusive models of what it means to be a mother today.

REFERENCES

Blumberg, Rae Lesser. 1984. A General Theory of Gender Stratification. In *Sociological Theory 1984,* edited by R. Collins. San Francisco: Jossey-Bass.

Bose, Christine. 1987. Dual Spheres. In *Analyzing Gender,* edited by B. Hess and M. M. Ferree. Newbury Park, CA: Sage.

Coltrane, Scott. 1998. *Gender and Families.* Thousand Oaks, CA: Pine Forge Press.

Coltrane, Scott, and Randall Collins. 2001. *Sociology of Marriage and the Family: Gender, Love, and Property.* 5th ed. Belmont, CA: Wadsworth.

Coontz, Stephanie. 1992. *The Way We Never Were.* New York: Basic Books.

Cowan, Carolyn P., and Phillip A. Cowan. 1992. *When Partners Become Parents.* New York: Basic Books.

Glenn, Evelyn Nakano. 1992. From servitude to service work: Historical continuities in the racial division of women's work. *Signs* 18:1–43.

Glenn, Evelyn Nakano. 1994. Social Constructions of Mothering. In *Mothering: Ideology, Experience, and Agency,* edited by E. N. Glenn, G. Chang, and L. R. Forcey. New York: Routledge.

Hays, Sharon. 1996. *The Cultural Contradictions of Motherhood.* New Haven: Yale University Press.

Oakley, Anne. 1974. *The Sociology of Housework.* New York: Pantheon.

Pogrebin, Letty C. 1983. *Family Politics.* New York: McGraw-Hill.

Romero, Mary. 1992. *Maid in the U.S.A.* New York: Routledge.

Rossi, Alice S. 1968. Transition to parenthood. *Journal of Marriage and the Family* 30:26–39.

Skolnick, Arlene. 1991. *Embattled Paradise: The American Family in an Age of Uncertainty.* New York: Basic Books.

SUGGESTED READINGS

Bassin, Donna, Margaret Honey, and Meryle Mahrer Kaplan, eds. 1994. *Representations of Motherhood.* New Haven, CT: Yale University Press.

Chodorow, Nancy. 1978. *The Reproduction of Mothering: Psychoanalysis and the Sociology of Gender.* Berkeley, CA: University of California Press.

Connolly, Deborah R. 2000. *Homeless Mothers: Face to Face with Women and Poverty.* Minneapolis: University of Minnesota Press.

Fineman, Martha Albertson, and Isabel Karpin, eds. 1995. *Mothers in Law: Feminist Theory and the Legal Regulation of Motherhood.* New York: Columbia University Press.

Garey, Anita. 1999. *Weaving Work and Motherhood.* Philadelphia, PA: Temple University Press.

Glenn, Evelyn Nakano, Grace Chang, and Linda Rennie Forcey, eds. 1994. *Mothering: Ideology, Experience, and Agency.* New York: Routledge.

Hrdy, Sarah Blaffer. 1999. *Mother Nature: Maternal Instincts and How They Shape the Human Species.* New York: Ballantine Books.

Johnson, Miriam M. 1988. *Strong Mothers, Weak Wives.* Berkeley: University of California Press.

Kaplan, E. Ann. 1992. *Motherhood and Representation: The Mother in Popular Culture and Melodrama.* London, UK: Routledge.

Ludtke, Melissa. 1997. *On Our Own: Unmarried Motherhood in America.* New York: Random House.

McMahon, Martha. 1995. *Engendering Motherhood: Identity and Self-Transformation in Women's Lives.* New York: Guilford Press.

Peters, Joan K. 1997. *When Mothers Work: Loving Our Children without Sacrificing Our Selves.* Reading, MA: Addison-Wesley.

Smart, Carol, ed. 1992. *Regulating Womanhood: Historical Essays on Marriage, Motherhood, and Sexuality.* London, UK: Routledge.

INFOTRAC® COLLEGE EDITION EXERCISES

The exercises that follow allow you to use the InfoTrac® College Edition online database of scholarly articles to explore the sociological implications of the selections in this chapter.

Search Keyword: Pronatalism. Using InfoTrac® College Edition, search for articles that talk about pronatalism. How do these selections describe pronatalism? What does it mean to be pronatalist? How is pronatalism reflected in the culture? Under what circumstances is pronatalism seen as a problem? How might pronatalism be used as a tool for social control?

Search Keyword: Cult of domesticity. Find articles that discuss the cult of domesticity or the cult of true womanhood. How is it defined? What are the ideal characteristics of women according to the cult of domesticity (or cult of true womanhood)? How have women been affected over the course of the nineteenth and twentieth centuries by this ideology? How have men been affected by it? Is there still evidence of the cult of domesticity today? If so, describe that evidence.

Search Keyword: Separate spheres. Use InfoTrac® College Edition to look for selections that talk about separate spheres ideology. What do these articles mean by "separate spheres"? Which sphere is associated with men? Which is associated with women? How are social power and social control implicated in separate spheres?

Search Keywords: Women and childcare. Do a keyword search in InfoTrac® College Edition looking for articles that discuss "women" and "childcare." What attitude is reflected in these selections about women and childcare? How is childcare seen to be a "woman's job"? Why do you think women are often portrayed as "ideal" childcare providers? Are patterns of childcare between women and men changing at all? If so, how?

Search Keyword: Surrogate mothers. Find articles that deal with surrogate mothers. What is a surrogate mother? In the articles that you find, what is the attitude toward surrogate mothers? How are surrogate mothers different than "real" mothers? How are they the same? What are the political controversies that arise over the issue of surrogacy? Why do these controversies arise?

Search Keywords: Parenting and Dr. Spock. Search for selections that talk about parenting advice manuals such as those written by Dr. Spock. What kinds of advice do these manuals give to mothers? What do the authors of these manuals think mothers should be like? What kinds of difficulties and contradictions do these manuals impose on mothers who also work outside of the home? What effect does this type of parenting advice have on mothers generally?

19

Motherhood

ELIZABETH CADY STANTON,

circa 1860

. . . Though motherhood is the most important of all the professions—requiring more knowledge than any other department in human affairs—there was no attention given to preparation for this office. If we buy a plant of a horticulturist we ask him many questions as to its needs, whether it thrives best in sunshine or in shade, whether it needs much or little water, what degrees of heat or cold; but when we hold in our arms for the first time a being of infinite possibilities, in whose wisdom may rest the destiny of a nation, we take it for granted that the laws governing its life, health, and happiness are intuitively understood, that there is nothing new to be learned in regard to it. Here is a science to which philosophers have as yet given but little attention. An important fact has only been discovered and acted upon within the last ten years; that children come into the world tired, and not hungry, exhausted with the perilous journey. Instead of being kept on the rack while the nurse makes a prolonged toilet and feeds it some nostrum supposed to have much-needed medicinal influence, the child's face, eyes, and mouth should be carefully washed, and the rest of its body thoroughly oiled, and then it should be slipped into a soft pillow case, wrapped in a blanket, and laid to sleep. Ordinarily, in the proper conditions, with its face uncovered in a cool, pure atmosphere, it will sleep twelve hours. Then it should be bathed, fed, and clothed in a high-neck, long-sleeved silk shirt and a blanket. As babies lie still most of the time for the first six weeks, they need no elaborate dressing. I think the nurse was a full hour bathing and dressing my first-born, who protested with a melancholy wail every blessed minute.

Ignorant myself of the initiative steps on the threshold of time, I supposed this proceeding was approved by the best authorities. However, I had been thinking, reading, observing, and had as little faith in the popular theories in regard to babies as on any other subject. I saw them, on all sides, ill half the time, pale and peevish, dying early, having no joy in life. I heard parents complaining of weary days and sleepless nights, while each child in turn ran the gauntlet of red gum, whooping cough, chicken pox, mumps, measles, and fits. Everyone seemed to think these inflictions were a part of the eternal plan—that Providence had a kind of Pandora's box, from which he scattered these venerable diseases most liberally among those whom he especially loved. Having gone through the ordeal of bearing a child, I was determined, if possible, to keep him, so I read

From Elizabeth Cady Stanton, 1922. "Motherhood" pp. 109–124 in Theodore Stanton and Harriot Stanton Blatch (eds.) *Elizabeth Cady Stanton: As Revealed in Her Letters, Diary, and Reminiscences*. New York: Harper and Brothers.

everything I could find on babies. But the literature on this subject was as confusing and unsatisfactory as the longer and shorter catechism and the Thirty-nine Articles of our faith. I had recently visited our dear friends, Theodore and Angelina Grimké-Weld, and they warned me against books on this subject. They had been so misled by one author, who assured them that the stomach of a child could only hold one tablespoonful, that they nearly starved their first-born to death. Though the child dwindled day by day, and, at the end of a month looked like a little old man, yet they still stood by the distinguished author. Fortunately, they both went off one day and left the child with "Sister Sarah," who thought she would make an experiment and see what a child's stomach could hold, as she had grave doubts about the tablespoonful theory. To her surprise the baby took a pint bottle full of milk, and had the sweetest sleep thereon he had known in his earthly career. After that he was permitted to take what he wanted, and "the author" was informed of his libel on the infantile stomach.

So here again I was entirely afloat, launched on the seas of doubt without chart or compass. The life and well-being of the race seemed to hang on the slender thread of such traditions as were handed down by ignorant mothers and nurses. One powerful ray of light illuminated the darkness; it was the work of Andrew Combe on *Infancy.* He had evidently watched some of the manifestations of man in the first stages of his development, and could tell at least as much of babies as naturalists could of beetles and bees. He did give young mothers some hints of what to do, and the whys and wherefores of certain lines of procedure. I read several chapters to the nurse. Although out of her ten children she had buried five, she still had too much confidence in her own wisdom and experience to pay much attention to any new idea that might be suggested to her. Among other things, Combe said that a child's bath should be regulated by the thermometer, in order to be always of the same temperature. She ridiculed the idea, and said her elbow was better than any thermometer, and, when I insisted on its use, she would invariably, with a smile of derision, put her elbow in first, to show how exactly it tallied with the thermometer.

When I insisted that the child should not be bandaged, she rebelled outright, and said she would not take the responsibility of caring for a child without a bandage. I said: "Pray, sit down, dear nurse, and let us reason together. Do not think I am setting up my judgment against yours, with all your experience. I am simply trying to act on the opinions of a distinguished physician, who says there should be no pressure on a child anywhere; that the limbs and body should be free; that it is cruel to bandage an infant from hip to armpit, as is usually done in America; or both body and legs, as is done in Europe; or strap them to boards, as is done by savages on both continents. Can you give me one good reason, nurse, why a child should be bandaged?" "Yes," she said emphatically, "I can give you a dozen." "I only asked for one," I replied. "Well," said she, after much hesitation, "the bones of a newborn infant are soft, like cartilage, and, unless you pin them up snugly there is danger of their falling apart." "It seems to me," I replied, "you have given the strongest reason why they should be carefully guarded against the slightest pressure. It is very remarkable that kittens and puppies should be so well put together that they need no artificial bracing, and the human family be left wholly to the mercy of a bandage. Suppose a child was born where you could not get a bandage, what then? Now, I think this child will remain intact without a bandage, and, if I am willing to take the risk, why should you complain?" "Because," said she, "if the child should die, it would injure my name as a nurse. I therefore wash my hands of all these new-fangled notions."

So she put a bandage on the child every morning, and I as regularly took it off. It has been fully proved since to be as useless an appendage as the vermiform. She had several cups with various concoctions of herbs standing in the chimney corner, ready for insomnia, colic, indigestion, etc., etc., all of which were spirited away when she was at her dinner. . . . I told her that if she would wash the baby's mouth with pure cold water morning and night, and give it a teaspoonful to drink occasionally during the day, there would be no danger of red gum; that if she would keep the blinds open and let in the air and sunshine, keep the temperature of the room at sixty-five degrees, leave the child's head uncovered so that it could breathe freely, stop rocking and trotting it, and singing such melancholy hymns as "Hark, from the tombs a doleful sound!" the baby and I would both be able to weather the cape. I told her I should nurse the child once in two hours, and that she must not feed it any of her nostrums in the meantime; that a child's stomach, being made on the same general plan as our own, needed intervals of rest as well as ours. She said it would be racked with colic if the stomach was empty any length of time, and that it would surely have rickets if it were kept too still. I told her if the child had no anodynes, nature would regulate its sleep. She said she could not stay in a room with the thermometer at sixty-five degrees, so I told her to sit in the next room and regulate the heat to suit herself; that I would ring a bell when her services were needed. . . .

Besides the obstinacy of the nurse, I had the ignorance of physicians to contend with. When the child was four days old we discovered that the collar bone was bent. The physician, wishing to get a pressure on the shoulder, braced the bandage round the wrist. "Leave that," he said, "ten days, and then it will be all right." Soon after he left I noticed that the child's hand was blue, showing that the circulation was impeded. "That will never do," said I; "nurse, take it off." "No, indeed," she answered, "I shall never interfere with the doctor." So I took it off myself, and sent for another doctor, who was said to know more of surgery. He expressed great surprise that the first physician called should have put on so severe a bandage. "That," said he, "would do for a grown man, but ten days of it on a child would make him a cripple." However, he did nearly the same thing, only fastening it round the hand instead of the wrist. I soon saw that the ends of the fingers were all purple, and that to leave that on ten days would be as dangerous as the first. So I took it off.

"What a woman!" exclaimed the nurse. "What do you propose to do?" "Think out something better myself; so brace me up with some pillows and give the baby to me." She looked at me aghast. "Now," I said, talking partly to myself and partly to her, "what we want is a little pressure on that bone; that is what both of those men have aimed at. How can we get it without involving the arm, is the question?" "I am sure I don't know," said she, rubbing her hands and taking two or three brisk turns around the room. "Well, bring me three strips of linen, four double." I then folded one, wet in arnica and water, and laid it on the collar bone, put two other bands, like a pair of suspenders over the shoulder, crossing them both in front and behind, pinning the ends to the diaper, which gave the needed pressure without impeding the circulation anywhere. As I finished she gave me a look of budding confidence, and seemed satisfied that all was well. Several times, night and day, we wet the compress and readjusted the bands, until all appearance of inflammation had subsided.

At the end of ten days the two sons of Æsculapius appeared and made their examination, and said all was right, whereupon I told them how badly their bandages worked, and what I had done myself. They smiled at each other, and one said, "Well, after all, a mother's instinct is better than a man's reason." "Thank you, gentlemen, there was no instinct about it. I did some hard thinking before I saw how I could get pressure on the shoulder without impeding the circulation, as you did." Thus, in the supreme moment of a young mother's life, when I needed tender care and support, the whole responsibility of my child's supervision fell upon me; but though uncertain at every step of my own knowledge, I learned another lesson in self-reliance. I trusted neither men nor books absolutely after this, either in regard to the heavens above or the earth beneath, but continued to use my "mother's instinct," if "reason" is too dignified a term to apply to a woman's thoughts. My advice to every mother is, above all other arts and sciences, study first what relates to babyhood, as there is no department of human action in which there is such lamentable ignorance.

20

Fictive Kin, Paper Sons, and Compadrazgo

BONNIE THORNTON DILL,
1988

Race has been fundamental to the construction of families in the United States since the country was settled. People of color were incorporated into the country and used to meet the need for cheap and exploitable labor. Little attention was given to their family and community life except as it related to their economic productivity. Upon their founding, the various colonies that ultimately formed the United States initiated legal, economic, political, and social practices designed to promote the growth of family life among European colonists. As the primary laborers in the reproduction and maintenance of families, White[1] women settlers were accorded the privileges and protection considered socially appropriate to their family roles. The structure of family life during this era was strongly patriarchal: denying women many rights, constraining their personal autonomy, and making them subject to the almost unfettered will of the male head of the household. Nevertheless, women were rewarded and protected within patriarchal families because their labor was recognized as essential to the maintenance and sustenance of family life. In addition, families were seen as the cornerstone of an incipient nation, and thus their existence was a matter of national interest.

In contrast, women of color experienced the oppression of a patriarchal society but were denied the protection and buffering of a patriarchal family. Although the presence of women of color was equally important to the growth of the nation, their value was based on their potential as workers, breeders, and entertainers of workers, not as family members. In the eighteenth and nineteenth centuries, labor, and not the existence or maintenance of families, was the critical aspect of their role in building the nation. Thus they were denied the societal supports necessary to make their families a vital element in the social order. For women of color, family membership was not a key means of access to participation in the wider society. In some instances racial-ethnic families were seen as a threat to the efficiency and exploitability of the work force and were actively prohibited. In other cases, they were tolerated when it was felt they might help solidify or expand the work force. The lack of social, legal, and economic support for the family life of people of color intensified and extended women's work, created tensions and strains in family relationships, and set the stage for a variety of creative and adaptive forms of resistance.

From Bonnie Thornton Dill, "Our Mothers' Grief: Racial Ethnic Women . . . ," *Journal of Family History,* vol. 13, no. 4, pp. 418–429,

AFRICAN AMERICAN SLAVES

Among students of slavery, there has been considerable debate over the relative "harshness" of American slavery, and the degree to which slaves were permitted or encouraged to form families. It is generally acknowledged that many slave owners found it economically advantageous to encourage family formation as a way of reproducing and perpetuating the slave labor force. This became increasingly true after 1807, when the importation of African slaves was explicitly prohibited. The existence of these families and many aspects of their functioning, however, were directly controlled by the master. Slaves married and formed families, but these groupings were completely subject to the master's decision to let them remain intact. One study has estimated that about 32 percent of all recorded slave marriages were disrupted by sale, about 45 percent by death of a spouse, about 10 percent by choice, and only 13 percent were not disrupted (Blassingame 1972). African slaves thus quickly learned that they had a limited degree of control over the formation and maintenance of their marriages and could not be assured of keeping their children with them. The threat of disruption was one of the most direct and pervasive assaults on families that slaves encountered. Yet there were a number of other aspects of the slave system that reinforced the precariousness of slave family life.

In contrast to some African traditions and the Euro-American patterns of the period, slave men were not the main providers or authority figures in the family. The mother-child tie was basic and of greatest interest to the slave owner because it was essential to the reproduction of the labor force.

In addition to the lack of authority and economic autonomy experienced by the husband–father in the slave family, use of rape of women slaves as a weapon of terror and control further undermined the integrity of the slave family.

> It would be a mistake to regard the institutionalized pattern of rape during slavery as an expression of white men's sexual urges, otherwise stifled by the specter of the white womanhood's chastity. . . . Rape was a weapon of domination, a weapon of repression, whose covert goal was to extinguish slave women's will to resist, and in the process, to demoralize their men. (Davis 1981: 23–24)

The slave family, therefore, was at the heart of a peculiar tension in the master–slave relationship. On the one hand, slave owners sought to encourage familiarities among slaves because, as Julie Matthaei (1982:81) states, "These provided the basis of the development of the slave into a self-conscious socialized human being." They also hoped and believed that this socialization process would help children learn to accept their place in society as slaves. Yet the master's need to control and intervene in the family life of the slaves is indicative of the other side of this tension. Family ties had the potential to become a competing and more potent source of allegiance than the master. Also, kin were as likely to socialize children in forms of resistance as in acts of compliance.

It was within this context of surveillance, assault, and ambivalence that slave women's reproductive labor took place. They and their menfolk had the task of preserving the human and family ties that could ultimately give them a reason for living. They had to socialize their children to believe in the possibility of a life in which they were not enslaved. The slave woman's labor on behalf of the family was, as Angela Davis (1971) has pointed out, the only labor in which the slave engaged that could not be directly used by the slave owner for his own profit. Yet, it was crucial to the reproduction of the slave owner's labor force, and thus a source of strong ambivalence for many slave women. Whereas some mothers murdered their babies to keep them from being slaves, many sought autonomy and creativity within the family that was denied them in other realms of the society. The maintenance of a distinct African American culture is testimony to the ways in which slaves maintained a degree of cultural autonomy and resisted the creation of a slave family that only served the needs of the master.

Herbert Gutman (1976) gives evidence of the ways which slaves expressed a unique African-American culture through their family practices. He provides data on naming patterns and kinship ties among slaves that fly in the face of the dominant ideology of the period, which argued that slaves were immoral and had little concern for or appreciation of family life. Yet Gutman demonstrates that within a system that denied the father authority over his family, slave boys were frequently named after their fathers, and many children were named after blood relatives as a way of maintaining family ties. Gutman also suggests that after emancipation a number of slaves took the names of former owners in order to reestablish family ties that had been disrupted earlier: On plantation after plantation, Gutman found considerable evidence of the building and maintenance of extensive kinship ties among slaves. In instances where slave families had been disrupted, slaves in new communities reconstituted the kinds of family and kin ties that came to characterize Black family life throughout the South. The patterns included, but were not limited to, a belief in the importance of marriage as a long-term commitment, rules of exogamy that excluded marriage between first cousins, and acceptance of women who had children outside of marriage. Kinship networks were an important source of resistance to the organization of labor that treated the individual slave, and not the family, as the unit of labor (Caulfield 1974). . . .

Perhaps most critical in developing an understanding of slave women's reproductive labor is the gender-based division of labor in the domestic sphere. The organization of slave labor enforced considerable equality among men and women. The ways in which equality in the labor force was translated into the family sphere is somewhat speculative. Davis (1981:18), for example, suggests that egalitarianism between males and females was a direct result of slavery: "Within the confines of their family and community life, therefore, Black people managed to accomplish a magnificent feat. They transformed that negative equality which emanated from the equal oppression they suffered as slaves into a positive quality; the egalitarianism characterizing their social relations."

It is likely, however, that this transformation was far less direct than Davis implies. We know, for example, that slave women experienced what has recently been called the "double day" before most other women in this society. Slave narratives (Jones 1985; White 1985; Blassingame 1977) reveal that women had primary responsibility for their family's domestic chores. They cooked (although on some plantations meals were prepared for all the slaves), sewed, cared for their children, and cleaned house after completing a full day of labor for the master. John Blassingame (1972) and others have pointed out that slave men engaged in hunting, trapping, perhaps some gardening, and furniture making as ways of contributing to the maintenance of their families. Clearly, a gender-based division of labor did exist within the family, and it appears that women bore the larger share of the burden for housekeeping and child care.

In contrast to White families of the period, however, the division of labor in the domestic sphere was reinforced neither in the relationship of slave women to work nor in the social institutions of the slave community. The gender-based division of labor among the slaves existed within a social system that treated men and women as almost equal, independent units of labor. Thus Matthaei (1982:94) is probably correct in concluding that

> whereas . . . the white homemaker interacted with the public sphere through her husband, and had her work life determined by him, the enslaved Afro-American homemaker was directly subordinated to and determined by her owner. . . . The equal enslavement of husband and wife gave the slave marriage a curious kind of equality, an equality of oppression.

Black men were denied the male resources of a patriarchal society and therefore were unable to turn gender distinctions into female subordination, even if that had been their desire. Black women, on the other hand, were denied support and protection for their roles as mothers and wives, and thus had to modify and structure those roles around the demands of their labor. Reproductive labor for slave women was intensified in several ways: by the demands of slave labor that forced them into the double day of work; by the desire and need to maintain family ties in the face of a system that gave them only limited recognition; by the stresses of building a family with men who were denied the standard social privileges of manhood; and by the struggle to raise children who could survive in a hostile environment.

This intensification of reproductive labor made networks of kin and fictive kin important instruments in carrying out the reproductive tasks of the slave community. Given an African cultural heritage where kinship ties formed the basis of social relations, it is not at all surprising that African American slaves developed an extensive system of kinship ties and obligations (Gutman 1976; Sudarkasa 1981). Research on Black families in slavery provides considerable documentation of participation of extended kin in child rearing, childbirth, and other domestic, social, and economic activities (Gutman 1976; Blassingame 1972; Genovese and Miller 1974).

After slavery, these ties continued to be an important factor linking individual household units in a variety of domestic activities. While kinship ties were also important among native-born Whites and European immigrants, Gutman (1976:213) has suggested that these ties

> were comparatively more important to Afro-Americans than to lower-class native white and immigrant Americans, the result of their distinctive low economic status, a condition that denied them the advantages of an extensive associational life beyond the kin group and the advantages and disadvantages resulting from mobility opportunities.

His argument is reaffirmed by research on African American families after slavery (Shimkin et al. 1978; Aschenbrenner 1975; Davis 1981; Stack 1974). Niara Sudarkasa (1981:49) takes this argument one step further, linking this pattern to the African cultural heritage.

> Historical realities require that the derivation of this aspect of Black family organization be traced to its African antecedents. Such a view does not deny the adaptive significance of consanguineal networks. In fact, it helps to clarify why these networks had the flexibility they had and why they, rather than conjugal relationships, came to be the stabilizing factor in Black families.

In individual households, the gender-based division of labor experienced some important shifts during emancipation. In their first real opportunity to establish family life beyond the controls and constraints imposed by a slave master, Black sharecroppers' family life changed radically. Most women, at least those who were wives and daughters of able-bodied men, withdrew from field labor and concentrated on their domestic duties in the home. Husbands took primary responsibility for the fieldwork and for relations with the owners, such as signing contracts on behalf of the family. Black women were severely criticized by Whites for removing themselves from field labor because they were seen to be aspiring to a model of womanhood that was considered inappropriate for them. The reorganization of female labor, however, represented an attempt on the part of Blacks to protect women from some of the abuses of the slave system and to thus secure their family life. It was more likely a response to the particular set of circumstances that the newly freed slaves faced than a reaction to the lives of their former masters. Jacqueline Jones (1985) argues that these patterns were "particularly significant" because at a time when industrial development was introducing a labor system that divided male and female labor, the freed Black family was establishing a pattern of joint work and complementarity of tasks between males and females that was reminiscent of preindustrial American families. Unfortunately, these former slaves had to do this without the institutional supports given white farm families and within a sharecropping system that deprived them of economic independence.

CHINESE SOJOURNERS

An increase in the African slave population was a desired goal. Therefore, Africans were permitted and even encouraged at times to form families, as long as they were under the direct control of the slave master. By sharp contrast, Chinese people were explicitly denied the right to form families in the United States through both law and social practice. Although male laborers began coming to the United States in sizable numbers in the middle of the nineteenth century, it was more than a century before an appreciable number of children of Chinese parents were born in America. Tom, a respondent in Victor Nee and Brett de Bary Nee's book, *Longtime Californ',* says: "One thing about Chinese men in America was you had to be either a merchant or a big gambler, have lot of side money to have a family here. A working man, an ordinary man, just can't!" (1973:80).

Working in the United States was a means of gaining support for one's family with an end of obtaining sufficient capital to return to China and purchase land. This practice of sojourning was reinforced by laws preventing Chinese laborers from becoming citizens, and by restrictions on their entry into this country. Chinese laborers who arrived before 1882 could not bring their wives and were prevented by law from marrying Whites. Thus, it is likely that the number of Chinese American families might have been negligible had it not been for two things: the San Francisco earthquake and fire in 1906, which destroyed all municipal records, and the ingenuity and persistence of the Chinese people, who used the opportunity created by the earthquake to increase their numbers in the United States. Since relatives of citizens were permitted entry, American-born Chinese (real and claimed) would visit China, report the birth of a son, and thus create an entry slot. Years later, since the records were destroyed, the slot could be used by a relative or purchased by someone outside the family. The purchasers were called "paper sons." Paper sons became a major mechanism for increasing the Chinese population, but it was a slow process and the sojourner community remained predominantly male for decades.

The high concentration of males in the Chinese community before 1920 resulted in a split household form of family. As Evelyn Nakano Glenn observes:

> In the split household family, production is separated from other functions and is carried out by a member living far from the rest of the household. The rest—consumption, reproduction and socialization—are carried out by the wife and other relatives from the home village. . . . The split household form makes possible maximum exploitation of the workers. . . . The labor of prime-age male workers can be bought relatively cheaply, since the cost of reproduction and family maintenance is borne partially by unpaid subsistence work of women and old people in the home village. (1983:38–39)

The Chinese women who were in the United States during this period consisted of a small number who were wives and daughters of merchants and a larger percentage who were prostitutes. Lucia Cheng Hirata (1979) has suggested that Chinese prostitution was an important element in helping to maintain the split household family. In conjunction with laws prohibiting intermarriage, it helped men avoid long-term relationships with women in the United States and ensured that the bulk of their meager earnings would continue to support the family at home.

The reproductive labor of Chinese women, therefore, took on two dimensions primarily because of the split household family. Wives who remained in China were forced to raise children and care for in-laws on the meager remittances of their sojourning husband. Although we know few details about their lives, it is clear that the everyday work of bearing and maintaining children and a household fell entirely on their shoulders. Those women who immigrated and worked as prostitutes performed the more nurturant aspects of reproductive labor, that is, providing emotional and sexual companionship for men who were far from home. Yet their role as prostitutes was more likely a means of supporting their families at home in China than a chosen vocation.

The Chinese family system during the nineteenth century was a patriarchal one and girls had little value. In fact, they were considered temporary members of their father's family because when they married, they became members of their husband's family. They also had little social value; girls were sold by some poor parents to work as prostitutes, concubines, or servants. This saved the family the expense of raising them, and their earnings became a source of family income. For most girls, however, marriages were arranged and families sought useful connections through this process. With the development of a sojourning pattern in the United States, some Chinese women in those regions of China where this pattern was more prevalent would be sold to become prostitutes in the United States. Most, however, were married to men whom they saw only once or twice in the twenty- or thirty-year period during which he was sojourning in the United States. A woman's status as wife ensured that a portion of the meager wages her husband earned would be returned to his family in China. This arrangement required considerable sacrifice and adjustment by wives who remained in China and those who joined their husbands after a long separation.

Maxine Hong Kingston tells the story of the unhappy meeting of her aunt, Moon Orchid, with her husband, from whom she had been separated for thirty years: "For thirty years she had been receiving money from him from America. But she had never told him that she wanted to come to the United States. She waited for him to suggest it, but he never did" (1977:144). His response to her when she arrived unexpectedly was to say: "'Look at her. She'd never fit into an American household. I have important American guests who come inside my house to eat.' He turned to Moon Orchid, 'You can't talk to them. You can barely talk to me.' Moon Orchid was so ashamed, she held her hands over her face" (1977:178).

Despite these handicaps, Chinese people collaborated to establish the opportunity to form families and settle in the United States. In some cases it took as long as three generations for a child to be born on U.S. soil.

> In one typical history, related by a 21 year old college student, great-grandfather arrived in the States in the 1890s as a "paper son" and worked for about 20 years as a laborer. He then sent for the grandfather, who worked alongside great-grandfather in a small business for several years. Great-grandfather subsequently returned to China, leaving grandfather to run the business and send remittance. In the 1940s, grandfather sent for father; up to this point, none of the wives had left China. Finally, in the late 1950s father returned to China and brought his wife back with him. Thus, after nearly 70 years, the first child was born in the United States. (Glenn 1981:14)

CHICANOS

Africans were uprooted from their native lands and encouraged to have families in order to increase the slave labor force. Chinese people were immigrant laborers whose "permanent" presence in the country was denied. By contrast, Mexican Americans were colonized and their traditional family life was disrupted by war and the imposition of a new set of laws and conditions of labor. The hardships faced by Chicano families, therefore, were the results of the U.S. colonization of the indigenous Mexican population, accompanied by the beginnings of industrial development. The treaty of Guadalupe Hidalgo; signed in 1848, granted American citizenship to Mexicans living in what is now called the Southwest. The American takeover, however, resulted in the gradual displacement of Mexicans from the land and their incorporation into a colonial labor force (Barrera

1979). Mexicans who immigrated into the United States after 1848 were also absorbed into the labor force.

Whether natives of northern Mexico (which became part of the United States after 1848) or immigrants from southern Mexico, Chicanos were a largely peasant population whose lives were defined by a feudal economy and a daily struggle on the land for economic survival. Patriarchal families were important instruments of community life, and nuclear family units were linked through an elaborate system of kinship and godparenting. Traditional life was characterized by hard work and a fairly distinct pattern of sex-role segregation.

> Most Mexican women were valued for their household qualities, men by their ability to work and to provide for a family. Children were taught to get up early, to contribute to their family's labor to prepare themselves for adult life. . . . Such a life demanded discipline, authority, deference—values that cemented the working of a family surrounded and shaped by the requirements of Mexico's distinctive historical pattern of agricultural development, especially its pervasive debt peonage. (Saragoza 1983:8)

As the primary caretakers of hearth and home in a rural environment, Chicanas' labor made a vital and important contribution to family survival. A description of women's reproductive labor in the early twentieth century may be used to gain insight into the work of the nineteenth-century rural women.

> For country women, work was seldom a salaried job. More often it was the work of growing and preparing food, of making adobes and plastering houses with mud, or making their children's clothes for school and teaching them the hymns and prayers of the church, or delivering babies and treating sickness with herbs and patience. In almost every town there were one or two women who, in addition to working in their own homes, served other families in the community as *curanderas* (healers); *parteras* (midwives), and schoolteachers. (Elasser et al. 1980:10)

Although some scholars have argued that family rituals and community life showed little change before Word War I (Saragoza 1983), the American conquest of Mexican lands, the introduction of a new system of labor, the loss of Mexican-owned land through the inability to document ownership, and the transient nature of most of the jobs in which Chicanos were employed resulted in the gradual erosion of this pastoral way of life. Families were uprooted as the economic basis for family life changed. Some people immigrated from Mexico in search of a better standard of living and worked in the mines and railroads. Others, who were native to the Southwest, faced a job market that no longer required their skills. They moved into mining, railroad, and agricultural labor in search of a means of earning a living. According to Albert Camarillo (1979), the influx of Anglo[5] capital into the pastoral economy of Santa Barbara rendered obsolete the skills of many Chicano males who had worked as ranch hands and farmers prior to the urbanization of that economy. While some women and children accompanied their husbands to the railroad and mining camps, many of these camps discouraged or prohibited family settlement.

The American period (after 1848) was characterized by considerable transiency for the Chicano population. Its impact on families is seen in the growth of female-headed households, reflected in the data as early as 1860. Richard Griswold del Castillo (1979) found a sharp increase in female-headed households in Los Angeles, from a low of 13 percent in 1844 to 31 percent in 1880. Camarillo (1979:120) documents a similar increase in Santa Barbara, from 15 percent in 1844 to 30 percent by 1880. These increases appear to be due not so much to divorce, which was infrequent in this Catholic population; as to widowhood and temporary abandonment in search of work. Given the hazardous nature of work in the mines and railroad camps, the death of a husband, father, or son who was laboring in these sites was not uncommon. Griswold del Castillo (1979) reports a higher death rate among men than women in Los Angeles. The rise in female-headed households, therefore, reflects the instabilities and insecurities introduced into women's lives as a result of the changing social organization of work.

One outcome, the increasing participation of women and children in the labor force, was primarily a response to economic factors that required the modification of traditional values. According to Louisa Vigil, who was born in 1890, "The women didn't work at that time. The man was supposed to marry that girl and take care of her. . . . Your grandpa never did let me work for nobody. He always had to work, and we never did have really bad times" (Elasser et al., 1980:14).

Vigil's comments are reinforced in Mario Garcia's (1980) study of El Paso. In the 393 households he examined in the 1900 census, he found 17.1 percent of the women to be employed. The majority of this group were daughters, mothers with no husbands, and single women. In Los Angeles and Santa Barbara, where there were greater work opportunities for women than in El Paso, wives who were heads of household worked in seasonal and part-time jobs, and lived from the earnings of children and relatives in an effort to maintain traditional females roles.

Slowly, entire families were encouraged to go to railroad work camps and were eventually incorporated into the agricultural labor market. This was a response both to the extremely low wages paid to Chicano laborers and to the preferences of employers, who saw family labor as a way of stabilizing the work force. For Chicanos, engaging all family members in agricultural work was a means of increasing their earnings to a level close to subsistence for the entire group and of keeping the family unit together. Camarillo provides a picture of the interplay of work, family, and migration in the Santa Barbara area in the following observation:

> The time of year when women and children were employed in the fruit cannery and participated in the almond and olive harvest coincided with the seasons when the men were most likely to be engaged in seasonal migratory work. There were seasons, however, especially in the early summer when the entire family migrated from the city to pick fruit. This type of family seasonal harvest was evident in Santa Barbara by the 1890s. As walnuts replaced almonds and as the fruit industry expanded, Chicano family labor became essential. (1979:93)

This arrangement, while bringing families together, did not decrease the hardships that Chicanas had to confront in raising their families. We may infer something about the rigors of that life from Jesse Lopez de la Cruz's description of the workday of migrant farm laborers in the 1940s. Work conditions in the 1890s were as difficult, if not worse.

> We always went to where the women and men were going to work, because if it were just the men working it wasn't worth going out there because we wouldn't earn enough to support a family. . . . We would start around 6:30 A.M and work for four or five hours, then walk home and eat and rest until about three-thirty in the afternoon when it cooled off. We would go back and work until we couldn't see. Then I'd clean up the kitchen. I was doing the housework and working out in the fields and taking care of two children. (Quoted in Goldman 1981:119–120)

In the towns, women's reproductive labor was intensified by the congested and unsanitary conditions of the barrios in which they lived. Garcia described the following conditions in El Paso:

> Mexican women had to haul water for washing and cooking from the river or public water pipes. To feed their families, they had to spend time marketing, often in Ciudad Juarez across the border, as well as long, hot hours cooking meals and coping with the burden of desert sand both inside and outside their

> homes. Besides the problem of raising children, unsanitary living conditions forced Mexican mothers to deal with disease and illness in their families. Diphtheria, tuberculosis, typhus and influenza were never too far away. Some diseases could be directly traced to inferior city services. . . . As a result, Mexican mothers had to devote much energy to caring for sick children, many of whom died. (1980:320–321)

While the extended family has remained an important element of Chicano life, it was eroded in the American period in several ways. Griswold del Castillo (1979), for example, points out that in 1845 about 71 percent of Angelenos lived in extended families, whereas by 1880, fewer than half did. This decrease in extended families appears to be a response to the changed economic conditions and the instabilities generated by the new sociopolitical structure. Additionally, the imposition of American law and custom ignored, and ultimately undermined, some aspects of the extended family. The extended family in traditional Mexican life consisted of an important set of family, religious, and community obligations. Women, while valued primarily for their domesticity, had certain legal and property rights that acknowledged the importance of their work, their families of origin, and their children. In California, for example,

> equal ownership of property between husband and wife had been one of the mainstays of the Spanish and Mexican family systems. Community-property laws were written into the civil codes with the intention of strengthening the economic controls of the wife and her relatives. The American government incorporated these Mexican laws into the state constitution, but later court decisions interpreted these statutes so as to undermine the wife's economic rights. In 1861, the legislature passed a law that allowed the deceased wife's property to revert to her husband. Previously it had been inherited by her children and relatives if she died without a will. (Griswold del Castillo 1979:69)

The impact of this and similar court rulings was to "strengthen the property rights of the husband at the expense of his wife and children" (Griswold del Castillo 1979:69)

In the face of the legal, social, and economic changes that occurred during the American period, Chicanas were forced to cope with a series of dislocations in traditional life. They were caught between conflicting pressures to maintain traditional women's roles and family customs, and the need to participate in the economic support of their families by working outside the home. During this period the preservation of traditional customs—such as languages, celebrations, and healing practices—became an important element in maintaining and supporting familial ties.

According to Alex Saragoza (1983), transiency, the effects of racism and segregation, and proximity to Mexico aided in the maintenance of traditional family practices. Garcia has suggested that women were the guardians of Mexican cultural traditions within the family. He cites the work of anthropologist Manuel Gamio, who identified the retention of many Mexican customs among Chicanos in settlements around the United States in the early 1900s.

> These included folklore, songs, and ballads, birthday celebrations, saints' days, baptisms, weddings, and funerals in the traditional style. Because of poverty, a lack of physicians in the barrios, and adherence to traditional customs, Mexicans continued to use medicinal herbs. Gamio also identified the maintenance of a number of oral traditions, and Mexican style cooking. (Garcia 1980:322)

Of vital importance to the integrity of traditional culture was the perpetuation of the Spanish language. Factors that aided in the maintenance of other aspects of Mexican culture also helped in sustaining the language. However, entry into English-language public schools introduced the children and their families to systematic efforts to erase their native tongue. Griswold del Castillo reports that in the early 1880s there was considerable pressure against speakers of Spanish in the public schools. He also found that some Chicano parents responded to this kind of discrimination by helping support independent bilingual schools. These efforts, however, were short-lived.

Another key factor in conserving Chicano culture was the extended family network, particularly the system of *compadrazgo* (godparenting). Although the full extent of the impact of the American period on the Chicano extended family is not known, it is generally acknowledged that this family system, though lacking many legal and social sanctions, played an important role in the preservation of the Mexican community (Camarillo 1979). In Mexican society, godparents were an important way of linking family and community through respected friends or authorities. Participants in the important rites of passage in the child's life, such as baptism, first Communion, confirmation, and marriage, godparents had a moral obligation to act as guardians, to provide financial assistance in times of need, and to substitute in case of the death of a parent. Camarillo (1979) points out that in traditional society these bonds cut across class and racial lines.

The rite of baptism established kinship networks between rich and poor, between Spanish, mestizo and American Indian, and often carried with it political loyalty and economic-occupational ties. The leading California patriarchs in the pueblo played important roles in the *compadrazgo* network. They sponsored dozens of children for their workers or poorer relatives. The kindness of the *padrino* and *madrina* was repaid with respect and support from the *pobladores* (Camarillo 1979:12–13).

The extended family network, which included godparents, expanded the support groups for women who were widowed or temporarily abandoned and for those who were in seasonal, part- or full-time work. It suggests, therefore, the potential for an exchange of services among poor people whose income did not provide the basis for family subsistence. Griswold del Castillo (1979) argues that family organization influenced literacy rates and socioeconomic mobility among Chicanos in Los Angeles between 1850 and 1880. His data suggest that children in extended families (defined as those with at least one relative living in a nuclear family household) had higher literacy rates than those in nuclear families. He also argues that those in larger families fared better economically and experienced less downward mobility. The data here are too limited to generalize to the Chicano experience as a whole, but they do reinforce the actual and potential importance of this family form to the continued cultural autonomy of the Chicano community.

CONCLUSION

Reproductive labor for African American, Chinese American, and Mexican American women in the nineteenth century centered on the struggle to maintain family units in the face of a variety of assaults. Treated primarily as workers rather than as members of family groups, these women labored to maintain, sustain, stabilize, and reproduce their families while working in both the public (productive) and private (reproductive) spheres. Thus, the concept of reproductive labor, when applied to women of color, must be modified to account for the fact that labor in the productive sphere was required to achieve even minimal levels of family subsistence. Long after industrialization had begun to reshape family roles among middle-class White families, driving White women into a cult of domesticity, women of color were coping with an extended day. This day included subsistence labor outside the family and domestic labor within the family. For slaves, domestics, migrant farm laborers, seasonal factory workers, and prostitutes, the distinctions between labor that reproduced family life and labor that economically sustained it were minimized. The expanded workday was one of the primary ways in which reproductive labor increased.

Racial-ethnic families were sustained and maintained in the face of various forms of disruption. Yet the women and their families paid a high price in the process. High rates of infant mortality, a shortened life span, and the early onset of crippling and debilitating disease give some insight into the costs of survival.

The poor quality of housing and the neglect of communities further increased reproductive labor. Not only did racial-ethnic women work hard outside the home for mere subsistence, they worked very hard inside the home to achieve even minimal standards of privacy and cleanliness. They were continually faced with disease and illness that resulted directly from the absence of basic sanitation. The fact that some African women murdered their children to prevent them from becoming slaves is an indication of the emotional strain associated with bearing and raising children while participating in the colonial labor system.

We have uncovered little information about the use of birth control, the prevalence of infanticide, or the motivations that may have generated these or other behaviors. We can surmise, however, that no matter how much children were accepted, loved, or valued among any of these groups of people, their futures were precarious. Keeping children alive, helping them to understand and participate in a system that exploited them, and working to ensure a measure—no matter how small—of cultural integrity intensified women's reproductive labor.

Being a woman of color in nineteenth-century American society meant having extra work both inside and outside the home. It meant being defined as outside of or deviant from the norms and values about women that were being generated in the dominant White culture. The notion of separate spheres of male and female labor that developed in the nineteenth century had contradictory outcomes for the Whites. It was the basis for the confinement of upper-middle-class White women to the household and for much of the protective legislation that subsequently developed in the workplace. At the same time, it sustained White families by providing social acknowledgment and support to women in the performance of their family roles. For racial-ethnic women, however, the notion of separate spheres served to reinforce their subordinate status and became, in effect, another assault. As they increased their work outside the home, they were forced into a productive labor sphere that was organized for men and "desperate" women who were so unfortunate or immoral that they could not confine their work to the domestic sphere. In the productive sphere, racial-ethnic women faced exploitative jobs and depressed wages. In the reproductive sphere, they were denied the opportunity to embrace the dominant ideological definition of "good" wife or mother. In essence, they were faced with a double-bind situation, one that required their participation in the labor force to sustain family life but damned them as women, wives, and mothers because they did not confine their labor to the home.

Finally, the struggle of women of color to build and maintain families provides vivid testimony to the role of race in structuring family life in the United States. As Maxine Baca Zinn points out:

> Social categories and groups subordinate in the racial hierarchy are often deprived of access to social institutions that offer supports for family life. Social categories and groups elevated in the racial hierarchy have different and better connections to institutions that can sustain families. Social location and its varied connection with social resources thus have profound consequences for family life. (1990:74)

From the founding of the United States, and throughout its history, race has been a fundamental criterion determining the kind of work people do, the wages they receive, and the kind of legal, economic, political, and social support provided for their families. Women of color have faced limited economic resources, inferior living conditions, alien cultures and languages, and overt hostility in their struggle to create a "place" for families of color in the United States. That place, however, has been a precarious one because the society has not provided supports for these families. Today we see the outcomes of that legacy in statistics showing that people of color, compared with whites, have higher rates of female-headed households, out-of-wedlock births, divorce, and other factors associated with family disruption. Yet the causes of these variations do not lie merely in the higher concentrations of poverty among people of color; they are also due to the ways race has been used as a basis for denying and providing support to families. Women of color have struggled to maintain their families against all of these odds.

SELECTED REFERENCES

Aschenbrenner, Joyce. 1975. *Lifelines: Black Families in Change.* New York: Holt, Rinehart, and Winston.

Baca Zinn, Maxine. 1990. "Family, Feminism and Race in America." *Gender and Society* 4 (1) (March): 6S-32.

Barrera, Mario. 1979. *Race and Class in the Southwest.* Notre Dame, Ind.: Notre Dame University Press.

Blassingame, John. 1972. *The Slave Community: Plantation Life in the Antebellum South.* New York: Oxford University Press.

———. 1977. *Slave Testimony: Two Centuries of Letters, Speeches, Interviews, and Autobiographies.* Baton Rouge: Louisiana State University Press.

Camarillo, Albert. 1979. *Chicanos in a Changing Society.* Cambridge, Mass.:Harvard University Press.

Caulfield, Mina Davis. 1974. "Imperialism, the Family, and Cultures of Resistance." *Socialist Review* 4 (2) (October): 67–85.

Davis, Angela. 1971. "Reflections on the Black Woman's Role in the Community of Slaves." *Black Scholar* 3 (4) (December): 2–15.

———. 1981. *Women, Race, and Class.* New York: Random House.

Elasser, Nan, Kyle MacKenzie, and Yvonne Tixier Y. Vigil. 1980. *Las Mujeres.* New York: The Feminist Press.

Garcia, Mario T. 1980. "The Chicano in American History: The Mexican Women of El Paso, 1880–1920—A Case Study." *Pacific Historical Review* 49 (2) (May): 315–358.

Genovese, Eugene D., and Elinor Miller, eds. 1974. *Plantation, Town, and County: Essays on the Local History of American Slave Society.* Urbana: University of Illinois Press.

Glenn, Evelyn Nakano. 1983. "Split Household, Small Producer, and Dual Earner: An Analysis of Chinese-American Family Strategies." *Journal of Marriage and the Family* 45 (1) (February): 35–46.

Goldman, Marion S. 1981. *Gold Diggers and Silver Miners.* Ann Arbor: University of Michigan Press.

Griswold del Castillo, Richard. 1979. *The Los Angeles Barrio: 1850–1890.* Los Angeles: University of California Press.

Gutman, Herbert. 1976. *The Black Family in Slavery and Freedom, 1750–1925.* New York: Pantheon.

Hirata, Lucia Cheng. 1979. "Free, Indentured, Enslaved: Chinese Prostitutes in Nineteenth Century America." *Signs* 5 (Autumn): 3–29.

Jones, Jacqueline. 1985. *Labor of Love, Labor of Sorrow.* New York: Basic Books.

Kingston, Maxine Hong. 1977. *The Woman Warrior.* New York: Vintage Books.

Matthaei, Julie. 1982. *An Economic History of Women in America.* New York: Schocken Books.

Nee, Victor G., and Brett de Bary Nee. 1973. *Longtime Californ'.* New York: Pantheon Books.

Saragoza, Alex M. 1983. "The Conceptualization of the History of the Chicano Family: Work, Family, and Migration in Chicanes." In *Research Proceedings of the Symposium on Chicano Research and Public Policy.* Stanford, Calif.: Stanford University, Center for Chicano Research.

Shimkin, Demetri, E. M. Shimkin, and D. A. Frate, eds. 1978. *The Extended Family in Black Societies.* The Hague: Mouton.

Stack, Carol S. 1974. *All Our Kin: Strategies for Survival in a Black Community.* New York: Harper & Row.

Sudarkasa, Niara. 1981. "Interpreting the African Heritage in Afro-American Family Organization." Pp. 37–53 in *Black Families,* edited by Harriette Pipes McAdoo. Beverly Hills, Calif.: Sage.

White, Deborah Gray. 1985. *Ar'n't la Woman? Female Slaves in the Plantation South.* New York: W. W. Norton.

21

The Meanings of Latina Transnational Motherhood

PIERRETTE HONDAGNEU SOTELO
AND ERNESTINE AVILA,
1997

While mothering is generally understood as practice that involves the preservation, nurturance, and training of children for adult life (Ruddick 1989), there are many contemporary variants distinguished by race, class, and culture (Collins 1994; Dill 1988, 1994; Glenn 1994). Latina immigrant women who work and reside in the United States while their children remain in their countries of origin constitute one variation in the organizational arrangements, meanings, and priorities of motherhood. We call this arrangement "transnational motherhood," and we explore how the meanings of motherhood are rearranged to accommodate these spatial and temporal separations. In the United States, there is a long legacy of Caribbean women and African American women from the South, leaving their children "back home" to seek work in the North. Since the early 1980s, thousands of Central American women, and increasing numbers of Mexican women, have migrated to the United States in search of jobs, many of them leaving their children behind with grandmothers, with other female kin, with the children's fathers, and sometimes with paid caregivers. In some cases, the separations of time and distance are substantial; 10 years may elapse before women are reunited with their children. In this article we confine our analysis to Latina transnational mothers currently employed in Los Angeles in paid domestic work, one of the most gendered and racialized occupations. We examine how their meanings of motherhood shift in relation to the structures of late-20th-century global capitalism.

Motherhood is not biologically predetermined in any fixed way but is historically and socially constructed. Many factors set the stage for transnational motherhood. These factors include labor demand for Latina immigrant women in the United States, particularly in paid domestic work; civil war, national economic crises, and particular development strategies, along with tenuous and scarce job opportunities for women and men in Mexico and Central America; and the subsequent increasing numbers of female-headed households (although many transnational mothers are married). More interesting to us than the macro determinants of transnational motherhood, however, is the forging of new arrangements and meanings of motherhood.

Central American and Mexican women who leave their young children "back home" and come to the United States in search of employment are in the process of actively, if not voluntarily, build-

From Pierrette Hondagneu Sotelo and Ernestine Avila, "I'm Here, but I'm There," *Gender and Society,* vol. 11, no. 5, pp. 548, 549, 550–553, 554–559, 560, 562, 564, 566–568, fig. 9.5.

ing alternative constructions of motherhood. Transnational motherhood contradicts both dominant U.S., White, middle-class models of motherhood, and most Latina ideological notions of motherhood. On the cusp of the millennium, transnational mothers and their families are blazing new terrain, spanning national borders, and improvising strategies for mothering. It is a brave odyssey, but one with deep costs. . . .

RETHINKING MOTHERHOOD

Feminist scholarship has long challenged monolithic notions of family and motherhood that relegate women to the domestic arena of private/public dichotomies and that rely on the ideological conflation of family, woman, reproduction, and nurturance (Collier and Yanagisako 1987, 36). "Rethinking the family" prompts the rethinking of motherhood (Glenn 1994; Thorne and Yalom 1992), allowing us to see that the glorification and exaltation of isolationist, privatized mothering is historically and culturally specific.

The "cult of domesticity" is a cultural variant of motherhood, one made possible by the industrial revolution, by breadwinner husbands who have access to employers who pay a "family wage," and by particular configurations of global and national socioeconomic and racial inequalities. Working-class women of color in the United States have rarely had access to the economic security that permits a biological mother to be the only one exclusively involved with mothering during the children's early years (Collins, 1994; Dill 1988, 1994; Glenn 1994). As Evelyn Nakano Glenn puts it, "Mothering is not just gendered, but also racialized" (1994, 7) and differentiated by class. Both historically and in the contemporary period, women lacking the resources that allow for exclusive, full-time, round-the-clock mothering rely on various arrangements to care for children. Sharing mothering responsibilities with female kin and friends as "other mothers" (Collins 1991), by "kin-scription" (Stack and Burton 1994), or by hiring child care (Uttal 1996) are widely used alternatives

Women of color have always worked. Yet, many working women—including Latina women—hold the cultural prescription of solo mothering in the home as an ideal. We believe this ideal is disseminated through cultural institutions of industrialization and urbanization, as well as from preindustrial, rural peasant arrangements that allow for women to work while tending to their children. It is not only White, middle-class ideology but also strong Latina/o traditions, cultural practices, and ideals—Catholicism, and the Virgin Madonna figure—that cast employment as oppositional to mothering. Cultural symbols that model maternal femininity, such as the Virgen de Guadalupe, and negative femininity, such as *La Llorona* and *La Malinche,* serve to control Mexican and Chicana women's conduct by prescribing idealized visions of motherhood.

Culture, however, does not deterministically dictate what people do. Many Latina women must work for pay, and many Latinas innovate income-earning strategies that allow them to simultaneously earn money and care for their children. They sew garments on industrial sewing machines at home (Fernandez-Kelly and Garcia 1990) and incorporate their children into informal vending to friends and neighbors, at swap meets, or on the sidewalks (Chinchilla and Hamilton 1996). They may perform agricultural work alongside their children or engage in seasonal work (Zavella 1987); or they may clean houses when their children are at school or alternatively, incorporate their daughters into paid house cleaning (Romero 1992, 1997). Engagement in "invisible employment" allows for urgently needed income and the maintenance of the ideal of privatized mothering. The middle-class model of mothering is predicated on mother-child isolation in the home, while women of color have often worked with their children in close proximity (Collins 1994), as in some of the examples listed above. In both cases, however, mothers are with their children. The long distances of time and space that separate transnational mothers from their children contrast sharply to both mother-child isolation in the home or mother-child integration in the workplace.

TRANSNATIONAL MOTHERS' WORK, PLACE, AND SPACE

Feminist geographers have focused on how gendered orientations to space influence the way we organize our daily work lives. While sociologists have tended to explain occupational segregation as rooted either in family or individual characteristics (human capital theory) or in the workplace (labor market segmentation), feminist geographers observe that women tend to take jobs close to home so that they can fulfill child rearing and domestic duties (Hanson and Pratt 1995; Massey 1994). Transnational mothers, on the other hand, congregate in paid domestic work, an occupation that is relentlessly segregated not only by gender but also by race, class, and nationality/citizenship. To perform child rearing and domestic duties for others, they radically break with deeply gendered spatial and temporal boundaries of family and work.

Performing domestic work for pay, especially in a live-in job, is often incompatible with providing primary care for one's own family and home (Glenn 1986; Rollins 1985; Romero 1992, 1997). Transnational mothering, however, is neither exclusive to live-in domestic workers nor to single mothers. Many women continue with transnational mothering after they move into live-out paid domestic work, or into other jobs. Women with income-earning husbands may also become transnational mothers. The women we interviewed do not necessarily divert their mothering to the children and homes of their employers but instead reformulate their own mothering to accommodate spatial and temporal gulfs.

Like other immigrant workers, most transnational mothers came to the United States with the intention to stay for a finite period of time. But as time passes and economic need remains, prolonged stays evolve. Marxist-informed theory maintains that the separation of work life and family life constitutes the separation of labor maintenance costs from the labor reproduction costs (Burawoy 1976; Glenn 1986). According to this framework, Latina transnational mothers work to maintain themselves in the United States and to support their children—and reproduce the next generation of workers—in Mexico or Central America. One precursor to these arrangements is the mid-20th-century Bracero Program, which in effect legislatively mandated Mexican "absentee fathers" who came to work as contracted agricultural laborers in the United States. Other precursors, going back further in history, include the 18th- and 19th-centuries' coercive systems of labor, whereby African American slaves and Chinese sojourner laborers were denied the right to form residentially intact families (Dill 1988, 1994).

Transnational mothering is different from some of these other arrangements in that now women with young children are recruited for U.S. jobs that pay far less than a "family wage." When men come north and leave their families in Mexico—as they did during the Bracero Program and as many continue to do today—they are fulfilling familial obligations defined as breadwinning for the family. When women do so, they are embarking not only on an immigration journey but on a more radical gender-transformative odyssey. They are initiating separations of space and time from their communities of origin, homes, children, and—sometimes—husbands. In doing so, they must cope with stigma, guilt, and criticism from others. A second difference is that these women work primarily not in production of agricultural products or manufacturing but in reproductive labor, in paid domestic work, and/or vending. Performing paid reproductive work for pay—especially caring for other people's children—is not always compatible with taking daily care of one's own family. All of this raises questions about the meanings and variations of motherhood in the late 20th century. . . .

Materials for this article draw from a larger study of paid domestic work in Los Angeles

County and from interviews conducted in adjacent Riverside County. The materials include in-depth interviews, a survey, and ethnographic fieldwork. We had not initially anticipated studying women who live and work apart from their children but serendipitously stumbled on this theme in the course of our research. . . . Transnational motherhood arrangements are not exclusive to paid domestic work, but there are particular features about the way domestic work is organized that encourage temporal and spatial separations of a mother-employee and her children. Historically and in the contemporary period, paid domestic workers have had to limit or forfeit primary care of their families and homes to earn income by providing primary care to the families and homes of employers, who are privileged by race and class (Glenn 1986; Rollins 1985; Romero 1992). Paid domestic work is organized in various ways, and there is a clear relationship between the type of job arrangement women have and the likelihood of experiencing transnational family arrangements with their children. To understand the variations, it is necessary to explain how the employment is organized. Although there are variations within categories, we find it useful to employ a tripartite taxonomy of paid domestic work arrangements. This includes live-in and live-out nanny-housekeeper jobs, and weekly housecleaning jobs.

Weekly house cleaners clean different houses on different days according to what Romero (1992) calls modernized "job work" arrangements. These contractual-like employee-employer relations often resemble those between customer and vendor, and they allow employees a degree of autonomy and scheduling flexibility. Weekly employees are generally paid a flat fee, and they work shorter hours and earn considerably higher hourly rates than do live-in or live-out domestic workers. By contrast, live-in domestic workers work and live in isolation from their own families and communities, sometimes in arrangements with feudal remnants (Glenn 1986). There are often no hourly parameters to their jobs, and as our survey results show, most live-in workers in Los Angeles earn below minimum wage. Live-out domestic workers also usually work as combination nanny-housekeepers, generally working for one household, but contrary to live-ins, they enter daily and return to their own home in the evening. Because of this, live-out workers better resemble industrial wage workers (Glenn 1986).

Live-in jobs are the least compatible with conventional mothering responsibilities. Only about half (16 out of 30) of live-ins surveyed have children, while 83 percent (53 out of 64) of live-outs and 77 percent (45 out of 59) of house cleaners do. As Table 1 shows, 82 percent of live-ins with children have at least one of their children in their country of origin. It is very difficult to work a live-in job when your children are in the United States. Employers who hire live-in workers do so because they generally want employees for jobs that may require round-the-clock service. As one owner of a domestic employment agency put it,

> They (employers) want a live-in to have somebody at their beck and call. They want the hours that are most difficult for them covered, which is like six thirty in the morning 'till eight when the kids go to

Table 1 Domestic Workers: Wages, Hours Worked and Children's Country of Residence

	Live-ins (n = 30)	Live-outs (n = 64)	House Cleaners (n = 59)
Mean hourly wage	$3.79	$5.90	$9.40
Mean hours worked per week	64	35	23
Domestic workers with children	(n = 16)	(n = 53)	(n = 45)
All children in the United States (%)	18	58	76
At least one child "back home"	82	42	24

school, and four to seven when the kids are home, and it's homework, bath, and dinner.

According to our survey, live-ins work an average of 64 hours per week. The best live-in worker, from an employer's perspective, is one without daily family obligations of her own. The workweek may consist of six very long workdays. These may span from dawn to midnight and may include overnight responsibilities with sleepless or sick children, making it virtually impossible for live-in workers to sustain daily contact with their own families. Although some employers do allow for their employees' children to live in as well (Romero 1996), this is rare. When it does occur, it is often fraught with special problems, and we discuss these in a subsequent section of this article. In fact, minimal family and mothering obligations are an informal job placement criterion for live-in workers. Many of the agencies specializing in the placement of live-in nanny-housekeepers will not even refer a woman who has children in Los Angeles to interviews for live-in jobs. As one agency owner explained, "As a policy here, we will not knowingly place a nanny in a live-in job if she has young kids here." A job seeker in an employment agency waiting room acknowledged that she understood this job criterion more broadly, "You can't have a family, you can't have a family, you can't have anyone (if you want a live-in job)."

The subminimum pay and the long hours for live-in workers also make it very difficult for these workers to have their children in the United States. Some live-in workers who have children in the same city as their place of employment hire their own nanny-housekeeper—often a much younger, female relative—to provide daily care for their children, as did Patricia, one of the interview respondents whom we discuss later in this article. Most live-ins, however, cannot afford this alternative; ninety-three percent of the live-ins surveyed earn below minimum wage (then $4.25 per hour). Many live-in workers cannot afford to bring their children to Los Angeles, but once their children are in the same city, most women try to leave live-in work to live with their children.

At the other end of the spectrum are the house cleaners that we surveyed, who earn substantially higher wages than live-ins (averaging $9.46 per hour as opposed to $3.79) and who work fewer hours per week than live-ins (23 as opposed to 64). We suspect that many house cleaners in Los Angeles make even higher earnings and work more hours per week, because we know that the survey undersampled women who drive their own cars to work and who speak English. The survey suggests that house cleaners appear to be the least likely to experience transnational spatial and temporal separations from their children.

Financial resources and job terms enhance house cleaners' abilities to bring their children to the United States. Weekly housecleaning is not a bottom-of-the-barrel job but rather an achievement. Breaking into housecleaning work is difficult because an employee needs to locate and secure several different employers. For this reason, relatively well-established women with more years of experience in the United States, who speak some English, who have a car, and who have job references predominate in weekly housecleaning. Women who are better established in the United States are also more likely to have their children here. The terms of weekly housecleaning employment—particularly the relatively fewer hours worked per week, scheduling flexibility, and relatively higher wages—allow them to live with, and care for, their children. So, it is not surprising that 76 percent of house cleaners who are mothers have their children in the United States.

Compared with live-ins and weekly cleaners, live-out nanny-housekeepers are at an intermediate level with respect to the likelihood of transnational motherhood. Forty-two percent of the live-out nanny-housekeepers who are mothers reported having at least one of their children in their country of origin. Live-out domestic workers, according to the survey, earn $5.90 per hour and work an average workweek of 35 hours. Their lower earnings, more regimented schedules, and longer workweeks than house cleaners, but higher earnings, shorter hours, and more scheduling flexibility than live-ins explain their intermediate incidence of transnational motherhood.

THE MEANINGS OF TRANSNATIONAL MOTHERHOOD

How do women transform the meaning of motherhood to fit immigration and employment? Being a transnational mother means more than being the mother to children raised in another country. It means forsaking deeply felt beliefs that biological mothers should raise their own children, and replacing that belief with new definitions of motherhood. The ideal of biological mothers raising their own children is widely held but is also widely broken at both ends of the class spectrum. Wealthy elites have always relied on others—nannies, governesses, and boarding schools—to raise their children (Wrigley 1995), while poor, urban families often rely on kin and "other mothers" (Collins 1991).

In Latin America, in large, peasant families, the eldest daughters are often in charge of the daily care of the younger children, and in situations of extreme poverty, children as young as five or six may be loaned or hired out to well-to-do families as "child-servants," sometimes called *criadas* (Gill 1994). A middle-aged Mexican woman that we interviewed, now a weekly house cleaner, homeowner, and mother of five children, recalled her own experience as a child-servant in Mexico: "I started working in a house when I was 8 . . . they hardly let me eat any food. . . . It was terrible, but I had to work to help my mother with the rent." This recollection of her childhood experiences reminds us how our contemporary notions of motherhood are historically and socially circumscribed, and also correspond to the meanings we assign to childhood (Zelizer 1994).

This example also underlines how the expectation on the child to help financially support her mother required daily spatial and temporal separations of mother and child. There are, in fact, many transgressions of the mother-child symbiosis in practice—large families where older daughters care for younger siblings, child-servants who at an early age leave their mothers, children raised by paid nannies and other caregivers, and mothers who leave young children to seek employment—but these are fluid enough to sustain ideological adherence to the prescription that children should be raised exclusively by biological mothers. Long-term physical and temporal separation disrupts this notion. Transnational mothering radically rearranges mother-child interactions and requires a concomitant radical reshaping of the meanings and definitions of appropriate mothering.

Transnational mothers distinguish their version of motherhood from estrangement, child abandonment, or disowning. A youthful Salvadoran woman at the domestic employment waiting room reported that she had not seen her two eldest boys, now ages 14 and 15 and under the care of her own mother in El Salvador, since they were toddlers. Yet, she made it clear that this was different from putting a child up for adoption, a practice that she viewed negatively, as a form of child abandonment. Although she had been physically separated from her boys for more than a decade, she maintained her mothering ties and financial obligations to them by regularly sending home money. The exchange of letters, photos, and phone calls also helped to sustain the connection. Her physical absence did not signify emotional absence from her children. Another woman who remains intimately involved in the lives of her two daughters, now ages 17 and 21 in El Salvador, succinctly summed up this stance when she said, "I'm here, but I'm there." Over the phone, and through letters, she regularly reminds her daughters to take their vitamins, to never go to bed or to school on an empty stomach, and to use protection from pregnancy and sexually transmitted diseases if they engage in sexual relations with their boyfriends.

Transnational mothers fully understand and explain the conditions that prompt their situations. In particular, many Central American women recognize that the gendered employment demand in Los Angeles has produced transnational motherhood arrangements. These new mothering arrangements, they acknowledge, take shape despite strong beliefs that biological mothers should care for their own children. Emelia, a 49-year-old woman who left her five children in Guatemala nine years ago to join her husband in Los Angeles explained this changing relationship between family arrangements, migration, and job demand:

> One supposes that the mother must care for the children. A mother cannot so easily throw her children aside. So, in all families, the decision is that the man comes (to the U.S.) first. But now, since the man cannot find work here so easily, the woman comes first. Recently, women have been coming and the men staying.

A steady demand for live-in housekeepers means that Central American women may arrive in Los Angeles on a Friday and begin working Monday at a live-in job that provides at least some minimal accommodations. Meanwhile, her male counterpart may spend weeks or months before securing even casual day laborer jobs. While Emelia, formerly a homemaker who previously earned income in Guatemala by baking cakes and pastries in her home, expressed pain and sadness at not being with her children as they grew, she was also proud of her accomplishments. "My children," she stated, "recognize what I have been able to do for them."

Most transnational mothers, like many other immigrant workers, come to the United States with the intention to stay for a finite period of time, until they can pay off bills or raise the money for an investment in a house, their children's education, or a small business. Some of these women return to their countries of origin, but many stay. As time passes, and as their stays grow longer, some of the women eventually bring some or all of their children. Other women who stay at their U.S. jobs are adamant that they do not wish for their children to traverse the multiple hazards of adolescence in U.S. cities or to repeat the job experiences they themselves have had in the United States. One Salvadoran woman in the waiting room at the domestic employment agency—whose children had been raised on earnings predicated on her separation from them—put it this way:

> I've been here 19 years, I've got my legal papers and everything. But I'd have to be crazy to bring my children here: All of them have studied for a career, so why would I bring them here? To bus tables and earn minimum wage? So they won't have enough money for bus fare or food?

WHO IS TAKING CARE OF THE NANNY'S CHILDREN?

Transnational Central American and Mexican mothers may rely on various people to care for their children's daily, round-the-clock needs, but they prefer a close relative: The "other mothers" on which Latinas rely include their own mothers, *comadres* (co-godmothers) and other female kin, the children's fathers, and paid caregivers. Reliance on grandmothers and comadres for shared mothering is well established in Latina culture, and it is a practice that signifies a more collectivist, shared approach to mothering in contrast to a more individualistic, Anglo-American approach (Griswold del Castillo 1984; Segura and Pierce 1993). Perhaps this cultural legacy facilitates the emergence of transnational motherhood.

Transnational mothers express a strong preference for their own biological mother to serve as the primary caregiver. Here, the violation of the cultural preference for the biological mother is rehabilitated by reliance on the biological grandmother or by reliance on the ceremonially bound comadres. Clemencia, for example, left her three young children behind in Mexico, each with their respective *madrina,* or godmother.

Emelia left her five children, then ranging in ages from 6 to 16, under the care of her mother and sister in Guatemala. As she spoke of the hardships faced by transnational mothers, she counted herself among the fortunate ones who did not need to leave the children alone with paid caregivers:

> One's mother is the only one who can really and truly care for your children. No one else can.... Women who aren't able to leave their children with their mother or with someone very special, they'll wire money to Guatemala and the people (caregivers) don't feed the children well. They don't buy the children clothes the mother would want. They take the money and the children suffer a lot....

New family fissures emerge for the transnational mother as she negotiates various aspects of the arrangement with her children, and with the "other mother" who provides daily care and supervision for the children. Any impulse to romanticize transnational motherhood is tempered by the sadness with which the women related their experiences and by the problems they sometimes encounter with their children and caregivers. A primary worry among transnational mothers is that their children are being neglected or abused in their absence. While there is a long legacy of child servants being mistreated and physically beaten in Latin America, transnational mothers also worry that their own paid caregivers will harm or neglect their children. They worry that their children may not receive proper nourishment, schooling and educational support, and moral guidance. They may remain unsure as to whether their children are receiving the full financial support they send home. In some cases, their concerns are intensified by the eldest child or a nearby relative who is able to monitor and report the caregiver's transgression to the transnational mother....

Milk, shoes, and schooling—these are the currency of transnational motherhood. Providing for children's sustenance, protecting their current well-being, and preparing them for the future are widely shared concerns of motherhood. Central American and Mexican women involved in transnational mothering attempt to ensure the present and future well-being of their children through U.S. wage earning, and as we have seen, this requires long-term physical separation from their children.

For these women, the meanings of motherhood do not appear to be in a liminal stage. That is, they do not appear to be making a linear progression from a way of motherhood that involves daily, face-to-face caregiving toward one that is defined primarily through breadwinning. Rather than replacing caregiving with breadwinning definitions of motherhood, they appear to be expanding their definitions of motherhood to encompass breadwinning that may require long-term physical separations. For these women, a core belief is that they can best fulfill traditional caregiving responsibilities through income earning in the United States while their children remain "back home."

Transnational mothers continue to state that caregiving is a defining feature of their mothering experiences. They wish to provide their children with better nutrition, clothing, and schooling, and most of them are able to purchase these items with dollars earned in the United States. They recognize, however, that their transnational relationships incur painful costs. Transnational mothers worry about some of the negative effects on their children, but they also experience the absence of domestic family life as a deeply personal loss. Transnational mothers who primarily identified as homemakers before coming to the United States identified the loss of daily contact with family as a sacrifice ventured to financially support the children....

Transnational mothers seek to mesh caregiving and guidance with breadwinning. While breadwinning may require their long-term and long-distance separations from their children, they attempt to sustain family connections by showing emotional ties through letters, phone calls, and money sent home. If at all financially and logistically possible, they try to travel home to visit their children. They maintain their mothering responsibilities not only by earning money for their children's livelihood but also by communicating and advising across national borders, and across the boundaries that separate their children's place of residence from their own places of employment and residence. . . .

As observers of late-20th-century U.S. families (Skolnick 1991; Stacey 1996) have noted, we live in an era wherein no one normative family arrangement predominates. Just as no one type of mothering unequivocally prevails in the White middle class, no singular mothering arrangement prevails among Latina immigrant women. In fact, the exigencies of contemporary immigration seem to multiply the variety of mothering arrangements. Through our research with Latina immigrant women who work as nannies, housekeepers, and house cleaners, we have encountered a broad range of mothering arrangements. Some Latinas migrate to the United States without their children to establish employment, and after some stability has been achieved, they may send for their children or they may work for a while to save money, and then return to their countries of origin. Other Latinas migrate and may postpone having children until they are financially established. Still others arrive with their children and may search for employment that allows them to live together with their children, and other Latinas may have sufficient financial support—from their husbands or kin—to stay home full-time with their children.

In the absence of a universal or at least widely shared mothering arrangement, there is tremendous uncertainty about what constitutes "good mothering," and transnational mothers must work hard to defend their choices. Some Latina nannies who have their children with them in the United States condemn transnational mothers as "bad women." One interview respondent, who was able to take her young daughter to work with her, claimed that she could never leave her daughter. For this woman, transnational mothers were not only bad mothers but also nannies who could not be trusted to adequately care for other people's children. As she said of an acquaintance, "This woman left her children (in Honduras) . . . she was taking care (of other people's children), and I said, 'Lord, who are they (the employers) leaving their children with if she did that with her own children!' "

Given the uncertainty of what is "good mothering," and to defend their integrity as mothers when others may criticize them, transnational mothers construct new scales for gauging the quality of mothering. By favorably comparing themselves with the negative models of mothering that they see in others—especially those that they are able to closely scrutinize in their employers' homes—transnational mothers create new definitions of good-mothering standards. At the same time, selectively developing motherlike ties with other people's children allows them to enjoy affectionate, face-to-face interactions that they cannot experience on a daily basis with their own children. . . .

In California, with few exceptions, paid domestic work has become a Latina immigrant women's job. One observer has referred to these Latinas as "the new employable mothers" (Chang 1994), but taking on these wage labor duties often requires Latina workers to expand the frontiers of motherhood by leaving their own children for several years. While today there is a greater openness to accepting a plurality of mothering arrangements—single mothers, employed mothers, stay-at-home mothers, lesbian mothers, surrogate mothers, to name a few—even feminist discussions generally assume that mothers, by definition, will reside with their children.

Transnational mothering situations disrupt the notion of family in one place and break distinctively with what some commentators have referred to as the "epoxy glue" view of motherhood (Blum and Deussen 1996; Scheper-Hughes 1992). Latina transnational mothers are improvising new mothering arrangements that are borne out of women's financial struggles, played out in a new global arena, to provide the best future for themselves and their children. Like many other women of color and employed mothers, transnational mothers rely on an expanded and sometimes fluid number of family members and paid caregivers. Their caring circuits, however, span stretches of geography and time that are much wider than typical joint custody or "other mother" arrangements that are more closely bound, both spatially and temporally.

The transnational perspective in immigration studies is useful in conceptualizing how relationships across borders are important. Yet, an examination of transnational motherhood suggests that transnationalism is a contradictory process of the late 20th century. It is an achievement, but one accompanied by numerous costs and attained in a context of extremely scarce options. The alienation and anxiety of mothering organized by long temporal and spatial distances should give pause to the celebratory impulses of transnational perspectives of immigration. Although not addressed directly in this article, the experiences of these mothers resonate with current major political issues. For example, transnational mothering resembles precisely what immigration restrictionists have advocated through California's Proposition 187 (Hondagneu-Sotelo 1995). While proponents of Proposition 187 have never questioned California's reliance on low-waged Latino immigrant workers, this restrictionist policy calls for fully dehumanized immigrant workers, not workers with families and family needs (such as education and health services for children). In this respect, transnational mothering's externalization of the cost of labor reproduction to Mexico and Central America is a dream come true for the proponents of Proposition 187.

Contemporary transnational motherhood continues a long historical legacy of people of color being incorporated into the United States through coercive systems of labor that do not recognize family rights. As Bonnie Thornton Dill (1988), Evelyn Nakano Glenn (1986), and others have pointed out, slavery and contract labor systems were organized to maximize economic productivity and offered few supports to sustain family life. The job characteristics of paid domestic work, especially live-in work, virtually impose transnational motherhood for many Mexican and Central American women who have children of their own.

The ties of transnational motherhood suggest simultaneously the relative permeability of borders, as witnessed by the maintenance of family ties and the new meanings of motherhood, and the impermeability of nation-state borders. Ironically, just at the moment when free trade proponents and pundits celebrate globalization and transnationalism, and when "borderlands" and "border crossings" have become the metaphors of preference for describing a mind-boggling range of conditions, nation-state borders prove to be very real obstacles for many Mexican and Central American women who work in the United States and who, given the appropriate circumstances, wish to be with their children. While demanding the right for women workers to live with their children may provoke critiques of sentimentality, essentialism, and the glorification of motherhood, demanding the right for women workers to choose their own motherhood arrangements would be the beginning of truly just family and work policies, policies that address not only inequalities of gender but also inequalities of race, class, and citizenship status.

SELECTED REFERENCES

Blum, Linda, and Theresa Deussen. 1996. Negotiating independent motherhood: Working-class African American women talk about marriage and motherhood. *Gender & Society* 10:199–211.

Burawoy, Michael. 1976. The functions and reproduction of migrant labor: Comparative material from Southern Africa and the United States. *American Journal of Sociology* 81:1050–87.

Chang, Grace. 1994. Undocumented Latinas: Welfare burdens or beasts of burden? *Socialist Review* 23:151–85.

Chinchilla, Norma Stoltz, and Nora Hamilton. 1996. Negotiating urban space: Latina workers in domestic work and street vending in Los Angeles. *Humbolt Journal of Social Relations* 22:25–35.

Collier, Jane Fishbume, and Sylvia Junko Yanagisako. 1987. *Gender and kinship: Essays toward a unified analysis.* Stanford, CA: Stanford University Press.

Collins, Patricia Hill. 1991. *Black feminist thought: Knowledge, consciousness, and the politics of empowerment.* New York: Routledge.

Dill, Bonnie Thornton. 1988. Our mothers' grief: Racial-ethnic women and the maintenance of families. *Journal of Family History* 13:415–31.

———. 1994. Fictive kin, paper sons and compadrazgo: Women of color and the struggle for family survival. In *Women of color in U.S. society,* edited by Maxine Baca Zinn and Bonnie Thornton Dill. Philadelphia: Temple University Press.

Fernandez-Kelly, M. Patricia, and Anna Garcia. 1990. Power surrendered, power restored: The politics of work and family among Hispanic garment workers in California and Florida. In *Women, politics & change,* edited by Louise A. Tilly and Patricia Gurin. New York: Russell Sage.

Gill, Lesley. 1994. *Precarious dependencies: Gender, class and domestic service in Bolivia.* New York: Columbia University Press.

Glenn, Evelyn Nakano. 1986. *Issei, Nisei, warbride: Three generations of Japanese American women in domestic service.* Philadelphia: Temple University Press.

———. 1994. Social constructions of mothering: A thematic overview. In *Mothering: Ideology, experience, and agency,* edited by Evelyn Nakano Glenn, Grace Chang, and Linda Rennie Forcey. New York: Routledge.

Griswold del Castillo, Richard. 1984. *La Familia: Chicano families in the urban Southwest, 1848 to the present.* Notre Dame, IN: University of Notre Dame Press.

Hanson, Susan, and Geraldine Pratt. 1995. *Gender, work and space.* New York: Routledge.

Hondagneu-Sotelo, Pierrette. 1995. Women and children first: New directions in anti-immigrant politics. *Socialist Review* 25:169–90.

Massey, Doreen. 1994. *Space, place and gender.* Minneapolis: University of Minnesota Press.

Rollins, Judith. 1985. *Between women: Domestics and their employers.* Philadelphia: Temple University Press.

Romero, Mary. 1992. *Maid in the U.S.A.* New York: Routledge.

———. 1996. Life as the maid's daughter: An exploration of the everyday boundaries of race, class and gender. In *Feminisms in the academy: Rethinking the disciplines,* edited by Abigail J. Steward and Donna Stanon. Ann Arbor: University of Michigan Press.

———. 1997. Who takes care of the maid's children? Exploring the costs of domestic service. In *Feminism and families,* edited by Hilde L. Nelson. New York: Routledge.

Ruddick, Sara. 1989. *Maternal thinking: Toward a politics of peace.* Boston: Beacon.

Scheper-Hughes, Nancy. 1992. *Death without weeping: The violence of everyday life in Brazil.* Berkeley: University of California Press.

Segura, Denise A., and Jennifer L. Pierce. 1993. Chicana/o family structure and gender personality: Chodorow, familism, and psychoanalytic sociology revisited. *Signs: Journal of Women in Culture and Society* 19:62–79.

Skolnick, Arlene S. 1991. *Embattled paradise: The American family in an age of uncertainty.* New York: Basic Books.

Stacey, Judith. 1996. *In the name of the family: Rethinking family values in the postmodern age.* Boston: Beacon.

Stack, Carol B., and Linda M. Burton. 1994. Kinscripts: Reflections on family, generation, and culture. In *Mothering: Ideology, experience, and agency,* edited by Evelyn Nakano Glenn, Grace Chang, and Linda Rennie Forcey. New York: Routledge.

Thorne, Barrie, and Marilyn Yalom. 1992. *Rethinking the family: Some feminist questions.* Boston: Northeastern University Press.

Uttal, Lynet. 1996. Custodial care, surrogate care, and coordinated care: Employed mothers and the meaning of child care. *Gender & Society* 10:291–311.

Wrigley. 1995. *Other people's children.* New York: Basic Books.

Zavella, Patricia. 1987. *Women's work and Chicano families: Cannery workers of the Santa Clara Valley.* Ithaca, NY: Cornell University Press.

Zelizer, Viviana. 1994. *Pricing the priceless child: The social value of children.* Princeton, NJ: Princeton University Press.

CHAPTER 7

Fatherhood

Fatherhood defines a biological and social relationship between a male parent and his offspring. "To father" means to impregnate a woman and beget a child, thus describing a kinship connection that facilitates the intergenerational transfer of wealth and authority (at least in patrilineal descent systems such as ours). Fatherhood also reflects a society's ideals about the rights, duties, and activities of men in families. In general, fatherhood demands that men support, protect, teach, discipline, and control their children, but in reality fathers may do only some of these things. The concept of fatherhood also generalizes to other social and symbolic relationships, as when Christians refer to "God The Father," Catholics call priests "Father," Germans speak of their native country as "The Fatherland," and Americans label George Washington "the Father" of their country. Fatherhood thus reflects a normative set of social practices and expectations that become institutionalized within religion, politics, law, and culture (Coltrane, 2001).

Fathering, in contrast to fatherhood, refers more directly to what men do with and for their children. As noted in the last chapter, although folk beliefs suggest that fathering and mothering entail behaviors that are fixed by reproductive biology, humans must learn how to parent much like they learn other social behaviors. Although women have been the primary caretakers of young children in all cultures, fathers' participation in child rearing has varied from virtually no direct involvement to active participation in all aspects of children's routine care, feeding, instruction, and discipline. Except for breast-feeding and the earliest care of infants, there are no cross-cultural universals in the tasks that mothers and fathers perform (Mead, 1949; Johnson, 1988). Anthropologists have identified two general patterns of fathers' family involvement—one intimate and the other aloof. In the intimate pattern, men eat and sleep with their wives and children, talk with them during evening meals, attend births, and participate actively in infant care. In the contrasting aloof pattern, men often eat and sleep apart from women, spend their leisure time in the company of other men, stay away during births, and seldom help with childcare (Whiting and Whiting, 1975). Research shows that about half of the world's known societies have exhibited close father–child relationships. Compared to societies with distant father–child relationships, those with involved fathers are more likely to be peaceful and to afford women opportunities to own property and be public leaders. Distant-father societies, in contrast, are more likely to encourage violent competition among men and to exclude women from leadership (Coltrane, 1996; Sanday, 1981).

Historical studies show that fatherhood has been linked to the exercise of authority both inside and outside of families. In America, the colonial economy of the seventeenth and eighteenth centuries was based on agriculture and productive family households. As Jessie Bernard notes in "The Good-Provider Role," both mothers and fathers were actively involved in the home-based work that allowed the family to survive. Because men's work as farmers, artisans, and tradesmen occurred in the family **household,** most fathers were a visible presence in their children's lives. Fathers introduced sons to farming or craft work, oversaw the work of others, and were responsible for maintaining harmonious household relations. The preindustrial family home was thus a system of control, as well as a center of production, and both functions reinforced the father's authority and shaped family relationships (Griswold, 1993; Mintz, 1998). Though mothers provided most direct care for infants and young children, men tended to be very active in the training and tutoring of children; so much so that most parental advice was addressed to fathers. Because they were moral teachers and family heads, fathers were thought to have greater responsibility for and influence on children than mothers (Pleck, 1987).

As Bernard's article describes, when market economies took over from home-based production in the nineteenth and twentieth centuries, the father's position as head of the household and moral instructor of his children was slowly transformed. Men were increasingly called upon to seek employment outside the home, and their direct contact with family members declined. As the wage labor economy developed, men's occupational achievement outside the household took on stronger moral overtones, and men came to be seen as fulfilling their family and civic duty not by teaching and interacting with their children as before, but by supporting the family financially. The **good provider** "set a good table, provided a decent home, paid the mortgage, bought the shoes, and kept his children warmly clothed." Bernard describes some of the costs and rewards associated with this male concentration on jobs and careers, including a strong emphasis on success and men's neglect of the emotional aspects of family life. Although most of Bernard's writings were published in the 1950s, 1960s, and 1970s, her comments and questions in this article highlight dilemmas and challenges that fathers are facing today.

In the next selection, Nicholas Townsend reports on an ethnographic study and interviews with fathers who identify their ethnicity as White, Latino, and Asian. Rather than focusing on their differences, however, Townsend finds that they embrace a similar conventional vision of a "package deal" of marriage, work, and fatherhood. Even though most of the wives in his study were employed, most couples were striving to maintain a division of labor based on "father **breadwinners**" and "mother **homemakers.**" The men talked about how they felt obligated to be primarily responsible for earning money, justified in part by their belief that children should be cared for by mothers. Such divisions of labor, as Bernard predicted, influence how parenting is organized and shape the emotional relationships that parents develop with their children. Townsend draws on the fathers' accounts to show how the parenting of men who are "enforcers," as well as those who are "fun dads," is mediated by mothers. Mothers and fathers thus collude in a gendered system of parenting in which mothers schedule and manage their children's activities, as well as monitor their social and emotional lives, with fathers remaining on the sidelines as helpers, playmates, and occasional disciplinarians. The resulting emotional distance between fathers and children is increasingly seen as a problem, insofar as popular cultural ideals now expect even **breadwinner** fathers to be emotionally connected to their children. Do you agree with Townsend that marriage, work, and fatherhood are viewed by most people as a package deal?

Susan Faludi further illuminates the tensions and contradictions of contemporary fatherhood in an excerpt from the book *Stiffed* that focuses on the Promise Keepers. Promise Keepers is a grassroots religious movement combining biblical teachings with stadium rallies and men's support groups. This evangelical Christian organization was founded in 1990 by former University of Colorado football Coach Bill McCartney and is now one of the largest men's organizations in the United States. Promise Keepers promotes conservative family values by teaching that God has ordained men to be heads and masters of their households, but at the same time exhorts them to be "servant leaders" who pay attention to their wives' feelings. The organization holds male-only rallies in sports stadiums and endorses the ideal of separate gender spheres, but also encourages its mostly middle-aged, middle-class members to confess their sins and express their emotions in men's "accountability" groups. Faludi lampoons some aspects of Promise Keepers, but simultaneously appreciates the predicament of men who are increasingly unable to feel that they are valued as fathers because their manhood and their jobs no longer guarantee them authority over women and children.

Although the selections in this chapter do not highlight an alternative vision of fathers who attempt to share breadwinning and homemaking equally with their partners, such practices are becoming more common in all ethnic groups, class levels, and geographic regions. Although the average American father still does only about half as much family work as his wife, absolute levels of involvement have increased, and the expectations for all types of fathers to be involved with their children have increased substantially (Coltrane, 1996; Deutsch, chapter 8, this volume; Parke, 1996). When men participate in the nurturing and supportive activities that serve children, they are also more likely to share in child and home maintenance activities. In contrast, when fathers enact fatherhood based on masculine recreation or family headship, they are less likely to share domestic work with wives. In an estimated 15 to 20 percent of two-parent families, men are almost as involved as mothers interacting with and being available to their children. When fathers share childcare and housework with their partners, employed mothers escape total responsibility for family work, evaluate the division of labor as more fair, are less depressed, and enjoy higher levels of marital satisfaction (Coltrane, 2000). When men care for young children on a regular basis, they emphasize verbal interaction, notice and use more subtle cues, and treat sons and daughters similarly, rather than focusing on play, giving orders, and sex-typing children. These types of father involvement have been found to encourage less gender stereotyping among young adults and to promote independence in daughters and emotional sensitivity in sons (Coltrane, 1996).

Not surprisingly, debates about the proper role of fathers are framed by an older political dichotomy: conservatives focus on fatherhood and family values and stress the importance of male headship and breadwinning, respect for authority, and moral leadership (Blankenhorn, 1995; Popenoe, 1996), whereas liberals focus on the importance of the economy and motherhood, highlighting the benefits that accrue to women when **patriarchal** family traditions are replaced by more individualistic and democratic family forms (Coontz, 1992; Silverstein and Auerbach, 1999, Stacey, 1996). As the Townsend and Faludi articles suggest, traditional assumptions about men's supposed "natural" role as primary breadwinners still have plenty of support. As the historical review by Bernard attests, however, our beliefs about what is natural and inevitable are also tied to changes in the economy and the society. It is likely that women's dramatic entry into the paid labor force will continue to transform images of ideal fathers to include more nurturing of children and performance of housework, but it remains to be seen whether cultural ideals and family practices will come to reflect those more egalitarian ideals.

REFERENCES

Blankenhorn, David. 1995. *Fatherless America: Confronting Our Most Urgent Social Problem.* New York: Basic Books.

Coltrane, Scott. 1996. *Family Man: Fatherhood, Housework, and Gender Equity.* New York: Oxford University Press.

Coltrane, Scott. 2000. Research on Household Labor: Modeling and Measuring the Social Embeddedness of Routine Family Work. *Journal of Marriage and the Family,* 62, 1208–1233.

Coltrane, Scott. 2001. Fatherhood. In *International Encyclopedia of the Social and Behavioral Sciences,* edited by Neil Smelser and Paul Baltes. Oxford, UK: Elsevier.

Coontz, Stephanie. 1992. *The Way We Never Were.* New York: Basic Books.

Griswold, Robert. 1993. *Fatherhood in America: A History.* New York: Basic Books.

Johnson, Miriam. 1988. *Strong Mothers, Weak Wives.* Berkeley: University of California Press.

Mead, Margaret. 1949. *Male and Female.* New York: William Morrow.

Mintz, Steven. 1998. From Patriarchy to Androgyny and Other Myths: Placing Men's Family Roles in Historical Perspective. In *Men in Families,* edited by A. Booth and A. C. Crouter. Mahwah, NJ: Erlbaum.

Parke, Ross D. 1996. *Fatherhood.* Cambridge, MA: Harvard University Press.

Pleck, Joseph. 1987. American Fathering in Historical Perspective. In *Changing Men,* edited by M. Kimmel. Newbury Park, CA: Sage.

Popenoe, David. 1996. *Life without Father.* New York: Martin Kessler/Free Press.

Sanday, Peggy. 1981. *Female Power and Male Dominance.* Cambridge, England: Cambridge University Press.

Silverstein, Louise B., and Carl F. Auerbach. 1999. Deconstructing the Essential Father. *American Psychologist* 54(6):397–407.

Stacey, Judith. 1996. *In the Name of the Family: Rethinking Family Values in the Postmodern Age.* Boston: Beacon.

Whiting, John, and Beatrice Whiting. 1975. Aloofness and Intimacy of Husbands and Wives. *Ethos* 3:183–207.

SUGGESTED READINGS

Collier, Richard. 1995. *Masculinity, Law, and the Family.* London; New York: Routledge.

Coltrane, Scott. 1996. *Family Man: Fatherhood, Housework, and Gender Equity.* New York: Oxford University Press.

Daniels, Cynthia R., ed. 1998. *Lost Fathers: The Politics of Fatherlessness in America.* New York: St. Martin's Press.

Dowd, Nancy E. 2000. *Redefining Fatherhood.* New York: New York University Press.

Gerson, Kathleen. 1993. *No Man's Land: Men's Changing Commitments to Family and Work.* New York: Basic Books.

Griswold, Robert L. 1993. *Fatherhood in America: A History.* New York: Basic Books.

Hobson, Barbara, ed. 2002. *Making Men into Fathers: Men, Masculinities, and the Social Politics of Fatherhood.* Cambridge; New York: Cambridge University Press.

Lamb, Michael E. 2003. *The Role of the Father in Child Development.* New York: Wiley.

Levine, J., and E. W. Pitt. 1995. *New Expectations: Community Strategies for Responsible Fatherhood.* New York: Families and Work Institute.

Levine, J., and T. L. Pittinsky. 1997. *Working Fathers: New Strategies for Balancing Work and Family.* New York: Harcourt Brace & Company.

Parke, Ross D. 1996. *Fatherhood.* Cambridge, MA: Harvard University Press.

INFOTRAC® COLLEGE EDITION EXERCISES

The exercises that follow allow you to use the InfoTrac® College Edition online database of scholarly articles to explore the sociological implications of the selections in this chapter.

Search Keyword: Fatherhood. Search for articles that discuss fatherhood. Examine the different ways the notion of fatherhood is used. How is it defined? What are its key components? What are some of the cultural images of fatherhood? To what degree does fatherhood appear to be biological, and to what degree is it a social construction? Is fatherhood considered to be similar to, or different from, motherhood? In what ways?

Search Keyword: Fatherlessness. Using InfoTrac® College Edition, find articles on fatherlessness. How is this notion defined? Is fatherlessness alleged to have increased or decreased over time? What do these articles suggest are some of the causes of fatherlessness? How does fatherlessness affect children? How does it affect families generally? How do these impacts differ by ethnicity and social class? Do the articles that you find suggest "solutions" to fatherlessness? If so, what are they?

Search Keyword: Breadwinner. Search for articles that describe the role of family breadwinner. What does it mean to be the family breadwinner? How is the notion of the family breadwinner related to gender? What does being the family breadwinner (or not) say about men and fathers? How does it relate to cultural images of men? How does it relate to cultural images of women? How is the good-provider role implicated in family power relationships?

Search Keyword: Promise Keepers. Using InfoTrac® College Edition, find articles that talk about the men's movement called Promise Keepers. How is this movement described? What are the goals of the Promise Keepers? How is this movement similar to, or different from, other men's movements, both current and past? How do Promise Keepers define their roles in families, as fathers and as husbands? What are some of the controversial aspects of the Promise Keepers? With whom are they controversial, and why?

Search Keyword: Shared parenting. Find articles that deal with shared parenting. What constitutes shared parenting? How is this phrase used to describe relationships between husbands and wives? How is it used to describe relationships between fathers and their children? What are some of the cultural images (movies, television programs) that are used to portray men in shared parenting roles? In what ways has parenting become more equally shared over time? In what ways has it become less equally shared? How does shared parenting affect relationships of power within the family?

22

The Good-Provider Role:

Its Rise and Fall

JESSIE BERNARD,
1981

. . . In our country in Colonial times women were still viewed as performing a providing role, and they pursued a variety of occupations. Abigail Adams managed the family estate, which provided the wherewithal for John to spend so much time in Philadelphia. In the 18th century "many women were active in business and professional pursuits. They ran inns and taverns; they managed a wide variety of stores and shops; and, at least occasionally, they worked in careers like publishing, journalism and medicine" (Demos, 1974, p. 430). Women sometimes even "joined the menfolk for work in the fields" (p. 430). Like the household of the proverbial virtuous woman, the Colonial household was a little factory that produced clothing, furniture, bedding, candles, and other accessories, and again, as in the case of the virtuous woman, the female role was central. It was taken for granted that women provided for the family along with men.

The good provider as a specialized male role seems to have arisen in the transition from subsistence to market—especially money—economies that accelerated with the industrial revolution. The good-provider role for males emerged in this country roughly, say, from the 1830s, when de Tocqueville was observing it, to the late 1970s, when the 1980 census declared that a male was not automatically to be assumed to be head of the household. This gives the role a life span of about a century and a half. Although relatively short-lived, while it lasted the role was a seemingly rock-like feature of the national landscape.

As a psychological and sociological phenomenon, the good-provider role had wide ramifications for all of our thinking about families. It marked a new kind of marriage. It did not have good effects on women: The role deprived them of many chips by placing them in a peculiarly vulnerable position. Because she was not reimbursed for her contribution to the family in either products or services, a wife was stripped to a considerable extent of her access to cash-mediated markets. By discouraging labor force participation, it deprived many women, especially affluent ones, of opportunities to achieve strength and competence. It deterred young women from acquiring productive skills. They dedicated themselves instead to winning a good provider who would "take care of" them. The wife of a more successful provider became for all intents and purposes a parasite, with little to do except indulge or pamper herself. The psychology of such dependence could become all but

From Jessie Bernard, "The Good-Provider Role: Its Rise and Fall," from *American Psychologist,* 36(1):1–12.

crippling. There were other concomitants of the good-provider role.

EXPRESSIVITY AND THE GOOD-PROVIDER ROLE

The new industrial order that produced the good provider changed not so much the division of labor between the sexes as it did the site of the work they engaged in. Only two of the concomitants of this change in work site are selected for comment here, namely, (a) the identification of gender with work site as well as with work itself and (b) the reduction of time for personal interaction and intimacy within the family.

It is not so much the specific kinds of work men and women do—they have always varied from time to time and place to place—but the simple fact that the sexes do different kinds of work, whatever it is, which is in and of itself important. The division of labor by sex means that the work group becomes also a sex group. The very nature of maleness and femaleness becomes embedded in the sexual division of labor. One's sex and one's work are part of one another. One's work defines one's gender.

Any division of labor implies that people doing different kinds of work will occupy different work sites. When the division is based on sex, men and women will necessarily have different work sites. Even within the home itself, men and women had different work spaces. The woman's spinning wheel occupied a different area from the man's anvil. When the factory took over much of the work formerly done in the house, the separation of work space became especially marked. Not only did the separation of the sexes become spatially extended, but it came to relate work and gender in a special way. The work site as well as the work itself became associated with gender; each sex had its own turf. This sexual "territoriality" has had complicating effects on efforts to change any sexual division of labor. The good provider worked primarily in the outside male world of business and industry. The homemaker worked primarily in the home.

Spatial separation of the sexes not only identifies gender with work site and work but also reduces the amount of time available for spontaneous emotional give-and-take between husbands and wives. When men and women work in an economy based in the home, there are frequent occasions for interaction. (Consider, for example, the suggestive allusions made today to the rise in the birth rate nine months after a blackout.) When men and women are in close proximity, there is always the possibility of reassuring glances, the comfort of simple physical presence. But when the division of labor removes the man from the family dwelling for most of the day, intimate relationships become less feasible. De Tocqueville was one of the first to call our attention to this. In 1840 he noted that

> almost all men in democracies are engaged in public or professional life; and . . . the limited extent of common income obliges a wife to confine herself to the house, in order to watch in person and very closely over the details of domestic economy. All these distinct and compulsory occupations are so many natural barriers, which, by keeping the two sexes asunder, render the solicitations of the one less frequent and less ardent—the resistance of the other more easy. (de Tocqueville, 1840, p. 212)

Not directly related to the spatial constraints on emotional expression by men, but nevertheless a concomitant of the new industrial order with the same effect, was the enormous drive for achievement, for success, for "making it" that escalated the provider role into the good-provider role. De Tocqueville (1840) is again our source:

> The tumultuous and constantly harassed life which equality makes men lead [becoming good providers] not only distracts them from the passions of love, by denying them time to indulge in it, but it diverts them from it by another more secret but more certain road.

> All men who live in democratic ages more or less contract ways of thinking of the manufacturing and trading classes. (p. 221)

As a result of this male concentration on jobs and careers, much abnegation and "a constant sacrifice of her pleasures to her duties" (de Tocqueville, 1840, p. 212) were demanded of the American woman. The good-provider role, as it came to be shaped by this ambience, was thus restricted in what it was called upon to provide. Emotional expressivity was not included in the role. One of the things a parent might say about a man to persuade a daughter to marry him, or a daughter might say to explain to her parents why she wanted to, was not that he was a gentle, loving, or tender man but that he was a good provider. He might have many other qualities, good or bad, but if a man was a good provider, everything else was either gravy or the price one had to pay for a good provider.

Lack of expressivity did not imply neglect of the family. The good provider was a "family man." He set a good table, provided a decent home, paid the mortgage, bought the shoes, and kept his children warmly clothed. He might, with the help of the children's part-time jobs, have been able to finance their educations through high school and, sometimes, even college. There might even have been a little left over for an occasional celebration in most families. The good provider made a decent contribution to the church. His work might have been demanding, but he expected it to be. If in addition to being a good provider, a man was kind, gentle, generous, and not a heavy drinker or gambler, that was all frosting on the cake. Loving attention and emotional involvement in the family were not part of a woman's implicit bargain with the good provider.

By the time de Tocqueville published his observations in 1840, the general outlines of the good-provider role had taken shape. It called for a hard-working man who spent most of his time at his work. In the traditional conception of the role, a man's chief responsibility is his job, so that "by definition any family behaviors must be subordinate to it in terms of significance and [the job] has priority in the event of a clash" (Scanzoni, 1975, p. 38). This was the classic form of the good-provider role, which remained a powerful component of our societal structure until well into the present century.

COSTS AND REWARDS OF THE GOOD-PROVIDER ROLE FOR MEN

There were both costs and rewards for those men attached to the good-provider role. The most serious cost was perhaps the identification of maleness not only with the work site but especially with success in the role. "The American male looks to his breadwinning role to confirm his manliness" (Brenton, 1966, p. 194). To be a man one had to be not only a provider but a *good* provider. Success in the good-provider role came in time to define masculinity itself. The good provider had to achieve, to win, to succeed, to dominate. He was a bread*winner.* He had to show "strength, cunning, inventiveness, endurance—a whole range of traits henceforth defined as exclusively 'masculine'" (Demos, 1974, p. 436). Men were judged as men by the level of living they provided. They were judged by the myth "that endows a money-making man with sexiness and virility, and is based on man's dominance, strength, and ability to provide for and care for 'his' woman" (Gould, 1974, p. 97). The good provider became a player in the male competitive macho game. What one man provided for his family in the way of luxury and display had to be equaled or topped by what another could provide. Families became display cases for the success of the good provider.

The psychic costs could be high:

> By depending so heavily on his breadwinning role to validate his sense of himself as a man, instead of also letting his roles as husband, father, and citizen of the community

> count as validating sources, the American male treads on psychically dangerous ground. It's always dangerous to put all of one's psychic eggs into one basket. (Brenton, 1966, p. 194)

The good-provider role not only put all of a man's gender-identifying eggs into one psychic basket, but it also put all the family-providing eggs into one basket. One individual became responsible for the support of the whole family. Countless stories portrayed the humiliation families underwent to keep wives and especially mothers out of the labor force, a circumstance that would admit to the world the male head's failure in the good-provider role. If a married woman had to enter the labor force at all, that was bad enough. If she made a good salary, however, she was "co-opting the man's passport to masculinity" (Gould, 1974, p. 98) and he was effectively castrated. A wife's earning capacity diminished a man's position as head of the household (Gould, 1974, p. 99).

Failure in the role of good provider, which employment of wives evidenced, could produce deep frustration. As Komarovsky (1940, p. 20) explains, this is "because in his own estimation he is failing to fulfill what is the central duty of his life, the very touchstone of his manhood—the role of family provider."

But just as there was punishment for failure in the good-provider role, so also were there rewards for successful performance. A man "derived strength from his role as provider" (Komarovsky, 1940, p. 205). He achieved a good deal of satisfaction from his ability to support his family. It won kudos. Being a good provider led to status in both the family and the community. Within the family it gave him the power of the purse and the right to decide about expenditures, standards of living, and what constituted good providing. "Every purchase of the family—the radio, his wife's new hat, the children's skates, the meals set before him—all were symbols of their dependence upon him" (Komarovsky, 1940, pp. 74–75). Such dependence gave him a "profound sense of stability" (p. 74). It was a strong counterpoise vis-à-vis a wife with a stronger personality. "Whether he had considerable authority within the family and was recognized as its head, or whether the wife's stronger personality . . . dominated the family, he nevertheless derived strength from his role as a provider" (Komarovsky, 1940, p. 75). As recently as 1975, in a sample of 3,100 husbands and wives in 10 cities, Scanzoni found that despite increasing egalitarian norms, the good provider still had "considerable power in ultimate decision-making" and as "unique provider" had the right "to organize his life and the lives of other family members around his occupation" (p. 38).

A man who was successful in the good-provider role might be freed from other obligations to the family. But the flip side of this dispensation was that he could not make up for poor performance by excellence in other family roles. Since everything depended on his success as provider, everything was at stake. The good provider played an all-or-nothing game.

DIFFERENT WAYS OF PERFORMING THE GOOD-PROVIDER ROLE

Although the legal specifications for the role were laid out in the common law, in legislation, in legal precedents, in court decisions, and, most importantly, in custom and convention, in real-life situations the social and social-psychological specifications were set by the husband or, perhaps more accurately, by the community, alias the Joneses, and there were many ways to perform it.

Some men resented the burdens the role forced them to bear. A man could easily vent such resentment toward his family by keeping complete control over all expenditures, dispensing the money for household maintenance, and complaining about bills as though it were his wife's fault that shoes cost so much. He could, in effect, punish his family for his having to perform the role. Since the money he earned belonged to him—was "his"—he could do with it what he pleased. Through extreme parsimony he could dole out his money in a mean, humiliating way, forcing his wife to come begging for pennies. By his reluctance and resentment he could make his family pay emotionally for the provisioning he supplied.

At the other extreme were the highly competitive men who were so involved in outdoing the Joneses that the fur coat became more important than the affectionate hug. They "bought off" their families. They sometimes succeeded so well in their extravagance that they sacrificed the family they were presumably providing for to the achievements that made it possible (Keniston, 1965).

The Depression of the 1930s revealed in harsh detail what the loss of the role could mean both to the good provider and to his family, not only in the loss of income itself—which could be supplied by welfare agencies or even by other family members, including wives—but also and especially in the loss of face.

The Great Depression did not mark the demise of the good-provider role. But it did teach us what a slender thread the family hung on. It stimulated a whole array of programs designed to strengthen that thread, to ensure that it would never again be similarly threatened. Unemployment insurance was incorporated into the Social Security Act of 1935, for example, and a Full Employment Act was passed in 1946. But there proved to be many other ways in which the good-provider role could be subverted.

ROLE REJECTORS AND ROLE OVERPERFORMERS

Recent research in psychology, anthropology, and sociology has familiarized us with the tremendous power of roles. But we also know that one of the fundamental principles of role behavior is that conformity to role norms is not universal. Not everyone lives up to the specifications of roles, either in the psychological or in the sociological definition of the concept. Two extremes have attracted research attention: (a) the men who could not live up to the norms of the good-provider role or did not want to, at one extreme, and (b) the men who overperformed the role, at the other. For the wide range in between, from blue-collar workers to professionals, there was fairly consistent acceptance of the role, however well or poorly, however grumblingly or willingly, performed.

First the nonconformists. Even in Colonial times, desertion and divorce occurred:

> Women may have deserted because, say, their husbands beat them; husbands, on the other hand, may have deserted because they were unable or unwilling to provide for their usually large families in the face of the wives' demands to do so. These demands were, of course, backed by community norms making the husband's financial support a sacred duty. (Scanzoni, 1979, pp. 24–25)

Fiedler (1962) has traced the theme of male escape from domestic responsibilities in the American novel from the time of Rip Van Winkle to the present:

> The figure of Rip Van Winkle presides over the birth of the American imagination; and it is fitting that our first successful home-grown legend should memorialize, however playfully, the flight of the dreamer from the shrew—into the mountains and out of time, away from the drab duties of home . . . anywhere to avoid . . . marriage and responsibility. One of the factors that determine theme and form in our great books is this strategy of evasion, this retreat to nature and childhood which makes our literature (and life) so charmingly and infuriatingly "boyish." (pp. xx–xxi)

Among the men who pulled up stakes and departed for the West or went down to the sea in ships, there must have been a certain proportion who, like their mythic prototype, were simply fleeing the good-provider role.

The work of Demos (1974), a historian, offers considerable support for Fiedler's thesis. He tells us that the burdens thrust on men in the 19th century by the new patterns of work began to show their effects in the family. When "the [spatial] separation of the work lives of husbands and wives made communication so problematic," he asks, "what was the likelihood of meaningful communication?" (Demos, 1974, p. 438). The answer is, relatively little. Divorce and separation increased, either formally or by tacit consent—or simply by default, as in the case of a variety of defaulters—tramps, bums, hoboes—among them.

In this connection, "the development of the notorious 'tramp' phenomenon is worth noticing," Demos (1974, p. 438) tells us. The tramp was a man who just gave up, who dropped out of the role entirely. He preferred not to work, but he would do small chores or other small-scale work for a handout if he had to. He was not above begging the housewife for a meal, hoping she would not find work for him to do in repayment. Demos (1974) describes the type:

> Demoralized and destitute wanderers, their numbers mounting into the hundreds of thousands, tramps can be fairly characterized as men who had run away from their wives. . . . Their presence was mute testimony to the strains that tugged at the very core of American family life. . . . Many observers noted that the tramps had created a virtual society of their own [a kind of counterculture] based on a principle of single-sex companionship. (p. 438)

A considerable number of them came to be described as "homeless men" and, as the country became more urbanized, landed ultimately on skid row. A large part of the task of social workers for almost a century was the care of the "evaded" women they left behind. When the tramp became wholly demoralized, a chronic alcoholic, almost unreachable, he fell into a category of his own—he was a bum.

Quite a different kettle of fish was the hobo, the migratory worker who spent several months harvesting wheat and other large crops and the rest of the year in cities. Many were the so-called Wobblies, or Industrial Workers of the World, who repudiated the good-provider role on principle. They had contempt for the men who accepted it and could be called conscientious objectors to the role. "In some IWW circles, wives were regarded as the 'ball and chain.' In the West, IWW literature proclaimed that the migratory worker, usually a young, unmarried male, was 'the finest specimen of American manhood . . . the leaven of the revolutionary labor movement' " (Foner, 1979, p. 400). Exemplars of the Wobblies were the nomadic workers of the West. They were free men. The migratory worker, "unlike the factory slave of the Atlantic seaboard and the central states, . . . was most emphatically 'not afraid of losing his job.' No wife and family cumbered him. The worker of the East, oppressed by the fear of want for wife and babies, dared not venture much" (Foner, 1979, p. 400). The reference to fear of loss of job was well taken; employers preferred married men, disciplined into the good-provider role, who had given hostages to fortune and were therefore more tractable.

Just on the verge between the area of conformity to the good-provider role—at whatever level—and the area of complete nonconformity to it was the non-good provider, the marginal group of workers usually made up of "the undereducated, the under-trained, the under-employed, or part-time employed, as well as the under-paid, and of course the unemployed" (Snyder, 1979, p. 597). These included men who wanted—sometimes desperately—to perform the good-provider role but who for one reason or another were unable to do so. Liebow (1966) has discussed the ramifications of failure among the black men of Tally's corner: The black man is

> under legal and social constraints to provide for them [their families] to be a husband to his wife and a father to his children. The chances are, however, that he is failing to provide for them, and failure in this primary function contaminates his performance as father in other respects as well. (p. 86).

In some cases, leaving the family entirely was the best substitute a man could supply. The community was left to take over.

At the other extreme was the overperformer. De Tocqueville, quoted earlier, was already describing him as he manifested in the 1830s. And as late as 1955 Warner and Ablegglen were adding to the considerable literature on industrial leaders and tycoons, referring to their "driving concentration" on their careers and their "intense focusing" of interests, energies, and skills on these careers, "even limiting their sexual activity" (pp. 48–49). They came to be known as workaholics or work-intoxicated men. Their preoccupation with their work even at the expense of their families was, as I have already noted, quite acceptable in our society.

Poorly or well performed, the good-provider role lingered on. World War II initiated a challenge, this time in the form of attracting more and more married women into the labor force, but the challenge was papered over in the 1950s with an "age of togetherness" that all but apotheosized the good provider, his house in the suburbs, his homebody wife, and his third, fourth, even fifth, child. As late as the 1960s most housewives (87%) still saw breadwinning as their husband's primary role (Lopata, 1971, p. 91). . . .

The present discussion began with the woman's part in the provider role. We saw how as more and more of the provisioning of the family came to be by way of monetary exchange, the woman's part shrank. A woman could still provide services, but could furnish little in the way of food, clothing, and shelter. But now that she is entering the labor force in large numbers, she can once more resume her ancient role, this time, like her male counterpart the provider, by way of a monetary contribution. More and more women are doing just this.

The assault on the good-provider role in the Depression was traumatic. But a modified version began to appear in the 1970s as a single income became inadequate for more and more families. Husbands have remained the major providers, but in an increasing number of cases the wife has begun to share this role. . . .

For some men the relief from the strain of sole responsibility for the provider role has been welcome. But for others the feeling of degradation resembles the feelings reported 40 years earlier in the Great Depression. It is not that they are no longer providing for the family but that the role-sharing wife now feels justified in making demands on them. The good-provider role with all its prerogatives and perquisites has undergone profound changes. It will never be the same again. Its death knell was sounded when . . . the 1980 census no longer automatically assumed that the male member of the household was its head.

THE CURRENT SCENE

Among the new demands being made on the good-provider role, two deserve special consideration, namely, (a) more intimacy, expressivity, and nurturance—specifications never included in it as it originally took shape—and (b) more sharing of household responsibility and child care.

As the pampered wife in an affluent household came often to be an economic parasite, so also the good provider was often, in a way, a kind of emotional parasite. Implicit in the definition of the role was that he provided goods and material things. Tender loving care was not one of the requirements. Emotional ministrations from the family were his right; providing them was not a corresponding obligation. Therefore, as de Tocqueville had already noted by 1840, women suffered a kind of emotional deprivation labeled by Robert Weiss "relational deficit" (cited in Bernard, 1976). Only recently has this male rejection of emotional expression come to be challenged. Today, even blue-collar women are imposing "a host of new role expectations upon their husbands or lovers. . . . A new role set asks the blue-collar male to strive for . . . deep-coursing intimacy" (Shostak, 1973). It was not only vis-à-vis his family that the good provider was lacking in expressivity. This lack was built into the whole male role script. Today not only women but also men are beginning to protest the repudiation of expressivity prescribed in male roles (David & Brannon, 1976; Farrell, 1974; Fasteau, 1974; Pleck & Sawyer, 1974).

Is there any relationship between the "imposing" on men of "deep-coursing intimacy" by women on one side and the increasing proportion of men who find marriage burdensome and restrictive on the other? Are men seeing the new emotional involvements being asked of them as "all burdens and restrictions"? Are they responding to the new involvements under duress? Are they feeling oppressed by them? Fearful of them?

From the standpoint of high-level pure-science research there may be something bizarre, if not even slightly absurd, in the growing corpus of serious research on how much or how little husbands of employed wives contribute to household chores and child care. Yet it is serious enough that all over the industrialized world such research is going on. Time studies in a dozen countries—communist as well as capitalist—trace the slow and bungling process by which marriage accommodates to changing conditions and by which women struggle to mold the changing conditions in their behalf. For everywhere the same picture shows up in the research: an image of women sharing the provider role and at the same time retaining responsibility for the household. Until recently such a topic would have been judged unworthy of serious attention. It was a subject that might be worth a good laugh, for instance, as when an all-thumbs man in a cartoon burns the potatoes or finds himself bumbling awkwardly over a diaper, demonstrating his—proud—male ineptness at such female work. But it is no longer funny.

The "politics of housework" (Mainardi, 1970) proves to be more profound than originally believed. It has to do not only with tasks but also with gender—and perhaps more with the site of the tasks than with their intrinsic nature. A man can cook magnificently if he does it on a hunting or fishing trip; he can wield a skillful needle if he does it mending a tent or a fishing net; he can even feed and clean a toddler on a camping trip. Few of the skills of the homemaker are beyond his reach so long as they are practiced in a suitably male environment. It is not only women's work in and of itself that is degrading but any work on female turf. It may be true, as Brenton (1966) says, that "the secure man can wash a dish, diaper a baby, and throw the dirty clothes into the washing machine—or do anything else women used to do exclusively—without thinking twice about it" (p. 311), but not all men are that secure. To a great many men such chores are demasculinizing. The apron is shameful on a man in the kitchen; it is all right at the carpenter's bench.

The male world may look upon the man who shares household responsibilities as, in effect, a slob. One informant tells the interviewer about a conversation on the job: "What, are you crazy?" his hard-hat fellow workers ask him when he speaks of helping his wife. "The guys want to kill me. 'You son of a bitch! You are getting us in trouble.' . . . The men get really mad" (Lein, 1979, p. 492). Something more than persiflage is involved here. We are fairly familiar with the trauma associated with the invasion by women of the male work turf, the hazing women can be subjected to, and the male resentment of admitting them except into their own segregated areas. The corresponding entrance of men into the traditional turf of women—the kitchen or the nursery—has analogous but not identical concomitants. . . .

A considerable amount of thought has been devoted to studying the effects of the large influx of women into the work force. An equally interesting question is what the effect will be if a large number of men actually do increase their participation in the family and the household. Will men find the apron shameful? . . .

The demise of the good-provider role also calls for consideration of other questions: What does the demotion of the good provider to the status of senior provider or even mere coprovider do to him? To marriage? To gender identity? What does expanding the role of housewife to that of junior provider or even coprovider do to her? To marriage? To gender identity? . . .

The good-provider role may be on its way out, but its legitimate successor has not yet appeared on the scene.

SELECTED REFERENCES

Bernard, J. Homosociality and female depression. *Journal of Social Issues,* 1976, *32,* 207–224.

Brenton, M. *The American male.* New York: Coward-McCann, 1966.

David, D. S., & Brannon, R. (Eds.). *The forty-nine percent majority: The male sex role.* Reading, Mass: Addison-Wesley, 1976.

de Tocqueville, A. *Democracy in America.* New York: J. & H. G. Langley, 1840.

Demos, J. The American family in past time. *American Scholar,* 1974, *43,* 422–446.

Farrell, W. *The liberated man.* New York: Random House, 1974.

Fasteau, M. F. *The male machine.* New York: McGraw-Hill. 1974.

Fiedler, L. *Love and death in the American novel.* New York: Meredith, 1962.

Foner, P. S. *Women and the American labor movement.* New York: Free Press, 1979.

Gould, R. E. Measuring masculinity by the size of a paycheck. In J. E. Pleck & J. Sawyer (Eds.), *Men and masculinity.* Englewood Cliffs, N.J.: Prentice-Hall, 1974. (Also published in *Ms.,* June 1973, pp. 18ff.)

Keniston, K. *The uncommitted: Alienated youth in American society.* New York: Harcourt, Brace & World, 1965.

Komarovsky, M. *The unemployed man and his family.* New York: Dryden Press, 1940.

Lein, L. Responsibility in the allocation of tasks. *Family Coordinator,* 1979, *28,* 489–496.

Liebow, E. *Tally's corner.* Boston: Little, Brown, 1966.

Lopata, H. *Occupation housewife.* New York: Oxford University Press, 1971.

Mainardi, P. The politics of housework. In R. Morgan (Ed.), *Sisterhood is powerful.* New York: Vintage Books, 1970.

Pleck, J. H., & Sawyer, J. (Eds.). *Men and masculinity.* Englewood Cliffs, N.J.: Prentice-Hall, 1974.

Scanzoni, J. H. *Sex roles, life styles, and childbearing: Changing patterns in marriage and the family.* New York: Free Press, 1975.

Scanzoni, J. H. An historical perspective on husband-wife bargaining power and marital dissolution. In G. Levinger & O. Moles (Eds.), *Divorce and separation in America.* New York: Basic Books, 1979.

Shostak, A. *Working class Americans at home: Changing expectations of manhood.* Unpublished manuscript, 1973.

Snyder, L. The deserting, non-supporting father. Scapegoat of family non-policy. *Family Coordinator,* 1979, *38,* 594–598.

Warner, W. L., & Ablegglen, J. O. *Big business leaders in America.* New York: Harper, 1955.

23

Marriage, Work, and Fatherhood in Men's Lives

NICHOLAS TOWNSEND,
2002

. . . In my interviews, the racial-ethnic category was not associated with different fundamental values about the place of fatherhood and family in men's lives, although some men did invoke their particular ethnic or cultural background to explain adherence to values that were in fact widely shared. I was struck by the way that men whose ancestry was Italian, Chinese, Irish, Portuguese, Filipino, or Mexican (none of whom spoke their ancestral language to their children) invoked, in almost identical terms, the "old country values" of their fathers as a support for the importance of family life, loyalty, and respect for the older generation. I found a remarkable degree of uniformity in men's depictions of the central elements of fatherhood, a discovery that mirrors the similar findings of a dominant image of motherhood by researchers who have studied diverse groups of mothers (Garey 1999; Hays 1996; Segura 1994; Walker 1990).

My study is not one of the racial-ethnic patterning of difference but is an investigation of the composition of, and internal contradiction within, a cultural model of successful male adulthood and fatherhood. While labels such as "Hispanic" and "Asian American" certainly point to important aspects of discrimination in the United States, they also accept a particular racializing division of people and obscure other crucial dimensions of difference. Denise Segura (1994), for instance, discovered that Latinas did not share a uniform orientation to motherhood, but rather that there were important distinctions between mothers who had immigrated from Mexico and Mexican American women who were themselves born and raised in the United States. To use the categories Hispanic or Latino to include Californians of Mexican descent whose families have been in the United States for several generations as well as immigrants from Mexico, Guatemala, Argentina, and other countries in Central or South America; or to classify as Asian American third-generation Chinese Americans as well as people who have themselves immigrated or whose parents immigrated from China, Japan, Vietnam, Korea, and the Philippines is not so much to recognize cultural difference as it is to impose dominant American patterns of discrimination and difference.

My interviews, which I quote in detail, revealed a basic pattern of goals and tensions that was shared by the men with whom I talked. I did not find that the Asian American or Hispanic men I talked to, or the men whose parents, grandpar-

From *The Package Deal: Marriage, Work, and Fatherhood in Men's Lives,* by Nicholas Townsend. Reprinted by permission of Temple University Press.

ents, or more remote ancestors had come from various parts of Europe, had different visions of what it means to be a successful man and a successful father in the contemporary United States. My conclusion is supported by the research of many others on parenthood in U.S. culture. Every specific study is, of course, subject to the limitations of its specificity, but, as the authors of so many of these specific studies point out, the findings of specific studies reveal a broadly shared and general picture.

Although I have not treated racial-ethnic categories as explanatory variables, I refer throughout this book to the pervasive racism and discrimination of U.S. culture and to the ways in which the dominant values of the culture exclude members of some groups from participation in the supposedly universal "American dream."

The men I talked with reflected the heterogeneity of the community, but they are a particular group, not a representative sample. The group does not include all types of men, it is class specific, all the men were avowedly heterosexual, and none of them was African American. Within the group, however, there was variation in opportunity and in outcomes. I deliberately chose men who were the same age and who went to school in the same town because it allowed me to learn about the circumstances of their lives in a way that would have been impossible if I had talked to a random sample of men from different parts of the country or of different ages.

My approach has been to examine the lives and words of people who accept the dominant values, who judge themselves by those standards, and who occupy social positions that enable them to realize those values to an extent that they can find acceptable. These men are "typical" in the sense that their lives and values represent the dominant cultural norms. It is worth nothing that for men in the United States who were born in 1950, 95 percent were in the labor force at the end of the twentieth century, more than 90 percent have married, and almost 90 percent of their wives have had at least one child. . . .

MEN AT WORK, WOMEN AT HOME: THE STRUCTURAL DIVISION OF LABOR

Men presented their own employment as primarily oriented toward providing the material resources for them to protect their children from danger and endow them with opportunities. They felt that their employment also contributed to the emotional closeness that is an integral facet of fatherhood, because these men interpreted providing as an expression of paternal love. In some dual-earner couples, husbands who wanted to emphasize that they were the primary providers for their families explained that their wives' incomes were used for "extras" or "luxuries." Other men said their wives worked to add variety to their lives, for social contacts, or to "get away from the kids." These men were all doing cultural work to interpret their arrangements as conforming to a hegemonic picture of the structural division of labor in marriage.

For these men one of the most important contributions of employment to fatherhood was that it enabled them to provide their children with a mother who was at home for them and could be close to them almost as a surrogate for their father. Jean Potuchek (1997) studied the gendered division of labor, identity, and responsibility in dual-earner couples. She described the distinction so many of these employed husbands and wives made between working (and earning an income) and breadwinning (and providing for a family). Some employment and income, that is, was described as more essential and basic than others. It is not that the couples Potuchek interviewed denied that both were employed, but that they described the husband's breadwinning as central to their family strategy while the wife's earnings were supplemental. The decline in the family wage (the ability for one worker to be able to earn enough to support a family) has made the simple division between "father breadwinner" and "mother homemaker" harder to maintain in its pure form, but many couples maintain and reinforce the division of responsibility that accompanies it. For the men I talked to, the division of labor in which they were the primary breadwinners reinforced not only their own identities as workers and providers but also enabled them to fulfill other facets of fatherhood by ensuring that there was an available parent in the home.

In support of this division of labor in their parenting, the men I talked to made three interlocking arguments: They liked or chose the arrangement, it was best for the children, and it was natural. Gordon, the engineer with three sons who felt "more married" once he had children, expressed very clearly the structural division of labor between parents: One parent should stay home to raise the children, and it should be the mother.

> I think it's wrong to have kids and then lock them in daycare centers while you're working. That's why I'm really grateful that my wife can stay home. And although at times we were real tight for money, and I told her she might have to start looking for a job if we were going to make ends meet, I was grateful when things worked out and she didn't have to. Because this is really the place the kids need a full-time mother, to watch them.

This arrangement worked, Gordon said, because "She's not the working type." This gendered division of labor between husband and wife was a reproduction of his parents' pattern. His father had been a skilled machinist; his mother, with a college degree, had stayed home and not worked outside the home until her children were in high school. Gordon explained the arrangement he had with his wife as the result of their "choice" and in accord with his wife's personality. Although Gordon described both the division of labor and the fact that he followed his father into working on machines as "natural," he and his wife's arrangement was an instance of a social fact: In the overwhelming number of cases where one of a couple works full time, it is the husband.

Like Gordon, Marvin attributed the division of labor in his marriage to his wife's preference. She had worked off and on, he said, selling products from the home and working as a teacher's aide for the local school district, which "gives her a lot of flexibility." When I asked him if she had ever wanted to work full time, he said,

> She seems to have wanted more to be a good mother. And she was the type of person that when we got married, she had this view of herself as not "Super Mom," but "Nice Mom" that does the things that moms do and takes the kids and gets involved in things. And that was a really big thing to her.

Marvin was articulating what Garey (1999) points to as a dominant cultural image of mothers as oriented either to work or to family. Garey argues that many employed mothers downplay their aspirations to "career" or to being "Super Mom" and, rather, practice "maternal visibility" by making a point of being seen as doing "the things that moms do."

Paul went a step beyond Gordon and Marvin in his defense of a structural gendered division of labor, turning it into a timeless and natural pattern. Paul was a serious, intense man who talked quietly but displayed a fierce protectiveness of his family. He worked a night shift with lots of overtime and shift differential pay. He and his wife, who was employed full time, lived with their two sons, ages six and eight, in a townhouse near his work. His mother-in-law cared for the children during the day, but she was about to move away and Paul's plan was for his wife to reduce her hours of employment and work part time:

> I was thinking about trying to buy a [single family] house over here, but if it's going to cause me to be away from the family, or cause [my wife] to have to work all the time, I think we're gonna back out. If I can't afford a house on my pay alone, and make it, if we can't do it on my paycheck alone, we're not gonna do it. Because that's just basic. It's just the way it's been since time began. Women

> stay home. I'm not trying to be chauvinist by any—but if you're gonna have a family, that's the way it works best.

The gendered division of labor, then, puts women in the home as the mothers of men's children, and this division of labor is reinforced by cultural work that emphasizes men's responsibility as providers and women's involvement in their children's lives. Such a division of labor is presented as natural and equal, but it is a product of a particular economic structure and social organization of work. The gendered division of labor at the structural level also has profound implications for the daily activity of parenting.

PROTECTING, ENDOWING, AND BEING THERE: GENDERED PARENTING

In the activities of parenting and child rearing, men who devote themselves to material providing once again place women between themselves and their children. Their interactions with their children are controlled, arranged, or supervised by their wives. Women have most of the responsibility for organizing and enforcing children's activities, with men exerting their influence through their wives. Some men do put a lot of energy into their children's activities, especially into their athletics, and even more express the desire to do so, especially to do more with their children than their fathers did with them. But studies of time use continue to find differences between working husbands and wives in the total number of hours worked when paid labor, childcare, and housework are combined. Studies in the 1980s concluded that employed husbands worked between ten to twenty fewer hours per week than their employed wives (Hochschild 1989: 3–4 and 271–73). More recent studies have found that men and women spend approximately equal amounts of time on the combination of housework and paid work (Ferree 1991; Pleck 1985; Schor 1991), but certain domestic tasks continue to be overwhelmingly women's work. (Coltrane 1996; Shelton 1992), and men's contribution to housework still tends to be thought of, by both husbands and wives, as "helping" (Coltrane 1989; Walzer 1998). When married couples have children, the division of domestic labor tends to become more traditionally gendered (Cowan and Cowan 2000) and women spend less time in the paid labor force and men spend more (Shelton 1992).

Not only is there a difference in the number of hours men and women spend in child rearing, but fathers and mothers approach parenting very differently. The men I talked to expressed the belief that mothers are the "default parent." They acted on this belief and by their actions made it true. Being the default parent means being the one to whom a child turns first, and being the one with the responsibility for knowing the child's needs and schedules. The default parent, ultimately, is the one who must be there, to whom parenting is in no sense optional (Walzer 1998). For example, even though fathers sometimes went to meetings at their children's schools or took their children to sports practice, it was usually mothers who kept track of the meeting and practice schedules. Lareau (2000), in her study of children's routines, reported that fathers were very vague and general in their accounts of their children's daily routine, in contrast to the detailed and specific responses of mothers. In general, the men I talked to indicated that their wives kept the mental and physical calendar. In parenting, as in so many areas, the person who keeps track of scheduling often has a great deal of control over what activities and events become scheduled.

Even in the area of discipline and punishment, where it would seem that the father's position as ultimate authority was secure, mothers are the gatekeepers or mediators. Consider the proverbial threat of mothers to their children: "Just wait until your father gets home!" A number of men used exactly this expression to show that they were deeply involved in their children's lives. It was meant to indicate that they were the source of discipline, even if they were not directly supervising most of the time. The words, however, indicate a very different relationship, for it is the mother who decides when and what the father is told, and thus when he can act. Rather than being in an immediate disciplinary relationship with his children, he is a resource to be mobilized by his wife in her dealings with the children, and thus in a relationship mediated by his wife.

The disciplinary dynamic in families could take several forms, but two were common. While they may seem very different, in both cases the wife and mother was ultimately responsible for discipline. In the first, the husband was an authority figure and disciplinarian who saw himself as supporting or backing up his wife. In the second, the husband was allowed to be the fun parent because his wife was the disciplinarian.

Both Ralph Colson and Terry Evans used the word "enforcer" to describe their role in their children's discipline. Neither of them liked this, though both accepted it as their responsibility to support their wives. Ralph told me that there had been one disciplinarian in his family when he was growing up, and that it was the same in his marriage:

> One parent seems like the disciplinarian and the other one is not. And in my family I am. And my wife doesn't understand: "Why won't the children listen to me?" Because it's always: "I'm gonna tell your father." She had to call me here [at work]. I've had to talk to them on the phone. And they straighten right up.

Ralph felt that his wife should be more consistent in her discipline so that the children would not ignore her threats and cause her to lose her temper. She, on the other hand, sometimes felt overwhelmed and told him that if she were to hit the children instead of threatening them, "I'd beat them to death. I'd be constantly hitting them."

Terry, the father of two boys, was also critical of his wife's treatment of the children, but he, too, accepted his role as enforcer:

> The thing is, you've gotta be the enforcer. The man has to be the enforcer and that's the only thing that sometimes irritates me. I come home from working a hard day and my wife right off: "Terry, he's done this, he's done that." And I get mad and I go in there and yell at him. That's where a lot of times I would like to say, "You're the mother. Handle it. If you want to restrict him, restrict him. If you want him to be whupped, do it." She's home every day. She knows exactly what's going on. I think she ought to handle it more herself.

For both Ralph and Terry, the structural division of labor, their position at work, and their wives' presence at home meant that their wives determined where the children were and what was expected of them. The women then decided what to tell their husbands and so shaped the types of interactions between the fathers and children. Terry yelled at his children and Ralph spanked his, but their wives, as the default parents, mediated the flow of information and expectation between fathers and children. Women's mediation should not be seen as deliberately manipulative. The gendered division of discipline is not simply an individual choice or decision, but part of a whole gendered system of division of labor.

Gordon's situation was a rather different manifestation of the same gendered system. In Gordon's family, he was the one who could relax and have fun with his sons because his wife was protective and strict. When it came to parenting, he said,

> My wife does a better job, although she is very protective of the kids. Like my eleven-year-old, she won't drop off at baseball practice. She'll stay and wait until it's over. And even though sitting in a car, she's always there. She won't leave the kids anywhere alone. . . . I think she's just worried about something happening to them. Not having an accident, like falling off of something, but with all the crazies out there, she's just worried about losing one of them. Which is—I mean, it's a real-life concern. I can't blame her for that, but it gets a little excessive sometimes. And she does discipline them bet-

> ter. They mind her better; she's more sensitive to their feelings and that kind of thing. It's the insensitive dad, sometimes. . . . I don't treat my kids the way [my father] treated us. He was a *very* heavy disciplinarian and we were afraid of him when we grew up. I don't want my kids to be afraid of me.

Partly because he did not want his children to fear him as he had feared his father, and partly because his wife was watching over them, protecting them, and disciplining them, Gordon felt he could relax and let them run a little wild: "When they're just goofing off and it's Friday night, I'm not going to crack the whip and put them to bed." He laughed: "It wouldn't work anyway."

Roy's wife, Sarah, also mediated Roy's position as a "fun dad" to his children. She had run a childcare business in her home when the youngest was a toddler and taught at the private school the children attended. Sarah was very involved in the lives of children in general and her own children in particular, and part of her involvement was in scheduling her husband's time with his children. Roy said, "When there are three there, it's tough. They're all vying for your attention." So Sarah intervened and Roy described to me what he would say to his children:

> "Your mom says it's your turn." So each time I do something I take a different child with me. And it works out two ways. It's a lot cheaper for one. And also I get that one-on-one with my kids.

Many of the things he did with his children, such as going to baseball and basketball games, were recreational activities he himself enjoyed. While Sarah was orchestrating this activity, Roy was the fun father, spending quality time with his children doing things they could enjoy together. Roy also was able to be spontaneous with the whole family. Several times he told me that he would, "on the spur of the moment," sweep the family up and drive to the beach:

> Like I get them going at seven in the morning up to Santa Cruz and I'll bring my camping stuff and we'll cook breakfast and we'll just have breakfast and when other people are coming, we're leaving and coming back home. We do stuff off the wall like that. Spur-of-the-moment-type things. On Friday afternoon I'll tell everyone to pack their suitcase and we'll go to Monterey for a night and things like that. I think that's pretty neat.

Overall, Roy emphasized the fun and spontaneity of his relations with his children: "Agewise, I'm probably considered an adult, but you talk to my kids and I'm probably the biggest kid around. I'm not kidding. I'm a big kid at heart. I love sports. I love my kids." It is important to notice that Roy's ability to be spontaneous and to have fun with his children, just like his one-on-one time with them, is dependent on the routine, day-by-day, planned, and conscientious work of his wife.

The gendered division of labor in parenting not only distributes work and fun differentially between fathers and mothers, it also distributes who gets taken for granted, and who gets the credit. Hochschild described how couples negotiate not only a material division of labor, but also an economy of gratitude (1989): People do not just want to be appreciated, they want to be appreciated for the contributions *they* think are important. Psychologists Carolyn and Philip Cowan (2000) observed that, for men, employment "counts" as childcare—both they and their wives interpret men's employment as doing something for their children. What is more, wives not only appreciate their husbands' work as something they do for their children, they also see their husbands' direct attention to their children as contributions to the marriage relationship. Husbands, on the other hand (and many wives), see women's employment as detracting from their mothering, and their husbands do not see the care mothers give their children as couple time or as building the marriage.

Mothers who are supervising and caring for their children may well know more about those children, about their hopes and insecurities, than fathers who are there to have occasional fun. This puts mothers in a position to relay or to hold back information about their children, and mothers are the ones who both fathers and children talk to about one another. Roy related a typical story of an incident between him and his daughter about which both of them had, independently, talked to his wife. Their communication about the event, and its resolution in Roy's mind, was very directly mediated by his wife:

> I just talked to my wife the other night. My oldest daughter, somehow I felt like she wasn't communicating with me lately, the last couple weeks. I was asking my wife if there was anything wrong. What particular things had happened at school? I went to pick up my son and she was gonna go somewhere else and I saw her and I know she saw me, but she didn't acknowledge me being there. So I was kind of hurt because usually they'll come up, "Hi dad!" And my wife goes, "It had nothing to do with her not wanting—" What it was, I guess, her friends were wearing makeup and she knows I'm against girls at this age wearing makeup and I guess that was why she didn't come and talk to me. So that's fine. I can see why she didn't want to talk to me.

These examples illustrate that the structural division of labor, in which men are seen as providers and women as homemakers, is connected to a gendered division of parenting.

Some people object to this assertion about the gendered division of labor between wives and husbands, between mothers and fathers. Surely, they say, the norm of breadwinning men and homemaking women is not a reality in a postfeminist world of gender equality. At the very least, they assert, this norm is outdated and both expectations and behavior are changing rapidly. Their objections, however, often are based on those familiar tendencies we examine in courses on social science methods: overgeneralization and selective observation. We tend to notice the exceptions. Women in traditionally male fields, for instance, or men who stay home with their children, stand out from their peers. They get noticed, they get remembered, and they get attention. Representations in popular culture and the news media concentrate on these exceptions, with the justification, if one is offered, that they are interesting or newsworthy. Selective observation is accompanied by overgeneralization and the assumption that what we have noticed is in fact representative of the whole.

Admittedly, in the cultural repertoire there is an attitude or image of sharing between couples, of equal participation in children's care, and of work being important to both women and men. But social and cultural institutions have an enduring persistence. In a review of the history of the family in Western Europe and North America, Martine Segalen (1986) concluded that, although norms about domestic life were fiercely debated during the 1960s and 1970s, behavior changed less than the representation of it, and marriage and parenthood remain key institutions and personal goals (cf. Modell 1985). We find a similar disjuncture between cultural attention to change and enduring patterns of behavior when we consider parental and household division of labor. Sara Harkness and Charles Super (1992) report that in a group of thirty-six middle- or professional-class fathers in intact marriages, all "committed parents," fathers spent only a little more than a quarter of their child's waking hours in the presence of their children aged one to four, and about 15 percent of their child's waking hours as primary caretaker, and were engaged in direct interaction with their children for less than half that time. The evidence for gendered parenting is overwhelming whether we look at cultural attitudes or actual behavior.

In impressive reviews of the recent literature, Coltrane (1998a: 64–74, 2000) described the basic inequality, and limited change in labor force participation, earnings, housework, and commitment to home and work between men and women in the United States. His summary picture is one in which changes have been more limited than attention to "new men" might suggest, and in which the transition to parenthood continues to be accompanied by an increase in men's hours of paid employment, a decrease in women's hours of paid employment, and a more traditional gender-based division of labor between husband and wife (cf. Cowan and Cowan 2000; LaRossa 1997; Walzer 1998).

Both structural reasons and deep cultural orientations work against the easy imposition or implementation of ideologies of equality. In *The Second Shift,* sociologist Arlie Hochschild distinguished between what she called "shallow" and "deep" gender ideologies in order to explain the contradiction between statements of belief in gender equality with very unequal divisions of domestic labor. She argued that many men and women profess an egalitarian gender ideology that is shallow because it is contradicted by their deep feelings (1989: 14–17). The behavioral reality is that only 2.5 percent of men in the labor force are the primary caregivers of a child under age fourteen (Marin 2000).

What *is* changing in the culture of the United States is the nature of the expectation of paternal closeness and involvement. It seems that mothers, children, and men themselves are less likely to assume paternal love and more likely to expect that it must be demonstrated and enabled through paternal practice. Successfully fulfilling the providing, protecting, and endowing facets of fatherhood is no longer enough to fulfill the emotional closeness facet as well. Fathers are experiencing a cultural pressure to contribute to the emotional closeness facet of fatherhood through direct action and personal interaction.

William Hughes, a married father of two children, who attributed his rise from the loading dock to a senior managerial position to the work ethic he had learned from his father, identified the change in cultural expectations of fatherhood. William also made the critical connection between changing visions of fatherhood and changing definitions of masculinity as he compared his own father with his peers. His swift transition from talking about domestic chores to talking about emotional expression indicates their close association in his mind. Changing diapers and showing affection seem to be linked as gendered activities:

> My father was lord and king of his realm and essentially he ruled the roost. And my mother accepted that and was glad for it and that's the way their relationship was set up. He handled all the money, he paid all the bills, he doled all the money out. I never saw my father wash a dish, sweep the floor, or do much of anything around the house. And my father didn't change diapers. My father was not extremely involved in all those little parts of our lives. And the men I know now, they change the diapers, they share the cooking, they help clean the house, they share in all the duties around the house. And I think relationships have really changed that way. It's more socially acceptable to be physically demonstrative and to show affection and love for your children. You're not considered weak or a pansy or anything along those lines. Or hugging and kissing your wife and children. Hugging friends. The image of a man I think has changed—What a man really is.

Despite his portrayal of an expansion in men's family activities and emotional expressiveness, William continued . . . to identify male responsibility in terms of work and providing and to devote his time to his job. He had often worked late into the evening and on many Saturdays over the years while his wife, a college graduate, had left her job when their first child was born. His embodiment of a new masculinity rested on achieving success by the old standards, and that relied on a gendered division of parenting.

Surveys of attitudes or aspirations indicate that a high percentage of men, and especially of young men, accept the notion that they should see themselves as fathers first and workers second, agree that they should aim for work schedules that allow time for family, and say they would give up pay to spend more time with their families (Radcliffe Public Policy Center 2000). However, there is overwhelming evidence not only that men's behavior has not changed to keep pace with their aspirations, but also that basic aspects of fatherhood have endured. Men still see being involved in the daily routines of their children as optional. They may be involved, but this is not constitutive of their fatherhood in the way that such involvement is constitutive of motherhood. And providing is still seen as fathers' primary responsibility.

SELECTED REFERENCES

Coltrane, Scott. 1989. "Household Labor and the Routine Production of Gender." *Social Problems* 36: 473–90.

———. 1996. *Family Man: Fatherhood, Housework, and Gender Equity.* New York: Oxford University Press.

———. 1998a. *Gender and Families.* Thousand Oaks, CA: Sage.

———. 2000. "Research on Household Labor: Modeling and Measuring the Social Embeddedness of Routine Family Work." *Journal of Marriage and the Family* 62 (November): 1208–33.

Cowan, Carolyn Pape, and Philip A. Cowan. 2000[1992]. *When Partners Become Parents: The Big Life Change for Couples.* Mahwah, NJ: Lawrence Erlbaum.

Ferree, Myra Marx. 1991. "The Gender Division of Labor in Two-Earner Marriages: Dimensions of Variability of Change." *Journal of Family Issues* 12:158–80.

Garey, Anita Ilta. 1999. *Weaving Work and Motherhood.* Philadelphia: Temple University Press.

Harkness, Sara, and Charles M. Super. 1992. "The Cultural Foundations of Fathers' Roles: Evidence from Kenya and the United States." Pp. 191–211 in *Father-Child Relations: Cultural and Biosocial Contexts,* ed. B. S. Hewlett. New York: Aldine De Gruyter.

Hays, Sharon. 1996. *The Cultural Contradictions of Motherhood.* New Haven, CT: Yale University Press.

Hochschild, Arlie. 1989. *The Second Shift: Working Parents and the Revolution at Home.* New York: Viking.

Lareau, Annette. 2000. "My Wife Can Tell Me Who I Know: Methodological and Conceptual Problems in Studying Fathers." *Qualitative Sociology* 23(4, July 21): 407–33.

LaRossa, Ralph. 1997. *The Modernization of Fatherhood: A Social and Political History.* Chicago: University of Chicago Press.

Marin, Rick. 2000. "At-Home Fathers Step Out to Find They Are Not Alone." *New York Times,* 2 January, pp. 1, 16.

Modell, John. 1985. "Historical Reflections on American Marriage." Pp. 181–96 in *Contemporary Marriage: Comparative Perspectives on a Changing Institution,* ed. K. Davis. New York: Russell Sage Foundation.

Pleck, Joseph H. 1985. *Working Wives/Working Husbands.* Beverly Hills, CA: Sage.

Potuchek, Jean. 1997. *Who Supports the Family? Gender and Breadwinning in Dual-Earner Marriages.* Stanford, CA: Stanford University Press.

Radcliffe Public Policy Center. 2000. *Life's Work: Generational Attitudes toward Work and Life Integration.* Cambridge, MA: Author. Available: http://www.radcliffe.edu/pubpol/publications.

Schor, Juliet B. 1991. *The Overworked American: The Unexpected Decline of Leisure.* New York: Basic Books.

Segalen, Martine. 1986. *Historical Anthropology of the Family,* trans. J. C. Whitehouse and S. Matthews. New York: Cambridge University Press.

Segura, Denise. 1994. "Working at Motherhood: Chicana and Mexican Immigrant Mothers and Employment." Pp. 211–33 in *Mothering: Ideology, Experience, and Agency,* ed. E. N. Glenn, G. Chang, and F. R. Forcey. New York: Routledge.

Shelton, Beth A. 1992. *Women, Men, and Time: Gender Differences in Paid Work, Housework, and Leisure.* New York: Greenwood.

Walker, Karen. 1990. "Class, Work, and Family in Women's Lives." *Qualitative Sociology* 13:297–320.

Walzer, Susan. 1998. *Thinking about the Baby: Gender and Transitions into Parenthood.* Philadelphia: Temple University Press.

24

A Christian Quest for Manhood

SUSAN FALUDI,

1999

. . . In the mid-1990s, the men of Promise Keepers had gathered most famously in football stadiums across the nation. Their weekend-long rallies were directed by born-again evangelist Bill McCartney, a former football coach from the University of Colorado who had announced in 1994 that he was quitting to devote himself to family and God. The scene at any stadium looked for all the world like a sports event; the Promise Keepers "fans" brought their coolers, wore their team hats, chanted slogans, and even did the wave. But these fans took to the field, tens of thousands of men amassed on the gridiron as if huddling for their next play. "This is the first time in my life I've been in a jam-packed stadium where every single person is cheering for the same team—team Jesus," a thirty-seven-year-old mall security director attending Promise Keepers told a reporter (Gushee, 1995). They were seeking to *participate* in this particular game, to earn for themselves laudable manly roles in their families. They intended to do it by reclaiming what the rally speakers described as "spiritual responsibility" and "servant leadership" in the home.

The group's promise of virtuous manhood had spawned a nationwide following almost overnight, culminating in 1997 in a massive convocation on the Washington Mall, broadcast live on C-SPAN. Promise Keepers wasn't alone in this enterprise. If anything, its Washington show of numbers had been upstaged by the celebrated "Million Man March" in the capital in 1995, a "day of atonement" organized by Nation of Islam leader Louis Farrakhan and attended by hundreds of thousands of black men. And both conclaves were preceded by a secular "men's movement," which had drawn tens of thousands of generally New Age-oriented men to its weekend drumming sessions and "wild man" retreats in the late eighties and early nineties. Poet Robert Bly, the men's movement's most prominent spokesman and author of the best-selling *Iron John,* packed auditoriums. But Promise Keepers overshadowed all these male gatherings, both in size and in longevity. At the organization's height, and in spite of the $60 admission ticket, its stadium events were drawing nearly a million men a year (and 1.1 million in 1996)—and the men returned home to organize a vast nationwide network of small groups, which continued the organization's quest for "servant leadership," in some cases for years.

The men who joined Promise Keepers were seeking to build something greater than the sum of their individually distressed lives. Their occupational

---

From Susan Faludi, *Stiffed: The Betrayal of the American Man,* pp. 227–236, 239–242.

betrayal and civic betrayal had been compounded by a powerful sense of domestic betrayal. It was a betrayal that Promise Keepers was at least trying to address. They hoped that from the ashes of one male institution, the football stadium that had once promised male communion but was now one big consumer billboard, might rise another, more traditional and thus more solid one. For what could be more sustaining than a religion that had united and nourished so many men for two millennia? They were looking to join a grand struggle that would restore to them not only the love of their wives and children but also the conviction that they were embarking on a mission, that they had at last found a purpose that would earn them the appreciation of society, and that this purpose would be backed by a brotherhood.

WELCOME TO TESTOSTERONE COUNTRY!

FROM HUGE STAGES ERECTED BY THE GOALPOSTS and draped with banners featuring the Promise Keepers emblem—three hands raising a flagpole, Iwo Jima style—speakers exhorted masses of male listeners to repent their domestic failings and take back the family helm through prayer and religious direction. They were to become the masters of their households' spiritual life, religious authorities who would take charge of the domestic circle through "submission" and "servitude." To much of the media and to many feminist groups, such counsel sounded suspiciously like sugar-coated instructions to plant patriarchal boots right back on wifely necks. It didn't help that Dallas preacher Chuck Swindoll liked to blaze onstage in the company of a tattooed Christian motorcycle gang and lead the crowds with chants of "Power, power, we got the power!" It didn't help that Promise Keepers' speakers were so fond of Saint Paul's famed domestic stricture, "Wives, submit to your husbands" (the same directive that the Southern Baptists would add to their official credo of belief in 1998) (Griffith and Harvey, 1998; Stammer, 1989). It didn't help that the retinue of onstage lecturers hailed uniformly from a religious right that had famously thrown itself with punitive zeal into a decades-long body block of women's progress. And most of all it didn't help that Promise Keepers founder Bill McCartney had proved himself a dedicated foe of women's reproductive rights, not to mention gay liberation. A year before he was urging men to commit themselves to family leadership, he was calling for the criminalization of abortion at pro-life rallies and decrying homosexuality as "an abomination against Almighty God" (Spolding, 1996). At press conferences during Promise Keepers events, his views on gender roles appeared to be fresh out of a Victorian marriage manual. He was eager to reiterate his theory, before open-mouthed journalists, that "the ladies" were "receptors" who needed to be "brought to splendor" by their men, the "initiators."

It's little wonder, then, that the advice Promise Keepers speaker and preacher Tony Evans gave men on "reclaiming your manhood" would be endlessly quoted by a dubious press:

> The first thing you do is sit down with your wife and say something like this: "Honey, I've made a terrible mistake. I've given you my role. I gave up leading this family, and I forced you to take my place. Now I must reclaim that role." Don't misunderstand what I'm saying here. I'm not suggesting that you *ask* for your role back, I'm urging you to *take it back*. (Evans, 1994).

Evans's words seemed like a smoking gun, incontrovertible evidence that under all the tributes to "servant leadership," Promise Keepers was promoting a stealth campaign of misogyny and macho dominance. Ignored was the rest of Evans's broadside, which focused almost entirely on deploring a "macho" mass culture that told men they were worthless unless they were constantly conquering beautiful babes and single-handedly gunning down bad guys. "Rambo is not only violent, but he's also uncaring," Evans protested (Evans, 1994). In the same passage, he urged men to reject material values; exhorted them to turn off the TV and talk to their wives; and assured them it was okay to cry. For biblical substantiation, he turned again and again to an Old Testament book that would seem an unlikely choice for a preacher endorsing triumphal male dominance: the book of Job.

As an organization Promise Keepers was, in fact, replete with such contradictions. After Chuck Swindoll barreled onstage in his Harley, speaker Gary Smalley trundled up the ramp on a Big Wheel, clutching a doll with, as he put it, "batteries not included"—a purposeful display of knee-high helplessness. "All of us are like this doll," he said, holding up the molded-plastic girl in pigtails, a pink dress, pink shoes, and barrettes. He confided how his bullying older brother "terrorized me" so much when he was a boy that "when I was twenty-five years old, I couldn't stay in a home alone because I was so afraid." For every "We got the power!" chant, there was an allegory delivered about the solace of relinquishing power, like the one pastor Greg Laurie of Harvest Christian Fellowship told about how unhappy a little bird was until he was returned to a cage, because each man yearns for the "restrictive" bars of God's commandments, "keeping him safe." The same organization that endorsed the submission of wives to husbands also ran first-person articles in its affiliated periodical *New Man* from men who had put their wives' careers first or discovered the challenges of homemaking. In "The Unexpected Choice," Robert V. Zoba, for instance, reported how he had reinterpreted Paul's stricture, "Wives, submit to your husbands," to mean that he should follow his wife to Chicago, where she had landed a prestigious job as an editor at *Christianity Today* and where he, under-employed, "ran chauffeur service for our boys, found the best bang-for-bucks deals at the local grocery store, and learned the difference between delicate and sturdy cycles on the washing machine" (Zoba, 1997; Wong, 1997). Even when it came to the "We got the power" chant, Swindoll revealed to his audience that he had borrowed it from "an old cheer my daughters used to lead in high school."

That the convention stage was more often than not occupied by speakers with alarming antiabortion records and Christian Coalition affiliations was indisputable. But on the field and in the bleachers, the men were not so easily categorized. While speakers thundered for a return to a "biblically sanctioned" patriarchal household and a "traditional" male order, the rank and file sat quietly—polite and attentive, but hardly clamorous. If they were plotting the overthrow of a feminist world, they showed no signs of it. Mostly they seemed intent on being mannerly and tidy. In an era when the sports spectators who were the bleachers' usual clientele left the stadiums littered and vandalized, the Promise Keepers were careful to throw away all their trash. They obediently took notes during the speeches and displayed at all times their Promise Keepers ID bracelets, which looked exactly like the identification bands worn by hospital patients. They stood dutifully and uncomplainingly in endless lines to pick up dreary box lunches, buy PK-branded coffee mugs and caps at the Product Tent, or use the ATM machines. They were willfully docile, as though, if they just obeyed enough, they would at last get their reward.

At a stadium event in Oakland, the longest lines inside the Product Tent were at a booth selling T-shirts with this inspirational biblical message emblazoned across the chest: I AM A WORM. Thus armored, the male shoppers flocked to the Christian Financial Concepts table to thumb through stacks of brochures and books with titles like *Debt-Free Living, Your Career in Changing Times,* and *Whatever Happened to the American Dream?* As I wandered around a rally at the Anaheim stadium in southern California, a man called out to me, "Welcome to Testosterone Country!" but his tone was wistful rather than cocky. More typical was the remark of Bill Moore, a Promise Keeper from Beaumont, California, whom I met at the same event. When I approached him and his seatmates for an interview, he told me forlornly, "If you're putting guys like us in your book, you should call it 'Men with Low Testosterone.' "

The men who attended Promise Keepers conferences were not biker outlaws; they were the "good sons." In the stands, Troy Barber, a thirty-two-year-old man from North Dakota who managed a shoe store, told me softly, "Last night, I was crying." Overhearing him, a nearby conferee chimed in with, "I've failed miserably." In the seat behind him, Chris Martinson, a security guard at a private college in St. Paul, Minnesota, said sorrowfully, "We know we aren't effective as men." A plane buzzed the stadium just then, hauling a banner paid for by a local women's group. Its big letters spelled out, PROMISE KEEPERS, LOSERS AND WEEPERS. The men shaded their eyes and followed its arc across the sky in silent bafflement. They didn't necessarily disagree with the assessment. As the plane vanished, a mystified Larry Coleman of Kansas City turned to me. "My wife is the furthest from someone you can dominate." His friend Chris Lopez was confused, too: "Guys out there, we're really lost. We need help."

What kind of help the men were getting from the stadium events was unclear. Bill McCartney's name was rarely invoked in conversations. More than a few men I spoke to got his name wrong—Mc*Carthy* or *Paul* McCartney were the two most common mistakes—and none said he was the reason they came. That McCartney junked a high-paying, flourishing career as a coach to spend more time with his family was unfathomable to them; most of the men in the stadium were there out of fear that their families would junk *them* because they *didn't* have high-paying, flourishing careers. While the coach's press conferences were packed, his customary pep speech at the rally's finale, full of overamped sports clichés ("It doesn't get better than this!"), generally provoked a mass exodus. The men took his ascendancy to the mike as their cue to head for the exits and beat the rush.

Nonetheless, they headed for the parking lots with a flushed enthusiasm. What had they seen that so enraptured them, I wondered at first, but that, it turned out, was the wrong way to frame my inquiry. Because the most commonly cited highlight of the weekend was not something seen. The private huddle, the chance to pray off to the side with a few other men, was what most enlivened them. "You get to cry," one man after another told me of their convention experience, their eyes lighting up at last.

Why did the huddle hold such appeal? and what were the men crying about? If you asked, a Promise Keeper would say he was "repenting," for sins about which he was always vague. He had been "disobedient" and "selfish" and needed "forgiveness." But what was he so remorseful about? What promises hadn't he kept?

The average Promise Keeper was hardly wayward. A Promise Keepers–commissioned survey of men attending stadium events in 1994 found the group of mostly middle-aged men to be dutiful, upright, and eager to comply with social expectations of propriety and judgment. (Sixty-three percent agreed with the statement, "I follow the 'letter' of the law," and 74 percent affirmed "How others perceive my spiritual life is very important to me") (National Center for Fathering, 1995). The typical conventiongoer was like the unsung brother in the parable of the Prodigal Son. He had commonly recoiled from the celebrated protest dramas of his baby-boom cohorts. While fellow students were staging sit-ins in the sixties and early seventies, many of the nascent Promise Keepers were joining, if anything at all, "movements" like the Inter-Varsity Christian Fellowship, Campus Ambassadors, and especially the Campus Crusade for Christ.

An organization founded by conservative California evangelist Bill Bright, Campus Crusade for Christ was consciously designed to counter student unrest by coopting it. The group launched "Revolution Now," a proselytizing blitzkrieg in 1967 at the University of California at Berkeley, specifically to assist Bright's close friend and then-governor Ronald Reagan in his efforts to quash collegiate antiwar activism. Campus Crusade targeted college athletes and frat boys, emphasizing "aggressive evangelism" and "brotherhood" (but not sisterhood); by the seventies, its young recruits far outnumbered those of the radical Students for a Democratic Society. While classmates raised fists for a student strike or hailed each other with *V* peace signs, a Campus Crusade recruit would be more likely to sport one of the Jesus movement's "One Way" pendants with a finger pointing heavenward—a logo created partly in response to the antiwar hand gestures of the time (Diamond, 1989; Bright, 1970; Quebedeaux, 1979; McDannell, 1995).

When Bill Bright became a Promise Keepers speaker and one of its biggest promoters, it seemed a theological quid pro quo: Bill McCartney had been converted in 1974 by a branch of Campus Crusade for Christ called Athletes in Action, relinquishing the institutional devotions of his father's Roman Catholicism for a "tangible, hands-on, heart-to-heart relationship with the person of Jesus Christ," as he would put it later. McCartney was seeking a personalized guardianship where he "surrendered control" and could "actually *feel* his overshadowing presence guiding me and helping me grow." What he and the other Promise Keepers in the making wanted wasn't to break with authority. "I needed to personally surrender my life," McCartney wrote. "What I needed was a *relationship*" (Campbell, 1996; McCartney, 1995, 1997).

* * *

"HERE'S WHERE WE HAVE TO START when you start talking about man," Bill McCartney shouted at me, though I was less than a foot away. I had come to talk to him at Promise Keepers headquarters in Denver, a block-long building that used to house a printing press and truck lobbying association. The ex-coach had gotten down on his knees by my chair with a Bible open to the gospel of Luke in his hand. "What is our purpose?" he asked. He was answering my question about why American men had responded so strongly to Promise Keepers. "The deepest longing of the human spirit *screams* for significance," he said, his pregame, let's-get-'em! delivery turning his face only a shade less purple than his PK-embossed golf shirt. "And the only way you can achieve real significance is to fulfill your *purpose,* okay?"

And what specifically was the purpose Promise Keepers was providing men?

"That they come in touch with how much he loves them. . . . You think of the guy that most closely approximates what it really means to be a loving father, and a loving husband, and he doesn't even scratch the surface when it comes to how much God loves us. See, God the Father loves us so much, with such a passion, with such an intensity, with such a comprehensive caring for us, concern for us, that he doesn't sleep. He never slumbers! He watches over us day and night. And he's always pulling for us, okay? Now—"

I tried to interject a few words into the verbal fusillade, without success. "Now don't stop me!" He waved an arm in the air, as if he were calling a play. "Don't stop me because I'm gonna tell you the whole story. Then you can ask me anything. Okay, so now, picture this: picture somebody being loved like that. He knows every hair on our head, he knows every thought before we think it, he remembers every word before we speak it, okay? He is *intensely* involved in our lives. . . . He loves this guy Jeff, okay? He loves Jeff with this kind of love."

I wasn't sure who Jeff was. The only other man in the room was named Steve—Steve Chavis, the media-relations director. I also didn't quite get how all of this added up to a "purpose" for men. If God was the one sacrificing sleep, working night and day to bring men his love, then wasn't McCartney describing a purpose for *God?*

"He sent his own son!" McCartney belted out enthusiastically when I tried to pin him down on this point.

I tried another tack. How did McCartney see his own "purpose" now?

After several attempts, I finally got this much out of him: "My life can be of some value now because I can help young guys." It seemed a curious remark, coming from a coach who had abandoned a team full of young men to lead an organization whose members were overwhelmingly over the age of thirty. And the young men he left behind, the University of Colorado Buffaloes, had been notoriously in need of guidance.

Under McCartney, the team had made a big splash in *Sports Illustrated*—for its stunning arrest record. The players' criminal activities were so profligate, the magazine reported, that the campus police carried copies of the Buffaloes' roster when they went to investigate a crime. In one three-year stretch, two dozen players were arrested for burglary, sexual assault, or rape. One athlete even turned out to be the community's Duct-Tape Rapist, who had taped shut the mouths and eyes of eight women and assaulted them. The player was charged with four rapes, pled guilty, and was sentenced to twenty-five years (Reilly, 1989; Abas, 1989).

McCartney was not a beloved coach in the eyes of many of his players, who had to endure his legendary temper tantrums, inflexibility, and religiosity. And some evidently exacted revenge the one way they knew how: they began a Spur Posse-like campaign of sexual conquest against McCartney's teenage daughter, according to an account published in the Denver alternative weekly *Westword*. McCartney fulminated against the *Westword* article as "sacrilege" and claimed that only divine intervention prevented him from killing its author. No disciplinary action was taken on the team, though McCartney could hardly overlook incontrovertible evidence of his daughter's seduction by certain players. She told him that the team's star quarterback and then, four years later, a defensive tackle *both* got her pregnant. McCartney, the chastity champion, had become the grandfather of two illegitimate babies (Abas, 1989; McCartney, 1995; Hoffer and Smith, 1995).

As McCartney now continued to testify to me at top volume, media-relations director Steve Chavis squirmed visibly in his chair. A former journalist for a radio station, Chavis no doubt recognized a media disaster in the making. He had been discreetly trying to shepherd McCartney out of the room almost from the moment the ex-coach had dropped to his knees before me. McCartney was feverishly whipping through his Bible once more, searching for an answer to my inquiry into his newfound "purpose." "Isaiah 38:19," he announced at last, an index finger pouncing on the verse. "The *father* to the children shall make known the truth." He looked up at me, triumphant, eyes moistly gleaming. "See, he charged the man with the responsibility of bringing the children up in the Lord, okay?"

In fact, as his wife, Lyndi McCartney, made clear in several commentaries he allowed her to insert in his books and in later comments in the press as well, her husband hadn't been much help bringing up baby. "In a way, being a coach's child is a dream," she wrote in McCartney's first book, *From Ashes to Glory*. "But it was a nightmare as well." The kids hardly saw their father. "Bill never saw the boys play football until they were in high school." Nor did she think much of Bill's husbandly "spiritual leadership" before or after he founded Promise Keepers. Mostly, she reported, he consigned her to a "bottom-shelf existence"; when he was off the field, he preferred bars, where he often drank himself into a stupor. When she miscarried, she couldn't find Bill—he had gone off to "tend bar" at a pub without telling her—and she had to take herself to the hospital. In desperation, Lyndi, who rarely drank, once tried to get his attention by drinking herself into a state of unconsciousness; he didn't even notice. After he launched Promise Keepers, nothing much changed. She described him in a *New York Times* interview as the plumber who was "always out fixing everybody else's plumbing." As Promise Keepers grew, so did Lyndi's distress; in 1993, as both she and her husband recounted in McCartney's memoirs, she came close to a breakdown and considered taking her life. She shut herself away, read "more than a hundred" self-help books, threw up daily for seven months, and ate almost nothing. She lost more than seventy pounds. Bill's observation: "I saw that she was losing weight, and I was proud of her" (McCartney, 1995, 1997; Goodstein, 1997).

I asked McCartney about his wife's account of domestic agony. "I don't sit here as someone you could point to as an example of anything other than a guy who's a sinner saved by grace. But I know something. You know what I know?" I confessed that I didn't. "*He* loves me." Anyway, McCartney said, reiterating a claim from his books, these days he was working harder to be his wife's "spiritual leader." I asked what that entailed. "It's my responsibility to get things ready," he said, "and to encourage others to let's take some time out now and shut off the television and let's come before the Lord. See, that's *my* responsibility to do that. I don't wait for my wife to do it. I don't. But when I suggest it and she says this isn't a good time, I say, 'Okay, well when do you think would be a good time?' But it's *mandated*. I'm the one. The only way the spiritual work's gonna get done is if the man takes responsibility."

McCartney paused for a breath and, seizing the moment, Steve Chavis jumped to his feet, announced that the coach had "other appointments" that couldn't be missed, and guided him decisively toward the door, leaving me as befuddled as before, if not more so.

And so I gave up on illumination from headquarters. McCartney was not the reason men were flocking to Promise Keepers anyway. If they weren't seeking enlightenment from him, maybe I shouldn't either. If the prayer huddle was the main attraction of Promise Keepers, as its constituents maintained, then neither the organization's executive suites nor the bleachers of Folsom Stadium were the place from which to observe it. And so I turned toward the smaller and stiller domain of a Glendora living room. It would prove a far more revealing listening post.

. . . The Promise Keepers group sat in a living room adorned with an empty birdcage and ceramic pots sprouting fake ivy. It was an aesthetic I would come to associate with Promise Keepers' households. We sometimes met at the home of Martin Booker, the group's de facto leader, and his living room reminded me of a nursery with its baby-blue coordinated drapes and couches, the latter heaped with an impressive lineup of decorative stuffed animals. Like many of the men in the group, Martin had countered his wife's pastel putsch the only way he knew how, with a steroid-sized home-entertainment system. A "high-end audiophile," as he described himself to me, Martin had installed six giant speakers in the living room to create a "surround-sound effect." He once demonstrated this to us by playing a recording of military jets whistling overhead. The effect was convincing; we all ducked, half expecting bombs to rain on the powder-blue sofas.

I would spend many evenings in the Atwaters' living room, listening as the men told their stories, studied scripture, and closed each meeting with a lengthy prayer circle.

But the first meeting I attended, on an October evening in 1995, convened a few miles away at the Bookers'. I arrived early; Martin was percolating the coffee and laying out pastries. . . .

"Millions of men are facing a loss of jobs," Martin said, a comment that elicited a chorus of "mmm-hmms," "and you are going to lose your identity if you rely on your job for it."

Martin wasn't telling many of these men anything they hadn't painfully learned for themselves. . . .

"And here's the great thing," Martin said. "It's not based on *performance*. You don't have to *do* to earn his acceptance. We're told 'Do the Nike thing,' when we could just be dependent on Him. It's like, I was talking to this guy from Arizona. He lost his job. He was on the verge of losing his marriage. He had a whole wall of self-help books, shelves and shelves of 'em, all about the power of positive thinking and do this and do that by yourself. But once he placed his identity in Jesus, he just threw them all in the trash. He didn't need to *do* anymore. Jesus was in control." Lost job, lost marriage, that was a familiar set of toppling dominoes to these men. The terror of being on the verge of divorce had driven most of them to the group. They were hoping the group would reverse the process; that somehow when the marriage domino was set up again, the other fallen pieces of their lives would pop up, too.

The solution that Promise Keepers offered to this work-marriage dilemma was masterful, in its own way. Once men had cemented their identity to Jesus, so the organization's theory went, they could reclaim a new masculine role in the family, not as breadwinners but as spiritual pathfinders. Promise Keepers proposed that men reimagine themselves as pioneers on the home front, Daniel Boones for Christ, hacking their way through a godless wilderness of broken marriages and homes lacking all spirituality to build a spiritually fortified bunker in which their families could settle for the long haul. The ingenuity of such a solution was that it slipped the traces of traditional male work identity without challenging the underlying structure of the American male paradigm; that paradigm was simply reformulated in religious-battle terms. Men's shared "mission" now became the spiritual salvation of their families; men's "frontier," the domestic front; men's "brotherhood," the Christian fraternity of Promise Keepers; and men's "provider and protector" role, offering not economic but religious sustenance and shielding their wives from the satanic forces lurking behind consumer culture.

This reconfiguration of the male role on a spiritual battlefield helped ease one of the group's greatest collective anxieties, . . . that their wives were really the well-armed ones. Their vision of women as the more powerful combatants was in itself, however, only another way to cloak in metaphor their painful domestic disputes, a way to make them look like so many heroic Davids before feminine Goliaths. "When a guy goes into the military," Mike Pettigrew said one evening, "you can simulate battle training. But when the bullets are flying at you at home . . ." He raised his arms, as if surrendering to an invisible army. War and battle metaphors were never far from the men's lips when they discussed their domestic situations, and they always presented themselves as the ones on the receiving end. "When you're facing one of those World War II fights from your wife" was the way Howard Payson phrased it, or "When the bombs are dropping on you from your family," as Mike Pettigrew put it.

Presented with a real example of their fear that a woman was attacking them, however, they tended to shield their eyes. A year earlier, Frank Camilla's daughter-in-law had been arrested and charged with murder—of her husband, Frank's son. But the group rarely discussed this horrific event or even seemed particularly curious about it. When I first joined the group and heard about the murder, few of the men knew anything beyond the barest details. Frank frequently offered the group updates on pretrial proceedings when it was his turn to talk, but the men tended to mumble polite condolences and nervously change the subject. They didn't know what to do with Frank's story. It was too literal for their purposes, plus his son hadn't won the battle.

The first meeting I attended, the men were still celebrating, or rather recycling, the group's two "success" stories. Both Mike Pettigrew and Howard Payson had been on the brink of marital collapse when they joined. Mike's wife of nine years, Margaret, had moved out and seen a divorce lawyer; Howard's wife of more than twenty years, Libby, had been loudly advertising her imminent departure. Now, Libby had rescinded her threat and the couple had recently renewed their wedding vows. At the same time, Margaret had given up her new apartment and moved back in with Mike.

"What happened to Mike, it's an inspiration," said Jeremy, whose wife was just then trying to get *him* to move out.

Bart agreed. "Mike's story gives me so much hope."

"We got to watch a miracle!" Martin Booker exclaimed.

Mike's "miracle" had unfolded at the group's first stadium event, in the summer of 1993. Nine of them had piled into a seven-seater van and driven twenty hours to Boulder. "The way the guys talk about it," Martin Booker told me, a bit starry-eyed, "it was like *Bury My Heart at Wounded Knee.*" I didn't quite get the analogy, but at any rate Mike's was definitely a wounded heart. He had no job and no wife. "When Margaret moved out, I went into crisis mode," he said, but he clung to the slim reed of hope that Margaret would reconsider. After all, she had paid for his admission ticket so he could attend the stadium conference. Mike was broke and she wasn't, and she figured that, since she had to get him something for his birthday, why not a weekend of repentance? Two other men in the group also had their fares paid by their wives.

"At the time," Mike recalled, recounting for the group a story they had heard many times but never tired of reliving, "my wife had rebelled against Christianity. She said the Bible was irrelevant." Worse, she was caught up in the "occult," he said, by which he meant her self-help books with a New Age bent. At the stadium event, "I got very charged up, because I learned how the world had tainted me and steered me wrong. Here I was, forty-one years old, trying to win my wife back by being somebody, being a success at work, all that, when what I should have been doing is praying for her release from bondage. Because that is what it is. My wife was really in bondage. She was in the Satan realm and it was up to me to pray for her return to the spiritual realm."

And so at the end of the first long day of stadium speeches, Mike Pettigrew and Martin Booker retreated to the van to pray for Margaret. "Martin told me about how Margaret had turned her back on God and she was living deceived by the world that she could make it on her own, that she didn't need God," Mike recalled. "Martin said what we need to do is pray to reclaim her in Jesus' name, to save her. We *attacked* Satan, that bond he had on her, to get him off of her." The two men huddled until two-thirty in the morning. "I confessed to Martin some of my sins, like 'being physical' with Margaret. We wept together." Two nights in a row they huddled and wept. And then the miracle happened. "When I got back to the dorm where we were staying." Mike recalled, "I called my [answering] machine and there were six or seven phone calls from my wife, saying, 'Please call me back! Please call me back!' "

"God flat-out answered Mike's prayers," Martin Booker said.

"That's why I say to guys in the group, just hang in there," Mike said. "Keep at the praying. Because God will save your marriage."

SELECTED REFERENCES

Abas, Bryan. Aug. 30–Sept. 5, 1989. "That Sinning Season," *Westward*.

Bright, Bill. 1970. *Come Help Change the World* (Old Tappan, N.J.: Fleming H. Revell).

Campbell, Matt. March 24, 1996. "Promise Keepers Praised, Panned and Set for KC." *Kansas City Star*, p. A1.

Diamond, Sara. 1989. *Spiritual Warfare: The Politics of the Christian Right* (Boston: South End Press).

Evans, Tony Evans. 1994. "Spiritual Purity" In *Seven Promises of a Promise Keeper* (Colorado Springs: Focus on the Family).

Goodstein, Laurie. Oct. 29, 1997. "A Marriage Gone Bad Struggles for Redemption." *New York Times*, p. A1.

Griffith, Marie, and Paul Harvey. July 1, 1998. "Wifely Submission: The SBC Resolution: Southern Baptist Convention and Marriage." *Christian Century*, p. 636.

Gushee, Steve. Aug. 8, 1995. "The Promise Keepers," *Palm Beach Post*, p. 1D.

Hoffer, Richard, and Shelley Smith. Jan. 16, 1995. "Putting His House in Order," *Sports Illustrated*.

McCartney, Bill with Dave Diles. 1995. *From Ashes to Glory* (Nashville: Thomas Nelson).

McCartney, Bill with David Halbrook. 1997. *Sold Out* (Nashville: Word Publishing).

McDannell, Colleen. 1995. *Material Christianity* (New Haven: Yale University Press).

National Center for Fathering. March, 1995. "Promise Keepers' Sample of 1994 National Survey on Men: Report on 1994 Conference Attendees."

Quebedeaux, Richard. 1979. *I Found It! The Story of Bill Bright and Campus Crusade* (New York: Harper & Row).

Reilly, Rick. Feb. 27, 1989. "What Price Glory?," *Sports Illustrated*.

Spalding, John D. March 6, 1996. "Bonding in the Bleachers: A Visit to the Promise Keepers," *Christian Century*, p. 260.

Stammer, Larry B. June 10, 1989. "A Wife's Role Is 'To Submit,' Baptists Declare," *Los Angeles Times*, p. A1.

Wong, Rholan. Sept. 1997. "Lessons from a Househusband," *New Man*, p. 58.

Zoba, Robert V. Nov./Dec., 1997. "The Unexpected Choice," *New Man*, p. 41.

CHAPTER 8

Families and Work

Work and family are often seen as separate things, but as earlier chapters have discussed, the two are inseparably linked. The work that people do to maintain families can be divided into two general types: paid and unpaid. Paid work provides resources for families to live on. In previous centuries, most people lived in households that were productive units, and men's and women's labor on farms or in family businesses provided resources that were used to sustain the family or traded or sold to provide the things that people needed to survive (food, shelter, etc.). As the market economy grew in the nineteenth and twentieth centuries, **households** moved from being productive units to being consumption units that sent family members (mostly men) out to be wage earners who earned money income that was spent on food, durable goods, shelter, and various services to support the family. Paid work is necessary to families because it provides resources for survival. Unpaid work also contributes to the survival of families. The unpaid cooking, cleaning, care giving, repairing, and maintenance work that goes on in households is also necessary for survival. This unpaid family work—or **social reproductive labor**—is just as important to the maintenance of society as the productive work that occurs in the formal market economy. Recent estimates suggest that the total amount of time spent in unpaid family work is about equal to the time spent in paid labor and that the bulk of the unpaid family work is performed by women (Robinson and Godbey, 1997).

This chapter considers the importance of both paid and unpaid labor for families. One way to consider the work–family linkage is to see how paid work affects the attitudes, values, behaviors, and life chances of parents and children. This "top-down" sociological approach assumes that the type of employment in which people are engaged, and fluctuations in that employment, can have profound implications for the organization of family life. The first two articles in this chapter rely primarily on this approach. Another way to consider the work–family linkage is to focus on the tasks that are done inside families and to examine how family members talk about dividing **breadwinning** and **household labor.** This "bottom-up" approach focuses on how individuals' commitments to paid and unpaid work are understood and negotiated. The last two articles in this chapter rely primarily on this second approach. Both approaches help us see how employment and family work are related, but not necessarily in the ways normally assumed.

As discussed earlier, **stratification** is a useful concept for understanding work–family linkages. Like a layer cake, society is divided into different levels or ranks, with those located in the lower "strata" possessing less wealth, **power,** or **status.** The most obvious basis for stratification is the amount of money that families have. Money income affects such things as food and nutrition, quality of housing, size of rooms and yards, types of toys in the home, number and types of books in the home, neighborhood safety, availability of community services, quality of schools, level of medical and dental care, style of recreation, frequency of travel, likelihood of college attendance, and a host of other factors. All these things influence how families live and how they bring up their children. Stratification researchers find that lack of income, not surprisingly, has detrimental effects on families and child development. People with less income have more health problems and more developmental difficulties. They have higher rates of birth defects, mental retardation, and infant mortality. They have shorter **life expectancy** and are more likely to die from accidents and most illnesses. Those facing economic hardship are more likely to experience child abuse, spouse abuse, divorce, and desertion. Because of the presence or absence of money along with its attendant resources, children's perceptual, cognitive, and motivational development will be different, even when two children start out the same or when their parents use the same child-rearing techniques (see Hofferth, Phillips, and Cabrera, chapter 13).

As Melvin Kohn's classic article on **social class** and parenting suggests, people from different class positions tend to parent differently. Kohn's basic thesis is that the types of work adults do in their everyday lives shapes the values they hold about children's behavior and the techniques they use to raise their children. Although parents from all class levels want the best for their children, Kohn observes that middle-class mothers and fathers are more likely to want their children to grow up happy, curious, interested in the world, and considerate of others. Working-class parents, in contrast, are more likely to want their children to be obedient, polite, neat, and clean. Kohn suggests that occupational conditions determine these class differences in child-rearing values.

Kohn argues that to understand the impact of **social class,** one must move beyond income and consider the influence of occupation (and to some extent, education). A person's position in the **stratification** system is associated with various job characteristics such as degree of supervision, amount of routinization, complexity of tasks, and possibilities for self-direction. Working-class jobs, whether dealing with people or things, tend to be repetitive and closely regulated. Middle-class jobs in contrast, tend to entail a greater degree of substantive complexity, independent thought, and self-direction. Kohn shows how these demands of the workplace shape people's worldviews, and thus their parenting styles. Working-class parents end up valuing children's outward appearance, conformity, and obedience to authority, whereas middle-class parents tend to value self-control and care more about their children's emotions and intentions. Parents thus tend to instill in their children traits that are likely to be functional for people who occupy specific types of occupations. Parenting values and practices therefore reproduce the class structure, insofar as they prepare the next generation to function best in the class (and occupations) into which they were born. Some researchers suggest that differences in value socialization across social class may be declining as Americans generally have come to place more emphasis on autonomy and self-direction (e.g., Cherlin 2002). Do you agree?

In an excerpt from the book *Families on the Fault Line,* Lillian Rubin reports on the results of her study of working-class families in America in the 1990s. She is well situated to comment on changes in family life brought on by economic changes because two decades earlier she wrote *Worlds of Pain: Life in the Working-Class Family.* In the selection reprinted here, Rubin focuses on how **economic class,** and especially job loss, influences family life and affects racial and ethnic tensions. She focuses on how men's identity, more than women's, is tied up with the jobs they do. When they lose their jobs, or when they cannot support their families, she finds that the quality of their family relationships often decline, even though they have more time to be around the house. She also finds that white men's anger at their shrinking paychecks often gets directed at men of color, who are frequently—but unjustly—perceived as having taken white men's jobs. She laments how some men focus on being victims and invoke white supremacy to compensate for their failed sense of masculinity. Her study highlights how economic downturn affects those near the bottom more harshly than those higher up. Although the changing job market could signal more racial and gender equality, she finds that outmoded ideas about white men's **entitlement** spell trouble for many families. As you read this selection, see if you can identify ideas or practices described by Townsend or Faludi in the last chapter.

In an excerpt from the book *Halving it All,* Francine Deutsch discusses how the sharing of family work between mothers and fathers is related to issues of parenting identity, negotiation, and job priorities. Although she finds that equally shared parenting benefits children and marriages, she shows how it entails negotiation on the part of women and some career sacrifices on the part of men. Instead of being buffered from the dual strains of work and family stress like sole-earner fathers, men in these families must juggle career and family as women have typically done. By prioritizing children and family (at least when the children are young) men have to sacrifice some career advancement opportunities. The men she interviewed valued the rewards of being emotionally connected to their children and involved in all aspects of family life far more than any raise or job promotion they might have otherwise received. Although these men did not benefit from the masculine **entitlements** of being sole **breadwinners,** they self-consciously invested in their "family careers," a choice undoubtedly aided by their middle-class status and by being in relationships with assertive women.

In *The Second Shift,* sociologist Arlie Hochschild presented compelling accounts of working parents trying to balance job and family responsibilities in the 1980s (Hochschild with Machung, 1989). According to Hochschild, American women put in fifteen hours more each week than their husbands on all types of work—both paid and unpaid—amounting to a full extra month of twenty-four-hour work days in a year. Trying to understand how and why this second shift falls on women, Hochschild and her assistants interviewed and observed 52 couples over an eight-year period. One of their most consistent findings was that women were "far more deeply torn" between the demands of work and family than their husbands. Since wives remained responsible for home management as well as performing the most time consuming and repetitive tasks, Hochschild found that they tended to talk intensely about being overtired, sick and emotionally drained. She labeled the common situation of men's favored position in the household economy as "the leisure gap." since most men enjoyed more free time than their wives and were less burdened by family obligations.

Hochschild's study acknowledged that economics play an important part in determining who does what around the house, but she found no simple trade-off between wages earned and household labor performed. Something else besides the bold assertion of power is at work when people divide up housework and child care. Hochschild used the term "gender strategies" to capture the interplay of ideology and practice that is continually and subtly negotiated as couples divide family labor and make assumptions about who should do which chores. The feelings of entitlement, guilt, obligation, or appreciation that are inevitably associated with the performance of certain activities reflect what Hochschild calls the "economy of gratitude." In general, the men in Hochschild's study did not consider the second shift to be "their issue." Nevertheless, about one in five shared housework roughly equally with their wives, and these men seemed to be just as pressed for time as the women.

Since Hochschild conducted her study, there has been more convergence in the work that men and women do. Compared to past decades, women are spending more time on the job and less time on family work. In contrast, men are spending slightly more time on family work and slightly less time on the job. Recent research confirms that family work is still sharply divided by gender, with women spending much more time on these tasks than men, and typically taking responsibility for monitoring and supervising the work, even when they pay for domestic services or delegate tasks to others. Research also shows that women perform more of the housework when they are married and when they become parents, whereas men tend to perform less housework when they marry and assume a smaller share of the household work after their wives have children. Because new mothers tend to reduce their employment hours, and new fathers often increase theirs, findings about housework are best understood within larger economic, social, and family contexts. When time spent on both paid and unpaid work is combined, most studies find that the total number of hours contributed by husbands and wives is much more equal than it used to be. Nevertheless, when women shoulder a disproportionate share of responsibility for housework, their perceptions of fairness and marital satisfaction decline, and depending on gender ideology and other mediating factors, marital conflict and women's depression increase.

Because gender is a major organizing feature of household labor, research has explored how men's and women's task performance differs and how their experience and evaluation of housework tend to diverge. In general, women have felt obligated to perform housework and men have assumed that domestic work is primarily the responsibility of mothers, wives, daughters, and low-paid female housekeepers. In contrast, men's participation in housework has appeared optional, with most couples—even those sharing substantial amounts of family work—characterizing men's contributions as "helping" their wives or partners (Coltrane 1996). Much recent research also attempts to isolate the conditions under which men and women might come to share more of the housework. Most studies show that women who are employed longer hours, earn more money, have more education, and endorse gender equity do less housework, whereas men who are employed fewer hours, have more education, and endorse gender equity do more of the housework. A preponderance of research also shows that when husbands do more, wives are likely to evaluate the division of labor as fair, which, in turn, is associated with various measures of positive marital quality (Coltrane, 2000). Although researchers have begun to isolate causes and consequences of various divisions of paid and unpaid labor in (mostly white middle-class) married couples, we know less about how labor is divided in other family types. The problems of juggling paid and unpaid work in single-parent families are particularly daunting, and may be growing even worse in the face of welfare-to-work reforms (see chapter 13).

In an excerpt from *No Place Like Home* called "The Political Economy of Constructing Family," sociologist Christopher Carrington shows how divisions of labor in lesbian and gay households are shaped both by family ideals and by occupational trajectories. Noting that lesbian and gay families with more economic resources maintain more extensive social connections, hold jobs offering more flexible time, and pay others to perform more domestic services, Carrington explores how they can invest more heavily in creating and maintaining a sense of family. Although all same-sex couples experience a lack of institutional support (if not outright hostility), those that are more prosperous are able to avoid some of the negative consequences of such prejudices. Carrington suggests that we should not idealize the supposed egalitarian arrangements of same-sex couples and documents how paid work influences the allocation and appreciation of **household labor** in these families. With sensitivity to class and race differences, Carrington explores how lesbian and gay families struggle to maintain a minimal notion of family as they face many of the same pressures to juggle career and family commitments that heterosexual couples face. Like the other selections in this chapter, this one reveals that paid employment and unpaid family work are not separate, but rather reflect and reproduce patterns of **stratification** found in the larger society.

REFERENCES

Cherlin, Andrew J. 2002. *Public and Private Families: An Introduction.* 3rd Edition. Boston: McGraw-Hill.

Coltrane, Scott. 1996. *Family Man: Fatherhood Housework and Gender Equity.* New York: Oxford University Press.

Coltrane, Scott. 2000. Research on Household Labor. *Journal of Marriage and the Family,* 62, 1208–1233.

Hochschild, Arlie Russell with Anne Machurg. 1989. *The Second Shift: Working Parents and the Revolution at Home.* New York: Viking.

Robinson, John, and G. Godbey. 1997. *Time for Life.* University Park: Pennsylvania State University Press.

SUGGESTED READINGS

DeVault, Marjorie. 1991. *Feeding the Family: The Social Organization of Caring as Gendered Work.* Chicago, IL: University of Chicago Press.

Ehrenreich, Barbara. 2001. *Nickel and Dimed: On (Not) Getting By in America.* New York: Henry Holt and Company.

Halberstam, Joshua. 2000. *Work: Making a Living and Making a Life.* New York: Perigee.

Hardill, Irene. 2002. *Gender, Migration and the Dual Career Household.* London; New York: Routledge.

Hertz, Roseanne, and Nancy Marshall, eds. 2001. *Working Families: The Transformation of the American Home.* Berkeley: University of California Press.

Hesse-Biber, Sharlene, and Gregg Lee Carter. 2000. *Working Women in America: Split Dreams.* New York: Oxford University Press.

Hochschild, Arlie Russell. 1997. *The Time Bind: When Work Becomes Home and Home Becomes Work.* New York: Metropolitan Books.

Jackson, Maggie. 2002. *What's Happening to Home: Balancing Work, Life, and Refuge in the Information Age.* Notre Dame, IN: Sorin Books.

Nelson, Margaret K., and Joan Smith. 1999. *Working Hard and Making Do: Surviving in Small Town America.* Berkeley: University of California Press.

Newman, Katherine S. 1999. *Falling from Grace: The Experience of Downward Mobility in the American Middle Class.* Berkeley: University of California Press.

Risman, Barbara. 1998. *Gender Vertigo: American Families in Transition.* New Haven, CT: Yale University Press.

Wharton, Amy. 1998. *Working in America: Continuity, Conflict, and Change.* Mountain View, CA: Mayfield Publishing Co.

INFOTRAC® COLLEGE EDITION EXERCISES

The exercises that follow allow you to use the InfoTrac® College Edition online database of scholarly articles to explore the sociological implications of the selections in this chapter.

Search Keyword: Dual-career family. Using InfoTrac® College Edition, search for articles that discuss dual-career families. How are dual-career families described? What are some of the advantages of being involved in a dual-career family? What are some of the disadvantages? How does being a dual-career family affect gender and power relationships in the home? How has this family type changed over time?

Search Keyword: Social class. Find articles that deal with social class. What are the different ways that social class is defined? What are some of the ways that social class is measured, i.e., how do things like income and education provide "clues" to a person's social class standing? What other ways do we signal social class standing? What are some of the cultural markers of social class? How are different social class values transmitted from one generation to the next? How does stratification by social class relate to power in society?

Search Keyword: Household labor. Search for articles that deal with household labor. What are the major tasks involved? How is performance of these tasks related to gender? How is household labor divided, in general, between men and women in families? Do women or men spend more time performing household labor? Does this relative distribution of household tasks differ by race/ethnicity and/or social class? If so, how does it differ? How does the division of household labor reflect power relationships and issues of entitlement? How does the division of household labor affect paid labor force participation? Has the division of household labor changed over time? If so, how and why has it changed?

Search Keywords: Labor force participation and men. Draw on InfoTrac® College Edition to explore the relationship of men to the paid labor force. What articles can you find that deal with men's participation in the paid workforce? What are the trends in men's labor force participation? Are wages for men generally going up or down? Are the types of jobs available to men changing? How do these trends differ by race/ethnicity and/or social class? What do cultural images project about the relationship of men's participation in the paid workforce to their family work? What are the implications of changing workplace trends for men and for families?

Search Keywords: Labor force participation and women. Now find articles that deal with women's labor force participation. Compare and contrast women's labor force participation with what you found about men in the previous exercise. What do trends show about how many women work in the paid labor force and how much they are paid? How has their participation changed over time? How have the jobs they do changed over time? What are the implications of these trends for women and for families?

25

The Effects of Social Class on Parental Values

MELVIN KOHN,
1979

My thesis is straightforward and relatively simple: that there are substantial differences in how parents of differing social-class position raise their children; that these differences in parental practices result chiefly from class differences in parents' values for their children; and that such class differences in parental values result in large measure from differences in the conditions of life experienced by parents at different social-class levels. This essay attempts to spell out this thesis more concretely and explicitly (Bronfenbrenner, 1958; Kohn, 1969; Gecas, 1977). Without getting into technical aspects of methodology, it also attempts to give some idea of the type of empirical evidence on which the generalizations are based.

SOCIAL CLASS

Since the heart of the thesis is that parents' social-class positions profoundly affect their values and child-rearing practices, it is well to begin by defining *social class.* I conceive of social classes as aggregates of individuals who occupy broadly similar positions in a hierarchy of power, privilege, and prestige (Williams, 1960; Barber, 1968). The two principal components of social class, according to most empirical evidence, are education and occupational position. Contrary to the impression of most laymen, income is of distinctly secondary importance, and subjective class identification is virtually irrelevant. The stratificational system of the contemporary United States is probably most accurately portrayed as a continuum of social class positions—a hierarchy, with no sharp demarcations anywhere along the line (Kohn, 1969, 1977). For convenience, though, most research on social class and parent-child relationships employs a somewhat over-simplified model, which conceives of American society as divided into four relatively discrete classes: a small "lower class" of unskilled manual workers, a much larger "working class" of manual workers in semiskilled and skilled occupations, a large "middle class" of white-collar workers and professionals, and a small "elite," differentiated from the middle class not so much in terms of occupation as of wealth and lineage. The middle class can be thought of as comprising two distinguishable segments: an upper-middle class of professionals, proprietors, and managers, who generally have at least some college training; and a lower-middle class of small shopkeepers, clerks, and salespersons, generally with less education.

It is probably unnecessary to underline education's importance for placing people in the social order, and it is self-evident that level of educational attainment can be treated as a quantitative variable: a college graduate unequivocally has higher educational credentials than does a

From Melvin L. Kohn, "The Effects of Social Class on Parental Values," from *The American Family: Dying or Developing,* pp. 45–68.

high school dropout. But it may be less apparent that occupational position is also a major criterion of ranking in this—and in all other—industrial societies. One of the most important and general findings in social science research is the relative invariance of people's ratings of occupational prestige, regardless of which country is studied. This finding is of great theoretical importance in its implication that the stratification system is much the same across all industrialized societies.

. . . People's positions in the class system are related to virtually every aspect of their lives: their political party preferences, their sexual behavior, their church membership, even their rates of ill health and death (Berelson and Steiner, 1964). Among these various phenomena, none, certainly, is more important than the relationship of social class to parental values and child-rearing practices. But it is well for us to be aware, when we focus on this relationship, that it is one instance of a much larger phenomenon: the wide ramifications of social stratification for people's lives. Any interpretation we develop of the relationship between social class and parental values and behavior must be applicable, at least in principle, to the larger phenomenon as well.

Social class has proved to be so useful a concept in social science because it refers to more than simply educational level, or occupation, or any of the large number of correlated variables. It is useful because it captures the reality that the intricate interplay of all these variables creates different basic conditions of life at different levels of the social order. Members of different social classes; by virtue of enjoying (or suffering) different conditions of life, come to see the world differently—to develop different conceptions of social reality, different aspirations and hopes and fears, different conceptions of the desirable.

The last is particularly important for our purposes, because conceptions of the desirable—that is, values—are a key bridge between position in the larger social structure and behavior. Of particular pertinence to our present interests are people's values for their children.

PARENTAL VALUES

By values, I mean standards of desirability—criteria of preference (Williams, 1968). By parental values, I mean those standards that parents would most like to see embodied in their children's behavior. Since values are hierarchically organized, a central manifestation of value is to be found in choice. For this reason, most studies of parental values require parents to choose, from among a list of generally desirable characteristics, those few that they consider most desirable of all, and, in some studies, those that they consider the least important, even if desirable (Kohn, 1969). Such a procedure makes it possible to place parents' valuations of each characteristic on a quantitative scale. We must recognize that parents are likely to accord high priority to those values that are not only important, in that failing to achieve them would affect the children's futures adversely, but also problematic, in that they are difficult of achievement. Thus, the indices of parental values used in most of the pertinent inquiries measure conceptions of the "important, but problematic" (Kohn, 1969).

There have been two central findings from these studies. One is that parents at all social-class levels value their children's being honest, happy, considerate, obedient, and dependable (Kohn, 1969). Middle- and working-class parents share values that emphasize, in addition to children's happiness, their acting in a way that shows respect for the rights of others. All class differences in parental values are variations on this common theme.

Nevertheless, there are distinct differences in emphasis between middle- and working-class parents' values. The higher a parent's social-class position, the more likely he is to value characteristics indicative of self-direction and the less likely he is to value characteristics indicative of conformity to external authority (Kohn, 1969). That is, the higher a parent's social-class position, the greater the likelihood that he will value for his children such characteristics as consideration, an interest in how and why things happen, responsibility, and self-control, and the less the likelihood that he will value such characteristics as manners, neatness and cleanliness, being a good student, honesty, and obedience. More detailed analyses show that the differential evaluation of self-direction and conformity to external authority by parents of varying social-class position obtains whatever the age and sex of the child, in families of varying size, composition, and functional pattern.

This essential finding has been repeatedly confirmed, both for fathers and for mothers. The original finding came from a small study in Washington, D.C., in the late 1950s, but it has since been confirmed in several other U.S. studies, including three nation-wide studies, one as recent as 1975. It has also been confirmed in studies in Italy, Germany, Great Britain, France, Ireland, and Taiwan. There are no known exceptions (Franklin and Scott, 1970; Campbell, 1978; LeMasters, 1975; Sennett and Cobb, 1973). . . .

PARENTAL VALUES AND PARENTAL PRACTICES

We would have little interest in parental values but for our belief that parents' values affect their child-rearing practices. The evidence here is much less definitive than on the relationship of class to parental values, but what evidence we do have is altogether consistent. Parents do behave in accord with their values in the two important realms where the question has been studied: in their disciplinary practices and in the allocation of parental responsibilities for imposing constraints on, and providing emotional support for, their children.

Disciplinary Practices

Most early research on class differences in disciplinary practices was directed toward learning whether working-class parents typically employ techniques of punishment different from those used by the middle class. In his definitive review of the research literature on social class and family relationships through the mid-1950s, Bronfenbrenner (1958, p. 424) summarized the results of the several relevant studies as indicating that "working-class parents are consistently more likely to employ physical punishment, while middle-class families rely more on reasoning, isolation, appeals to guilt, and other methods involving the threat of loss of love." This conclusion has been challenged in later research (Erlonger, 1974). Whether or not it is still true, the difference in middle- and working-class parents' propensity to resort to physical punishment certainly never has been great.

For our purposes, in any case, the crucial question is not which disciplinary method parents prefer but when and why they use one or another method of discipline. The early research tells us little about the when and why of discipline; most investigators had relied on parents' generalized statements about their usual or their preferred methods of dealing with disciplinary problems, irrespective of what the particular problem might be. But surely not all disciplinary problems evoke the same kind of parental response. In some sense, after all, the punishment fits the crime. Under what conditions do parents of a given social class punish their children physically, reason with them, isolate them—or ignore their actions altogether?

Recent studies have shown that neither middle- nor working-class parents resort to punishment as a first recourse when their children misbehave (Kohn, 1969). It seems instead that parents of both social classes initially post limits for their children. But when children persist in misbehavior, parents are likely to resort to one or another form of coercion. This is true of all social-class levels. The principal difference between the classes is in the specific conditions under which parents—particularly mothers—punish children's misbehavior. Working-class parents are more likely to punish or refrain from punishing on the basis of the direct and immediate consequences of children's actions, middle-class parents on the basis of their interpretation of children's intent in acting as they do (Gecas and Nye, 1974). Thus, for example, working-class parents are more likely to punish children for fighting than for arguing with their brothers and sisters and are also more likely to punish for aggressively wild play than for boisterousness—the transgression in both instances being measured in terms of how far the overt action transgresses the rules. Middle-class parents make no such distinction. But they do distinguish, for example, between wild play and a loss of temper, tolerating even excessive manifestations of the former as a childish form of emotional expression, but punishing the latter because it signifies a loss of mastery over self.

To say that working-class parents respond more to the consequences of children's misbehavior and middle-class parents more to their own interpretation of the children's intent gets dangerously close to implying that while middle-class parents act on the basis of long-range goals for children, working-class parents do not. On the contrary, the evidence suggests that parents of both social classes act on the basis of long-range goals—but that the goals are different. The interpretive key is provided by our knowledge of class differences in parental values. Because middle- and working-class parents differ in their values, they view children's misbehavior differently; what is intolerable to parents in one social class can be taken in stride by parents in the other. In both social classes, parents punish children for transgressing important values, but since the values are different, the transgressions are differently defined. If self-direction is valued, transgressions must be judged in terms of the reasons why the children misbehave. If conformity to external authority is valued, transgressions must be judged in terms of whether or not the actions violate externally imposed proscriptions.

The Allocation of Parental Responsibilities for Support and Constraint

The connection between values and punishment of disvalued behavior is direct: punishment is invoked when values are transgressed. There are also less direct but broader behavioral consequences of class differences in parental values. In particular, class differences in parental values have important consequences for the overall patterning of parent-child interaction.

In common with most investigators, I conceive of parent-child relationships as structured along two principal axes: support and constraint. This conception is derived in part from Parsons and Bales's (1955) theoretical analysis of family structure and in part from Schaefer's (1959) empirical demonstration that the findings of several past studies of parent-child relationships could be greatly clarified by arraying them along these two dimensions.

Because their values are different, middle- and working-class parents evaluate differently the relative importance of support and constraint in child rearing. One would expect middle-class parents to feel a greater obligation to be supportive, if only because of their concern about children's internal dynamics. Working-class parents, because of their higher valuation of conformity to external rules, should put greater emphasis upon the obligation to impose constraints. We should therefore expect the ratio of support to constraint in parents' handling of their children to be higher in middle-class than in working-class families. And this, according to Bronfenbrenner (1958, p. 425), is precisely what has been shown in those studies that have dealt with the overall relationship of parents to child:

> Parent-child relationships in the middle class are consistently reported as more acceptant and equalitarian, while those in the working class are oriented toward maintaining order and obedience. Within this context, the middle class has shown a shift away from emotional control toward freer expression of affection and greater tolerance of the child's impulses and desires.

Whatever relative weight parents give to support and constraint, the process of child rearing requires both. These responsibilities can, however, be apportioned between mother and father in any of several ways. Mothers can specialize in providing support, fathers in imposing constraints; both parents can play both roles more-or-less equally; mothers can monopolize both roles, with fathers playing little part in child rearing; and there are other possible, but less likely, arrangements. Given their high valuation of self-direction, middle-class parents—mothers and fathers both—should want fathers to play an important part in providing support to the children. It would seem more appropriate to working-class parents' high valuation of conformity to external authority that fathers' obligations should center on the imposition of constraints.

The pertinent studies show that in both the middle class and the working class, mothers would prefer to have their husbands play a role that facilitates children's development of valued characteristics (Kohn, 1969). To middle-class mothers, it is important that children be able to decide for themselves how to act and that they have the personal resources to act on these decisions. In this conception, fathers' responsibility for imposing constraints is secondary to their responsibility for being supportive; in the minds of some middle-class mothers, for fathers to take a major part in imposing constraints interferes with their ability to be supportive. To working-class mothers, on the other hand, it is more important that children conform to externally imposed rules. In this conception, the fathers' primary responsibility is to guide and direct the children. Constraint is accorded far greater value than it has for the middle class.

Most middle-class fathers seem to share their wives' views of fathers' responsibilities toward sons and act accordingly. They accept less responsibility for being supportive of daughters—apparently feeling that this is more properly the mothers' role. But many working-class fathers do not accept the obligations their wives would have them assume, either toward sons or toward daughters. These men do not see the constraining role as any less important than their wives do, but many of them see no reason why fathers should have to shoulder this responsibility. From their point of view, the important thing is that children be taught what limits they must not transgress. It does not particularly matter who does the teaching, and since mothers have primary responsibility for child care, the job should be theirs. Of course, there will be occasions when fathers have to backstop their wives. But there is no ideological imperative that makes it the fathers' responsibility to assume an important part in child rearing. As a consequence, many working-class fathers play little role in child rearing, considering it to be their wives' proper responsibility.

Theories of personality development, including Parsons and Bales's (1955) sociological reinterpretation of the classical Freudian developmental sequence, have generally been based on the model of a family in which the mothers' and fathers' intrafamily roles are necessarily differentiated, with mothers specializing in support and fathers in constraint. However useful a first approximation this may be, both middle- and working-class variations on this general theme are sufficiently great to compel a more precise formulation.

The empirical evidence is partly consistent with the mother-supportive, father-constraining formulation, for even in middle-class families, almost no one reports that fathers are more supportive than mothers. Yet, in a sizable proportion of middle-class families, mothers take primary responsibility for imposing constraints on sons, and fathers are at least as supportive as mothers. And although middle-class fathers are not likely to be as supportive of daughters as their wives are, it cannot be said that fathers typically specialize in constraint, even with daughters.

It would be a gross exaggeration to say that middle-class fathers have abandoned the prerogatives and responsibilities of authority in favor of being friends and confidants to their sons. Yet the historical drift is probably from primary emphasis on imposing constraints to primary emphasis on support (Bronson, Katten, and Livson, 1958; Bronfenbrenner, 1961). In any event, mothers' and fathers' roles are not sharply differentiated in most middle-class families; both parents tend to be supportive. Such division of functions as exists is chiefly a matter of each parent's taking special responsibility for being supportive of the children of the parent's own sex.

Mothers' and fathers' roles are more sharply differentiated in working-class families, with mothers almost always being the more supportive. Yet, despite the high valuation put on the constraining function, fathers do not necessarily specialize in setting limits, even for sons. In some working-class families, mothers specialize in support, fathers in constraint; in many others, the division of responsibilities is for the mothers to raise the children, the fathers to provide the wherewithal. This pattern of role allocation probably is and has been far more prevalent in American society than the formal theories of personality development have recognized.

SOCIAL CLASS, VALUES, AND CONDITIONS OF LIFE

There are, then, remarkably consistent relationships between social class and parental values and behavior. But we have not yet touched on the question: Why do these relationships exist? In analytic terms, the task is to discover which of the many conditions of life associated with class position are most pertinent for explaining why class is related to parental values. Since many of the relevant conditions are implicated in people's occupational lives, our further discussion is focused on one crucial set of occupational conditions: those that determine how much opportunity people have to exercise self-direction in their work.

The principal hypothesis that has guided this line of research is that class-correlated differences in people's opportunities to exercise occupational self-direction—that is, to use initiative, thought, and independent judgment in work—are basic to class differences in parental values. Few other conditions of life are so closely bound up with social class position as are those that determine how much opportunity, even necessity, people have for exercising self-direction in their work. Moreover, there is an appealing simplicity to the supposition that the experience of self-direction in so central a realm of life as work is conducive to valuing self-direction, off as well as on the job, and to seeing the possibilities for self-direction not only in work but also in other realms of life.

Although many conditions of work are either conducive to or deterrent of the exercise of occupational self-direction, three in particular are critical.

First, a limiting condition: people cannot exercise occupational self-direction if they are closely supervised. Not being closely supervised, however, does not necessarily mean that people are required—or even free—to use initiative, thought, and independent judgment; it depends on how complex and demanding is their work.

A second condition for occupational self-direction is that work allow a variety of approaches; otherwise the possibilities for exercising initiative, thought, and judgment are seriously limited. The organization of work must not be routinized; it must involve a variety of tasks that are in themselves complexly structured.

The third and most important determinant of occupational self-direction is that work be substantively complex. By the *substantive complexity* of work I mean, essentially, the degree to which performance of that work requires thought and independent judgment. All work involves dealing with things, with data, or with people; some jobs involve all three, others only one or two of these activities. Work with things can vary in complexity from ditchdigging to sculpturing; similarly, work with people can vary in complexity from receiving simple instructions to giving legal advice; and work with data can vary from reading instructions to synthesizing abstract conceptual systems. Although, in general, work with data or with people is likely to be more complex than work with things, this is not always the case, and an index of the overall complexity of work should reflect its degree of complexity in each of these three types of activity. What is important about work is not whether it deals with things, with data, or with people, but its complexity.

No one of these occupational conditions—freedom from close supervision, nonroutinization, and substantive complexity—is definitional of occupational self-direction. Nevertheless, each of these three conditions tend to be conducive to the exercise of occupational self-direction, and the combination of the three both enables and requires it. Insofar as people are free of close supervision, work at nonroutinized tasks, and do substantively complex work, their work is necessarily self-directed. And insofar as they are subject to close supervision, work at routinized tasks, and do work of little substantive complexity, their work does not permit self-direction.

THE RELATIONSHIP OF OCCUPATIONAL SELF-DIRECTION TO PARENTAL VALUES

Since most of the research on the relationship between occupational self-direction and parental values deals only with men's occupational conditions and men's values, I shall first discuss fathers' values, then broaden the discussion to include mothers' values as well. All three occupational conditions that are determinative of occupational self-direction prove to be empirically related to fathers' values (Kohn, 1969). Men who are free from close supervision, who work at nonroutinized tasks, and who do substantively complex work tend to value self-direction rather than conformity to external authority for their children. This being the case, it becomes pertinent to ask whether the relationship between social class and fathers' values can be explained as resulting from class differences in the conditions that make for occupational self-direction.

It must be emphasized that in dealing with these occupational conditions, we are concerned not with distinctions that cut across social class but with experiences constitutive of class. The objective is to learn whether these constitutive experiences are pertinent for explaining the class relationship. To achieve this objective, we statistically control occupational dimensions that have proved to be related to values and orientation, to determine whether this reduces the correlation between class and fathers' valuation of self-direction or conformity for their children. This procedure is altogether hypothetical, for it imagines an unreal social situation: social classes that did not differ from one another in the occupational conditions experienced by their members. But it is analytically appropriate to use such hypothetical procedures, for it helps us differentiate those occupational conditions that are pertinent for explaining the relationship of class to parental values from those occupational conditions that are not pertinent. In fact, statistically controlling the conditions that make for occupational self-direction reduces the correlation of class to fathers' valuation of self-direction or conformity by nearly two thirds. The lion's share of the reduction is attributable to the substantive complexity of the work, but closeness of supervision and routinization are relevant too. By contrast, though, statistically controlling numerous other occupational conditions has a much weaker effect—reducing the class correlation by only one-third. And controlling both sets of occupational conditions reduces the correlation of class to fathers' values by no more than does controlling occupational self-direction alone. Thus, other occupational conditions add little to the explanatory power of the three that are determinative of occupational self-direction. . . .

Much less is known about the relationship between occupational conditions and mothers' values. My colleagues and I are currently analyzing data on the relationship between employed mothers' occupational conditions and their values for their children (Kohn, 1977). Preliminary results indicate that women's occupational conditions affect their values in much the same way as do men's occupational conditions. . . .

Because the relationship between exercising self-direction on the job and valuing self-direction for children is so direct, one might conclude that parents are simply preparing their children for the occupational life to come. I believe, rather, that parents come to value self-direction or conformity as virtues in their own right, not simply as means to occupational goals. One important piece of evidence buttresses this impression: studies in both the United States and Italy show that the relationship between men's occupational experiences and their values is the same for daughters as for sons, yet it is hardly likely (especially in Italy) that most fathers think their daughters will have occupational careers comparable to those of their sons. It would thus seem that occupational experience helps structure parents' views not only of the occupational world but of social reality in general....

It could be argued that the empirical interrelationships of social class, occupational self-direction, and parental values reflect the propensity of people who value self-direction to seek out jobs that offer them an opportunity to be self-directed in their work and, once in a job, to maximize whatever opportunities the job allows for exercising self-direction. But we know that occupational choice is limited by educational qualifications, which in turn are greatly affected by the accidents of family background, economic circumstances, and available social resources. Moreover, the opportunity to exercise self-direction in one's work is circumscribed by job requirements. Thus, an executive must do complex work with data or with people; he cannot be closely supervised; and his tasks are too diverse to be routinized—to be an executive requires some substantial self-direction. Correspondingly, to be a semi-skilled factory worker precludes much self-direction. The substance of one's work cannot be especially complex; one cannot escape some measure of supervision; and if one's job is to fit into the flow of other people's work, it must necessarily be routinized. The relationship between being self-directed in one's work and holding self-directed values would thus seem to result not just from self-directed people's acting according to their values but also from job experiences affecting these very values....

Education matters for parental values in part because it is an important determinant of occupational conditions. A major reason for looking to such occupational conditions as substantive complexity, closeness of supervision, and routinization as possible keys to understanding the relationship between social class and parental values is that few other conditions of life are so closely related to educational attainment. This explanation has been confirmed in further analyses that have assessed the effects of education on occupational conditions at each stage of career. Education is a prime determinant, for example, of the substantive complexity of the job; and the substantive complexity of the job, in turn, has an appreciable effect on parental values. It is precisely because education is crucial for the very occupational conditions that most strongly affect parental values that education is so powerfully related to parental values (Kohn and Schooler, 1973).

Education also has important direct effects on parental values, quite apart from its indirect effects mediated through occupational conditions. Education matters, aside from its impact on job conditions, insofar as education provides the intellectual flexibility and breadth of perspective that are essential for self-directed values. Thus education has both direct and indirect effects upon parental values, both types of effect contributing importantly to the overall relationship between social class and parental values....

The facts and interpretations reviewed in this paper have many implications....

These findings are pertinent to our conception of what is normal and what is not in family functioning. I have tried in this paper to show that there are considerable variations in normal family functioning and that these variations are to be understood in terms of the actual conditions of life that families encounter. The values and child-rearing practices of American parents must be seen in terms of the realities parents face.

SELECTED REFERENCES

Barber, B. Social stratification. In D. L. Sills (Ed.), *International Encyclopedia of the Social Sciences,* Vol. 15. New York: Macmillan Company and Free Press, 1968, pp. 288–296.

Berelson, B., and Steiner, G. A. *Human Behavior: An Inventory of Scientific Findings.* New York: Harcourt, Brace and World, 1964.

Bronfenbrenner, U. Socialization and social class through time and space. In E. E. Maccoby, T. M. Newcomb, and E. L. Hartley (Eds.), *Readings in Social Psychology.* New York: Holt, Rinehart and Winston, 1958, pp. 400–425.

Bronfenbrenner, U. The changing American child—A speculative analysis. *Journal of Social Issues,* 1961, 17, 6–18.

Bronson, W. C., Katten, E. S., and Livson, N. Patterns of authority and affection in two generations. *Journal of Abnormal and Social Psychology,* 1959, 58 (March), 143–152.

Campbell, J. D. The child in the sick role: Contributions of age, sex, parental status, and parental values. *Journal of Health and Social Behavior,* 1978, 19 (March), 35–51.

Erlanger, H. S. Social class and corporal punishment in childrearing: A reassessment. *American Sociological Review,* 1974, 39 (February), 68–85.

Franklin, J. I., and Scott, J. E. Parental values: An inquiry into occupational setting. *Journal of Marriage and the Family,* 1970, 32 (August), 406–409.

Gecas, V. The influence of social class on socialization. In W. R. Burr, R. Hill, I. L. Reiss, and F. I. Nye (Eds.), *Theories about the Family.* New York: Free Press, 1977.

Gecas, V., and Nye, F. I. Sex and class differences in parent-child interaction: A test of Kohn's hypothesis. *Journal of Marriage and the Family,* 1974, 36 (November), 742–749.

Kohn, M. L. *Class and Conformity: A Study in Values.* Homewood, Ill.: Dorsey, 1969. (Second edition, 1977, published by the University of Chicago Press.)

Kohn, M. L. Occupational structure and alienation. *American Journal of Sociology,* 1976, 82 (July), 111–130.

Kohn, M. L., and Schooler, C. Occupational experience and psychological functioning: An assessment of reciprocal effects. *American Sociological Review,* 1973, 38 (February), 97–118.

Le Masters, E. E. *Blue-Collar Aristocrats: Life-Styles at a Working-Class Tavern.* Madison: University of Wisconsin Press, 1975.

Parsons, T, and Bales, R. F. *Family, Socialization and Interaction Process.* Glencoe, Ill.: Free Press, 1955.

Schaefer, E. S. A circumplex model for maternal behavior. *Journal of Abnormal and Social Psychology,* 1959, 59 (September), 226–235.

Sennett, R., and Cobb, J. *The Hidden Injuries of Class.* New York: Knopf, 1973.

Williams, R. M. Jr. *American Society: A Sociological Interpretation,* 2nd ed. New York: Knopf, 1960.

Williams, R. M. Jr. The concept of values. In D. L. Sills (Ed.), *International Encyclopedia of the Social Sciences,* Vol. 16. New York: Macmillan Company and Free Press, 1968, pp. 283–287.

26

Families on the Fault Line

LILLIAN RUBIN,
1995

. . .Two decades ago I did a study of white working-class family life that led to the publication of a book entitled *Worlds of Pain: Life in the Working-Class Family* (Rubin, 1976). In the intervening years, enormous social, political, and economic changes have been at work, defining and redefining family and social life, relations between women and men, between parents and children, and among the various ethnic and racial groups that make up the tapestry of American life. The time had come, therefore, to take another look at working-class family life, to compare then and now, to examine how the changes of these past years have affected life inside working-class families and, in turn, how these changes have helped to form their responses to the world outside, to the social and political issues of the day.

Twenty years ago it was reasonable to study white working-class families without reference to race or ethnicity. Given the heightened ethnic and racial strains with which we live now, it's unthinkable to leave these out. There was plenty of racial anger around then, too, of course. The rancorous conflicts about school busing all across the land made it obvious that racial tensions were an intensely felt part of our national life (Rubin, 1972; Lukas, 1986). Nevertheless, race wasn't high on the list of issues the people I interviewed wanted to talk about then—an impression confirmed by a recent *New York Times* article that traced the results of forty-five years of Gallup polls asking people what they thought were "the most important problems facing the country." Race relations, which was seen as a significant issue from the mid-1950s through the early years of the 1960s—the years when the civil rights movement was in full swing—was replaced by concerns about the cost of living toward the end of that decade. By the beginning of the 1970s, the period when I was doing the research for *Worlds of Pain,* race wasn't even mentioned. Two decades later, racial issues were prominent again. In the research for this book, race was a recurrent theme, most of the time arising spontaneously as people aired their grievances and gave vent to their wrath. . . .

A word about *class*—always a difficult problem in a society that lives with the fiction that everyone is middle class. In the research for *Worlds of Pain,* I interviewed only people who were clearly working class—that is, those who worked in *blue-collar* occupations and had no more than a high school education. Even then, I had questions about whether to leave out lower-middle-class white-collar families, since, as I wrote at the time, "many acute observers argue that the major division in our advanced industrial society is between the upper middle class and the combined working and lower middle class—an argument with which I concur" (Rubin, 1976). Nevertheless, given the aims of that research, which was to specify class differences in culture and life-style, it made sense to draw the line more tightly.

It's still true, as I argued then, that a salaried white-collar worker is less susceptible to economic and seasonal fluctuations than a blue-collar worker who's paid by the hour. But it's also true that the boundary between the working class and the lower level of the middle class is even less clear today than it was twenty years ago. This time, therefore, the line is looser, and the families I write about in this book are a somewhat more varied group. Most are blue-collar, pink-collar, and service workers; some earn their keep in the menial sales and office jobs that characterize the low-level white-collar world.

None had a college degree at the time of the interview; a few had some college, usually courses taken at a two-year community college. But whatever their differences, they all have one central thing in common: They are the women and men who keep this country's wheels in motion and who—whether they work in the manufacturing or the service sector—are so little rewarded for the work they do. Although they have much to teach us about life in these United States, they are all too often the invisible ones, those whose voices rarely are heard in all their complexity, their ambivalence, and their contradictions. . .

THE FAMILY AND THE ECONOMY

Nearly 15 percent of the men in the families I interviewed were jobless when I met them (U.S. Bureau of the Census 1992). Another 20 percent had suffered episodic bouts of unemployment—sometimes related to the recession of the early 1990s, sometimes simply because job security is fragile in the blue-collar world, especially among the younger, less experienced workers. With the latest recession, however, age and experience don't count for much; every man feels at risk (Ehrenreich, 1989; Newman, 1988).

Tenuous as the situation is for white men, it's worse for men of color, especially African-Americans. The last hired, they're likely to be the first fired. And when the axe falls, they have even fewer resources than whites to help them through the tough times. "After kicking around doing shit work for a long time, I finally got a job that paid decent," explains twenty-nine-year-old George Faucett, a black father of two who lost his factory job when the company was restructured—another word that came into vogue during the economic upheaval of the 1990s. "I worked there for two years, but I didn't have seniority, so when they started to lay guys off, I was it. We never really had a chance to catch up on all the bills before it was all over," he concludes dispiritedly.

I speak of men here partly because they're usually the biggest wage earners in intact families. Therefore, when father loses his job, it's likely to be a crushing blow to the family economy. And partly, also, it's because the issues unemployment raises are different for men and for women. For most women, identity is multifaceted, which means that the loss of a job isn't equivalent to the loss of self. No matter how invested a woman may be in her work, no matter how much her sense of self and competence are connected to it, work remains only one part of identity—a central part perhaps, especially for a professional woman, but still only a part. She's mother, wife, friend, daughter, sister—all valued facets of the self, none wholly obscuring the others. For the working-class women in this study, therefore, even those who were divorced or single mothers responsible for the support of young children, the loss of a job may have been met with pain, fear, and anxiety, but it didn't call their identity into question.

For a man, however, work is likely to be connected to the core of self. Going to work isn't just what he does, it's deeply linked to who he is. Obviously, a man is also father, husband, friend, son, brother. But these are likely to be roles he assumes, not without depth and meaning, to be sure, but not self-defining in the same way as he experiences work. Ask a man for a statement of his identity, and he'll almost always respond by telling you first what he does for a living. The same question asked of a woman brings forth a less predictable, more varied response, one that's embedded in the web of relationships that are central to her life (Rubin, 1976, 1986).

Some researchers studying the impact of male unemployment have observed a sequenced series of psychological responses (Hill, 1978; Rees, 1981). The first, they say, is shock, followed by denial and a sense of optimism, a belief that this is temporary, a holiday, like a hiatus between jobs rather than joblessness. This period is marked by heightened activity at home, a burst of do-it-yourself projects that had been long neglected for lack of time. But soon the novelty is gone and the projects wear thin, ushering in the second phase, a time of increasing distress, when inertia trades places with activity and anxiety succeeds denial. Now a jobless man awakens every day to the reality of unemployment. And, lest he forget, the weekly trip to the unemployment office is an unpleasant reminder. In the third phase, inertia deepens into depression, fed by feelings of identity loss, inadequacy, hopelessness, a lack of self-confidence, and a general failure of self-esteem. He's tense, irritable, and feels increasingly alienated and isolated from both social and personal relationships.

This may be an apt description of what happens in normal times. But in periods of economic crisis, when losing a job isn't a singular and essentially lonely event, the predictable pattern breaks down (Newman, 1988). During the years I was interviewing families for this book, millions of jobs disappeared almost overnight. Nearly everyone I met, therefore, knew someone—a family member, a neighbor, a friend—who was out of work. "My brother's been out of a job for a long time; now my brother-in-law just got laid off. It seems like every time I turn around, somebody's losing his job. I've been lucky so far, but it makes you wonder how long it'll last."

At such times, nothing cushions the reality of losing a job. When the unbelievable becomes commonplace and the unexpected is part of the mosaic of the times, denial is difficult and optimism impossible. Instead, any layoff, even if it's defined as temporary, is experienced immediately and viscerally as a potentially devastating, cataclysmic event.

It's always a shock when a person loses a job, of course. But disbelief? Denial? Not for those who have been living under a cloud of anxiety—those who leave work each night grateful for another day of safety, who wonder as they set off the next morning whether this is the day the axe will fall on them. "I tell my wife not to worry because she gets panicked about the bills. But the truth is, I stew about it plenty. The economy's gone to hell; guys are out of work all around me. I'd be nuts if I wasn't worried."

It's true that when a working-class man finds himself without a job he'll try to keep busy with projects around the house. But these aren't undertaken in the kind of holiday spirit earlier researchers describe (Hill, 1978; Rees, 1981). Rather, building a fence, cleaning the garage, painting the family room, or the dozens of other tasks that might occupy him are a way of coping with his anxiety, of distracting himself from the fears that threaten to overwhelm him, of warding off the depression that lurks just below the surface of his activity. Each thrust of the saw, each blow of the hammer helps to keep the demons at bay. "Since he lost his job, he's been out there hammering away at one thing or another like a maniac," says Janet Kovacs, a white thirty-four-year-old waitress. "First it was the fence; he built the whole thing in a few days. Then it was fixing the siding on the garage. Now he's up on the roof. He didn't even stop to watch the football game last Sunday."

Her husband, Mike, a cement finisher, explains it this way: "If I don't keep busy, I feel like I'll go nuts. It's funny," he says with a caustic, ironic laugh, "before I got laid off my wife was always complaining about me watching the ball games; now she keeps nagging me to watch. What do you make of that, huh? I guess she's trying to make me feel better."

"Why didn't you watch the game last Sunday?" I ask.

"I don't know, maybe I'm kind of scared if I sit down there in front of that TV, I won't want to get up again," he replies, his shoulders hunched, his fingers raking his hair. "Besides, when I was working, I figured I had a right."

His words startled me, and I kept turning them over in my mind long after he spoke them: "When I was working, I figured I had a right." It's a sentence any of the unemployed men I met might have uttered. For it's in getting up and going to work every day that they feel they've earned the right to their manhood, to their place in the world, to the respect of their family, even the right to relax with a sporting event on TV. . . .

RACE AND THE RISE OF ETHNICITY

Two decades ago, when I began the research for *Worlds of Pain,* we were living in the immediate aftermath of the civil rights revolution that had convulsed the nation since the mid-1950s. Significant gains had been won. And despite the tenacity with which this headway had been resisted by some, most white Americans were feeling good about themselves. No one expected the nation's racial problems and conflicts to dissolve easily or quickly. But there was also a sense that we were moving in the right direction, that there was a national commitment to redressing at least some of the worst aspects of black-white inequality.

In the intervening years, however, the national economy buckled under the weight of three recessions, while the nation's industrial base was undergoing a massive restructuring. At the same time, government policies requiring preferential treatment were enabling African-Americans and other minorities to make small but visible inroads into what had been, until then, largely white terrain. The sense of scarcity, always a part of American life but intensified sharply by the history of these economic upheavals, made minority gains seem particularly threatening to white working-class families.

It isn't, of course, just working-class whites who feel threatened by minority progress. Wherever racial minorities make inroads into formerly all-white territory, tensions increase. But it's working-class families who feel the fluctuations in the economy most quickly and most keenly. For them, these last decades have been like a bumpy roller coaster ride. "Every time we think we might be able to get ahead, it seems like we get knocked down again," declares Tom Ahmundsen, a forty-two-year-old white construction worker. "Things look a little better; there's a little more work; then all of a sudden, boom, the economy falls apart and it's gone. You can't count on anything; it really gets you down."

This is the story I heard repeatedly: Each small climb was followed by a fall, each glimmer of hope replaced by despair. As the economic vise tightened, despair turned to anger. But partly because we have so little concept of class resentment and conflict in America, this anger isn't directed so much at those above as at those below. And when whites at or near the bottom of the ladder look down in this nation, they generally see blacks and other minorities.

True, during all of the 1980s and into the 1990s, white ire was fostered by national administrations that fanned racial discord as a way of fending off white discontent—of diverting anger about the state of the economy and the declining quality of urban life to the foreigners and racial others in our midst. But our history of racial animosity coupled with our lack of class consciousness made this easier to accomplish than it might otherwise have been.

The difficult realities of white working-class life not withstanding, however, their whiteness has accorded them significant advantages—both materially and psychologically—over people of color. Racial discrimination and segregation in the workplace have kept competition for the best jobs at a minimum. They do, obviously, have to compete with each other for the resources available. But that's different. It's a competition among equals; they're all white. They don't think such things consciously, of course; they don't have to. It's understood, rooted in the culture and supported by the social contract that says they are the superior ones, the worthy ones. Indeed, this is precisely why, when the courts or the legislatures act in ways that seem to contravene that belief, whites experience themselves as victims. . . .

No matter how far down the socioeconomic ladder whites may fall, the one thing they can't lose is their whiteness. No small matter because, as W. E. B. DuBois observed decades ago, the compensation of white workers includes a psychological wage, a bonus that enables them to believe in their inherent superiority over nonwhites (DuBois, 1935).

It's also true, however, that this same psychological bonus that white workers prize so highly has cost them dearly. For along with the importation of an immigrant population, the separation of black and white workers has given American capital a reserve labor force to call upon whenever white workers seemed to them to get too "uppity." Thus, while racist ideology enables white workers to maintain the belief in their superiority, they have paid for that conviction by becoming far more vulnerable in the struggle for decent wages and working conditions than they might otherwise have been.

Politically and economically, the ideology of white supremacy disables white workers from making the kind of interracial alliances that would benefit all of the working class. Psychologically, it leaves them exposed to the double-edged sword of which I spoke earlier. On one side, their belief in the superiority of whiteness helps to reassure them that they're not at the bottom of the social hierarchy. But their insistence that their achievements are based on their special capacities and virtues, that it's only incompetence that keeps others from grabbing a piece of the American dream, threatens their precarious sense of self-esteem. For if they're the superior ones, the deserving ones, the ones who earned their place solely through hard work and merit, there's nothing left but to blame themselves for their inadequacies when hard times strike.

In the opening sentences of *Worlds of Pain* I wrote that America was choking on its differences. If we were choking then, we're being asphyxiated now. As the economy continues to falter, and local, state, and federal governments keep cutting services, there are more and more acrimonious debates about who will share in the shrinking pie. Racial and ethnic groups, each in their own corners, square off as they ready themselves for what seems to be the fight of their lives. Meanwhile, the quality of life for all but the wealthiest Americans is spiraling downward—a plunge that's felt most deeply by those at the lower end of the class spectrum, regardless of color (Greenstein, 1991).

As, more and more mothers of young children work full-time outside the home, the question of who will raise the children comes center stage. Decent, affordable child care is scandalously scarce, with no government intervention in sight for this crucial need. In poor and working-class families, therefore, child care often is uncertain and inadequate, leaving parents apprehensive and children at risk. To deal with their fears, substantial numbers of couples now work different shifts, a solution to the child-care problem that puts its own particular strains on family life.

In families with two working parents, time has become their most precious commodity—time to attend to the necessary tasks of family life; time to nurture the relationships between wife and husband, between parents and children; time for oneself, time for others; time for solitude, time for a social life (Schor, 1991). Today more than ever before, family life has become impoverished for want of time, adding another threat to the already fragile bonds that hold families together.

While women's presence in the labor force has given them a measure of independence unknown before, most also are stuck with doing two days' work in one—one on the job, the other when they get home at night. Unlike their counterparts in the earlier era, today's women are openly resentful about the burdens they carry, which makes for another dimension of conflict between wives and husbands.

Although the men generally say they've never heard of Robert Bly or any of the other modern-day gurus of manhood, the idea of men as victims has captured their imagination (Bly, 1990; Keen, 1991). Given the enormous amount of publicity these men's advocates have garnered in the last few years, it's likely that some of their ideas have filtered into the awareness of the men in this study, even if they don't know how they got there. But their belief in their victimization is also a response to the politics of our time, when so many different groups—women, gays, racial minorities, the handicapped—have demanded special privileges and entitlements on the basis of past victimization. And once the language of victimization enters the political discourse, it becomes a useful tool for anyone wanting to resist the claims of others or to stake one of their own.

As the men see it, then, if their wives are victims because of the special burdens of women, the husbands, who bear their own particular hardships, can make the claim as well. If African-American men are victims because of past discrimination, then the effort to redress their grievances turns white men into victims of present discrimination.

To those who have been victimized by centuries of racism, sexism, homophobia, and the like, the idea that straight white men are victims, too, seems ludicrous. Yet it's not wholly unreal, at least not for the men in this study who have so little control over their fate and who so often feel unheard and invisible, like little more than shadows shouting into the wind.

Whether inside the family or in the larger world outside, the white men keep hearing that they're the privileged ones, words that seem to them like a bad joke. How can they be advantaged when their inner experience is that they're perched precariously on the edge of a chasm that seems to have opened up in the earth around them? It's this sense of vulnerability, coupled with the conviction that their hardships go unseen and their pain unattended, that nourishes their claim to victimhood.

Some analysts of family and social life undoubtedly will argue that the picture I've presented here is too grim, that it gives insufficient weight to both the positive changes in family life and the gains in race relations over these past decades. It's true that the social and cultural changes we've witnessed have created families that, in some ways at least, are more responsive to the needs of their members, more democratic than any we have known before (Stacey, 1990; Coontz, 1992; Skolnick, 1991). But it's also true that without the economic stability they need, even the most positive changes will not be enough to hold families together.

Certainly, too, alongside the racial and ethnic divisions that are so prominent a part of American life today is the reality that many more members of these warring groups than ever before are living peaceably together in our schools, our factories, our shops, our corporations, and our neighborhoods. And, except for black-white marriages, many more are marrying and raising children together than would have seemed possible a few decades ago.

At the same time, there's reason to fear. The rise of ethnicity and the growing racial separation also means an escalating level of conflict that sometimes seems to threaten to fragment the nation. In this situation, ethnic entrepreneurs like Al Sharpton in New York and David Duke in Louisiana rise to power and prominence by fanning ethnic and racial discord. A tactic that works so well precisely because the economic pressures are felt so keenly on all sides of the racial fissures, because both whites and people of color now feel so deeply that "it's not fair."

As I reflect on the differences in family and social life in the last two decades, it seems to me that we were living then in a more innocent age—a time, difficult though it was for the working-class families of our nation, when we could believe anything was possible. Whether about the economy, race relations, or life inside the family, most Americans believed that the future promised progress, that the solution to the social problems and inequities of the age were within our grasp, that sacrifice today would pay off tomorrow. This is perhaps the biggest change in the last twenty years: The innocence is gone.

But is this a cause for mourning? Perhaps only when innocence is gone and our eyes unveiled will we be able to grasp fully the depth of our conflicts and the sources from which they spring.

We live in difficult and dangerous times, in a country deeply divided by class, race, and social philosophy. The pain with which so many American families are living today, and the anger they feel, won't be alleviated by a retreat to false optimism and easy assurances. Only when we are willing to see and reckon with the magnitude of our nation's problems and our people's suffering, only when we take in the full measure of that reality, will we be able to find the path to change. Until then, all our attempts at solutions will fail. And this, ultimately, will be the real cause for mourning. For without substantial change in both our public and our private worlds, it is not just the future of the family that is imperiled but the very life of the nation itself.

SELECTED REFERENCES

Bly, Robert. 1990. *Iron John.* Reading, MA: Addison-Wesley.

Coontz, Stephanie. 1992. *The Way We Never Were: American Families and the Nostalgia Trap.* New York: Basic Books.

DuBois, W. E. B. 1935. *Black Reconstruction in the United States, 1860–1880.* New York: Harcourt Brace.

Ehrenreich, Barbara. 1989. *Fear of Falling.* New York: Pantheon Books.

Greenstein, Robert. Fall, 1991. "The Kindest Cut." *The American Prospect:* 49–57.

Hill, John. 1978. "The Psychological Impact of Unemployment." *New Society* 43:118–120.

Keen, Sam. 1991. *Fire in the Belly.* New York: Bantam Books.

Lukas, Anthony J. 1986. *Common Ground.* New York: Alfred Knopf.

Newman, Katherine S. 1988. *Falling from Grace.* New York: Free Press.

Rees, Linford W. 1981. "Medical Aspects of Unemployment." *British Medical Journal* 6307: 1630–1631.

Rubin, Lillian B. 1972. *Busing & Backlash: White Against White in an Urban School District.* Berkeley: University of California Press.

Rubin, Lillian B. 1976. *Worlds of Pain: Life in the Working-Class Family.* New York: Basic Books.

Rubin, Lillian B. 1986. *Women of a Certain Age: The Midlife Search for Self.* New York: Harper Perennial.

Schor, Juliet B. 1991. *The Overworked American: The Unexpected Decline of Leisure.* New York: Basic Books.

Skolnick, Arlene. 1991. *Embattled Paradise: The American Family in an Age of Uncertainty.* New York: Basic Books.

Stacey, Judith. 1990. *Brave New Families.* New York: Basic Books.

U.S. Bureau of the Census. 1992. *Statistical Abstract.* Washington, D.C.: U.S. Government Printing Office.

27

Equality Works

FRANCINE DEUTSCH,
1999

. . . There is no doubt that men resist doing half of the work at home. Yet male resistance is often fueled by women's ambivalence about equality. Fears may be preventing women from enthusiastically encouraging their husbands to share fully in the joys and burdens of parenthood, but these fears are unfounded.

Imagine for a moment that we could listen to the kinds of internal monologues that women construct when they think about equal parenting.

I don't want to become just like a man, focused more on career than on family. I love my job, but my child is more important.

In the equally sharing couples I interviewed, mothers were intensely involved with their children. I had to change the first question of my interview from "Would you describe your children?" to "Would you *briefly* describe your children?" The difference between the equal sharers and other couples was not that mothers cared less, but that fathers cared more. Like all mothers, the equally sharing mothers compromised their careers for the good of their children, but so did their husbands. Recall that in several of these families, *both* fathers and mothers took leaves from their jobs or cut back to part-time paid work to care for their children. Yet paternal care may not always assuage women's worries.

No one can take care of my child the way I can. Especially during infancy, my baby needs me.

Mothers who believe that no substitute will suffice for their infants' well-being can take the primary role early in their children's lives, and still establish equality later on when they return to paid work. A majority of the equally sharing mothers were *the* parent in their family who took time off from the job to care for infants. When I began my study, I believed that if parents didn't share equally from birth, they were doomed to be forever unequal because it would be impossible to undo the patterns set up in the early days of parenting. I was wrong. Parenting continually changes during children's lives.

It was also true that in families committed to equality from the start, men made just as good "mothers" as their wives. These dads carved out significant roles for themselves even in the earliest days of parenting when their wives were nursing. Equally sharing fathers got up in the middle of the night to comfort their babies, diapered them, and lovingly handed them over to their wives to be fed. They encouraged their wives to use supplemental bottles so that they too could feed their sons and daughters. The intensity of attachment between these men and their infants was as strong as the bond we usually associate with mothers.

One father joked that when his son started childcare at seven months, he and his wife each visited an hour every day. They paid for five hours of childcare and stayed with their baby two out of those five hours. That same father introduced solid food to his child. He fretted over how the baby was eating, and read articles about what was important for his nutrition: "I'm not Jewish, but I'm kind of a Jewish mother." When fathers equally shared infant care, they worried about all the little details. Fathers *can* meet the needs of infants. Nevertheless, even when the needs of children are met, equality may be threatening, because of the needs of mothers.

> I want to be a real mother. Isn't being number one to your child the essence of motherhood? I don't want to give that up.

Motherhood is central to the identities of most women with children. For many, as we have seen, being number one, at least when children are small, is part of the definition of being a good mother. For some, that belief precluded equal sharing when they first became parents. One self-perceptive mother admitted that she lacked confidence in her ability to nurture the way she thought a mother should. She told me she quit her high-powered job to be home with her two daughters for four years to prove to herself that she could do it. Married to a highly nurturant man, she made room for him to be an equal parent only after she felt secure about her own special relationship with her children.

But we also saw equally sharing women who transformed the meaning of motherhood. Some completely rejected the notion that mothers should perform a unique role in the family, and shared the central spot with their husbands in all ways. Other mothers, reluctant to relinquish all of the special meanings of motherhood, retained some particular part of parenting for themselves. Equal parenting doesn't mean that men and women had identical roles or relationships with their children. To some women, motherhood meant tucking their children in every night, or staying home when they were sick, or taking the lead when discussing problems with them. Most of their husbands honored these maternal prerogatives and found other ways to contribute. The mothers' view is:

> I like being in control. The truth, which I am embarrassed to admit even to myself, is that I like things done my way with my children. I can just imagine the kind of clothes he would put on my daughter!

Yes, equal sharing does mean that women have to give up a certain kind of control. Most men are not willing to do half the work of raising children without a say in how it is to be done. And who would blame them? Yet, even though the mothers ceded control overall, a few domains remained predominantly under their influence. Funnily enough, fashion headed the list. On average in the equally sharing families, mothers did over 80 percent of the clothes-buying.

It always seemed ironic to me when the supermoms I interviewed talked about wanting to keep control. Like jugglers with too many balls in the air, women who did the lioness's share of the work at home while handling the demands of paid work had tenuous control at best. Although these supermoms usually did get to plan the meals, the children's activities, and the daily schedule of the household, their descriptions of the stress created by relentless demands sounded like anything but control. One mother working a full-time job and managing three fourths of the work at home described the burnout she felt: "When I go to sleep at night, I'm not tired. I'm really exhausted. A lot of that is mental energy, it's keeping all the balls in the air." At times, she told me, it all "unravels." No woman, no matter how organized, efficient, loving, or self-sacrificing, can work full time and care for children with minimal help from her husband and really be in control. But these supermoms are worried about dividing things equally:

> I don't know if I want to share everything equally with my husband. It would be so hard to negotiate every little decision. I know a couple who did that and they ended up getting divorced.

Equal parents don't negotiate every little decision. Sharing a profound commitment to their children, equal parents trust each other to do what is best for them. Of course they negotiate about who's going to do what, but so do all dual-earner couples today. We live in an age that lacks consensus about the roles of mothers and fathers. In one sense, couples who agree on the principle of equality have less to negotiate than other couples. They simply have to figure out how to put the 50-50 principle into practice. Equal or not, couples vary dramatically in how much time they spend negotiating and discussing decisions, but these variations reflect their personal styles more than how they divide childcare.

There is nothing to fear from equality. Equally shared parenting doesn't mean giving up the commitment to children's needs, or adopting identical roles, or endless negotiating. It does mean sharing it all instead of doing it all. There is nothing to lose from equality and a world to gain (Marx, 1950).

A WORLD TO GAIN

I wrote this book because all around me I saw women compromising their dreams. Growing up at a time when the barriers to women's achievement were crumbling at an incredible rate, I was shocked to see what was happening in the family. When motherhood hit, egalitarian ideals went out the window. Women were either cutting back substantially on their careers (in the absence of husbands who did so as well) or working full time, trying to manage the unmanageable dual burdens of the supermom. It seemed to me that there must be a better way.

I admit that after having a child myself, I realized I had underestimated the intense pull of a fragile new person. The day my husband and I took our newborn son home from the hospital I sat on my bed with him in my arms. All of a sudden, I realized I was in love. It is an indescribable love, comparable to nothing else in life. I never would have believed it before becoming a parent, but when my child had surgery at the age of three, I would have changed places with him in a second. Now twelve, he knows exactly how to exasperate me, yet the love I feel for him today is as intense and passionate as when I held him in my arms on that beautiful fall day over a decade ago.

The fallacy we have accepted is that this uncompromising love for our children means that we have to give up on equality. The stories of the equal sharers I interviewed show that we don't. Love doesn't diminish when it is shared. Equality is good for children. It is simply easier for two devoted parents to meet children's needs than for one to do it (Deutsch, Servis, and Payne, 1998). Ironically, by doing less childcare, mothers can do more for their children and feel better about what they are doing. "It's just giving the kids more . . . They still have their mother as much as they would otherwise, but they also have their father." Two fully involved parents means neither is "burned-out, over-involved, fried or depressed." Mothers and fathers buffer each other. "Children benefit in that they have two adults. If one happens to be sick or in a bad mood on a particular day, they have someone else to rely on." The endlessly patient, totally nurturing mother is a fiction that exists neither in the families where mothers do it all nor in those where parenting is shared. The difference is that in the equally sharing families when Mom has had it, Dad is right there to comfort, help with homework, or answer the 7,000th "Why?" that day (Risman and Myers, 1997).

Equally shared parenting benefits not just children, but women and men, and their marriages (Okin, 1989). Women obtain the most obvious advantages. Secure in the knowledge that their children are getting the best of both parents, equally sharing mothers have room in their lives to invest in work outside the home without suffering the excessive burdens of an unshared double day. Some even make time to exercise or see a friend. They report that their marriages benefit because "there's a whole lot less conflict if you aren't always mad at your husband about how you're being a martyr" (see Suitor, 1991). Equal sharers don't fall into the syndrome that this mother observed: "Oh God, the number of marriages that seem to have difficulties because the husband is going full steam ahead (in his career) and the wife has kind of been left behind in the diaper pail."

Equal sharing can strengthen marriage because family work, when shared, becomes a bond rather than a barrier to intimacy (Schwartz, 1994). The stuff of everyday life with children can seem pretty mundane to less involved fathers, but it is these seeming trivialities that connect parent and child, and parents to each other. Children's questions, their fumbling attempts to tie their shoes, their willingness to try a new food after months of resistance—all these ordinary details of life with children can be intensely interesting to a parent who is seriously involved in their care. Sometimes the hardest part of inequality was the loss of intimacy because the wonder of parenting wasn't shared. As one unequal mother confided: "There's a loneliness sometimes I feel. . . There are times when I feel alone in raising them." When parents share equally, interest and pleasure in the details become part of what they share as a couple. Sharon laughed that when she and Peter spend a rare night alone together, they go out to dinner and entertain each other with stories about their daughter, Nicole.

In truth, of course there is a price for men to pay when they fully shoulder the dual demands of home and paid work. Until now, juggling has been a woman's problem. When men pick up the slack at home, however, it becomes their problem as well. The equally sharing fathers were well aware of the cost to their careers entailed by their involvement in childcare. But just as so-called career women were amazed to discover the depth of their connection to their children, men who shared responsibility for daily childcare were surprised by the depth of their relationships with their children. One father, a factory worker who alternated work shifts with his wife, described the transformation that occurred after he started taking care of his infant son:

> Until I actually stayed home with him . . . (I knew) nothing. I mean in the deep-seated sense of knowing what he's like during the day, what he wants, what different looks are . . . I don't think our relationship developed to anything substantial until I started taking care of him . . . It was just a different closeness.

Perhaps these fathers cherish their relationships with their children all the more because they know how easy it would have been to miss out. Their own fathers missed what they are experiencing, and even now they know other men are missing it: "Now watching my children develop is just amazing. It's fantastic! It's really mind boggling! I can't imagine a parent who could miss that."

Poignantly, the fathers on the sidelines sometimes understood that they were missing something that could never be regained:

> I feel sad that I don't have the opportunity to do more with them, spend more time with them. I also know how quickly this time is passing by. Ten years from now my son won't be in the house more than likely and that's a scary feeling.

Equality means some sacrifices from men, but the men I interviewed told me, each in his own way, that the rewards reaped were well worth it. The bond they forged with their wives, the special relationships with their children, and the development they saw in themselves were priceless (Barnett and Rivers, 1996). This equally sharing father of two teenage daughters expressed what many of them felt: "A lot of things I would change in my life. (Parenting) I wouldn't consider changing. It's the best thing I've done in my life."

HOW COUPLES BECOME EQUAL SHARERS

Couples create equality by the accumulation of large and small decisions and acts that make up their everyday lives as parents. Couples become equal or unequal in working out the details: who makes children's breakfasts, washes out their diaper pails, kisses their boo-boos; takes off from work when they are sick, and teaches them to ride bikes. Like all dual-earner families today, the equal sharers grapple daily with how to manage the demands of work and family. It is not easy to create equality. Sometimes it seems that everything is stacked against it—from gender discrimination in the workplace and the seeming normalcy of the inequality in friends' relationships to the serious obstacles inside the marriage. Equal sharers must squarely confront all the complex feelings that arise when trying to change the "normal" course of family life: anger, resentment, frustration, and guilt.

The stories of Mary and Paul, Jonathan and Ruth, Donna and Kevin, Jake and Susan, and Rita and Charles show that equal sharers are not an elite group of gender radicals. They are ordinary couples grappling with how to manage two jobs and kids. Remember that over half did not start out sharing equally. Couples come to equality along diverse paths. A majority of the equal sharers in my study had comparable careers; some were self-consciously feminist; and many lived in social circles that touted gender equality. But although comparable careers, egalitarian ideology, and liberal friends facilitate equal sharing, they don't ensure it. And the couples we've met show that these things are not prerequisites to equality.

Almost all previous studies that examine gender inequality in the division of labor at home search for and examine its "causes." Inequality has been blamed on pay inequities between the sexes, men's longer hours in the workforce, old-fashioned ideas about men's and women's roles, maternal instinct, traditional social circles, and deep-seated male and female identities developed through childhood socialization. If only these obstacles could be overcome, then equality would prevail. So goes the implicit reasoning in much of current research.

This reasoning is fundamentally flawed, however, because it ignores the role that couples themselves play. The so-called causes of inequality are as likely to be consequences of decisions as they are to be forces driving inequality. For example, ... researchers and couples themselves often invoke inequality in paid work to explain inequality at home. Fathers work more hours and earn higher incomes, they argue, so it just makes sense that fathers would do less than mothers at home. But recall that both equal and unequal couples had often begun with very equal work schedules. Equal mothers were able to pursue jobs comparable to their husbands' by resisting the gendered decisions about careers often made by their unequal counterparts. Couples who created equality at home made choices and changed in other ways that supported equality. They found new friends, rejected the belief in maternal instinct, and developed nontraditional identities, resisting the pressures of conventional motherhood and conventional male careers.

FROM MALE CAREERS TO FAMILY CAREERS

Careers are designed for men. Privileged men in traditional relationships with wives backstage have lives that best promote conventional careers, which, when closely examined, are really two-person careers (Papanek, 1973). In my first teaching job, over twenty years ago, I witnessed firsthand the advantages of the backstage support entailed in a two-person career. The wives of my male colleagues brought hot lunches to the office for them, entertained for them, kept the children out of their way, and sometimes even worked for them as unpaid research assistants. When our department meetings lasted until six o'clock, I knew those men would soon be home eating a nicely prepared dinner, while the other lone single female colleague and I would be rummaging around in the supermarket trying to find something quick to fix before we settled in to prepare the next day's lecture. It didn't seem quite fair.

Conventional careers demand the willingness to put in long working hours, to relocate for good job opportunities, to shield work from personal responsibilities, and to give work priority over family. Career building at its most intense occurs during childbearing years. Is it any wonder that these requirements mean that careers are gendered? (Fowlkes, 1987). Men in traditional relationships are best situated to meet these requirements; mothers in traditional relationships are least able to meet them.

Dual-earner spouses without children, although disadvantaged as compared to men with wives at home, can still hope to compete, especially given that fewer and fewer men have the luxury of a wife backing them up at home. But when children appear, the picture changes entirely. As we have seen, in many unequal families, women drop out. When parents accept the conventional definition of career but reject the model of a "two-father" family, they have few options but to create an asymmetrical division of labor. Given the gender pressures, mothers give up on careers. No matter what their title and how educated they are, their paid jobs become jobs rather than careers.

In equally sharing families something very different happens on the way to equality. Parents degender careers by contesting the conventional career and turning to alternatives that fit primary parenting for two. What does it mean to degender career? It means limiting work hours, passing up opportunities, altering the career clock, and allowing family obligations to intrude on work, while still maintaining a commitment to work as a significant part of identity. When both husbands and wives made symmetrical adjustments, they avoided the spiral in which only the father ends up with a real career, but those adjustments meant that neither had a conventional male career. Equal sharers created family careers, one couple at a time with choices, compromises, and costs, some aided by the luxury of high-powered, well-paid work with plenty of room to cut back. Maybe, just maybe, by recreating themselves as professionals, the equal sharers can teach us a different model of family and professional life. Rather than pulling us back and forth, family and profession could work together like the two oars of a rowboat, each necessary to row us in the right direction.

SELECTED REFERENCES

Barnett, R. C., & Rivers, C. (1996). *She works he works: How two-income families are happier, healthier, and better-off.* San Francisco: Harper.

Deutsch, F. M., Servis, L. J., & Payne, J. D. (1998). Paternal participation in childcare and its effects on children's self-esteem. Manuscript submitted for publication.

Fowlkes, M. R. (1987). The myth of merit and male professional careers: The roles of wives. In N. Gerstel & H. E. Gross (eds.), *Families and work* (pp. 347–360). Philadelphia: Temple University Press.

Marx, K. (1950). *Communist Manifesto.* Chicago: Regency.

Okin, S. M. (1989) *Justice, gender, and the family.* New York: Basic Books.

Papanek, H. (1973). Men, women, and work: Reflections on the two-person career. In J. Huber (ed.), *Changing women in a changing society* (pp. 90–110). Chicago: University of Chicago Press.

Risman, B. J., & Myers, K. (1997). As the twig is bent: Children reared in feminist households. *Qualitative Sociology,* 20, 229–252.

Schwartz, P. (1994). *Love between equals: How peer marriage really works.* New York: Free Press.

Suitor, J. J. (1991). Marital quality and satisfaction with the division of household labor across the family life cycle. *Journal of Marriage and the Family,* 53, 221–230.

28

The Political Economy of Constructing Family

CHRISTOPHER CARRINGTON,

2000

. . . Constructing a fulfilling and durable family takes resources. Approximately one in three of the households studied here have achieved something approaching their ideal. Among more educated and more affluent subjects I heard many speak glowingly of their lesbigay families, their chosen families (Weston 1991). Those families who feel content with their chosen families also tend to fall toward the higher end of the socioeconomic ladder. The greater availability of resources, including time, makes the investment in domesticity—the building blocks of family—much more possible among the affluent, and consequently expands the opportunities to do family with satisfaction. These are the lesbigay families that marketers portray to the lesbigay communities themselves and to the broader public. Clearly, domesticity becomes more comprehensive and ample among those with greater socioeconomic resources. Such families maintain more relationships, engage in more extensive social interaction and community participation, hold jobs offering more flexible time, possess more money to spend and, consequently, can invest more effort in the construction and maintenance of family. Those lesbigay families who possess these things lead richer, more fulfilling family lives. I have taken quite a bit a flak for this assertion, but I refuse to romanticize the struggles of lesbigay people to build and sustain family, and the reality is that they often fail, not because of some inherent defect but due to social and economic realities that pare down their capacity to do family. Too much recent analysis of disempowered groups, including lesbian, gay, bisexual, and transgendered people, emphasizes the successful efforts by these groups to sustain family, religious identities, and cultural traditions in the face of oppression. Left out of such analysis is the brute fact that many such groups get trounced economically, which has consequences for their ability to build stable identities, families, and communities.

The resources available to these more prosperous families has enabled them to construct a somewhat alternative conception and practice of family, one that is organized more strongly around "choice."

DOMESTICITY AND FAMILY: WHO GETS TO DO FAMILY?

One of the critical lapses in the contemporary debate over lesbigay family life concerns the socioeconomic inequalities that prevail between different lesbigay families. I find it striking how differentiated the ability to do family is among lesbigay "families." Even the assertion that lesbigay relationships constitute family seems linked to socioeconomic resources. As resources accumulate, participants begin to think in familial terms about their lesbian- and gay-defined household lives. They have the resources required to create more elaborate domestic regimens, and more stable families. Family construction requires a combination of time, money, ideology, and relationships (friends and/or biolegal relatives). For instance, some of the more affluent lesbigay families had the money, the relationships, and even a commitment to notions of "chosen family," but insufficient time. Their work hours, commutes, and travel obligations made a more elaborate family life something they perhaps envisioned but rarely experienced. In these households there were well-furnished dining rooms, fully equipped kitchens, and inviting spaces for entertaining, but they were rarely used. They sat idle most of the time because in order to obtain them, their owners had to work for wages constantly, and when they were not working for wages they were exhausted and preferred to grab a burrito and rent a movie rather than to entertain themselves or guests with a big production. This scenario was most common among the affluent egalitarians.... As Susan Posner, a twenty-nine-year-old lesbian, put it: "We bought this dining room table, and now we are too tired to use it because we spend all of our time trying to earn the money to pay for it." In contrast to the few affluent families without time, there are a multitude of lesbigay families of more moderate means who lack time, money, kin networks, and the energy to create family.

THE MINIMAL LESBIGAY FAMILIES

The general sense of satisfaction among the more affluent households contrasts markedly with the isolation and disappointment I heard in the many other lesbigay households about family life. Most participants spoke of desiring more friends and more community, and of hopes of healing soured or rotted relationships with biolegal relatives, and even with former lesbigay family members. Many lived with an uneasy sense of how they would secure themselves financially, expressing doubt about ever owning their own homes, obtaining adequate healthcare, or having a secure retirement. These families live in walk-up or basement apartments that have no space for guests, or they live in suburban apartments where they have minimal interaction with their neighbors and only occasional interactions with other lesbigay people or with biolegal relations. Their employers do not offer domestic partnership benefits, and even when they do, those benefits are often not extensive enough to consider them "family friendly," assuming that means that such policies enable the recipients to put more emphasis on constructing and maintaining family. Their employers expect constant overtime and undivided attention to work when they are at work—a common experience among the service/working-class participants, as well as a majority of the middle-class participants in this study. When they arrive home they do not have the energy to plan birthday parties, cultivate gardens, have conversations with their partners, cook interesting food, create and maintain photo albums, visit friends, or participate in a host of other family-building activities. Many participants were too tired from exhausting workdays to spend much time cultivating chosen families, and instead collapsed in front of televisions seeking relaxation. When their relationships do falter, these families cannot afford a therapist to adjudicate domestic conflicts. Most of these lesbigay families find themselves struggling to do family against a set of social and economic conditions that impede their efforts.

Whether due to extensive and exhausting work schedules, inflexible work hours, small homes, high housing costs, soured relations with biolegal relatives, or a limited sense of obligation to significant others, most participants live within an attenuated family, or what Dizard and Gadlin (1990) call a "minimal family." Most lesbigay families experience at least some isolation from biolegal relatives and have yet to develop and fortify alternative family structures. Of course, many heterosexual families increasingly face a similar predicament. Jan Dizard and Howard Gadlin (1990), in a recent monograph concerning the status of family in our society, use the term *familism* to refer to the processes and functions of what to them constitute the essential core of family. They argue that familism consists of

> a reciprocal sense of commitment, sharing, cooperation, and intimacy that is taken as defining the bonds between family members. These bonds represent the more or less unconstrained acknowledgment of both material and emotional dependency and obligation. They put legitimate claims on one's own material and emotional resources and put forth a set of "loving obligations" that entitles members of the family to expect warmth and support from fellow family members. In addition, these bonds are assumed to be deeper and more lasting than those that exist in other, non-familial relationships. Familism embraces solicitude, unconditional love, personal loyalty and willingness to sacrifice for others. Familism makes the home a base to which you can always return when your independent endeavors fail or prove unsatisfactory. (1990, 6–7)

Dizard and Gadlin note, however, that many families increasingly have difficulty performing these functions and meeting these basic human needs. The economic demand that most families now include two full-time wage earners, along with the separation of paid work from family life, makes it difficult for many families to act in familistic ways, including many lesbigay families. Recognizing the difficulties more and more families face, Dizard and Gadlin call for the adoption of a public familism (1990, 204–19) in which the state and/or employers provide some of the basic needs previously found in families. The provision of comprehensive healthcare or publicly funded preschool would be beneficial to heterosexual families as well as to lesbigay families. It remains an open question as to whether such basic needs will be met. Even if such benefits are achieved, they will only partially assist in the creation and maintenance of family life. In order for more lesbigay families to invest more in the work that constructs and sustains them, we need to work toward political goals that reduce the risks and lower the costs of making the sacrifices that domesticity currently entails.

CONTEMPLATING THE RISKS OF DOING FAMILY WORK

Given the fact that domestic work doesn't pay much of a wage, engaging in it necessitates risks, the kind that are clearly seen in the life of Henry Zamora. These risks vary from person to person, depending on one's gender, race, educational level, and other factors. As we have seen in earlier chapters, lesbian families do less domesticity than their gay-male counterparts. Of course, notable exceptions exist (recall the affluent Kirbo-Pendleton household described in the housework chapter). In the main, lesbian families, as well as individual lesbians, earn significantly less money than do men (straight or gay). For many lesbians, achieving economic security precludes the option of making a significant investment in family work. Many of the lesbian families in this research face unique disadvantages in the struggle to create family. Due to the persistent nature of gender inequality in access to socioeconomic resources, lesbian families have fewer such resources (money, time, living space) to construct a durable family life. Paradoxically, this partially explains the greater degree of equality within the lesbian relationships studied here—there was less domesticity to do and a greater sacrifice entailed in doing it. For many of the lesbian families of moderate means, the choice to invest a great deal of energy into domesticity would mean a sacrifice of wages that they cannot afford to lose. In contrast, the longer-term affluent lesbian families studied here are thick in domesticity, and they are either specialized in their division of labor or they rely on domestic service workers to provide some domestic labor.

Similar dynamics are at work in the family lives of many people of color. A disproportionate number of African-, Asian-, and Latino-American participants dwell in working/service-class families. . . I do not think this is an artifact of sampling procedure, but rather a reflection of the greater economic constraints that operate in the lives of these participants. This lack of resources partially explains the stronger tendency of African-, Asian-, and Latino-American participants to conceive of family in biolegal terms. . . Barbara Cho is a Chinese-American woman who could not conceive of her partner as part of her family. Barbara's proximity to her biolegal Chinese-American relations (her parents lived less than a mile away), along with her economic ties to them (her parents owned the apartment building in which she lived), partially prevented Barbara from constructing a lesbigay family. Barbara's biolegal relations know that Barbara is a lesbian but they prefer to keep the matter concealed. Part of this reflects the legacy of compartmentalizing sexual matters as private in many Asian cultures (Chan 1989; Murray 1996, 264–65), but part also reflects the legacy of the economic ties that bind people (particularly women) to the patriarchal family.

Moreover, the legacy of economic marginality for people of color in the United States, combined in some cases with sociocultural traditions that conceive of the patriarchal family as the only means of economic viability, reinforces an ideology linking individuals to biolegal relations, particularly in periods of economic crisis (Amott and Matthaei 1996). Many people of color have not had the freedom to construct chosen families, although many have had to turn kith into kin, often a strategy of resisting poverty (Stack 1974). This economic marginality continues and prevents many lesbigay people of color from creating the kinds of family life (chosen and/or biolegal) that they prefer to lead. To make a greater investment in domesticity entails economic and social risks. Barbara Cho could insist on a stronger acknowledgment of her lesbian life, say, through inviting everyone to an event honoring her anniversary with her partner, Sandy. She might also get thrown out of her apartment. Moreover, the hours that Barbara Cho and Sandy Chao work for wages, along with the frequent swing shifts—Barbara works at a hotel covering night shifts three days a week and Sandy works at a daycare center, arriving at 6:30 A.M. to meet the first parents with their kids—prevent either of them from investing a great deal of energy into their domestic lives. For either or both of them to reduce work hours would entail a significant economic risk that neither is really prepared to take, not to mention that Barbara has a strong commitment to making herself economically independent of her biolegal family.

For affluent, often Euro-American, lesbigay families, economic independence and a reliance on service workers diminish the risks to members of the family; each can pursue employment with less chance of exploitation of other family members. Certainly, as indicated earlier, many of the affluent lesbigay families create a greater sense of equality between the partners through reliance on the service economy, or in other words, upon the poorly paid labors of others, notably women of color and younger, less-educated gay men and lesbians. As indicated in the housework chapter, financial resources enable many of the affluent lesbigay families to purchase several forms of domestic labor (laundry, housecleaning, gardening, routine meals) in the marketplace, and this allows such families to invest more time and energy in the more pleasant forms of family work (for example, kin work, ceremonial cooking, and the more pleasing forms of consumption work). This parallels patterns that exist among affluent heterosexual families (Ostrander 1984; Hertz 1986). These expenditures create greater equality within the family.

But is equality that is premised on the exploitation of outsiders a worthy ideal? Even if some find it so, this leaves the majority of lesbigay families who cannot afford such services in a dilemma. For them, creating family demands domesticity, while at the same time the reigning ideals demand egalitarianism. The two are not so easily merged. Conflict is inevitable given these contradictory expectations. One of the potential outcomes of this conflict is the diminution of domesticity in the lives of lesbigay families, which contributes to the emergence of minimal lesbigay families. I suspect this diminution frequently involves diminishing emotional needs as well as domestic activities. Lesbigay people are already experts at emotional asceticism, trained to conceal our most fundamental feelings, needs, and sentiments by the persistent potential of physical and psychological violence against us (Garnets, Herek, and Levy 1990). It seems quite plausible that lesbigay people feel less entitled to caregiving from other family members than do many heterosexuals. Is it really a sign of strength that lesbigay people live with reduced expectations of care from significant others?

Potentially, the combination of domestic needs and egalitarian expectations can lead to the demise of the relationship. Those who would place lesbigay families on the egalitarian pedestal need to come to terms with the fragility of lesbigay relationships (Blumstein and Schwartz 1983; Kurdek 1992). Fully aware of this fragility, many lesbigay family members who might invest more energy in domesticity probably do not, instead putting more energy into paid work and consequently reducing the quality and the long-term durability of their family relationships. . . .

Too many scholars have pinpointed gender as the crucial variable in creating egalitarianism in lesbian and gay relationships. Many have argued that because there are not both men and women in lesbigay families, the traditional assignment of domesticity on the basis of gender cannot take place, and therefore lesbigay families must negotiate domesticity, which leads to a more equitable division. This is simply not the case among longer-term families. As we have seen, lesbigay families carefully stage information about domesticity, as well as the performance of domesticity itself, in order to prevent stigma. Recall Rich Chesebro's effort to prevent people, including me, from thinking of his partner, Bill, as a "housewife, or something" by constantly emphasizing Bill's artistic career, or Dolores Bettenson's strategic emphasis on the "feminine bed linens" chosen by her partner, Arlene, an emphasis that partially attempts to conceal the reality that Arlene works as a prosecuting attorney seventy hours a week and has little time for things domestic. Recall the nearly universal assertion of participants claiming they split domestic work "fifty-fifty" when the empirical reality is that they do not. . . .

Many scholars of late have argued that lesbigay families carry the potential to undo the gendered ideology and inequality of midtwentieth-century family life because of their supposed egalitarianism (Stoddard 1989; Hunter 1991; Schwartz 1994, I; Okin 1997, 54–56). These scholars offer the lesbigay family as a model for the future. My research seriously challenges the effort to place the lesbigay family in the vanguard of social change, a model of equality for others to emulate. Such assertions are based on the ideology of egalitarianism, not on its actual existence, and on the invisibility, devaluation, and diminishment of domesticity. The minimal family and the familized corporation (Hochschild 1997) may well be the path of the future but it is not a path to equality, fulfillment, and happiness for most lesbigay people, nor for most other people. . . .

The effects of paid work upon lesbigay family life remain largely invisible and undocumented. I noted in the introduction the dearth of research on this matter. In part this invisibility occurs due to an overemphasis on individual volition when thinking about family life. We too often assume that the hours that someone works for wages, the energy they put into that work, and their subsequent capacity to make themselves available to family and domestic life is simply a question of individual values or personality. Such an emphasis fails to acknowledge the structures of constraint (Folbre 1994, 51) at work in people's lives, particularly in the lives of working/service-class families, where choices are greatly constrained by economic necessity. Neglecting the effects of paid work on family life leaves pressing political and economic questions unaddressed. Questions like, Why should most salaried workers be exempted from the forty-hour work week? Or, Why should household work go uncompensated, or when compensated, without benefits like healthcare?

In sum, lesbigay families are not as distinct from heterosexual families as many seem to believe. Taking the wider view, looking at domesticity in its myriad forms, examining it, and depicting it in its detail, as I have tried to do, creates a portrait of longer-term lesbian and gay family life that more often than not resembles patterns seen within heterosexual families. This raises a set of provocative questions about how lesbigay families and communities either resist or accommodate the broader cultural context in which they live. A debate rages about what constitutes gay and lesbian culture, and whether that culture does or should accommodate or resist the social practices and discourses of the broader society (Faderman 1989; Herdt 1991). My research and observations about domesticity suggest a significant pattern of accommodation to the predominant social structure of American society among lesbigay families. Much of that accommodation goes unrecognized, notably the seclusion and devaluation of domesticity. Some of that accommodation is denied, as in the case of the many gay men who conceal domesticity performed by themselves or other family members, or in the case of the domesticity that contributes to the creation and maintenance of social-class distinctions. Some of the accommodation is recognized by lesbigay people but undocumented and untheorized by researchers, as exemplified by the dearth of research exploring the effect of paid work upon the organization and character of lesbigay domestic life.

SELECTED REFERENCES

Amott, T., and J. Matthaei. 1996. *Race, Gender and Work: A Multi-Cultural Economic History of Women in the United States.* Boston: South End Press.

Blumstein, P., and P. Schwartz. 1983. *American Couples.* New York: Morrow.

Chan, C. 1989. Issues of sexual identity in an ethnic minority: The case of Chinese-American lesbians, gay men, and bisexual people. In A. D'Augelli and C. Patterson, eds., *Lesbian, Gay, and Bisexual Identities over the Lifespan.* New York: Oxford.

Dizard, J., and H. Gadlin. 1990. *The Minimal Family.* Amherst: University of Massachusetts Press.

Folbre, N. 1994. *Who Pays for the Kids: Gender and the Structures of Constraint.* New York: Routledge.

Garnets, L., G. Herek, and B. Levy. 1990. Violence and victimization of lesbians and gay men: Mental health consequences. *Journal of Interpersonal Violence* 5(3): 366–83.

Herdt, G. 1991. *Gay Culture in America.* Boston: Beacon Press.

Hertz, R. 1986. *More Equal than Others.* Berkeley: University of California Press.

Hochschild, A. 1997. *The Time Bind: When Work Becomes Home and Home Becomes Work.* New York: Metropolitan Books.

Hunter, N. 1991. Marriage, law and gender: A feminist inquiry. *Law and Sexuality* 1 (1): 9–30.

Kurdek, L. 1992. Relationship stability and relationship satisfaction in cohabiting gay and lesbian couples: A prospective longitudinal test of the contextual and interdependence models. *Journal of Social and Personal Relationships* 9: 125–42.

Ostrander, S. 1984. *Women of the Upper Class.* Philadelphia: Temple University Press.

Schwartz, P. 1994. *Peer Marriage: How Love Between Equals Really Works.* New York: Free Press.

Stack, C. 1974. *All Our Kin: Strategies for Survival in a Black Community.* New York: Harper and Row.

Stoddard, T. 1989. Why gay people should seek the right to marry. In S. Sherman, ed., *Lesbian and Gay Marriage.* Philadelphia: Temple University Press.

Weston, K. 1990. *Families We Choose: Lesbians, Gays, Kinship.* New York: Columbia University Press.

CHAPTER 9

Divorce

In an average year in the United States, over two million couples marry and over a million get divorced. Whether we consider these high or low numbers for a country with a population of almost 290 million people depends on how we perceive divorce and marriage and what comparisons we invoke. The **divorce rate** rose gradually during the twentieth century, but went down sharply during the Depression and then up following World War II. The rate almost doubled between 1965 and 1975 then stabilized and began dropping slightly after the early 1980s (Cherlin, 1992). In 1890 the chance that a marriage would end in divorce was only about one in ten. By 1930 the chance was about one in four, and by the 1970s, about one in two. That is the rate where we have more or less stabilized today, with predictions suggesting that between four and five out of every ten marriages will end in **divorce** (National Center for Health Statistics, 2000). With divorce rates dropping slightly and estimates of the proportion of children affected by divorce similarly declining, we might assume that people would be less likely to define divorce as a social problem than they were in the 1970s and the 1980s. In fact, the opposite seems to be occurring, with new political and religious efforts launched in the 1990s to reinstitute a fault basis into legal divorce proceedings and to require mandatory waiting periods and counseling before a divorce could be approved. The readings in this chapter introduce issues underlying debates about divorce and focus attention on variation in the meaning of divorce by historical period, gender, and **social class.**

The excerpts from essays by John Stuart Mill written in 1832 and in 1869 place debates about marriage, divorce, and women's rights in historical perspective. Many of the ideas in these essays were developed jointly by John Mill and Harriet Taylor Mill and provide a rare public endorsement of gender equality by one of England's leading intellects. The ideas that the Mills advanced about liberalizing marriage and divorce laws and giving women access to education and allowing them to vote were considered radical in their day. Harriet Taylor argued that there should be no laws regulating marriage and that a woman should take responsibility for her own children following divorce (at the time, fathers received **custody**). John Mill was more cautious than Harriet Taylor, but he also called for changes in laws that would allow women and men to be equals in marriage. Unlike Taylor, who advocated for women's right to pursue careers, John Mill continued to believe that women's true calling was to marry and share in her husband's occupations and interests. Nevertheless, he advocated delaying marriage and postponing having children so that couples could test their compatibility, an idea that was well ahead of its time, especially because birth control was still relatively ineffective (Rossi, 1970).

The following excerpts from John Stuart Mill's essays focus on marriage, divorce, and the position of women in society. According to Mill, the question of divorce ("the dissolubility of marriage") could not be determined until women gained equal status with men. The excerpt from the first essay comments on possible historical advantages to women of marriage being permanent ("indissoluble"), but notes the "absurdity and immorality" of requiring that women depend on marriage to achieve basic human rights. Here Mill suggests that the real question is "whether marriage is to be a relation between two equal beings, or between a superior and an inferior." In the second and more often cited essay, published at the urging of his daughter eleven years after the death of Harriet Taylor Mill, John Mill summarizes how the legal rules of the day treated women as little better than slaves. Mill suggests, often implicitly, that many marriages were unsuccessful and that marriage laws forced some women into "being the personal body-servant of a despot." In summary, he argues that divorce should be an acceptable way for wives to end oppressive marriages.

The John Stuart Mill selections show that **divorce** has been used in public discourse to symbolize other things. Religious and political leaders have exploited fears about divorce at least since the founding of our country, with orators and pamphleteers equating marriage with social order and portraying divorce as a revolutionary threat to society and a recipe for the destruction of civilization (Basch, 1999). Victorian American moralists championed the self-sacrificing communitarianism of marriage against the supposed selfish individualism of divorce. We hear similar arguments today, with attention paid to the decline of community and the rise of individualism. Then as now, such criticisms suggest that Americans (especially women) have become too self-absorbed and that they are no longer willing or able to make the personal sacrifices that are supposedly necessary to sustain families and build strong communities. The high divorce rate is typically cited as evidence of rising individualism and a decline in "family values" (Coltrane and Adams, 2003).

How should we evaluate claims about whether individualism is promoting divorce and undermining society or whether it is, instead, maintaining the institution of marriage by providing a safety valve for people who end up in bad marriages? Marriage is definitely less obligatory than it once was, and both women and men are now able to end bad marriages relatively easily compared to past eras. How are such choices affecting society? Opinion polls show that almost all Americans continue to believe in marriage and nine out of ten do marry, but a large majority also thinks that married couples should divorce if they don't get along (General Social Survey, 1994). These conflicting views influence marital relationships in ways that social scientists are just beginning to understand. In "Wives' Marital Work in a Culture of Divorce," Karla Hackstaff draws on interviews with married couples to explore how the possibility of divorce might influence marriage. We live in a "divorce culture," according to Hackstaff, reflecting three basic assumptions: (1) marriage is optional—we assume that every adult is free to marry or not marry; (2) marriage is contingent—we do not expect that all marriages will last a lifetime; and (3) divorce is a gateway—we now accept that getting divorced can lead to other opportunities in life for both men and women. "Divorce culture" can be contrasted with "marriage culture," which assumes that marriage is obligatory, lifelong, and not subject to dissolution for personal reasons.

Marriage culture and divorce culture coexist uneasily in our society, with one or the other sometimes dominant in a particular era, region, life stage, family, or individual. Hackstaff uses case studies to explore how the simple presence of divorce culture transforms the meaning and practice of marriage. She focuses on an emerging "marital work ethic," which refers to the things that couples do "for the marriage" to counter the possibility of divorce. She explores how couples sometimes cooperate in the work of maintaining the marriage, sometimes engage in conflict over it, and sometimes fail to share it, thus possibly precipitating divorce. Hackstaff explains that the concept of individualism is theoretically and practically different for women than for men, and focuses our attention on how gender is implicated in the maintenance of marriage as well as in possibilities for divorce.

Most people who divorce do so only for strong and pressing reasons. In the selection "How Marriages End" from the book *For Richer, For Poorer: Mothers Confront Divorce,* Demie Kurz discusses the principal reasons the women in her study gave for their divorces. She discusses four different groups: (1) those who reported separating from their husbands because of personal dissatisfaction or mutual consent; (2) those who reported leaving their marriages because of violence by their husbands; (3) those who reported ending their marriages because of "hard-living" (including husband's difficulty holding a job, his drinking or drug problems, or his absences from home); and (4) those who reported that their marriages ended because their husbands were involved with other women. These women's accounts do not support the idea that wives are putting their own interests ahead of other family members and leaving their families because of weak commitments to the institution of marriage. In fact, most of the women felt forced to live apart from their husbands to protect their children or to maintain the possibility of having a positive relationship with a man. Many described bleak relationships in which husbands had substance abuse problems, used threats and intimidation, were excessively controlling, and all too often resorted to violence. Others described being virtually abandoned or having noncommunicative husbands who were not willing to work on the marriage or contribute to the household.

Research shows that men provide somewhat different accounts based on their divorce experiences, typically reflecting a greater sense of **entitlement** and less awareness of problems in the marriage (Arendell, 1995). Although breaking up a marriage is difficult for everyone, women are more likely to be the initiators of divorce. This reflects the divergent "his" and "hers" version of marriage first described by Jessie Bernard (1972). Because wives have traditionally been held responsible for the well-being of family members and are more likely to define themselves in relation to others, they are more likely to sense and respond to deteriorating relationships. Women tend to be more attuned to the stresses and strains associated with emotion work, kin work, and care work (Hochschild, 1989). As Hackstaff notes, when a relationship has low levels of **reciprocity,** women tend to notice it and initiate private talks or therapy to improve the situation. Men, in contrast, are more likely to be satisfied with communication in the marriage and often refuse, resist, or withdraw when wives want to talk about problems in the marriage. If such talks fail to improve the relationship, women then become more likely than men to begin thinking about breaking up.

Although women, on average, monitor problems and begin thinking about a breakup before men, these are only general tendencies and men sometimes play this role. Both men and women become unhappy with relationships and anyone can initiate the actual breakup. In both heterosexual and homosexual relationships, the person who initiates the breakup is better prepared to handle the separation. The initiator typically has more alternatives and begins the process sooner, so can often more easily forge new identities not dependent on the former partner (Vaughn, 1986). Whoever initiates the breakup, divorce can provide opportunities for learning about oneself and making decisions about the future, but the process of redefining oneself and forging new relationships can be very disheartening, especially for the person who did not initiate the breakup.

Additional concerns enter into divorce when children are involved, and this includes about half of all divorces. Because most divorcing men do not seek (or are not awarded) **child custody** following divorce, the number of divorced men who are uninvolved fathers has risen (Eggebeen, 2002). At the same time, recent research shows that the actual involvement of fathers with children after divorce, and their payment of child support varies enormously, sometimes without regard to official post-divorce court orders (Seltzer, 1998). The number of men with **joint physical custody** has grown, though **joint legal custody** is still a more common post-divorce parenting arrangement. There is not enough space to evaluate scholarly evidence about the impact of divorce on children, in part because research findings are incomplete and sometimes contradictory. Children often initially blame themselves for the divorce, and their reactions to the breakup depend on such factors as the amount of conflict before and after the divorce and the quality of parent–child relationships over time. Amicable child custody agreements and practices and realistic **child support** orders and payments are also important to children's and adults' post-divorce adjustment.

Over three-fourths of children from divorced parents are virtually indistinguishable from children with parents who did not divorce (Hetherington and Kelly, 2002). At the same time, children of divorce are, on average, more likely to drop out of school, marry early, and become single parents themselves (Amato, 2000). These tendencies are also related to poverty, and about half of the difference between children with divorced or nondivorced parents can be traced to the differing economic conditions associated with living in a single-parent household. Although estimates of the significance of divorce on the economic circumstances of divorced men and women vary, research shows that women's average income declines more than men's following the breakup and recovers more slowly, principally because they are paid less than men and they are more likely to be awarded physical custody of any children. In what ways do you think that the possibility of divorce influences relationships with children as well as spouses when couples are still together? Do you think social policy can change the prospects of divorce? How?

REFERENCES

Amato, Paul. 2000. Diversity within Single-Parent Families. In *Handbook of Family Diversity,* edited by D. H. Demo, K. R. Allen, and M. A. Fine. New York: Oxford University Press.

Arendell, Terry. 1995. *Fathers and Divorce.* Berkeley: University of California Press.

Basch, Norma. 1999. *Framing American Divorce: From the Revolutionary Generation to the Victorians.* Berkeley, CA: University of California Press.

Bernard, Jessie. 1972. *The Future of Marriage.* New York: World.

Cherlin, Andrew. 1992. *Marriage, Divorce, Remarriage.* Cambridge, MA: Harvard University Press.

Coltrane, Scott, and Michele Adams. 2003. The Social Construction of the Divorce "Problem": Morality, Child Victims, and the Politics of Gender. *Family Relations.*

Eggebeen, D. 2002. The Changing Course of Fatherhood. *Journal of Family Issues* 23:486–506.

General Social Survey (GSS). 1994. University of Michigan, Inter-University Consortium for Political and Social Research. Available: http//:www.icpsr.umich.edu/

Hetherington, E. Mavis, and John Kelly. 2002. *For Better or For Worse: Divorce Reconsidered.* New York: W.W. Norton & Co.

Hochschild, Arlie, with Anne Manning. 1989. *The Second Shift.* New York: Viking.

National Center for Health Statistics. 2000. Available: http://www.cdc.gov/nchs/.

Rossi, Alice S., ed. 1970. *Essays on Sex Equality.* Chicago: University of Chicago Press.

Seltzer, J. A. 1998. Father by Law: Effects of Joint Legal Custody on Nonresident Fathers' Involvement with Children. *Demography* 35:135–46.

Vaughn, Diane. 1986. *Uncoupling: Turning Points in Intimate Relationships.* New York: Oxford University Press.

SUGGESTED READINGS

Ahrons, Constance R. 1994. *The Good Divorce: Keeping Your Family Together When Your Marriage Comes Apart.* New York: HarperCollins.

Arendell, Terry. 1995. *Fathers and Divorce.* Thousand Oaks, CA: Sage Publications.

Arendell, Terry. 1986. *Mothers and Divorce: Legal, Economic, and Social Dilemmas.* Berkeley: University of California Press.

Basch, Norma. *Framing American Divorce: From the Revolutionary Generation to the Victorians.* Berkeley: University of California Press.

Fineman, Martha. 1991. *The Illusion of Equality: The Rhetoric and Reality of Divorce Reform.* Chicago: University of Chicago Press.

Hartog, Hendrik. 2000. *Man and Wife in America: A History.* Cambridge, MA: Harvard University Press.

Hetherington, E. Mavis, and John Kelly. 2002. *For Better or For Worse: Divorce Reconsidered.* New York: W. W. Norton.

Riley, G. 1991. *Divorce: An American Tradition.* Lincoln: University of Nebraska Press.

Smart, Carol, and Bren Neale. 1999. *Family Fragments?* Cambridge, UK: Polity Press.

Thompson, Ross A., and Paul R. Amato, eds. 1999. *The Postdivorce Family: Children, Parenting, and Society.* Thousand Oaks, CA: Sage Publications.

Vaughan, Diane. 1986. *Uncoupling: Turning Points in Intimate Relationships.* New York: Oxford University Press.

INFOTRAC® COLLEGE EDITION EXERCISES

The exercises that follow allow you to use the InfoTrac® College Edition online database of scholarly articles to explore the sociological implications of the selections in this chapter.

Search Keyword: Divorce rates. Use InfoTrac® College Edition to examine trends in divorce rates over time. What is meant by the divorce rate? What is the present divorce rate in the United States? How has that rate changed over time? Is that trend similar to, or different from, trends in other parts of the world? What are some of the reasons given for the changes in the divorce rate in the United States? What are the implications of the present trend in divorce, according to the authors of the articles that you find? Do they feel that the divorce rate needs to be reduced? If so, what "solutions" do they offer to lower the present divorce rate?

Search Keyword: Divorce ceremony. What is a divorce ceremony? See if you can find any articles that discuss this modern phenomenon. How does a divorce ceremony differ from a wedding or a marriage ceremony? What does a divorce ceremony symbolize for the couple involved? What are the advantages and/or disadvantages to society of a couple's enacting a divorce ceremony at the end of a marriage?

Search Keyword: No-fault divorce. Using InfoTrac® College Edition, search for articles dealing with no-fault divorce. What is no-fault divorce? How does no-fault divorce compare to the "fault-based" divorce procedure? How did "no-fault" come about in the United States? What do the authors of the articles you find see as the benefits and/or disadvantages of no-fault divorce? What proposals are afoot to roll back no-fault divorce and reintroduce fault into divorce proceedings? In the articles that you've found, what are the prevailing attitudes toward changing no-fault divorce statutes? What would be the implications of such a change on relationships within the family?

Search Keyword: Feminization of poverty. Search for articles having to do with the feminization of poverty. What is meant by the term *feminization of poverty*? How has the phenomenon evolved over time? Why is poverty becoming more specific to women? What is the relationship of divorce to the occurrence of the feminization of poverty? How does divorce policy, including spousal support, child support, and property distribution, contribute to the greater impoverishment of women?

Search Keyword: Divorce and children. How many articles can you find that deal with children and divorce? According to these articles, about what percentage of the divorces that occur each year involve minor children? What do they have to say about the impacts of divorce on children? Describe some of the impacts; generally, how severe or long lasting are these effects? Are there ever times when children are better off having their parents divorce? If so, when is that? Under what conditions do children suffer least when their parents divorce; i.e., how can the effects of divorce on children be minimized by parents and by society? What policy proposals are currently being suggested to address issues related to the effects of divorce on children? How successful are such proposals likely to be?

29

Early Essays on Marriage and Divorce

JOHN STUART MILL,
1832 and 1869

There can . . . be no doubt that for a long time the indissolubility of marriage acted powerfully to elevate the social position of women. The state of things to which in almost all countries it succeeded, was one in which the power of repudiation existed on one side but not on both: in which the stronger might cast away the weaker, but the weaker could not fly from the stronger. To a woman of impassioned character, the difference between this and what now exists, is not worth much; for she would wish to be repudiated, rather than to remain united only because she could not be got rid of. But the aspirations of most women are less high. They would wish to retain any bond of union they have ever had with a man to whom they do not prefer any other, and for whom they have that inferior kind of affection which habits of intimacy frequently produce. Now, assuming what may be assumed of the greater number of men, that they are attracted to women solely by sensuality, or at best by transitory *taste;* it is not deniable, that the irrevocable vow gave to women, when the passing gust had blown over, a permanent hold upon the men who would otherwise have cast them off. Something, indeed *much,* of a community of interest, arose from the mere fact of being indissolubly united: the husband took an interest in the wife as being *his* wife, if he did not from any better feeling: it became essential to his respectability that his wife also should be respected; and commonly when the first revulsion of feeling produced by satiety, went off, the mere fact of continuing together if the woman had anything lovable in her and the man not wholly brutish, could hardly fail to raise up some feeling of regard and attachment. She obtained also, what is often far more precious to her, the certainty of not being separated from the children.

Now if this be all that human life *has* for women, it is little enough: and any woman who feels herself capable of great happiness, and whose aspirations have not been artificially checked, will claim to be set free from *only* this, to seek for more. But women in general, as I have already remarked, are more easily contented, and this I believe to be the cause of the general aversion of women to the idea of facilitating divorce. They have a habitual belief that their power over men is chiefly derived from men's sensuality; and that the same sensuality would go elsewhere in search of gratification, unless restrained by law and opinion. They on their part, mostly seek in marriage, a home, and the state or condition of a married woman, with the addition or not as it may happen, of a splendid establishment etc. etc. These things once obtained, the indissolubility of marriage renders them sure of keeping. And most women, either because these things give them all the happiness they are capable of, or from the artificial barriers which curb all

From John Stuart Mill. Excerpts from "Early Essays on Marriage and Divorce" (1832) and "The Subjection of Women" (1869).

spontaneous movements to seek their greatest felicity, are generally more anxious not to peril the good they have than to go in search of a greater. If marriage were dissoluble, they think they could not retain the position once acquired; or not without practicing upon the attention of men by those arts, disgusting in the extreme to any woman of simplicity, by which a cunning mistress sometimes established and retains her ascendancy.

These considerations are nothing to an impassioned character; but there is something in them, for the characters from which they emanate—is not that so? The only conclusion, however, which can be drawn from them, is one for which there would exist ample grounds even if the law of marriage as it now exists were perfection. This conclusion is, the absurdity and immorality of a state of society and opinion in which a woman is at all dependent for her social position upon the fact of her being or not being married. Surely it is wrong, wrong in every way, and on every view of morality, even the vulgar view—that there should exist any motives to marriage except the happiness which two persons who love one another feel in associating their existence.

The means by which the condition of married women is rendered artificially desirable, are not any superiority of legal rights, for in that respect single women, especially if possessed of property, have the advantage: the civil disabilities are greatest in the case of the married woman. It is not law, but education and custom which make the difference. Woman are so brought up, as not to be able to subsist in the mere physical sense, without a man to keep them: they are so brought up as not to be able to protect themselves against injury or insult, without some man on whom they have a special claim, to protect them: they are so brought up, as to have no vocation or useful office to fulfill in the world, remaining single; for all women who are educated to *be* married, and what little they are taught deserving the name useful, is chiefly what in the ordinary course of things will not come into actual use, unless nor until they are married. A single woman therefore is felt both by herself and others as a kind of excrescence on the surface of society, having no use or function or office there. She is not indeed precluded from useful and honorable exertion of various kinds: but a married woman is *presumed* to be a useful member of society unless there is evidence to the contrary; a single woman must establish what very few either women or men ever do establish, an *individual* claim.

All this, though not the less really absurd and immoral even under the law of marriage which now exists, evidently grows out of that law, and fits into the general state of society of which that law forms a part, nor could continue to exist if the law were changed, and marriage were not a contract at all, or were an easily dissoluble one: The indissolubility of marriage is the keystone of woman's present lot, and the whole comes down and must be reconstructed if that is removed.

And the truth is, that this question of marriage cannot properly be considered by itself alone. The question is not what marriage ought to be, but a far wider question, what woman ought to be. Settle that first, and the other will settle itself. Determine whether marriage is to be a relation between two equal beings, or between a superior and an inferior, between a protector and a dependent; and all other doubts will easily be resolved. . . .

. . . By the old laws of England, the husband was called the *lord* of the wife; he was literally regarded as her sovereign, inasmuch that the murder of a man by his wife was called treason (*petty* as distinguished from *high* treason), and was more cruelly avenged than was usually the case with high treason, for the penalty was burning to death. Because the various enormities have fallen into disuse (for most of them were never formally abolished, or not until they had long ceased to be practised) men suppose that all is now as it should be in regard to the marriage contract; and we are continually told that civilization and Christianity have restored to the woman her just rights. Meanwhile the wife is the actual bond-servant of her husband: no less so, as far as legal obligation goes, than slaves commonly so called. She vows a lifelong obedience to him at the altar, and is held to it all through her life by law. . . . She can do no act whatever but by his permission, at least tacit. She can acquire no property but for him; the instant it becomes hers, even if by inheritance, it becomes *ipso facto* his. In this respect the wife's position under the common law of England is worse than that of slaves in the laws of many countries. . . .

I am far from pretending that wives are in general no better treated than slaves; but no slave is a slave to the same lengths, and in so full a sense of the word, as a wife is. Hardly any slave, except one immediately attached to the master's person, is a slave at all hours and all minutes; in general he has, like a soldier, his fixed task, and when it is done, or when he is off duty, he disposes, within certain limits, of his own time, and has a family life into which the master rarely intrudes. "Uncle Tom" under his first master had his own life in his "cabin," almost as much as any man whose work takes him away from home, is able to have in his own family. But it cannot be so with the wife. Above all, a female slave has (in Christian countries) an admitted right, and is considered under a moral obligation, to refuse to her master the last familiarity. Not so the wife: however brutal a tyrant she may unfortunately be chained to—though she may know that he hates her, though it may be his daily pleasure to torture her, and though she may feel it impossible not to loathe him—he can claim from her and enforce the lowest degradation of a human being, that of being made the instrument of an animal function contrary to her inclinations. While she is held in this worst description of slavery as to her own person, what is her position in regard to the children in whom she and her master have a joint interest? They are by law *his* children. He alone has any legal rights over them. Not one act can she do towards or in relation to them, except by delegation from him. Even after he is dead she is not their legal guardian, unless he by will has made her so. He could even send them away from her, and deprive her of the means of seeing or corresponding with them, until this power was in some degree restricted by Serjeant Talfourd's Act. This is her legal state. And from this state she has no means of withdrawing herself. If she leaves her husband, she can take nothing with her, neither her children nor anything which is rightfully her own. If he chooses, he can compel her to return, by law, or by physical force; or he may content himself with seizing for his own use anything which she may earn, or which may be given to her by her relations. It is only legal separation by a decree of a court of justice, which entitles her to live apart, without being forced back into the custody of an exasperated jailer—or which empowers her to apply any earnings to her own use, without fear that a man whom perhaps she has not seen for twenty years will pounce upon her some day and carry all off. This legal separation, until lately, the courts of justice would only give at an expense which made it inaccessible to any one out of the higher ranks. Even now it is only given in cases of desertion, or of the extreme of cruelty; and yet complaints are made every day that it is granted too easily. Surely, if a woman is denied any lot in life but that of being the personal body-servant of a despot, and is dependent for everything upon the chance of finding one who may be disposed to make a favourite of her instead of merely a drudge, it is a very cruel aggravation of her fate that she should be allowed to try this chance only once. The natural sequel and corollary from this state of things would be, that since her all in life depends upon obtaining a good master, she should be allowed to change again and again until she finds one. I am not saying that she ought to be allowed this privilege. That is a totally different consideration. The question of divorce, in the sense involving liberty of remarriage, is one into which it is foreign to my purpose to enter. All I now say is, that to those to whom nothing but servitude is allowed, the free choice of servitude is the only, though a most insufficient, alleviation. Its refusal completes the assimilation of the wife to the slave—and the slave under not the mildest form of slavery: for in some slave codes the slave could, under certain circumstances of ill usage, legally compel the master to sell him. But no amount of ill usage, without adultery superadded, will in England free a wife from her tormentor. . . .

Whether the institution to be defended is slavery, political absolution, or the absolutism of the head of a family, we are always expected to judge of it from its best instances; and we are presented with pictures of loving exercise of authority on one side, loving submission to it on the other—superior wisdom ordering all things for the greatest good of the dependents, and surrounded by their smiles and benedictions. All this would be very much to the purpose if any one pretended that there are no such things as good men. Who doubts that there may be great goodness, and great happiness, and great affection, under the absolute government of a good man? Meanwhile, laws and institutions require to be adapted, not to good men, but to bad. Marriage is not an institution designed for a select few. Men are not required, as a preliminary to the marriage ceremony, to prove by testimonials that they are fit to be trusted with the exercise of absolute power. The tie of affection and obligation to a wife and children is very strong with those whose general social feelings are strong, and with many who are little sensible to any other social ties; but there are all degrees of sensibility and insensibility to it, as there are all grades of goodness and wickedness in men, down to those whom no ties will bind, and on whom society has no action but through its *ultima ratio,* the penalties of the law. In every grade of this descending scale are men to whom are committed all the legal powers of a husband. The vilest malefactor has some wretched woman tied to him, against whom he can commit any atrocity except killing her, and, if tolerably cautious, can do that without much danger of the legal penalty. And how many thousands are there among the lowest classes in every country, who, without being in a legal sense malefactors in any other respect, because in every other quarter their aggressions meet with resistance, indulge the utmost habitual excesses of bodily violence towards the unhappy wife, who alone, at least of grown persons, can neither repel nor escape from their brutality; and towards whom the excess of dependence inspires their mean and savage natures, not with a generous forbearance, and a point of honour to behave well to one whose lot in life is trusted entirely to their kindness, but on the contrary with a notion that the law has delivered her to them as their thing, to be used at their pleasure, and that they are not expected to practise the consideration towards her which is required from them towards everybody else. The law, which till lately left even these atrocious extremes of domestic oppression practically unpunished, has within these few years made some feeble attempts to repress them. But its attempts have done little, and cannot be expected to do much, because it is contrary to reason and experience to suppose that there can be any real check to brutality, consistent with leaving the victim still in the power of the executioner. Until a conviction for personal violence, or at all events a repetition of it after a first conviction, entitles the woman *ipso facto* to a divorce, or at least to a judicial separation, the attempt to repress these "aggravated assaults" by legal penalties will break down for want of a prosecutor, or for want of a witness.

30

Wives' Marital Work in a Culture of Divorce

KARLA HACKSTAFF,

1999

Ever since Jessie Bernard (1972) discovered that there is a "his" and "her" to every marriage—and that "his" marriage is generally better, researchers have examined inequalities between women's and men's experiences in families. Family scholars have analyzed this inequality, in part, by examining the work that wives and mothers do, beyond paid labor, to sustain marriages and families: kin work, emotion work, reproductive labor, the "second shift," and marriage work.

. . . Most researchers have found that women have been responsible for monitoring the emotional quality of marriage—a responsibility deriving from the legacy of "separate spheres" (. . . Thompson and Walker 1989). Blaisure and Allen (1995) found that beyond sharing housework and paid work, reflective assessment and emotional involvement are crucial to egalitarian marriages. No one, however, has examined how the context of divorce might be influencing women's work as wives. . . .

Divorce rates have been relatively stable from the late 1980s into the 1990s; half of all marriages contracted during the 1970s and thereafter are projected to end in divorce.

. . . How has the prevalence of divorce influenced the labor of sustaining marriages? Are women still monitoring the emotional quality of marriages, sharing the work, or falling down on the job? Some scholars imply that wives are becoming less self-sacrificing and increasingly individualistic and that they are abandoning the work of monitoring marriages (. . . Glenn 1987; Popenoe 1988 . . .). But one could argue alternatively that a context of prevalent divorce compels more work to sustain marriages. Indeed, monitoring a marriage may be more demanding than ever because we live in a culture of divorce (Hackstaff, 1999).

Divorce culture is a cluster of symbols, beliefs, and practices that anticipate and reinforce divorce and, in the process, redefine marriage. Divorce culture is marked by three key premises: marrying is an option, marriage is contingent, and divorce is a gateway. To believe that marriage is optional is to envision adulthood separate from marriage. To believe that marriage is contingent is to dilute the lifetime promise of "till death do us part." To say that divorce is a gateway is to suggest that divorce is not simply an option to be shunned but may be a means to a more fulfilling life. As an incipient cultural construction, divorce culture challenges an older ideology of marriage culture. However, the tenets of marriage culture—that marrying is given, marriage is forever, and divorce is a last resort—endure, even as its hegemony has lapsed.

Today, spouses feel vulnerable to divorce. In the early 1990s I interviewed 17 married couples to find out how they were coping with the rise of divorce and gender equality. I found that couples

From Karla Hackstaff, "Wives' Marital Work in a Culture of Divorce," from *Families in the US: Kinship and Domestic Politics,* Ed. Karen V. Hansen and Anita Ilta Garey, pp. 459–473. Reprinted by permission of Temple University Press.

devise various strategies to cope with divorce and that the "marital work ethic" is the most important of these. Nearly everyone I interviewed referred to the idea of a marital work ethic: the widespread *belief* that marital work is crucial to cope with the option of divorce. "Marital work" is the ongoing work necessary to create, sustain, and reproduce an emotionally gratifying relationship. The marital work ethic expresses divorce culture by assuming that marital endurance is contingent upon a gratifying marriage. Indeed, in their large surveys, Kitson and Holmes found that "relational complaints," such as the absence of communication or affection, are increasingly seen, by divorced and married alike, as legitimate grounds for marital dissolution (1992, 341). Despite spouses' shared beliefs in a marital work ethic, more often than not, the work itself is divided by gender. Like the literature documenting women's disproportionate responsibility for family work, my research finds that women are more likely to do or to initiate the work of taking care of the marriage.

Because wives have been responsible for monitoring marriages, they also take responsibility for redistributing marital work. I found that wives try to enlist husbands in the work of monitoring marriage, not to abandon the work, but to share it. However, sometimes the means are paradoxical. For many wives, achieving equality in the relational work of monitoring a marriage requires exercising the power of independence. Divorce threats allow wives to set limits. For example, after one wife threatened her husband with "recovery or the marriage," her drug-dependent husband complied. Her reservations about her ultimatum concerned relationships: "It was very hard; you've got two children who love their father—it's not just me." However, her concern for individual integrity and independence were apparent when she explained that she wanted her daughters to "know that their needs are important and should be honored"—that one needs to learn "to compromise without giving yourself away."

The wives I interviewed are not simply concerned with either their own individual rights or their relational responsibilities but with both. Increasing financial independence, made possible by women's increasing labor force participation, enables wives to take their own needs and rights into account and, in effect, rely upon individualism. Yet, it is "relationality" rather than individualism that motivates wives' desire to share the work of monitoring marriage. In contrast to individualism—a stance that puts the self first—relationality "is a stance which emphasizes expressivity and takes others into account not as 'other' but as important in themselves" (Johnson 1988, 68–69). While relationality informs many a wife's ideal of marriage, individualism can be crucial for wives to produce relational ends. Concerns about divorce that accuse wives of individualism or of a decreasing willingness to sacrifice for families overlook the complexity of wives' emotional desires and interactional limits. Wives may want an equality of sacrifice based on relational connection, yet relationality cannot be forced. Thus, wives may find it easier to secure an individualistic equality in which neither spouse sacrifices.

In this chapter, I document wives' and husbands' belief in a marital work ethic and suggest that this belief has been intensified by widespread divorce. I illustrate that wives are more likely than their husbands to do marital work, and how they attempt to recruit their husbands in the work of monitoring marriages. The dynamics and results of wives' efforts are represented in three case studies.

RESEARCH DESIGN AND METHOD

. . . I designed my research to interview heterosexual couples marrying after 1970. . . .

In order to hold age relatively constant, I primarily selected individuals from a twentieth-year high school reunion list. One spouse from each couple was born in 1953, so the average age of respondents was 39. This is basically a quota sample. I aimed to vary race-ethnicity and religion among respondents in order to capture diverse meanings regarding marriage, divorce, and gender. I did not intentionally pursue quotas on religious affiliation, but I obtained variation. The couples were mostly dual career couples, and all were middle-class, married parents from the San Francisco Bay area.

I interviewed 17 couples (34 individuals) in total in the early 1990s. The average year of marriage was 1979. At the time of the interviews, the mean number of years married was 11; six couples had been married less than 10 years and 11 couples had been married 10 or more years. Of the 34 spouses, 25 were in their first marriage, and the remaining nine were in their second marriage. All couples had children and averaged two children each.

I designed my study to conduct three indepth interviews with every couple. The wife and husband were first interviewed separately; the individual interviews averaged three hours each. Then a joint interview was scheduled; the joint interviews averaged two and a half hours. The interview tapes were transcribed and the text was coded and analyzed according to grounded theory procedures (Corbin and Strauss 1990). My research aim was to discover how spouses' talked about marriage, divorce, and gender, with the rise of both divorce and gender equality.

THE MARITAL WORK ETHIC

As Craig Morris pondered changes in marriage and divorce since his parents' marriage in the late 1940s, he articulated a relationship between divorce and work in marriage.

> I guess there were a lot of people who were stuck in marriages, and very unhappy at that time, and didn't feel like they had a way out. And yet, on the other hand, um, you look at today and you say—I think maybe a lot of people are leaving too soon, before they really look for ways to . . . save their marriage. But then again people are working harder today to save their marriages than ever before. Um, because you have to work hard to save it 'cause it's so easy to let it slip away, compared to 30 years ago. (European American, age 45, remarried 5 years)

Craig's discussion reflects two interpretations of divorce. Divorce rates seem to suggest that spouses are "leaving too soon" and not working on marriages. Yet a context of prevalent divorce, alternatively, compels marital work. Indeed, nearly everyone I interviewed confirmed Craig's belief that to maintain a marriage in a context of divorce, spouses are "working harder today."

I found that a majority of spouses believed in a marital work ethic. About 80 percent of this sample (28 of 34 spouses) talked about the need to work on marriage.

> And it was work. That's the other thing. It meant working really hard at what we had, and if we had something. And I think I took it more seriously than he did. (Pamela Jordan, European American, age 38, married 7 years)

> Mostly it involves effort, work, communication, and tolerance. (Naomi Rosenberg, European American, age 42, remarried 9 years)

> To make it work. Maybe that's the largest decision. It's so easy to say, "I quit." It's more difficult to stay and make it work. (Rosemary Gilmore, African American, age 38, married 13 years)

> [To] do custodial things, you know, to keep myself in a sound frame of mind. And not go overboard in voicing my way of doing things. Sometimes, you know, it does go overboard and I have to reel that in. Sometimes she'll point those out to me, and sometimes I'm aware of them. And sometimes the kids bring points and stuff up—it's about my being willing to, you know, make those changes necessary to get back on course. (Bill Gilmore, African American, age 39, married 13 years)

These husbands and wives talk about a work ethic—assuming, implying, or asserting that coping with the divorce option requires periodic effort. It is interesting that the six spouses who did not talk about the marital work ethic consider themselves "traditional"—traditional denoting both a strong belief in "marriage as forever" and in a male-dominant marriage. But for most couples, widespread divorce means husbands are talking about the need to work on marriage fully as much as wives are.

THE WORK OF MARRIAGE

The belief in the marital work ethic is not the same as actually *doing* the work of monitoring a marriage. The work of marriage should be understood as an extension of Hochschild's (1983) concept of "emotion work." Emotion work entails inducing or suppressing "feeling in order to sustain the outward countenance that produces the proper state of mind in others" (Hochschild 1983, 7). As Hochschild observes, "We put emotion to private use. . . . We continually try to put together things that threaten to pull apart—the situation, an appropriate way to see and feel about it, and our own real thoughts and feelings" (1983, 85). Among other private relationships, spouses certainly "try to put together things" within the marital situation—and hold them together despite forces that threaten to pull a marriage apart. Here, I focus on the aspect of emotion work that attempts to induce feelings of intimacy, connection, and attunement. The code words of marital work are: adjusting, adapting, trying, learning, growing, fulfilling needs, caring, and, above all, communicating.

For example, Jane and Gordon Walker's 17-year marriage has not been gratifying throughout, and Jane emphasized communication and sharing feelings as she discussed a marital crisis:

> We weren't communicating. I guess I wasn't really expressing my feelings. He wasn't expressing his. It just seems like the door was closing out, and I realized that—and he realized that as well—but he wouldn't identify it as much as I would. And say something's really wrong here—you have to let me in. (African American, age 39)

As Gordon also attests, Jane initiated the repair work. Although Jane identified the problem, Gordon was willing to reflect, respond, and work.

> I think we both recognize—I'm pretty sure she'd tell you the same thing—we worked at it. We understand that you just don't: "okay, we love each other, we got married, okay, fine that's it, let's go on." Because if you are an individual who is growing, you change. And if you care about the relationship that you're in, you're constantly communicating with your mate so that they adjust with you. Or if the adjustments are uncomfortable, you are aware of those changes, and you make them fit if you're concerned about the relationship. (African American, age 40)

Perhaps reflecting the view of a longstanding tradition of egalitarianism and relationality among African Americans (Collins 1990; Davis 1983; Nobles 1976; Taylor et al. 1990), Gordon was not a resistant recruit to the work of monitoring marriage. For other wives, however, the process of recruiting husbands to marital work has meant more conflict.

When 35-year-old Iris Sutton asserted that the biggest challenge of her 13-year marriage is "probably just staying married" and added "I mean there are so many reasons to get divorced, I mean that are very good reasons," she not only was referring to marital difficulties, but also was implying that marital endurance requires active effort. Under the influence of egalitarian ideals, wives like Iris increasingly attempt to share the work of monitoring marriage. Iris has threatened to divorce her husband, Ben Yoshida because of her frustrations with his lack of emotional attunement. Ben explained that he realized he must pay attention to the marriage.

> I think I realized I'm not dealing with a Japanese woman; in the sense that when Iris was threatening me to break up—at one time

> or another I really had the realization that she really, really means it—if I don't behave she really would not hesitate to leave me and that really scared the hell out of me. (Japanese, age 40)

"Behaving" in this case means attuning to his wife and his children. For Ben, Iris's European American ethnicity explains her seriousness. But this seriousness appears to be relatively recent for European American women, too. Women's use of divorce threats to produce emotional equality requires them to have financial independence, egalitarian ideals, and accessible divorce.

Wives' efforts to recruit men into the relational work of marriage, in essence, can be understood as efforts to "defeminize" love. By "defeminizing" love, I mean fostering men's valuation of nurturance and expressiveness, rather than Cancian's meaning, for example, which emphasizes redefining love. Cancian (1987) has argued that since the nineteenth century, love has been exceedingly "feminized," in its emphasis on tenderness, nurturance, and the expression of feelings. She emphasizes the need to expand the definition of love to include qualities that have been considered "masculine," not only sex, but the practical giving of help or sharing of activities that are absent from dominant cultural understandings of love in the West. But we may miss a more important point if we simply expand our definition of love rather than redistribute the emotion work of love. My data suggest that it is not wives who undervalue the practical giving of help or the sharing of activities—in fact, they dearly value the sharing of the "second shift" (Hochschild with Machung 1989); rather it is husbands who resist the nurturance and expression of feelings required by marital work.

Like their efforts to redistribute reproductive and productive labor. I found that wives' attempts to get men to do more emotion work in marriage often met with resistance. The following three case studies reveal this process and the uneven results. One shows cooperation in the work, the second shows a conflicted process, and the third shows a failed attempt. Whether wives' efforts are met with cooperation or resistance, understanding these gendered dynamics is crucial in a culture of divorce. In the past when there was a sense of "no alternative," it may have taken more effort to leave the marriage; today, it may take more effort to stay in the marriage.

THE CLEMENT-LEONETTIS: COOPERATION IN MARITAL WORK

The Clement-Leonetti marriage illustrates how an active work ethic serves to sustain a gratifying marriage and arises in response to "unexpected divorces." Dana Clement and Robert Leonetti are in a 10-year marriage, the first for both, and have three children. They are struggling to maintain a middle-class lifestyle as Robert works two jobs as a professional policy analyst and a consultant, while Dana works part-time in a biological laboratory and is the primary caretaker of their children. Both spouses are European American, age 32, and are lapsed Catholics.

For Robert, the first unexpected divorce was that of his parents; for Dana, it was that of her best friend from kindergarten. Children, homes, and marital length all figure into the unexpected quality of these divorces for Dana. The topic arises when I open the interview by asking Dana "what's good or happy about your marriage?" She replies by reviewing their brushes with others' divorces:

> I think that Robert and I have certain understandings that—I think a big part of what has shaped our marriage and the reason we're still together and all that is because he comes from a family that is divorced. His parents are divorced. And when my first child was born, my best friend from kindergarten went through a divorce. And I was totally, you know, unaccustomed to divorces in a family. And one of the things that helped us, I think, is that we both shared our feelings about

> both of those sets of divorces. For me, it was terribly shocking that my friend who had been married 15 years and had two kids, and a car, and a house, and a yard, and a dog, and you know the whole American picture—to be getting divorced. At that time, this was eight years ago already, we agreed that we would never stay together with one of us unhappy, that if something was going wrong, we would at least have the respect for the other person to tell them what was happening.

Dana assumed that she should give a reason for still being together. When marriage culture prevailed, this kind of explanation was unnecessary except under "last resort" conditions.

As Dana discussed the array of divorces surrounding Robert and herself, she turned a potential vulnerability to divorce on its head. They pledged to share their feelings with one another and to keep one another abreast of encroaching unhappiness or doubts. That the goal of the work ethic is a "gratifying" and not simply an "enduring" marriage is captured in Dana's statement that "we agreed we would never stay together with one of us unhappy." To avert an unhappy marriage and subsequent divorce, they "work" at it. But how exactly do they do this?

Part of the work is managing emotions and their expression. When I asked Robert about how they have changed over the course of their marriage, he explained that in the early years of their relationship their differences were more "pronounced" and that there were many more arguments. When I ask Robert how they had gotten through those years of conflict, he begins by laughingly reporting "years of arguing," and then elaborates.

> And we struggled with it. I don't know maybe we grew up—I grew up, she grew up—together.
>
> . . . And I wasn't, clearly, I really wasn't asking what she was feeling, I was just telling her what she was feeling. And as I watched that in myself, and I had to 'cause she forced me to—and I was also kind of growing . . . I was really forced to deal with it. I either had to deal with it or leave. [A lesson was learned] which is that there are ways that are different than what you want, and that I really wasn't—I was expressing my own disappointment, my own anger in a way that didn't leave enough room for her.

As Robert learned to monitor himself and avoid imposing his "version" of reality—learned to suppress his anger and redefine his disappointment—he also learned to recognize Dana's feelings. He felt forced, as he repeated twice, to adjust. He reiterated later that he had to "let go of his version of resolution as right" and remind himself that "her way of expression was okay and mine was not better." In this way, he feels he has "grown."

In addition to these struggles and their pledge to keep one another informed of their feelings about the marriage, Dana contended that her friend's divorce was "one of the reasons I went to therapy at the time I did." She observed that these divorces were "what I talked about a lot." Both Dana and Robert portray therapy as part of the work they did, not only to deal with the divorces around them, but also to further the relational work in their own marriage. While Robert and Dana simultaneously went to individual, as opposed to couples, therapy, their reliance upon therapy reflected a widespread generation-based pattern.

The number of couples who attend "couples therapy" serves as one indicator of marital monitoring and was prevalent in my sample. Seven of the 17 couples had gone to couples therapy at some point during their relationship—some briefly and some at length. Other research, also, suggests that spouses rely on therapy more often than in the past (Bellah et al. 1985; Philipson 1993, 58).

Therapy helps couples monitor their relationships, but the men's and women's attitudes toward therapy differed. Although some of the wives criticized or resisted therapy, wives were more likely to be dissatisfied with marital communication, to advocate therapy, and to want husbands to go to therapy with them. The wives were more likely to talk the "vocabulary" of therapeutic culture, using such terms as "the inner child" and "codependency." In contrast, husbands were less likely to use such words. They were less likely to express dissatisfaction with communication and more likely to refuse, resist, deride, or downplay the role of therapy in their marriages. Therapists are, in some sense, allies to many women not because they necessarily support their agendas but because they legitimate women's relational concerns, validate women's desires for men to take active responsibility for their marriages, and suggest that men may have to contribute to the "emotion work" of marriage. While some scholars have argued that therapeutic culture has been a vehicle for individualism, it can also serve as a vehicle for relationality. Therapy often requires emotion work that can nurture relationality in marriage.

Robert and Dana's mutual efforts to monitor the marriage are less typical in this sample. However, they illustrate the key facets of marital work by responding to a culture of divorce with the aims of growth, fulfillment, and happiness, and by using therapy. They provide a portrait of a successful redistribution of relationship work in marriage.

THE TURNERS: CONFLICT OVER MARITAL WORK

The Turners engaged in a more overt struggle over the marital work. Mina and Nick Turner lived together for three years and have been married for 10. They are one of four interracial couples I interviewed; Mina is Japanese American and Nick is African American. While this is Mina's second, it is Nick's first marriage; she is 41, he is 38. They are currently raising three children. Both Mina and Nick are college-educated, middle class, and full-time workers.

When I opened my interview with Mina, I asked what was good about her marriage. She initially stated: "Everything." She elaborates that it is good "sexually," provides "emotional support as far as career-wise, . . . there's no really set role of who does what, you know, everyone does everything." The latter comment represents their egalitarian marriage; the Turners share work and home responsibilities.

Mina's previous marriage has affected her view of marriage with Nick. When I asked Mina about the biggest surprise of her marriage, she replied in a way that reflects divorce culture.

> One of the surprises is that it's lasted this long. (Laughs.) No really, it's like I feel like "who is gonna put up with me?" You know? Because at this point in my life, it's like, you take me like I am or else forget it, because I'm not going to be in the same situation as I was in my first, where I did everything to please a person and lost myself, you know. So I'm just going to be the bitch I am and you take me, or else you're gone.

As Mina suggested by appropriating the pejorative term "bitch," criticisms of her aggressiveness or selfishness will not sway Mina to cede her independence. Perhaps concluding that "offense is the best defense," she refused to "lose herself" again. Her marriage is undeniably contingent upon her independence. The strength of her independence is verified in a story both Mina and Nick tell of a crisis in their marriage.

Two years ago the Turner's youngest child was hit by a car. While their daughter has generally recovered, the crisis represented a marital communication problem to both spouses. One evening, during this traumatic period, a new crisis emerged when, as they both attested, he "erupted"—got drunk and violent. When I asked about the biggest surprise of "his" marriage, Nick began to relate his version of the same event. He replied, "I guess the—we're still together after I erupted one evening, you know, I had too much to drink and one thing set me off and I went on a screaming rampage." This "near divorce" was averted because, with some resistance, Nick conformed to Mina's condition. Mina's conditions were, as she stated:

> I go, "Okay," you know, "you go see the therapist or I'm out of here." . . . You know, I wasn't gonna wait around for him to put his mitts on me. That was—that's all it took for me, you know. And uh—but you know, it was just so out of character for him. . . . like I said, I think it was all that internalizing.

Mina went on to frame the "violence" as not only pivotal but also exceptional; it was, as she put it, "Dr. Jeckyll and Mr. Hyde." When Mina demanded that Nick go into therapy, she was trying to share the work of monitoring marriage. Mina not only forged a direct link between Nick's noncommunication and his explosive behavior that night but also she expects Nick to share the emotion work of marriage. She wants Nick to learn the expressive and relational skills for the survival of the marriage. Or else.

Independence empowers Mina to achieve her goal of a relational marriage. She is defining the terms of the marriage contract—and using the option of divorce to redefine the division of emotion work in the marriage. "Communication" and "talk" are marks of her relational criteria of marital happiness. She does not simply want independence, she also wants to feel connected.

A dilemma for many wives is how to procure equality without exercising the power of independence—without, in effect, relying upon individualism. Mina Turner's individualism is apparent in her interview and appears to be a reaction to her first marriage. Women can use their income and occupational status to improve their standing and power in relationships; yet, their desire for independence should not be viewed apart from their relational concerns. Wives like Mina have a second agenda based on relationality; they want to share not only individual rights but also relational responsibilities—including the "emotion work" of monitoring marriage. Although some women are using independence to bring about relational ideals in their marriages, they need husbands who are willing to reciprocate; that is, willing to value and to do the marital work entailed in sustaining heterosexual marriage.

The final case study, the Kason-Morris marriage, suggests that not all husbands are willing to reciprocate—and not all wives are as willing as Mina to exercise their power.

THE KASON-MORRISES: FAILURE TO REDISTRIBUTE THE WORK

Roxanne and Craig have been married for five years and are raising two daughters. They are European American, nonreligious, and are college educated. Craig is an accountant and Roxanne is a part-time word processor. Both have been married before and were wary of remarriage, but they asserted that the theme of their marriage is "practice makes perfect." Still, this marriage is not perfect. While Craig clearly articulated that spouses are "working harder today" because they must if they do not want it to "slip away," he was not as clear about what that "work" entails.

Whether speaking of child care, household, or market labor, Roxanne did not feel that responsibilities are shared in her marriage. When I ask Roxanne about the biggest compromise of her marriage, she asserted: "Housework. It's my compromise, not his." Later, she described the choice not to work full-time as the biggest "concession" of her marriage: "It's really a powerless situation in a marriage when you're barely working." She explained: "you have no monetary power" and "you're at their beck and call." Finally, when I asked what's not good about the marriage, she stated, "He complains that I strap him with the load of child care when he comes home from work . . . which isn't true, but that's what he thinks."

It might seem that Roxanne's claimed sacrifices, compromises, and concessions are an array of separate issues; however, in many ways her account suggests that productive labor, reproductive labor, and emotion work are all of a piece. Roxanne reports that in her first marriage she was repeatedly venting her anger about the "second shift." In this marriage, she resolved, "I'm not going to do that anymore, I'm not going to lower myself and I'm not going to get that angry." She has learned, as she puts it, to "press down my anger when he won't do the dishes." Roxanne's efforts illustrate Hochschild's point that "across the nation at this particular time in history, this emotion work is often all that stands between the stalled revolution on the one hand, and broken marriages on the other" (with Machung 1989, 56).

In addition to the spouses divergent expectations of marital labors, the Kason-Morris marriage also revealed tension in their visions of what constitutes a fulfilling marriage. Craig wants Roxanne to be more appreciative of things as they are. In contrast, Roxanne perceives "talk" as constituting emotional intimacy and wants more of it. When I ask what's missing in the marriage, Roxanne replies: "One thing that I would want more of is more intimacy, just on a, on a daily basis. More visiting. More communication. More talking. Sharing." Roxanne, is among a number of wives who feel that their husbands refuse to express thoughts and feelings and do not care about women's emotional lives (Thompson and Walker 1989, 846).

The Kason-Morris marriage reflects conventional gender differences found in heterosexual marriages (Rubin 1983). If these conflicts are not new, the ways in which wives contend with them are: wives are increasingly likely to press for their vision of a fulfilling marriage, a vision that is informed by connection and an increasing awareness that the emotion work of marriage needs to be shared. Wives' desire for relationality have them searching for strategies to recruit husbands into the emotion work of marriage and family and, as I have suggested, couples therapy has become an increasingly common vehicle for sharing this work.

Roxanne has tried to get Craig to go to therapy with her, and she "would still like to," but "he hated it." She reported, "we got in our worst arguments after each session." Most of these arguments concerned family responsibilities, housework, and sharing feelings. She "decided it would be the ruin of our marriage if we continued."

Craig would also like to see improvement in their communication dynamics—not more talk but more understanding. In this way, Craig's view reflects Cancian's (1987) insight that his practical giving of help and sharing of activities go unrecognized as love. For Craig, therapy was not a means to overcome these difficulties but a setting to reenact, them. He states that he "doesn't care for" therapy and that he is uncomfortable with a lot of talk. Moreover, Craig felt manipulated into therapy—he felt that Roxanne was hauling him into therapy to focus on his problems. In ways he was right: Roxanne wants him to take more time and responsibility for the emotional contours of marital and parental relationships. Yet, ultimately, Roxanne is more focused on how the marriage needs change. Craig may be aware that therapy legitimates Roxanne's relational concerns, but his resistance blinds him to Roxanne's larger aims and to therapy's potential for averting a growing breach between himself and Roxanne. Craig's refusal to participate in therapy doesn't change the degree to which therapeutic culture informs Roxanne's vision of marital relationships. What it does change, however, is whether she attends to the vying messages in therapeutic culture of relationality or individualism. Relationality cannot be wrested from another. By choice or default, Roxanne is attuned to the message of individualism. She wants to conquer her "codependency," that is, her tendency to take care of others and neglect herself.

According to Roxanne's talk, her marriage is contingent not only on a sense of autonomy but also on a sense of connection; in actuality, she is neither very autonomous nor connected. Roxanne wanted connection, but I sensed that she might "choose" separation. She asserted that it would be his "moodiness, pouting, and childlike behavior" that would break her marriage. She was tired of Craig's refusal to answer her and she was tired of doing all the emotion work in their marriage. Roxanne's initiation of a divorce could suggest that her individualism split the marriage. Craig's resistance to the emotion work would not be reflected in most statistics; yet, he could become one of those ex-husbands who "didn't know what happened."

Kitson and Holmes discovered a notable gender difference in their research on the marital complaint "not sure what happened": For ex-husbands this complaint ranked third, for ex-wives it ranked twenty-eighth (1992, 123). These results suggest that these husbands were not attuned to the well-being of their wives or their marriages. Other research also finds that "in marriages that eventually ended in separation or divorce, women usually knew the relationship was in trouble long before their partners did" (Thompson and Walker 1989, 848). In short, Craig may be missing the implication of not sharing the marital work with Roxanne. In response, Roxanne gravitates toward the ideal of individualism.

WIVES AND MARITAL WORK: INDIVIDUALISM OR RELATIONALITY?

In the past decade, family scholars have debated whether we should be optimistic or pessimistic about family life in the United States (Glenn 1987, 349). Optimistic theorists have argued that the family is not falling apart, but simply changing and adapting (Riley 1991; Scanzoni 1987; Skolnick 1991). These theorists emphasize the oppression that has attended women's self-sacrifices in marriage and point to the potential for greater self-determination and happier relationships today (Cancian 1987; Coontz 1992; Reissman 1990; Skolnick 1991; Stacey 1990). They tend to downplay individualism and troubles in family bonds.

Pessimistic theorists have argued alternatively that the institution of marriage is a cause for concern—that divorce rates signify an unraveling of social bonds (Bellah et al. 1985; Glenn 1987; . . . Popenoe 1988 . . .). Given that marital dissolution by divorce, rather than death, entails individual choice, these theorists focus on the role played by "individualism" in divorce. Pessimists fear we have forsaken nurturance, commitment, and responsibility in our aggregate rush to divorce. And because these are the very virtues that have been traditionally valorized in women, these "divorce debates" are always implicitly, if not explicitly, about gender. As one scholar has observed, "when commentators lament the collapse of traditional family commitments and values, they almost invariably mean the uniquely female duties associated with the doctrine of separate spheres for men and women" (Coontz 1992, 40).

We know that as heads of the household, even when not primary breadwinners, most husbands, historically, have had greater authority, and therefore greater freedom, to be independent, compared to wives. Women's greater labor force participation, increased activity in the political sphere, and greater initiation of divorces suggest that women are appropriating the masculinist model of individualism. However, if we focus on these social changes, we obscure women's enduring responsibility for family labors—including monitoring marriages.

While marital monitoring was important in the past, when a culture of marriage prevailed, it was more likely to be done by and expected of wives and was more likely to lead to wives' accommodation within the marriage. Economic constraints and cultural disincentives pressed for accommodation. In a culture of divorce, however, marital monitoring by wives in heterosexual marriage takes on new meanings.

In a climate of marital contingency, the work of monitoring marriage demands to be shared. Wives' efforts to redistribute the monitoring of marriage may be successful, as in the Clement-Leonetti case, because of the spouses' shared receptivity to the marital work; or these efforts may be successful, as in the Turner case, because the wife engages the power of independence and the husband responds. However, the wife's efforts to redistribute the monitoring of marriage may falter, as in the Kason-Morris case, because her husband resists and because she hesitates to engage her power. My research leads me to conclude that more wives are less interested in conducting a marriage on old terms, terms that denied their autonomy and devalued their relational concerns.

Women such as Dana, Roxanne, and Mina are alert to the meaning of sacrificing self to others. Given an awareness of the subordinate role of wives and the devaluation of "women's work," they are rightfully suspicious of the wife role. Many refuse to sacrifice themselves.

While women may want an equality of sacrifice based on relational connection, they may have to settle for an equality of nonsacrifice based on individualism. It is easier for women to appropriate the male model of individualism than it is to compel relationality from men. Because relationality cannot be wrested from husbands, wives' individualism is a predictable default in a cultural context that valorizes individual choice.

Optimistic scholars have been quick to deny women's increasing individualism—because it has been used against women. Rather than denying it, women's individualism needs to be understood in a political, structural, and historical context. Because we proceed from a history of male-dominant marriages, individualism does not *mean* the same thing for women as it does for men. Wives who are more individualistic are often trying to counter male-dominance in marriage. As one family scholar succinctly observed about the egalitarian marriage: "What most women seek is not power but the absence of domination" (Johnson 1988, 261). Wives fear being subordinated in a way that husbands do not. Moreover, independence is not just an end but a means for some wives to counter old terms and to secure a relational agenda. Wives are not necessarily forgoing relational responsibilities, but are combining them with individual rights. Yet individualism does mean that a wife can use the lever of the divorce option to secure her marital vision.

In sum, even though wives want a relational marriage, they may need to draw on individualism to secure it; and if secured, they may change the power dynamics of their marriages. If wives are unable to secure their vision of the marriage, they may choose to pass through the gateway called divorce. Yet, ultimately, what many wives want is not freedom from but freedom within a relational marriage.

SELECTED REFERENCES

Bellah, Robert N., Richard Madsen, William Sullivan, Ann Swidler, and Steven Tipton. 1985. *Habits of the Heart.* Berkeley: University of California Press.

Bernard, Jessie. 1972/1982. *The Future of Marriage.* New Haven, Conn.: Yale University Press.

Blaisure, Karen R., and Katherine R. Allen. 1995. "Feminists and the Ideology and Practice of Marital Equality." *Journal of Marriage and the Family* 57:5–19.

Cancian, Francesca. 1987. *Love in America: Gender and Self-Development.* New York: Cambridge University Press.

Collins, Patricia Hill. 1990. *Black Feminist Thought: Knowledge, Consciousness, and the Politics of Empowerment.* Boston: Unwin Hyman.

Coontz, Stephanie. 1992. *The Way We Never Were: American Families and the Nostalgia Trap.* New York: Basic Books.

Corbin, Juliet, and Anselm Strauss. 1990. "Grounded Theory Research: Procedures, Canons, and Evaluative Criteria." *Qualitative Sociology* 13(1): 3–21.

Davis, Angela. 1983. *Women, Race, and Class.* New York: Vintage Books.

Glenn, Norval D. 1987. "Continuity Versus Change, Sanguineness Versus Concern." *Journal of Family Issues* 8 (4): 348–354.

Hackstaff, Karla B. 1999. *Marriage in a Culture of Divorce.* Philadelphia, PA: Temple University Press.

Hochschild, Arlie R. 1983. *The Managed Heart: The Commercialization of Human Feeling.* Berkeley: University of California Press.

Hochschild, Arlie R., with Anne Machung. 1989. *The Second Shift: Working Parents and the Revolution at Home.* New York: Viking Press.

Johnson, Miriam. 1988. *Strong Mothers, Weak Wives: The Search for Gender Equality.* Berkeley: University of California Press.

Kitson, Gay C, with William Holmes. 1992. *Portrait of Divorce: Adjustment to Marital Breakdown.* New York: Guilford Press.

Nobles, Wade W. 1976. "Extended Self: Rethinking the So-Called Negro Self-Concept." *Journal of Black Psychology* 2(2): 15–24.

Philipson, Ilene. 1993. *On the Shoulders of Women: The Feminization of Psychotherapy.* New York: Guilford Press.

Popenoe, D. 1988. *Disturbing the Nest: Family Change and Decline in Modern Societies.* New York: Aldine De Gruyter.

Riessman, Catherine Kohler. 1990. *Divorce Talk: Women and Men Make Sense of Personal Relationships.* New Brunswick, N.J.: Rutgers University Press.

Riley, Glenda. 1991. *Divorce: An American Tradition.* New York: Oxford University Press.

Rubin, Lillian. 1983. *Intimate Strangers: Men and Women Together.* New York: Harper & Row.

Scanzoni, John. 1987. "Families in the 1980s: Time to Refocus Our Thinking." *Journal of Family Issues* 8(4): 394–421.

Skolnick, Arlene. 1991. *Embattled Paradise.* New York: Basic Books.

Stacey, Judith. 1990. *Brave New Families: Stories of Domestic Upheaval in Late Twentieth Century America.* New York: Basic Books.

Taylor, Robert J., Linda M. Chatters, M. Belinda Tucker, and Edith Lewis, 1990. "Developments in Research on Black Families: A Decade Review." *Journal of Marriage and the Family* 52:993–1014.

Thompson, Linda, and Alexis Walker. 1989. "Gender in Families: Women and Men in Marriage, Work, and Parenthood." *Journal of Marriage and the Family* 51:844–871.

31

How Marriages End

DEMIE KURZ,
1995

Some social scientists believe that a key factor shaping contemporary divorce is "individualism"—husbands and wives, influenced by corporate culture, personal growth movements, and the women's movement, now place a high priority on their own personal interests and needs, and so are more ready to divorce when they encounter difficulties in their marriages. Despite the widespread interest in the topic of divorce, however, we don't really understand why people end their marriages. While research shows that women typically initiate the divorce process, we don't know why this is so.

In order to form accurate images and assessments of those who divorce, we need to know more about why couples are leaving their marriages. Are divorced women "selfish" and "misguided"? Are they "victims"? If women are leaving marriages simply to "see if they can do better," this phenomenon creates a different understanding of marriage than if they are leaving because of dissatisfaction with traditional sex roles, domestic violence, or their husbands' drug abuse or domestic violence.

Our views of divorced women in turn influence the policies we choose to respond to divorce. Depending on whether we believe that divorce is caused by the rise of individualism and the abandonment of commitments to the family, by men's unemployment and inability to support a family, or by women's wishes to escape a violent marriage, we will adopt one policy while rejecting another.

In this research I explored the reasons that women themselves give for their marriages ending. These women's views challenge contemporary perspectives on divorce which stress the role that individualism plays in a couple's discontent. Their stories illustrate how their decisions about marriage and divorce were rooted in family contexts that were significantly shaped by factors related to gender, race, and class. The domestic violence in these women's marriages, for example, was a major influence on their experiences of marriage and their decision to divorce. Economic hardship and poverty also placed heavy burdens on the marriages of some women. Their accounts demonstrate the disadvantages that divorced women can experience in marriage, and the ways that their marriages can be not only fraught with problems, but also destructive.

Women in this study were asked to describe the circumstances of their separation. Most women stated, "I left when/because . . ." or "He left when/because. . . ." These data do not tell us what "causes" divorce. An answer to that complex question would require an in-depth analysis of historical and demographic data, as well as more comprehensive data from women themselves. Nor did these women give extended accounts of the process of ending their marriages such as are found in some other studies. These data represent what a random sample of women chose to report in interviews an average of five years after their separation. Their descriptions of why their marriages ended relate to their understanding of their divorces at a particular moment in time; they would very likely differ at different points in the divorce process, with different interviewers, and at different historical periods. These women's accounts in many cases would also differ from those of their ex-husbands.... There are "his" and "her" versions of marriage and divorce based on the different social positions of men and women in the family.

What is important about these accounts is that they provide women's perspectives on what shaped their marriage and divorce. These women's views of the ending of their marriage strongly influenced their entire experience of divorce. While being interviewed, the women referred repeatedly to these reasons for their divorces, indicating their enduring importance to their lives. They often reported that the circumstances of their separation had a strong impact on their experience of divorce as well, including their ability to negotiate for resources. For example, if a couple parts mutually and amicably, a woman may be more likely to obtain child support from her ex-husband than if they had serious conflicts. How a marriage ends also affects a woman's assessment of her divorce. A woman who is left by her husband against her wishes may see herself as having lost major opportunities because of divorce, while a woman whose ex-husband was abusive or violent may feel divorce gives her the opportunity to pursue a more fulfilling life. These women also reported problems in their marriages that have been widely noted as problems for families in the United States today, particularly domestic violence. And finally, their responses to questions about the circumstances of their separation are consistent with responses women have given other researchers who have asked similar questions about the ending of their marriages. (Goode, 1956; Kitson and Holmes, 1992).

REASONS FOR DIVORCE

Based on their accounts of why their marriages ended, I have placed the women into four different groups. One group includes those women who reported having left their husbands because of personal dissatisfaction and those who termed their separation "mutual." A second group includes those women who said their marriages ended because their husbands were involved with other women. In some cases the men left their wives, in others wives said they left because of their husbands' involvement with other women. I placed in a third category those women who reported having left their marriages because of violence they experienced at the hands of their husbands. In a fourth category are those women who state that their marriages ended for what I call "hard-living" reasons, after Joseph Howell's description of certain types of blue-collar families in which the men's difficulties holding jobs led to alcoholism, extended absences from home, and related behaviors that threatened their marriages (Howell, 1973)....

There were significant differences between reports of women in different classes. The higher their class position, the more likely the women were to cite personal dissatisfaction as a reason for their divorce, while the lower their class position, the more they reported leaving because of violence. "Hard-living" divorces were more prevalent among working-class and poverty-level women and were significantly more likely to occur among white women. There were no other statistically significant differences between women in terms of their race.

Personal Dissatisfaction

I begin my discussion of the ending of these marriages with the category of what I have called "personal dissatisfaction," since this category is closest to the kinds of dissatisfaction that contemporary observers have said motivate Americans to divorce. The 19 percent of women who left their marriages for reasons of personal dissatisfaction stated that: they didn't love their husbands anymore; the communication in their marriages was not good; they fought too often with their husbands; their husbands had been too controlling; or they were tired of carrying the emotional load of the marriage. At the same time that they cited a variety of factors as critical to the ending of their marriages, however, women also identified certain patterns to the separation, particularly the importance of gender roles.

Typically, there is a division of family labor by gender, with women assuming emotional and caretaking functions and men expected to be the primary breadwinners. In addition, . . . despite norms favoring equality in marriage, men often still control decision-making in the family. Gender, of course, permeates all social interactions. . . .

First, women mentioned their ex-husbands' controlling behavior. One 34-year-old white middle-class woman, an administrator who had been married for eleven years, felt that because of her ex-husband's controlling behavior, she could not participate in decision-making during the marriage:

> I left him . . . he wasn't physically abusive, but he was emotionally abusive. . . . I didn't like how he made all the decisions. He always argued very logically with what I said. So it always seemed like what I said didn't make sense and what he said was right. I just didn't have respect for him anymore. I thought long and hard about leaving, for over a year. I tried to get him into counseling. He went but he didn't really participate. And when we got home it was just the same thing.

Several working-class women also spoke of their ex-husbands as being too controlling.

> I married him when I was young and dumb. I had a scholarship at [local university] I could have taken but I married him instead. He also wanted to control everything about my life. He wanted to control my friends, my time. [36-year-old black nurse, married for fifteen years with two children]

> One problem with my husband, he was like "women have their place." He would provide but the woman was to keep her mouth shut. (Now) I would look for somebody who would treat their partner as an equal. [39-year-old black part-time temporary worker, married for seventeen years with three children]

A few women spoke in both general terms and in more gendered terms about their marital relationships. One middle-class white woman whose ex-husband was an artist initially spoke of her ex-husband's "irresponsibility." She described how she felt guilty when she left "because my child was very young and she was very confused but at the same time I couldn't be with someone so irresponsible." They had tried counseling, but according to this woman, it was too late. "I had been up for that a year and a half before and he kept saying there was nothing wrong. Only when I got him out the door was he ready." However, later in the interview this woman gave a more gendered account of her marriage. She stated that what she liked about the divorce was "moving more in the direction I want." During her marriage, she said, "I'd been running in the direction he wanted."

Second, women stated that they did not get enough emotional support from their husbands. A 33-year-old white psychiatrist, who had been married nine years, spoke about the issue of emotional support in very gendered terms.

> I had to do all the emotional work in the relationship and I wasn't getting anything back. I really tried to make it work for a couple of years. I kept thinking that it would change, that if I did the right thing he would become more emotionally active and responsive in the relationship. We went to therapy but things didn't seem to change. Finally I told him I wanted him to leave. He was very angry.
>
> I knew I couldn't stay with someone who was so completely unsupportive. He was only looking out for what was good for him. I don't think all men are the same, but

> I believe it was very male behavior. I guess he was angry that he wasn't getting enough emotional attention. First, because we had recently had a baby. And then because I went back to school. . . .

A 34-year-old black poverty-level woman, a part-time classroom aide, spoke in a similar vein:

> I want a supportive relationship. One where I can give support and get support. My husband thought a man was supposed to be a good provider and that was it. I want more than that. I want a real relationship. . . .

Third, a small number of women voiced discontent with their husband's failure to be adequate enough providers. These women are discontented because their husbands are not working, or are not working hard enough. Those who are themselves working for pay are particularly unhappy. One white middle-class woman, a registered nurse, stated she left her husband because he remained unemployed throughout the marriage. She claimed that he accumulated debts and smoked a lot of marijuana, and that he didn't spend time with the children or help her take care of them. One woman, a 36-year-old white middle-class mother of two, who was married for twelve years, felt that her husband did not try hard enough to be successful in his work.

> He is bitter. He feels he got kicked out. It was me that wanted the divorce. I didn't love him anymore. I didn't respect him anymore. He didn't have any ambition. I had to really push him to improve his business. . . .

Several women commented that they were dissatisfied with the subordinate roles they adopted in their marriages and indicated that they would not assume these roles again.

> I've learned from this experience [the divorce] not to give all of myself. Women have a tendency to love more than a man, to make him happy, to make things right for him. But then women get the blunt of the blow. Women get hurt. [36-year old black pharmacy technician, married thirteen years]

> I'm not going to be somebody's servant. That's the way I was in my marriage, picking up all the time. My ex-husband was a slob. Love is not being a maid. Men don't know the meaning of love. My ex-husband treated me like his mother. [34-year-old unemployed working-class white woman, married eleven years]

> My ex-husband was very impressive when we met. Anything that money could buy he got us—clothes, furniture. But he changed when we got married. Men do that. First they have to impress you. But when you marry them, they feel they own you. And when he says jump, you have to jump. [24-year-old poverty-level black woman, married five years]

These women's dissatisfaction with male control raises the question of a gender gap between men's and women's expectations of marital roles. As noted in the previous chapter, polling data show women want more equality in marriage, while men think there is enough. One group of researchers believes that the divorced population includes a higher percentage of men who have more traditional gender-role attitudes and women who have nontraditional views. They argue that particularly those women who are in the paid labor force are now in a position to leave a relationship in which they view their husbands' demands as unfair (Finlay, Starnes, and Alvarez, 1985).

Violence

Although the home is spoken of as a site of love and caring, we know now that the prevalence of violence is alarmingly high in contemporary marriages. . . .

In this sample, 70 percent of women of all classes and races experienced violence at the hands of their husbands at least once. Fifty percent of women experienced violence at least two to three times. . . .

The accounts of these women resonate with fear and pain. One can easily imagine how the children who witnessed this violence were very fearful and deeply saddened. At the same time, a quiet courage runs through the accounts of these women as they assessed the costs and benefits of staying in their marriages.

While many women in the sample experienced violence, those who stated that they separated because of the violence had experienced much more serious physical violence than women who gave other reasons for separating. Many of these women stated that they left their husbands after a particularly serious fight. Some women, such as this 41-year-old white working-class mother of two, felt that their lives could have been in danger.

> It took a year for the separation to come through. I filed. We separated for the last time after he beat me up. It was Mother's Day. He beat me up in front of the kids and his parents. I was really scared then. I thought, "if he'll do this in front of them, what could he do next?" I had to get a protection order at the time and that cost $300.
>
> We had been going to a counselor. . . . The therapist called me one night and said to come right over. He said, "Your husband doesn't know right from wrong. He only thinks he is right. You'd better get away. He could kill you." I now believe that. At the time I thought he would still change.

This 31-year-old mother of one, a black woman living at poverty level, left an eleven-year marriage after a particularly serious incident of physical abuse. She addressed the question most frequently put to battered women, the question of why she didn't leave sooner, despite experiencing a lot of very serious violence throughout her marriage.

> We separated after a big fight where he was physically abusive. First I went to the Emergency Room. Then I went to the Police Roundhouse. The police came to the house and made him leave. . . . I got a restraining order. It lasted for a whole year.
>
> There was violence constantly for the 10 years. It would usually happen on the weekends. We would fight over small things like, if he would go out on Friday, I would say I want to go out on Saturday. But slowly over the years something was clicking inside. I said to myself, "Are you going to let someone else run your life?"
>
> Interviewer: So what made you realize that you wanted to get out of this relationship?
>
> I have thought about that a lot. Somebody told me people scheme on others but you scheme on yourself. First, I don't want to hurt or disappoint other people. So instead I get hurt. Also, there would be repercussions. Where would I go with three children? I didn't want to go back to my mother's. I get along with my mother but if I went home it would be like I had been a failure.

Another poverty-level woman also described why she didn't leave sooner.

> He degraded me constantly and he injured me. I've had a broken jaw. And see this scar on my face—it's not a dimple. We had violent fights three times a week. It started when the honeymoon was over. He said, "Now we're going to change all this shit." I had thought everything was fine.
>
> Interviewer: Did you think of leaving?
>
> That's what everyone asked, "Why don't you leave?" I said, where was I going to go? People who ask that question don't understand. [31-year-old black part-time classroom aide]

Like this woman, some other women also described injuries.

> There was no violence for the first five years, before the kids were born. Before they were born he was violent maybe only three times. But after the kids were born, it was often. He seemed to especially resent time I spent with

> the kids. I'd have to hurry them through their baths.
>
> The night he broke my rib and my jaw I was giving the kids a bath. I guess he thought it was too long. He broke my rib because he was trying to throw me down the stairs. I was resisting and I guess he cracked my rib over the banister. . . . I would say there was a violent incident once a month. [42-year-old white secretary, mother of two, married sixteen years]

Some women left when the violence affected their children. One woman left when her husband sexually assaulted her son by a former marriage. She filed a criminal charge against her ex-husband, who is now serving time in prison for this crime. This poverty-level black woman left because of the effect of the violence on her children and on herself.

> All the violence was hard on my son. He saw me injured when he was two years old. He saw blood, he saw a lot. It's affected my son. He's mixed up.
>
> I left because I was afraid of what this was doing for my son. I left because of what it was doing to me. I realized I could have shot my ex-husband. But I couldn't do that for my son's sake.
>
> I finally realized the marriage wasn't working. I really wanted a marriage. I wanted that marriage to work. But I finally realized it just wasn't working. [28-year old black woman living at the poverty-level, mother of one, married six years]

Some women also watched their husbands destroy property and found this frightening. This 33-year-old white middle-class mother of two, who owned and ran a business with her husband, described his violence:

> I was the one who left. My ex-husband had a terrible temper. He used violence a lot. He didn't hurt me physically very much but he destroyed property a lot. He flew off the handle a lot. He was also an alcoholic. He had an explosive angry temper.

In conclusion, many of the divorced women in this sample were "battered women." We rarely think of divorce and battering together, instead viewing "battered women" and "divorced women" as two different categories. We must incorporate an understanding of violence into our portrayal of all aspects of family life, including divorce. There undoubtedly have always been women who left marriages because of domestic violence. Perhaps there are more at the present time because of increased publicity about wife battering. The problem is, because we do little to alleviate the hardships of divorce for women, we make them pay a heavy price for leaving marriages because of domestic violence.

Hard-Living

Seventeen percent of women gave "hard-living" as the reason for the ending of their marriage. As noted in the previous chapter, "hard-living" is the term Joseph Howell used some time ago to describe certain behaviors, such as heavy drinking and frequent absences from the family, which men, particularly working-class and poverty-level men, exhibit when they feel frustrated by unstable work conditions and high unemployment. Since Howell first described hard-living, alcohol abuse has continued to plague men with weak employment histories. Hard-living is usually associated with male behavior. Lillian Rubin argues that some men continue to use alcohol as a way of coping with unemployment and Katherine Newman stresses that alcohol abuse is associated with the downward mobility and job insecurity of male workers at all socio-economic levels (Howell, 1973; Newman, 1988; Rubin, 1994, 1976). It is important, however, to highlight the role of gender in hard-living; more often than not, it is women who are the victims of these behaviors.

Women in this sample who said they left because of hard-living behavior mentioned most often drug abuse, then alcohol abuse, followed by accounts of their husbands' absence from the home. Drug use has become a particularly serious problem during recent decades. It has devastated minority populations in urban neighborhoods, and—as is reported less frequently—it also has posed a threat to white blue-collar families. Judith Stacey relates the recent upsurge in drug use to the unstable employment conditions created by economic restructuring. Most of the women who left because of their ex-husbands' use of drugs or alcohol found their husbands' behavior very troubling and particularly disruptive of family life (Stacey, 1990). The majority of women who reported leaving their marriages for reasons of hard-living are working-class and poverty-level women, although a few middle-class women reported that their ex-husbands abused alcohol, and one that her ex-husband used drugs. Said one woman living at the poverty level:

> He's still an alcoholic and he's still into drugs. That was the problem in our marriage. At first he'd be gone one night a month. I would stay up all night worrying and he would come in at 6:30 in the morning and take a shower and leave. I didn't realize what it was at first. Then it became more and more frequent and he was gone a whole lot. . . . It started getting to the kids. [34-year-old white book-keeper, mother of two, married twelve years]

In some of these cases the husbands also used violence.

> He kicked me out of the house, violently. It was our house. By then there was a lot of violence. Things hadn't been that way to start. Initially there was none. But around the time my son was born I noted he was acting funny. I didn't know what it was. I thought maybe he was jealous of the baby because I was paying the baby more attention. But then I came to realize it was drugs. [33-year-old white secretary, mother of one, married seven years]

Evelyn . . . said that drugs made her ex-husband behave in strange and frightening ways.

> The reason for the divorce was drugs. I would have stuck with my ex-husband for anything else, including just plain robbery. But not drugs, because of what it was doing to the kids and me.
>
> He's not a bad guy at all. But when people are on drugs they are fearless. It's scary. You could see all the neighbors stay out of his way. Nobody would cross him. I wouldn't cross him. No way. I think he liked that feeling. The neighbors called him a crazy man. When he was on drugs he would throw things around here. He wouldn't harm me physically very much but he destroyed property. He also liked guns. He would put a gun in my mouth, to my throat. When he started fights he would rip the phone off the wall.

The women were particularly concerned about the effect of drugs on their children.

> I was married a long time. Things were fine for six years. We got along pretty well. Then he decided he wanted other women. He wanted a wife and a girl-friend. But I won't be number two. I told him good-bye and good luck. I also found out during the last eighteen months of our marriage that he was selling drugs. Then I knew I wanted out. That's no way to raise a child. [38-year-old black middle-class office manager, mother of one, married fifteen years]

And of course, men who use drugs or alcohol are more likely to have employment problems and thus not be able to support their families.

> During the marriage we lived off my unemployment insurance. My ex-husband didn't keep jobs. He used alcohol a lot and was a drug addict. When I was eight months pregnant he still didn't have a job. I said, "That's it. We're having a baby in one month and you can't take any responsibility." [33-year-old white poverty-level homemaker with a year-old child] . . .

Other Women

Nineteen percent of the women reported that their ex-husbands were involved with other women. These women described two different kinds of experiences. One group of women stated they were left by their ex-husbands for other women. These women suffered a lot of emotional pain. A 51-year-old white middle-class woman with two college-age children who had been married for twenty-seven years said:

> After he said he was leaving for another woman I was in very, very bad shape. I went into a depression. It continued into the separation. We were in counseling together and I was in alone at the same time. I personally suffered a lot and it can't be measured or weighed. It's taken a toll.

A 31-year-old black teacher with a two-year-old child explained how it was when her husband said he was leaving:

> It was a horrible experience. I was on an unpaid maternity leave. I was a new mother. I had been hoping things would work out in the marriage. We had been having problems, but I thought we could work them out. But he was running around with another woman. And he had been, even while I was pregnant. I felt a lot of hurt and a lot of frustration. . . . The separation was the worst period of my life.

One woman, a 44-year-old white teacher, said that after two children and twenty-one years of marriage, her husband left for a younger woman. She described the experience as "devastating." She developed panic attacks, insomnia, and alcoholism and has seen a therapist for several years to cope with her emotional problems. She is now on friendly terms with her ex-husband, who has broken up with the woman he fell in love with. She finds she still loves her ex-husband and recently went on a date with him. She thinks she would like to remarry her ex-husband but she believes he is in a mid-life crisis.

> I would like to be married to my ex-husband again, if he would only straighten himself out. . . . The therapist says maybe he's learned now and he wants to come back. But the other possibility is that he'll just keep going off and having affairs. He won't go to a therapist. He says it wouldn't do any good. He talks like he hasn't been a success, he hasn't done the work he wanted to do. I don't know what the problem is because he did fine in his job.

Men who leave their marriages for other women are also presumably experiencing some kind of "personal dissatisfaction" with those marriages. They are in a different situation from their former wives, however. They are freer than women to leave marriage and pursue new relationships because unlike women they rarely have to consider that they will have full-time care of their children. Three women did report having affairs after they had decided to separate. For one woman, the affair resulted in remarriage. However, while most of the women who left for reasons of personal dissatisfaction said they would like to remarry, they are not currently in a serious relationship, and unlike their husbands they did not leave their marriages for another relationship.

Women of all classes expressed the same feelings of rejection and emotional pain when they were left by husbands for other women.

> I took my husband's finding another woman as rejection. I felt I failed. The other thing was money. I got really down. I was a candidate for [a local mental institution]. [33-year-old working-class black woman, part-time retail clothing, married seven years, one child]
>
> There was a period of about nine months . . . I knew he was leaving . . . for the first six months I was okay but then I wasn't okay. By not okay, I mean not functioning. I was not a functioning parent, not a functioning daughter, nothing. It's an upheaval and then a tremendous adjustment. [41-year-old white

working-class woman, legal secretary, married twelve years, three children]

> I'd like to stay married to him . . . I still love him. He's the man I married. He will never change. He lives with that lady and he goes to bars. He's a macho man. He still doesn't want me to talk to any man and he finds out if I go out with any man. The year he left I tried to kill myself. I was really depressed cause I never thought this would happen. [40-year-old poverty-level Hispanic woman, married nineteen years, three children]

All of these women experienced great emotional pain, and spoke of being "devastated" and very depressed.

> Emotionally it ripped me apart. I depended on him and when he left me I felt rejected and lonely. The kids played a major part in my life. I lived for them. I didn't want them to get hurt by the divorce. I wish I had gone to counseling earlier to face the reality sooner. [33-year-old white working-class woman, part-time accountant, married twelve years, three children]

A few women mentioned that their ex-husbands wanted to come back when their affairs didn't work out, but these women had lost trust in their husbands and were not willing to try and reconstruct the marriage.

Within the category of women who reported that their marriages ended because their husbands became involved with other women, some women reported that they left their husbands because their husbands were seeing or "fooling around" with other women. Particularly poverty-level and African-American women reported having left for this reason, in contrast to the majority of white women, whose husbands left them. These women whose husbands were "fooling around" were upset not only because of their husbands' involvement with other women, but also because these men were rarely at home and did not take any responsibility for the household. Thus "fooling around" is related to the hard-living category of never being around. . . .

On the whole, . . . we don't know under what circumstances men and women choose to have affairs. Recent data show that 25 percent of men and 15 percent of women in marriages have affairs. Presumably they have affairs when they experience personal dissatisfaction with their partners and their marriages. In this sample, however, when women left for reasons of personal dissatisfaction, they did not leave for a man. Thus, according to these women, when they left they were becoming single parents, while their ex-husbands left to begin relationships with other woman.

Other Reasons for Ending Marriage

I categorized 26 percent of women as having "other reasons" for ending their marriages. Six percent of these women gave no reason for ending the marriage. An additional 10 percent said they were left by their husbands but did not say why. Three percent were left by their husbands but didn't know why. Of the remaining 7 percent, 4 percent said they left their husbands because of their husband's sexual orientation. One woman left her husband because he came out as a transvestite, another because her husband had come out as a gay man. One woman left because of her husband's mental illness. Three percent of women said their husbands had left them for assorted reasons. One woman said her husband had left her because she decided to go back to school for a graduate degree.

According to their accounts of their divorces, the experiences of these women do not support the idea that women are putting their own interests ahead of other family members and leaving their families because of weak commitments to the institution of marriage. These women reported marriages embedded in different gender and class contexts. There was a lot of violence and a lot of hard-living in many of the marriages, and many women reported being left for other women. Even those women who reported leaving for reasons of personal dissatisfaction say that gender issues were part of these reasons and that they were very discontented with male controlling behaviors.

A major factor in these men's behavior, in addition to gender, was class. A higher level of hard-living and violence was evident among working-class and poverty-level men, particularly poverty-level men. Some of these men come from disadvantaged educational backgrounds and employment histories. Forty-one percent of the ex-husbands of poverty-level women did not have a high school diploma. These men, whose median income was $19,500, worked as security guards, cab drivers, truck drivers, or factory workers, or they held jobs in construction, roofing, or maintenance. Some poverty-level women reported that their ex-husbands were unemployed or in prison. Several women spoke sympathetically about the difficulties their ex-husbands had faced making a living. . . .

These women sound like the working-class couples described by Rubin and others, who live at home until they marry and use marriage as way to leave their family of origin. They believe that marriage will provide them with the autonomy they desire. Instead, they are disappointed to find that early marriage and child-bearing quickly create new burdens (Rubin, 1976).

SELECTED REFERENCES

Finlay, Barbara, Charles E. Starnes, and Fausto B. Alvarez. 1985. "Recent Changes in Sex-Role Ideology Among Divorced Men and Women: Some Possible Causes and Implications." *Sex Roles* 12:637–653.

Goode, William J. 1956. *After Divorce.* New York: Free Press.

Kitson, Gay C. and William Holmes. 1992. *Portrait of Divorce: Adjustment to Marital Breakdown.* New York: The Guilford Press.

Howell, Joseph T. 1973. *Hard Living on Clay Street.* Garden City, NY: Anchor Books.

Newman, Katherine S. 1988. *Falling From Grace.* New York: Free Press.

Rubin, Lillian B. 1976. *Worlds of Pain.* New York: Basic Books.

———. 1994. *Families on the Fault Line.* New York: Harper.

Stacey, Judith. 1990. *Brave New Families: Stories of Domestic Upheaval in Late Twentieth Century America.* New York: Basic Books.

CHAPTER 10

Diversity in Family Forms

Observing trends described in earlier chapters, we should not be surprised at the increasing diversity in family types and styles in the United States today. The selections in this chapter focus on a few of these, including remarriage, stepparent families, single-parent families, and adoptive families. Although some scholars and politicians have worried that families are going to disappear, from a sociological perspective, the changes we are witnessing are modifications in the **institution** of the family. The new family systems that are emerging differ from older ones in several ways. Increases in premarital sex, cohabitation, divorce, and nonmarital birth mean that what formerly occurred only inside of marriage now also occurs outside of it. These formerly nonlegalized sexual and parenting arrangements follow predictable patterns like the marriage-based forms that preceded them and thus should be understood as new versions of family structure rather than as nonfamily arrangements. We used to consider it scandalous to have sex before marriage, to live with someone outside of marriage, to have a child without getting married, to get a divorce, or to live with a romantic partner in a same-sex relationship. Though not necessarily accepted by everyone, these activities are now common and considered by most to be normal.

The latest forms of sexual and parenting arrangements have not replaced the conventional **nuclear family** of father, mother, and minor children living together, but these households now make up 24 percent of all households, whereas they were 39 percent of all households in 1970. Married couple households without minor children stayed fairly constant at around 30 percent of all households, but single-parent households increased from just 5 percent of households in 1970 to 8 percent in 2000. The largest growth, however, was in other family and nonfamily households, jumping from 7 percent to 13 percent, and in single-person households, jumping from 18 percent to 25 percent of all households. What we are seeing is the emergence of a permanent diversity of types of families rather than one form taking over from the others. The nuclear family is not disappearing because it keeps on being reformed, though it is not the dominant form it once was. People put off getting married longer these days; hence, there are more of them to count as single-person households, shared-with-roommate households, or cohabiting arrangements. The rate at which people eventually marry is still high, and even though almost half of marriages end in divorce, the rate of remarriage is almost as high as for first marriage.

One interesting result of these patterns of cohabitation, marriage, divorce, and remarriage is that the **kinship structure** is becoming more complicated. In the tradition of the nuclear family, each child has one father and one mother and lives in the same family household until he or she is old enough to leave home and establish another household. With a high rate of **cohabitation,** some children begin living with two parents who are not married to each other (though most do end up marrying). With high rates of nonmarital birth, many children begin living in a single-parent household and some transition to living with both birth parents or with one birth parent and a stepparent. With high rates of divorce, many children transition from living in two-parent households to single-parent households. With high rates of subsequent cohabitation and remarriage, many children have two sets of parents and two or more households. Instead of having one set of brothers and sisters, they have two or more sets, often living in different places. As those who have lived through such transitions are well aware, this makes social etiquette for things such as celebrating holidays or attending wedding receptions considerably more complicated than they once were. And it makes life inside the new family forms considerably less scripted and predictable than our image of the conventional nuclear family from the 1950s.

The first reading in this chapter is from the 1956 book *Remarriage* by sociologist Jessie Bernard. The subtitle of the book, *A Study of Marriage,* hints that readers were likely to feel that the topic was unusual and that such families should be considered as different from "normal" marriages. Bernard first lays out the complexity of remarriage, specifying how remarried families differ according to the previous marital status of the spouses, the presence or absence of children from previous marriages, and the relative significance of earlier marriages. After discussing such variation, she comments on how both the public and family scholars of the time tended to treat remarriage after divorce as a "problem" and how community norms were undergoing change. Finally, she comments on larger issues of technological change and families. Although she notes that adherence to older normative prescriptions and Victorian ideals creates problems for modern families, she also observes that without the old "institutional props," families based on love and personal satisfaction are difficult to maintain. Many of Bernard's observations about modern families are still relevant today, even as remarried families have moved into the mainstream and other multiple family forms based on love and individual commitment have emerged.

The selection by psychologist Jeanne Brooks-Gunn is a review of the literature on stepparenting from psychology, sociology, and economics. This article reports technical details about the studies that have been conducted as well as identifying omissions in previous research. Like the Bernard selection on remarriage from four decades earlier, this one reacts to previous findings that focused on the shortcomings of stepparenting families and presents a typology of the major differences among stepfamily types. Brooks-Gunn argues for using a process model to understand the transitions that families face and suggests that researchers control for preexisting family conditions and problems associated with financial limitations. One of the major puzzles about research on stepfamilies is why children from stepparent families resemble children from single-parent families more than they do children from original-parent families. Although stepfamilies tend to have more resources and parental time than single-parent households, they share some important characteristics with the latter. Brooks-Gunn links disparate literatures on parental commitment, income and poverty, neighborhood and schools, and family systems and concludes with recommendations for more and better research on the many types of stepfamilies that are becoming increasingly prevalent.

In the next article, legal scholar Stephen Sugarman reviews both academic and popular conceptions of single-parent families and considers various policies designed to aid or discourage them. The selection begins with a categorization of the demographic, legal, and relationship status of single parents. Sugarman describes several different types of single-parent households and shows how definitional boundaries between them are somewhat fluid and overlapping. Single mothers rarely conform to the myths that satirize and rebuke them, though more than a third do live below the poverty level (even after government support is taken into account). Sugarman provides an historical overview of policies toward single parents beginning with the first White House Conference on

Children convened in 1909. He chronicles changing attitudes toward divorce, "illegitimacy," child support, and cohabitation, as well as summarizing policies directed toward each of these issues. Sugarman profiles recent policy reforms and current initiatives in child support and divorce law, concluding that to be effective, policy proposals should be designed to address the needs of children.

The last reading in this chapter focuses on adoptive families in the United States. Christine Ward Gailey summarizes research in this area and discusses how adoptive families are similar to, and different from, other families. She shows how informal adoption has been common in America since the beginning of the country and how various forms of adoption are growing today. Gailey differentiates among stepparent adoption, independent adoption, international adoption, and public agency adoption, pointing out trends and issues associated with each type. In discussing how adoption is influenced by global politics as well as the race, class, gender, and sexuality of both adopters and adoptees, Gailey highlights the importance of social stratification in our understanding of adoptive families. In discussing the rhetoric and ideology of "natural" and "unnatural" kinship, she shows how all families are socially constructed, whether parents and children are genetically related or not. At the end of the article, the author suggests that looking at adoptive families can tell us something important about the ways that all families work. Why do you think she makes this claim?

All of the readings in this chapter focus on how diverse family forms are becoming more common. Most also advocate a focus on how children in these families thrive or face challenges. Do you think that the authors treat all families equally? Do you think that all family forms should be valued?

SUGGESTED READINGS

Amato, Paul R. 1994. The Implications of Research Findings on Children in Stepfamilies. In A. Booth and J. Dunn (Eds.), *Stepfamilies: Who Benefits? Who Does Not?* (pp. 81–87). Hillsdale, NJ: Erlbaum.

Cath, Stanley H., and Moisy Shopper, eds. 2001. *Stepparenting: Creating and Recreating Families in America Today.* Hillsdale, NJ: Analytic Press.

Cherlin, Andrew J. 1992. *Marriage, Divorce, Remarriage.* 2nd ed. Cambridge, MA: Harvard University Press.

Dowd, Nancy E. 1997. *In Defense of Single-Parent Families.* New York: New York University Press.

Ganong, Lawrence H., and Marilyn Coleman. 1999. *Changing Families, Changing Responsibilities: Family Obligations Following Divorce and Remarriage.* Mahwah, NJ: Lawrence Erlbaum Associates.

Ganong, Lawrence H., and Marilyn Coleman. 1994. *Remarried Family Relationships.* Thousand Oaks, CA: Sage Publications.

Ludtke, Melissa. 1997. *On Our Own: Unmarried Motherhood in America.* New York: Random House.

Stiers, Gretchen A. 1999. *From this Day Forward: Commitment, Marriage, and Family in Lesbian and Gay Relationships.* New York: St. Martin's Press.

VanEvery, Jo. 1995. *Heterosexual Women Changing the Family: Refusing to be a 'Wife'!* Bristol, PA: Taylor & Francis.

Weeks, Jeffrey, Brian Heaphy, and Catherine Donovan. 2001. *Same Sex Intimacies: Families of Choice and Other Life Experiments.* London; New York: Routledge.

Wu, Zheng. 2000. *Cohabitation: An Alternative Form of Family Living.* New York: Oxford University Press.

INFOTRAC® COLLEGE EDITION EXERCISES

The exercises that follow allow you to use the InfoTrac® College Edition online database of scholarly articles to explore the sociological implications of the selections in this chapter.

Search Keyword: Family diversity. Using InfoTrac® College Edition, how many articles can you find that discuss diversity in family structure? What types of families do these articles discuss? How are they diverse? What is the trend in diversification of family structures? Why do these authors believe there are so many different types of families today? What are the benefits and opportunities associated with each of the types of family discussed? What are the drawbacks and problems associated with each? What laws and/or public policy proposals are being advanced to address issues of different family forms? What is the projection for diversification of family structures in the future?

Search Keyword: Grandparents and custody. For a number of reasons, grandparents are increasingly being asked to assume custody of their minor grandchildren. Use InfoTrac® College Edition to search for articles that have to do with the issue of custodial grandparents. Why is this an increasingly prevalent phenomenon? What are the benefits of custodial grandparenting for the child? for the custodial grandparents? for society? What are some of the problems faced by grandparents assuming custody of their grandchildren? How are these problems being addressed by the custodial grandparents and by society as a whole? What institutional supports and rights (laws, funding, etc.) do custodial grandparents have? What supports and rights do they lack? What are the implications of custodial grandparenting as a new "type" of family form?

Search Keyword: Foster families. Using InfoTrac® College Edition, look up the key phrase *foster families.* What is a foster family? How does a foster family differ from a family in which the child(ren) have been formally adopted? What are the institutional supports for foster families? What institutional supports are unavailable to foster families; i.e., how does society fail to support foster families? What are the benefits of the foster family situation for the child? What are the disadvantages? What are some of the problems associated with the foster family system in the United States?

Search Keyword: Blended families. Search for articles that deal with blended families. What is a blended family? What are the different ways that blended families can be created? What are some of the opportunities associated with living in a blended family? What are some of the problems? What are some of the transitions that children moving into blended families must make? How can parents avoid potential problems associated with blended families? What public policies can help address issues related to blended families?

Search Keywords: Cohabiting and children. An increasing number of children are born to parents who are cohabiting without being legally married. Using the search keywords *cohabiting and children,* find articles that address the issue of children raised in homes with cohabiting parents. What similarities exist between these cohabiting home environments and the home environments of children brought up with married parents? What differences are there between the two types of child-rearing situations? What problems do cohabiting parent homes present for children? Are there benefits of living in this type of family? What institutional supports (laws, policies, funding, etc.) are available to cohabiting parents? What supports are lacking? Are there any public policy programs proposed to address issues related to children in cohabiting parent homes?

32

Remarriage

JESSIE BERNARD,
1956

DIFFERENCES AMONG REMARRIAGES

Remarriage is by no means a simple sociological phenomenon, and remarried persons are by no means easily classifiable. First marriage itself involves many kinds and types of relationships, among many kinds and types of people, with many kinds and types of problems; to these variations in relationships and people and problems, the institution of remarriage adds still others which are peculiar to it and which differ from those of first marriage in both degree and kind. The differences among remarriages and among remarried people may, indeed, be even greater than the differences between remarriages and first marriages.

We shall discuss briefly here three major sources of differences among remarriages. . .

1. the previous marital status of the spouses;
2. the presence or absence of children by previous marriages; and
3. the relative significance of the first and subsequent marriages.

Previous Marital Status

The first and most significant distinction among remarriages is based upon the previous marital status of the spouses. Remarriages that take place after one or both partners have been bereaved differ so greatly from those involving divorced partners as virtually to warrant separate study. As we shall see throughout this book, there is a difference in incidence of almost every variable we scrutinize between the remarried divorced, as a group, and the remarried widowed.

These two types of remarriage differ especially in four respects: (1) in the kinds of people involved; (2) in the impact of the previous spouse; (3) in the community evaluation of the manner in which the first marriage was terminated; and (4) in the amount of guilt feeling which may be present. Two additional factors must be considered: (5) the complexity introduced by the mating of a man and a woman with different marital histories; and (6) the heterogeneity of the divorced population itself.

The Kinds of People Involved. The men and women who marry more than once are not selected at random. They do not constitute a cross section of all adults. Rather clear-cut forces, as well as many more or less obscure ones, are at work selecting them. First of all, their first marriage selected them into the ever-married population. Then death or divorce se-

From Jessie Bernard, *Remarriage*. pg. 5–9, 26–37, and 331–333. New York: Dryden. Copyright © 1956.

lected them into the population of persons *formerly* married—those now eligible for second marriages. But death and divorce select different kinds of people. . . . The people who remarry after the death of a spouse are different, then, from those who remarry after divorce.

The Impact of a Previous Spouse. The impact of the previous spouse on the new marriage is generally not the same in remarriage after divorce as in remarriage after death. In both cases, the impact of the former partner may be favorable or unfavorable for the new marriage—but not in the same way. The major difference is, of course, that in remarriage after divorce the former spouse is still living and may continue to intervene actively in the new relationship. . . .

Even when the former partner does not actively intervene in a remarriage, he or she may nevertheless exert an influence. The first spouse may, for example, be used in a second marriage as an ally against a present husband or a wife: "No wonder John couldn't live with you!" shouts the angry spouse, strategically reminding his mate of an earlier marital failure. The coalition may also operate in the other direction, the first spouse remaining a perennial rival to the present one, and the previous relationship thus constituting a threat to the second marriage.

In the remarriage of a widowed person, also, the first spouse may be a rival to the new partner, but with the additional and "unfair" advantage of being no longer present to remind the living spouse of the frailties and foibles of his former partner. To criticize a deceased person constitutes a breach of taste. The first spouse may be so idealized and memories of him may become so selective that he becomes virtually sanctified. In retrospect, the past often seems better than it really was. There is also, for the person who marries a widowed man or woman, the ego-threatening knowledge that the present mate did not voluntarily terminate or destroy his or her previous match—that but for death the new marriage would not have taken place.

On the other hand, the former spouse may have a favorable impact on the second marriage, whether the remarriage takes place after divorce or after bereavement. But even in this case remarriage after divorce typically differs from marriage after bereavement.

When one member of a marriage has been divorced, the present spouse may derive support and comfort from the fact that comparisons between spouses are likely to be in his or her favor. "My husband used to tell amusing stories about the foibles of his first wife," one informant reports, "and this made me feel secure, because I knew that so far as the outside world was concerned she was far superior to me. It was nice to feel that *in his eyes* I was superior."

When remarriage occurs after one partner has been bereaved, the impact of the first spouse may also be favorable, but in a quite different way: the second spouse may feel that he would not have been chosen unless he met the standard set by the first husband or wife.

In at least one respect, the comparison between first and second spouses may be the same for the remarried widowed and the remarried divorced: the second marriage may be better for both. One informant, a widowed woman who had always considered her first marriage highly satisfactory, found that her second marriage was so far superior that, by comparison, the first seemed less successful than she had believed. She had not known her own capacity for happiness, and, with no standard of comparison, she had been satisfied with her first marriage. But her second husband was so much more congenial, so much better suited to her, so much more responsive to her needs that, in retrospect, her first husband seemed far less exemplary than he had in his lifetime. This is a reaction more characteristic of remarriage after divorce, but it may also occur in remarriage after bereavement.

Community Evaluation. Since marriages must function in a social matrix, the community evaluation of the manner in which the previous marriage was terminated can be an extremely important factor. In this respect also, remarriages after death are likely to differ from those after divorce. . . . Bereavement elicits sympathy from the community; divorce may evoke opprobrium. Despite increasing recognition of the necessity for divorce in certain circumstances, there is still much opposition to it. To some, it is a "release from intolerable bonds, an end to the most dreadful suffering, the advancement of good morals by enabling unhappy husbands and wives to contract respectable instead of adulterous relationships and to terminate pretended and barren unions for more lasting and fruitful ones." But to others, it is "the basest form of betrayal, evasion, and compromise, a breeder of sexual promiscuity, disruptive of the family, an interference with God's will, undermining the very foundation upon which our society is erected." With regard to its effects on children, "divorce is approved because it liberates children from hostile domestic environments"; or "it is condemned because it deprives them of home and family life and subjects them to a relentless-tug-of-war by parents."

These divergent attitudes toward divorce reflect a great moral conflict of all time, one that has especially wide ramifications throughout our culture today—that of the individual against the institution. Traditionally, in part because of our pioneer background, we have tended in our culture to support the individual in this conflict with the institution; we have tolerated, even encouraged, nonconformity. We have tended to accommodate the rules of society to the wishes of the individual. . . .

But so long as the public attitude toward divorce varies from community to community, the status of the divorced, as contrasted with that of the widowed, will be equivocal. It is often difficult for divorced persons to know how they will be judged, especially after remarriage; what their precise status is in the community; whether they do or do not have community support; whether they must prepare to fight a hostile world or can rely on a sympathetic one. . . . Discrimination against divorced people is no longer so institutionalized as it once was, but neither is any other policy toward them.

Guilt Feelings. The two factors we have just considered—the impact of a former spouse and the equivocal attitude of the community—may combine to inflict on those who remarry a heavy emotional tax in the form of feelings of guilt. The difference in this respect between remarriage after death and remarriage after divorce, or even between remarriage and first marriage, is a matter of degree rather than of kind. Partners in a first marriage may feel guilty if, for example, they have married against the will of their parents or after a previous broken engagement. Partners in a remarriage after death may feel similar pangs of guilt if they had ever wished for the death of their first spouse or had ever experienced even a fleeting desire for freedom from the bonds of marriage. But guilt feeling in the case of first marriage or remarriage after death is not likely to be so severe as in remarriage after divorce.

The occasions for guilt feeling in remarriage after divorce are numerous. A man who had forced a divorce on his partner, for example, may feel guilty, especially if his second marriage is a happy one, even more so if his first spouse is not remarried or is unhappy. Guilt-feelings may have protean manifestations. And the compensatory devices that psychologically enable the individual to tolerate the guilt may produce difficulties in the new marriage. Too great solicitude for a former spouse, or too great hostility; too great solicitude for children, or too little—whatever means the individual uses to deal with his problem of guilt may have hazardous consequences.

Sometimes even the children become enmeshed. Also feeling guilt, fearing that they were the cause of their parents' divorce, they may find it difficult to be at ease in a new family constellation.

Thus, remarriage after divorce differs from remarriage after death both in the likelihood that guilt feelings will be present and in the degree of guilt feeling experienced by the remarrying spouse.

Combinations of Previous Marital Status. The fact that there are at least eight possible kinds of mating in remarriage makes the remarried population extremely heterogeneous with respect to previous marital status. A remarriage may take place between (1) a widowed man and a widowed woman, (2) a widowed man and a divorced woman, (3) a widowed man and a single woman, (4) a divorced man and a widowed woman, (5) a divorced man and a divorced woman, (6) a divorced man and a single woman, (7) a single man and a widowed woman, and (8) a single man and a divorced woman. . . .

SOME INSTITUTIONAL PATTERNS OF REMARRIAGE IN OUR CULTURE

With respect to the remarriage of widowed persons, the institutional pattern in our culture lies between the two extremes of prohibition and prescription. . . . We no longer uphold the romantic, one-love ideology. (Indeed, some people even argue that remarriage is the greatest possible compliment one can pay to the deceased spouse.) Nor are there any legal or religious obstacles to the remarriage of the widowed. But our social conventions require that a "decent" interval—usually a year—intervene between bereavement and remarriage and that "due respect" should be paid the deceased. Emily Post once formulated the conventional pattern for a widow as follows: she should never remain in mourning after she has decided to remarry; it is quite acceptable for a widow to remarry but she should not accept attentions from a suitor before the "year of respect" has passed, especially if her first marriage had been a happy one (Post, 1945).

The regulations with respect to divorced persons are more restrictive. Indeed, they are often of a punitive or repressive nature, their fundamental objective being to discourage divorce by making it difficult for the divorced to remarry. Many religious groups do not permit their ministers to perform the marriage ceremony for divorced persons, especially for the "guilty" parties. . . .

The legal restrictions against remarriage after divorce in our culture are perhaps less drastic than these religious ones, but they have essentially the same purpose; to discourage divorce. Some states require that an interval of time elapse between the final divorce decree and remarriage. Laws in several states require the "guilty" party in a divorce granted on grounds of adultery to wait longer than the plaintiff before remarrying. In New York, for example, the "guilty" party may not remarry for three years, and then only if he can prove good behavior. . . .

Among the impediments to the remarriage of divorced persons are social conventions—perhaps less binding now than formerly—whose effect is similar to that of religious and legal restrictions. Thus, according to Emily Post, a divorcee may send out invitations to or announcements of her remarriage only when she is so clearly "innocent" that public opinion will sanction her right to remarry (Post, 1945). Interestingly enough, however, the bride of a divorced man may send out announcements of the marriage regardless of the groom's "guilt" or "innocence." It would appear, then, that the divorced woman is required to demonstrate her probity and consequent right to happiness before she can obtain community sanction for her remarriage, but the divorced man is not.

In general, however, although legal and religious restrictions concerning remarriage of the divorced are more stringent than those governing remarriage of the widowed, there appear to be fewer conventional rules bearing on remarriage of the divorced. What conventions there are tend to be up-to-date, simple, and much less standardized than those that apply to remarriage of the widowed, varying widely from one community to another according to the prevailing attitude toward divorce.

CHANGING COMMUNITY ATTITUDES TOWARD REMARRIAGE IN OUR CULTURE

The controls—religious, legal, and conventional—discussed above are by no means fixed and unchanging, as we have pointed out. I referred briefly to the change in attitude toward divorce which is now in process. Accompanying this change is a change in attitude toward the remarriage of divorced persons.

Until recently, most laymen and even most students of the family tended to consider all remarriage as a "problem." They took an especially disapproving view of remarriage after divorce, which they pictured in terms of quarrels and frustration, suffering children, and ruthless spouses fleeing from their legitimate responsibilities. It has been suggested that some of the disapproval may have stemmed from repressed envy masked as moral reprobation. Even remarriage after widowhood suggested to these critics the inevitability of such problems as "stepchild-stepparent" hostilities resulting from opposition of the children to the remarriage of the surviving parent and their resentment that the bereaved could be so quickly consoled. This attitude also perhaps stems from envy on the part of some that bereavement had honorably released another for a venture that, for them, could be made possible only by divorce, with its attendant scandal. But the conditions of remarriage are now in process of redefinition in our culture; community attitudes seem to be incorporating the most recent "facts of life" and to be stamping remarriage with acceptance if not with positive approval. . . .

THE FAMILY AND TECHNOLOGICAL CHANGE

Even if remarriage were becoming less prevalent, it would still be an interesting and important aspect of social life in our country. But the phenomenon is not waning; it is probably going to have an increasing influence on family life, at least in the immediate future. And increasingly, it is going to take place after divorce rather than after bereavement. Like divorce itself, remarriage is part of the accommodation now taking place in all aspects of society to the style of living demanded by modern technology.

We are so close to the family and its problems that we often lose sight of the long-term trend. The family has a long history. For the thousands, perhaps the hundreds of thousands, of years during which human beings supplied their wants by hunting, the mother and her children constituted the central, stable core of family life. Relatively recently—roughly ten thousand years ago, with the introduction of the plow into agriculture—patriarchy became the typical scheme in Western societies, for it seemed well suited to an agricultural economy. To millions of people today, it seems still to be the most "natural" way in which to organize family relationships. A strong father, with authority, prestige, and, concomitantly, serious responsibilities, a compliant wife, and obedient children—these constituted an efficient human grouping for an agricultural or domestic manufacturing economy. The costs to individual members of the family as thus constituted in terms of initiative, freedom, and individualization did not seem excessive, for everyone paid the same price. The reference-group values were the same for all. The advantages were many—security, stability, and a sense of belonging to a cohesive group, with a specific function in the group. Individuals might have been lonely for friends who could share personal interests, but there was little social isolation. Kinship ties were strong and binding. One always belonged. In recent years, the so-called Victorian family has come to stand for the archetype of the patriarchally organized family.

But even in the nineteenth century the heyday of the patriarchally organized family was past. The foundation of this type of family structure was already beginning to crumble under the weight of new technological advances, as revolutionary as the discovery of agriculture. Factories were growing up, and, around them, industrial cities. The production of goods and services was moving, little by little, out of the home into the factory. The development of new technical processes was reshaping industry, and the economic organization of society was slowly accommodating itself to the new modes of production. Political organization too was also being reshaped to fit the new industrial pattern, for neither the economic nor the political structure of a predominantly agricultural economy could withstand the impact of the new industrial system.

Nor was the previously satisfactory pattern of family organization able to survive in this new culture. Factories, not families, became the producing units. Except in agriculture, the family has ceased to be a functioning economic unit in production. Although a larger proportion of adults than ever before are now in the working force, they work as discrete individuals, at jobs that have no necessary relationship to the jobs of other members of the family, making products that do not directly satisfy the wants of the family. Work, in short, is no longer a family enterprise. The world of work is wholly separate from the world of the family.

And so for almost two centuries—a very brief time in human history—we have been trying to develop a model for organizing the family suitable to the conditions of modern living. The significance of the change in the producing unit to the institution of marriage and the family is that members of the family are no longer totally dependent upon one another for economic and social necessities. That is, life is possible outside of families. Clothing, shelter, food, and social activities are available to people—even to women—outside the framework of marriage. In pre-industrialized society, people were kept within the family group not only by institutional sanctions but also by the absence of any alternative way to live.

Alternatives, even pleasant ones, now exist. And institutional sanctions are slipping in their hold on people. We have come to depend to a large extent if not wholly on personal ties of love and affection to maintain family stability. . . .

Modern life, then, has knocked the institutional props from under the old type of family; it has attenuated the concept of duty as related to family life. It has left little but the slender support of love and satisfaction to sustain the family as a unit.

In many cases this support is not enough, for modern life does not prepare everyone for the kind of discipline that living in a modern family entails. A marriage that must constantly generate its own support is much more difficult to maintain than one upheld by a sturdy underpinning of institutional props. For all its freedom—indeed, perhaps because of it—the current form of family organization is difficult for the persons involved. Little by little, a body of knowledge about modern family life is being built up, and courses are being introduced into schools and colleges; it is conceivable that in time we may succeed in training people for the kind of marriage and family life required by our present social structure. But as yet we can offer relatively little guidance.

SELECTED REFERENCES

Post, Emily. 1945. "Etiquette." *Funk & Wagnalls.*

33

Research on Stepparenting Families

JEANNE BROOKS-GUNN,
1994

Much of the stepfamily research to date focuses on marital or parental states, rather than on changes in these states, as *transition* implies. Consequently, most studies focus on residence in various family forms. Much of the work is descriptive and comparative in nature, rather than process-oriented. The three family forms most often compared are: original-parent or traditional families (defined as a family with the biological mother and father residing with their children), ever-married single-parent families (in most cases, comprised of the mother and her children), and stepparent families (a biological mother who has been married previously, and a stepfather who reside with the children).

The benefits or costs of any particular household or parenting arrangement are not only studied by comparing different family forms, but also investigated by looking at changes in these arrangements. Marital transitions have become the object of longitudinal study, which allows for an understanding of how changes in parental marital status and household residence influence changes in children's adjustment. . . .

A transition-oriented perspective allows for an examination of the characteristics of parents and families (and, to a lesser extent, children) that undergo transformations versus those who do not. . . .

Most of the research in this tradition focuses on the two primary transitions that result in the three most studied family configurations. These are the transitions of divorce and remarriage.

DEFINING STEPPARENTS

The most frequent stepparent family form studied is the custodial mother, her children, and her spouse. Other family forms are not typically studied, even though several include stepparents. These include at least five types of families.

1. Stepparent families may have the father as the custodial parent and the new wife as the stepparent. Recent research. . . does address this stepfamily type, which is briefly discussed in this chapter.

2. Stepfamilies may also be formed when the custodial mother has not been married prior to her marriage to the stepfather. Although there is a voluminous literature on the life circumstances of the never-married single mother and the consequences for her children—especially those who give birth in their teenage years or in their early 20s. . ., little work follows these mothers and their children into stepparent situations. These stepparent families are also not included in most comparative analyses of

the effects of residing in different family types, although recent work is separating out never-married and ever-married single mothers into two groups when comparisons among single, original-parent and remarried families are made. . . .

3. Never-married single fathers also may marry and become stepparents. This group received no research attention.

4. Custodial parents who marry a third or fourth time bring new stepparents into the household. The same is true for noncustodial parents who remarry several times. Little research has focused on multiple remarriages, or on transitions from one stepparent family to another. Additionally, almost nothing is known about the continued relationships of previous stepfathers (or stepmothers) and children. However, children will sometimes list a previous stepfather as someone "who is like a father to me." Research is needed on these more complex family relationships, and these larger but perhaps, more diffuse social networks.

5. The final stepparent family form to be mentioned involves cohabitating adults with children from a previous marriage. Although not a stepparent family in the legal sense, these families present similar challenges to remarried families in that the mother, her partner, and her children must forge new relationships, renegotiate old ones, and develop a new family system. While little research addresses this family form, a small but significant number of families consist of cohabitating adults with children from either the man or the woman. . . .

This family form may evolve into a more traditional stepparent family. Its existence may confound some of the research findings for remarried families. Family systems seem to go through a period of disequilibrium in the several years following a remarriage (increased conflict, decline in parental responsibility, difficulties in acceptance of the stepparent/stepchild, decreased monitoring and supervision, and altered individual alliances), after which the family system seems to stabilize (Hetherington, 1993). The advent of cohabitation needs to be studied, since disequilibrium may occur when the adult partner enters the household and not when a remarriage occurs. Research on remarried family forms may be confounding length of time in the new family form with the actual remarriage, such that periods of disequilibrium might actually be periods of stabilization. Alternatively, it is possible that a remarriage after a period of cohabitation signals a second reorganization of family relationships (e.g., the stepparent takes on a more active parenting role, the stepchild accepts the inevitability of the stepparent staying in the household, the mother actively promotes a more active parenting role for her spouse, and so on). Neither the condition of cohabitation prior to remarriage nor the condition of cohabitation without eventual remarriage has been addressed. Cohabitation without remarriage probably confers different roles and responsibilities upon stepparents than does remarriage. Cohabitating partners may be less involved in family life than are stepparents, although one recent study did not support this premise (Thomson et al., 1992).

Cohabitation also may occur in families where the single mother has never married. Little information is available on this family form.

What is clear from this brief description of various stepparent forms is that remarriage and stepparenthood are not equivalent. Unfortunately, almost all of the research on stepparenting focuses on a first remarriage rather than a first marriage, cohabitation, or multiple marriages. . . .

Models Used in the Study of Stepparenting

Many different models are used to frame the discourse on stepparents. It is beyond the scope of this chapter to summarize these models regarding their use and influence in studying stepparent families. Economic models consider the decisions on income distribution within the family, the provision of income to children living outside one's household, compliance with child support awards, and changes in economic behavior as a function of marital status changes (e.g., the remarriage of the noncustodial parent). Effects of income upon children's outcomes are also examined as a function of family type.

A series of family systems or parental behavior models focus on how individual family members, as well as the family as a whole, respond to changes in the family. Much of this research takes a transition-oriented approach, postulating that moves of parents and stepparents in and out of the child's household trigger a reorganization and renegotiation of relationships, roles, and responsibilities. Hetherington (1993; Hetherington & Clingempeel, 1992) has made the important distinction between the phase immediately following a parental transition and subsequent phases. The early phase is marked by disequilibrium in the family system; the system stabilizes after a period of several years.

Another conceptual approach is based in risk and resilience models, as championed by Rutter and Garmezy (1983). This approach takes as its starting point the fact that reactions to parental disruptions are quite variable. Individual differences in adaptation to events such as parental changes are the focus. Factors that promote positive adaptation to such an event are explored (protective factors) as well as those that impede adaptations (risk factors). The individual difference approach considers a wide range of characteristics within the child, the parent(s), and the stepparent, as well as the relationships among them. Risk and resilience models are often used in conjunction with family systems models. While comparisons among family types are not the focus, comparative questions arise regarding the pattern of associations among variables within each family type: for example, whether mother–child conflict operates similarly with regard to child outcomes in original-parent, single-mother or stepfather families, or whether certain children, by virtue of characteristics such as temperament or reactivity, respond differentially to such conflict across family types.

More contextual models attempt to place the child of a stepparent family in a broader context. Bronfenbrenner's ecological model is the prime exemplar of this approach (Bronfenbrenner, 1986; Bronfenbrenner & Crouter, 1983). How the peer, neighborhood, school, and kin networks are altered by a parental transition is the object of study. Whether these contextual changes influence the family and the child is explored, as well as whether the characteristics of contexts interact to influence children. . . .

Social network and role theories are also useful in studying stepparent families. As White (1994) beautifully demonstrated, contact and support between stepparents and stepchildren are relatively infrequent after the stepchildren become adults and no longer live with the stepparent. The demonstration of such loose step-ties leads us to question the stepparents' role in the family (Rossi & Rossi, 1990; Thoits, 1992). Loose step-ties also raise questions about the primacy of biological sameness in the commitment of parents to offspring.

Prevention models focus on approaches to altering behavior in families. Little prevention work has focused on stepparent families. However, programs have targeted high conflict families or families in which the spouses have just become or are about to become, parents (Cowan & Cowan, 1990. . .). Via social skills and conflict resolution training, and identification of common stressors on families, goals of such programs are to reduce conflict and potential marital disruption. Such approaches could easily be applied to stepparent families.

All of these models provide useful frames for the study of stepfamilies. They often overlap in interesting ways. For example, all models speak to the investment and commitment of stepparents (or lack thereof). However, different measures and theory are garnered to support the premise that stepparents are, on average, less committed to their stepchildren than are original parents. Much more integrative research across models is needed.

PROCESSES UNDERLYING NEGATIVE OUTCOMES OF CHILDREN IN STEPPARENT FAMILIES

The outcomes of children in two-parent families differ as a function of the identity of the second parent in the household. Children in stepparent families often have lower school-related achievement scores and more adjustment problems than children in original-parent families. They are similar to children in ever-married single-parent households, although achievement decrements are somewhat less likely to be seen in children of stepparent families than in single-parent families (Thomson et al., 1992 . . .). However, variations are seen with respect to timing of parental disruptions in the child's and parents' life, duration of different family structures, and intensity of initial responses to such disruptions. These critical topics are not discussed here.

Single parenthood is known to confer risk for children, in part due to low income, drops in income (compared to family incomes prior to divorce), moves to new neighborhoods and schools, and possibly, reduced amount of time in parent-child interaction (due in part to the availability of one parent rather than two parents in the household, and to employment patterns in single-parent families). Stepparent households have several advantages over single-parent households, which might be expected to translate into better adjustment for the children residing in them, compared to single-parent households. Stepparent households are not as poor, on average, as ever-married single-parent households. Additionally, two parents reside in the household, rather than one parent, making adult time for child care and interaction more available.

Given these possible advantages, why do children in stepparent families look so similar to children in single-parent families? Economic, psychological, and family system perspectives all speak to these findings. Data from each of these perspectives are discussed briefly, in order to elucidate the mechanisms underlying the poorer adjustment of children in stepparent than in original-parent households. Economic perspectives focus on income, moves to new neighborhoods or schools, and adult time available for children in the household (see McLanahan et al., 1991). The psychological perspectives considered here focus on parental commitment to children in stepparent families. Family systems perspectives address relationships among family members. Comparisons are typically made among original-parent, ever-married single-parent, and stepparent families. Outcomes are school achievement (grades, high school drop-out, postsecondary schooling) or emotional problems (behavior problems, leaving home early). The most comprehensive comparative work has been conducted by McLanahan and her colleagues (. . . McLanahan et al., 1991; McLanahan et al., in press; Thomson et al., 1992; 1993) on the first two perspectives and Hetherington (1993) . . . on the third perspective. Research across perspectives, however, still does not adequately address whether or not the negative effects of stepparent families are really due to the first description—the divorce of the original parents.

Economic Perspectives

Income Differences in Original-Parent, Single-Parent, and Stepparent Families. A marital dissolution results in many children living below the poverty threshold. This is because mothers are usually the custodial parents and because fathers earn more income and are more likely to be employed than children. As incomes for the custodial parent declines, it increases for the noncustodial parent. The incomes in households a year following a marital dissolution tell the story—single mothers have 67% of the pre-divorce family income and fathers have 90% of the pre-divorce income (based on data from the PSID; Duncan & Hoffman, 1985). . . .

Family incomes rise with remarriage, but not always to levels comparable with those of original marriages. McLanahan et al. (1991) documented that stepfamilies have higher incomes than single-parent families, but not always as high as those reported by original-parent families (race/ethnic and sample differences are found for these comparisons).

Even if the national studies uniformly documented that stepparent families have incomes comparable to those of original-parent families, children in stepparent families are more likely to spend time in low-income, single-parent families. Persistence of poverty conditions is associated with child outcomes: longer bouts of poverty are associated with decrements in well-being, compared to more transient spells of poverty (Duncan et al., in press; Haveman, Wolfe, & Spaulding, 1991). Consequently, stepparent families may have higher incomes than single-parent families, but children in both families are more likely to have spent some childhood years in poverty than children in original-parent families. Effects of living in stepparent families need to be modeled, taking into account income levels and changes in income levels over time, including the years preceding the formation of a stepparent family (see Knox & Bane, in press).

Income volatility is also more likely in single-parent households than in two-parent households. Women experience a sharp drop in income following a divorce or separation. Some gains are made for those who may respond to the loss of income by entering the work force or increasing the number of hours of work. Changes in labor force participation, or difficulty in securing a stable job, can precipitate further drops and possible rises. Remarriage results in a rise in income for many families and perhaps short-term income volatility. No studies accounted for income volatility in the study of stepparent effects upon children.

Income Differentials and Outcomes in Stepparent Families. Research has addressed whether income or family configuration explains more variance in child and youth outcomes. In an analysis of the outcomes of adolescent girls from the PSID, my research group found that income was a more potent predictor of high school dropout and nonmarital teenage childbearing than was family configuration (Brooks-Gunn et al., 1993). However, income did not account for all of the negative effects of single parenthood following divorce or separation. The question may be framed: How much do income differentials among original-parent, single-parent and stepparent families account for the between-group variance in youth outcomes? PSID analyses by McLanahan et al. (1991) suggest that between two fifths to four fifths of the single-parent family effect on school-related outcomes is accounted for by family income. In contrast to the findings for single-parents, where income differences accounted for most of the between-group variance (original-parent vs. single-parent families) for youth outcomes, income accounted for very few of the differences between original-parent and stepparent families. (These analyses did not take into account the income differentials occurring during the years of single motherhood prior to a remarriage.) However, the risk of poor adolescent outcomes was high in stepparents, often as high as in single-parent families (McLanahan et al., 1991; Thomson & McLanahan, 1993).

Family Income and Neighborhood. While the line of research just reviewed suggests that family income explains many of the effects of residing in a single-parent family, it still does not provide any explanation of how poverty actually operates, nor why poverty does not explain differences between children in original-parent and stepparent families. One obvious possibility is that single parenthood and stepparenthood, operating through income, influences the stability of the family in several ways, including moves to a new neighborhood. Youth in single-parent households are more likely to live in poor neighborhoods than youth in original-parent households. This is not true of youth in stepparent families (McLanahan et al., 1991).

McLanahan and colleagues (1991) also examined the current school climate of youth in single-parent, stepparent and original-parent families.

Using the HSB, they reported that White youth are more likely to attend a school with a high drop-out rate if they are in a stepparent household than if they are in an original-parent family. This effect was not significant for African Americans, but in the same direction (McLanahan et al., 1991). The same was true for youth in single-parent families. Youth in stepparent families also reported that their peers had lower academic aspirations than did the youth from original-parent families. Is this because stepparent families are less likely to invest in the children than are original-parent families, the result being residence in neighborhoods with lower quality schools?

The schools were rated as having more problems by the offspring in stepparent compared to original-parent households. Single-parent households were similar to original-parent households. Why are neighborhood or school characteristics more negative for youth in stepfamilies, as compared to original-parent families, than for youth in single-parent families?

Psychological Perspectives

A number of research lines suggest that parents who are remarried may invest less psychologically (and economically) in their children than do parents in first marriages. Investment is inferred from time spent with the child, supervision and monitoring of the child's activities, and participating in adolescent decision-making.

Stepparent households differ from single-parent households in that the former have more adults in the household. Higher adult-to-child ratios were associated with better child outcomes. However, it is not clear that children in stepparent households spend more time with parents than children in single-parent households. Newly married couples may spend more time together, rather than time with the child. And stepfathers may not spend much time with their stepchildren, as Hetherington's results on the high rates of stepparent disengagement suggest (1993; Hetherington & Clingempeel, 1992; see also Thomson & McLanahan, 1993). More time-use data in different family configurations, and possible links between time and child outcome, would be welcome additions to the field (Lazear & Michael, 1988).

Remarried, single-parent mothers seemed to exhibit lower degrees of social control in comparison to mothers in original-parent families. This was expressed by less routinized household schedules, less supervision of homework, and less assistance in academic plans (Dornbush et al., 1985; Hetherington, Cox, & Cox, 1978; McLanahan et al., 1991; Thomson et al., 1992). These findings were more pronounced for single-parent families. Disagreements about parental plans for their adolescents' further schooling were equally likely in stepparent and single-parent households, as compared to original-family households, at least for the White HSB sample (McLanahan et al., 1991).

How do such indicators of supervision relate to outcomes? In comparisons using the HSB data set, McLanahan and colleagues (1991) reported lower levels of involvement in homework and in planning of the high school program in stepparent families than in original-parent families, with this effect consistent across ethnic groups (stepparent families being one half as likely to be involved, similar to the findings for single-parent families). However, supervision, involvement, and aspirations did not alter differences in youth outcomes between stepparent and original-parent families, controlling for social and economic status. The question remains as to whether supervision is really an explanatory mechanism, or whether or not the measures used in national data sets are reliable (youth report vs. multiple respondents; single items vs. scales). Other types of analyses might be helpful, especially those that examine the effects of supervision as a function of poverty, maternal employment, time spent with child, and so on.

Investment also has been studied regarding the salience of the stepparent role. Thoits (1992) examined the importance of 17 roles to adults, rated using a 4-point scale. Parenthood was rated as the most important of all roles to individuals; stepparenthood (for those individuals who were stepparents) was rated as one of the least important roles. These findings suggest that the role of stepparent may not be seen as particularly salient to most stepparents, which would partially explain the findings on disengagement and low monitoring.

Family Systems Perspectives

Family systems perspectives chart the changes in relationships within the family as a function of parental transitions. The research clearly describes a number of disruptions or disequilibrium in families when a stepparent enters the system. Children are often very resistant to the entrance of another adult into the household. Consequently, when stepparents do try to establish a relationship, many children do not respond. In such cases, as Hetherington (1993) showed, stepparents become disengaged—exhibiting little warmth, control, or monitoring. Stepparents are not necessarily negative, they are just distant. Relationships with the custodial parent often change as well, with conflict increasing.

Distant parenting is not solely the result of children's feelings and responses to the intrusion of another adult—a person who also has ties to the parent. Stepparents are simply less involved with stepchildren, regardless of the children's reactions. They are less likely to exhibit authoritative parenting, for example. And the custodial parent in remarriages may respond with less monitoring, supervision, and less general support (Thomson et al., 1992).

Clearly, the entire family system is altered by a remarriage. Hetherington (1993; Chase-Lansdale & Hetherington, 1990) pinpointed the developmental stages at which remarriage may have the most negative consequences for relationships and for children's outcomes—early adolescence. More developmentally oriented work on the interaction of child's life phase and parental transitions is needed.

Hetherington and others charted the initial child (and, to a lesser extent, the family) characteristics that might moderate the effects of single parenthood and remarriage on families and children. This work is in the tradition of risk and resilience models. One finding is illustrative: Children of parents who divorced or separated were often quite different than children of parents who did not divorce prior to the actual marital disruption (Baydar, 1988; Block et al., 1986; Cherlin et al., 1991). It is believed that these differences arise because families that divorce actually experience more conflict and less cohesion prior to the transition than those who do not divorce. Some of the negative effects of marital disruptions upon children are accounted for by these pre-existing differences, and such findings lead to a re-examination of the ways in which marital disruptions are conceptualized. Children are not just influenced by the actual disruption, but by the entire process leading up to the disruption. Family system perspectives, in the tradition of Hetherington, could offer a way to study family reorganization and stability prior to a disruption, if samples existed where families who do and do not divorce are studied intensively.

Longer-term effects also need to be studied. Kiernan's (1992) recent analysis of the National Child Development Study in Britain suggests that children in stepparent families leave home earlier than do children in other family configurations (single-parent and original-parent families). Youth from stepparent families were more likely to say that conflict was the reason for their leaving home early. Thus, conflict in stepparent families may have significant effects on life course trajectories.

OTHER FAMILY STRUCTURES

Thus far, discussion has centered upon the three most studied family structures—original-parent, ever-married single-parent, and the stepfather families. The effects of divorce and remarriage upon children tend to be confounded with the gender of the custodial parent, as most research focuses upon the mother. A few studies explored both parenting behavior and child outcomes in single-parent and stepparent families including the original father rather than the mother. Using the NSFH, Thomson et al. (1992) looked at parental activities for several family structures, including original-parent families, single-mother, single-father, stepmother, and stepfather families to see whether gender of the parent was associated with parenting behavior within family structure types. Generally, stepmother and stepfather families were similar to one another and differed from original-parent families, being less positive towards the children and engaging in less frequent activities. Single-parent families, whether headed by a mother or father, were also similar to one another. They differed from all of the two-parent families in terms of control and supervision, suggesting that parental monitoring is facilitated by having more than one adult in the household.

While the gender of the parent did not explain differences between original-parent families on the one hand and single-parent and stepparent families on the other, the gender of the parent was associated with parental behavior. Across family structure types, mothers were more involved with their children's lives than were fathers. This is true for custodial and noncustodial mothers alike (see also, Hetherington, 1993).

Thomson and her colleagues examined parental behavior in households where the mother is cohabitating but is not married. In general, parents in this family structure reported behavior that was quite similar to that reported by parents in stepparent families.

These lines of research need to be extended. Additionally, more work on children's responses to these various family structure forms is recommended. Although custodial mothers and fathers are quite similar, the within-family processes, or reactions to the stepparent, may differ by gender of the parent as well as by gender of the child. . . .

CONCLUSION

This chapter focused on several research and policy agendas. Research needs to focus on how remarriage operates in families for whom the remarriage has positive income effects, versus those for whom it does not. How stepparenting and remarriage operate in different cultural or ethnic groups (using economic, psychological, and family systems perspectives) also needs to be considered.

The connection of the noncustodial father is still an understudied topic in remarried and stepparent families. Different custody and child support mandates might influence the process of adapting to the new stepparent family form. Much more research is recommended on these topics, particularly work integrating economic and family system perspectives. In addition, almost nothing has been done regarding policies related to family processes. All policy initiatives address family economics, although policy scholars are considering the potential impact of economic policies on parent–child contact and family conflict.

Programs could focus on helping stepparent families adapt to this sometimes difficult family setting. Programs focus on new parents, and some ventures have universal home visiting during the neonatal period. The programs probably meet with widespread acceptance because health and safety are considered paramount (rather than family conflict or adaptation to the new family form), and because parenthood is a transition experienced by 90% of all adults. The formation of a stepparent family is not normative. Research such as that by Hetherington suggests potential directions for prevention research programs.

Another vexing issue involves the low commitment of many stepparents to their stepchildren. We have not yet begun to grapple with the question of whether or not it is desirable to increase stepparents' commitment (i.e., would the commitment of noncustodial parents decrease with stepparents' commitment?). Even if the answer is yes, it is not clear how to alter parental commitment and concern.

Stepparenting is a family form for many children in the United States. If stepparenting is defined to include never-married parents after they marry a first time, multiple remarriages, and perhaps even cohabitating parents with children from previous unions, even more children would be living with stepparents than the statistics suggest. At the same time, it is important to remember that stepparent families are not the most frequent living arrangements for children who are not in original-parent family situations. Children living in nonoriginal-parent families are most likely to be residing in parent-only families, not stepparent families. The study of stepparent families might be embedded in a larger contextual framework that considers the effects of all types of parental transitions upon children and the family system.

Such a framework would also allow for a more precise explanation of the effects of stepparents upon children. Currently, it is not clear whether the untoward effects are due to the residual effects of the original marital disruption and move to a single-parent household, or to the entrance of a new and unrelated adult into the household, or to a second disruption (independent of the identity of the new parental figure). Looking at various parental transitions would help untangle these possibilities. What happens to children and families when the father moves back into the household after a separation—and many separations do not end in divorce (Baydar & Brooks-Gunn, in press). In what ways are the disequilibrium and restabilization processes similar and different to those in families where a stepfather enters the family? Another example may come from the work on cohabitation. Does remarriage to a cohabitating adult alter the family systems already in place?

A final example involves grandmothers, since so many never–married single mothers co-reside and co-parent with their mothers or other female kin (Brooks-Gunn & Chase-Lansdale, 1991; Chase-Lansdale, Brooks-Gunn, & Zamsky, 1994). The move out of the grandmother's house is a common occurrence for many children in the preschool years. These children are losing a co-parent; how do the effects on the family system compare to those in a divorce situation? And how do these children and their mothers respond to a marriage or cohabitation situation? Although answers to these questions do not yet exist, they point to research agendas for the next decade.

SELECTED REFERENCES

Baydar, N. (1988). Effects of parental separation and reentry into union on the emotional well-being of children. *Journal of Marriage and the Family, 50,* 967–981.

Baydar, N., & Brooks-Gunn, J. (in press). The dynamics of child support and its consequences for children. In I. Garfinkel, S. McLanahan, & P. Robins (Eds.), *Child support reform and child well-being.* Washington, DC: Urban Institute Press.

Block, J. H., Block, J., & Gjerde, P. F. (1986). The personality of children prior to divorce: A prospective study. *Child Development, 57,* 827–840.

Bronfenbrenner, U. (1986). Ecology of the family as a context for human development: Research perspectives. *Developmental Psychology, 22*(6), 723–742.

Bronfenbrenner, U., & Crouter, A. C. (1983). The evolution of environmental models in developmental research. In P. Mussen (Ed.), *Handbook of child psychology* (Vol. 4, pp. 357–414). New York: Wiley.

Brooks-Gunn, J., & Chase-Lansdale, P. L. (1991). Children having children: Effects on the family system. *Pediatric Annals, 20*(9), 467–481.

Brooks-Gunn, J., Duncan, G. J., Klebanov, P. K., & Sealand, N. (1993). Do neighborhoods influence child and adolescent behavior? *American Journal of Sociology, 99*(2), 353–395.

Chase-Lansdale, P. L., Brooks-Gunn, J., & Zamsky, E. S. (1994). Young multigenerational families in poverty: Quality of mothering and grandmothering. *Child Development, 65*(2), 373–393.

Chase-Lansdale, P. L., & Hetherington, E. M. (1990). The impact of divorce on life-span development: Short and long term effects. In P. B. Baltes, D. L. Featherman, & R. M. Lerner (Eds.), *Life-span development and behavior* (pp. 105–150). Hillsdale, NJ: Lawrence Erlbaum Associates.

Cherlin, A. J., Furstenberg, F. F., Jr., Chase-Lansdale, P. L., Kiernan, K. E., Robins, P. K., & Morrison, D. R. (1991). Longitudinal studies of effects of divorce on children in Great Britain and the United States. *Science, 252,* 1386–1389.

Cowan, P. A., & Cowan, C. P. (1990). Becoming a family: Research and intervention. In I. Sigel & A. Brody (Eds.), *Family research* (pp. 1–51). Hillsdale, NJ: Lawrence Erlbaum Associates.

Dornbusch, S. M., Carlsmith, J. M., Bushwall, S. J., Ritter, P. L., Leiderman, N., Hastorf, A. H., & Gross, R. T. (1985). Single parents, extended households and the control of adolescents. *Child Development, 56,* 326–341.

Duncan, G. J., & Hoffman, S. D. (1985). A reconsideration of the economic consequences of marital dissolution. *Demography, 22,* 485–498.

Duncan, G. J., Klebanov, P. K., & Brooks-Gunn, J. (in press). Economic deprivation and early-childhood development. *Child Development, 65*(2), 296–318.

Haveman, R., Wolfe, B., & Spaulding, J. (1991). Childhood events and circumstances influencing high school completion. *Demography, 25*(1), 133–157.

Hetherington, E. M. (1993). An overview of the Virginia Longitudinal Study of Divorce and Remarriage: A focus on early adolescence. *Journal of Family Psychology, 7,* 39–56.

Hetherington, E. M., & Clingempeel, W. G. (1992). Coping with marital transitions: A family systems perspective. *Monographs of the Society for Research in Child Development, 57* (2–3, Serial No. 227).

Hetherington, E. M., Cox, M., & Cox, R. (1978). The aftermath of divorce. In. H. J. Stevens, Jr. & M. Mathews (Eds.), *Mother-child, father-child relations.* Washington, DC: National Association for the Education of Young Children.

Kiernan, K. (1992). The impact of family disruption in childhood on transitions made in young adult life. *Population Studies, 46,* 218–234.

Knox, V. W., & Bane, M. J. (in press). The effects of child support payments on educational attainment. In I. Garfinkel, S. McLanahan, & P. Robins (Eds.), *Child support reform and child well-being.* Washington, DC: Urban Institute Press.

Lazear, E. P., & Michael, R. T. (1988). *Allocation of income within the household.* Chicago, IL: University of Chicago Press.

McLanahan, S., Astone, N. M., & Marks, N. F. (1991). The role of mother-only families in reducing poverty. In A. C. Huston (Ed.), *Children in poverty: Child development and public-policy* (pp. 51–78). Cambridge, MA: Cambridge University Press.

McLanahan, S., Seltzer, J. A., Hanson, T. L., Thomson, E. (in press). Child support enforcement and child well-being: Greater security or greater conflict? In I. Garfinkel, S. McLanahan, & P. Robins (Eds.), *Child support reform and child well-being.* Washington, DC: Urban Institute Press.

Rossi, A. S., & Rossi, P. H. (1990). *Of human bonding: Parent-child relations across the life course.* Hawthorne, NY: Aldine de Gruyter.

Rutter, M., & Garmezy, N. (1983). Developmental Psychopathology. In P. Mussen & M. Hetherington, (Eds.), *Handbook of child psychology: Socialization, personality and social development* (pp. 775–911). New York: John Wiley and Sons.

Thoits, P. (1992). Identity structures and psychological well-being: Gender and marital status comparisons. *Social Psychology Quarterly, 55,* 236–256.

Thomson, E., & McLanahan, S. (1993, August). *Family structure and child well-being: Economic resource versus parental behavior.* Paper presented at the annual meeting of the American Sociological Association, Washington, DC.

Thomson, E., McLanahan, S., & Curtin, R. B. (1992). Family structure, gender, and parental socialization. *Journal of Marriage and Family, 54,* 368–378.

White, L. (1994). Stepfamilies over the life course: Social support. In A. Booth & J. Dunn, (Eds.), *Stepfamilies: Who Benefits? Who Does Not?* (pp. 109–137). Hillsdale, NJ: Lawrence Erlbaum Associates, Publishers.

34

Single-Parent Families

STEPHEN D. SUGARMAN,
1998

What do the former First Lady Jackie Kennedy, Princess Diana, the movie star Susan Sarandon, and the TV character Murphy Brown have in common? They all are, or at one time were, single mothers—unmarried women caring for their minor children. This chapter concerns public policy and the single-parent family, a family type dominated by single mothers.

Because these four women are fitting subjects for *Lifestyles of the Rich and Famous,* they are a far cry from what most people have in mind when the phrase "single mother" is used. Many picture, say, a nineteen-year-old high school dropout living on welfare in public housing. Hence, just mentioning these four prominent women vividly demonstrates the diversity of single mothers. These four also illustrate the major categories of single mothers—the widowed mother, the divorced or separated mother, and the single woman who bears her child outside of marriage. (Women in this last category are often misleadingly called "never married" even though approximately one-fourth of the women who are unmarried at the birth of their child had been married at an earlier time.)

One further distinction should also be made here. The usual picture of the single mother is of a woman living *alone* with her children—Jackie Kennedy, Princess Diana, and Murphy Brown. But those we call "cohabitants" are also single mothers as a legal matter, even though their children are living in two-adult households. Indeed, where the woman, like Susan Sarandon, is cohabiting with the father of her child (Tim Robbins), although the mother is single, from the child's perspective it is an intact family.

As with single mothers generally, these four prominent women arouse a wide range of feelings, from support to dismay, in the public at large. Jackie Kennedy surely gained the maximum empathy of our foursome when her husband was murdered in her presence. Even those women who are widowed in less horrifying ways have long been viewed as victims of cruel fate and strongly deserving of community compassion.

Not too long ago, having been divorced was by itself thought to disqualify those seeking public office or other positions of public prominence. That no longer holds, as Ronald Reagan's presidency made clear. As a result, once Prince Charles and Princess Diana split up, she probably appeared to most Americans as facing the challenging task confronting many divorced women of having to balance the pursuit of her career with raising children on her own. Yet, despite Diana's wealth and fame, her marital breakup brought to the fore our society's general uneasiness about how well children fare in these settings, as well as our uncertainty about the appropriate roles of divorced fathers as providers of both cash and care.

In Sweden today, Susan and Tim's family structure is commonplace. There, a very large number of men and women live together and have children together, but do not go through the formalities of marriage—or at least have not done so at the time their first child is born. Lately, in America as well, the cohabitation category, long ignored by the census, is rapidly growing. This is not to say that most Americans, unlike the Swedes, accept cohabitation as though it were marriage. Indeed, American public policy, as we will see, treats cohabitation very differently from marriage.

Perhaps because Murphy Brown is a fictional character, this has allowed those who are on the rampage against unmarried women who bear children to be candid about their feelings without having to be so openly nasty to a "real" person. Yet Murphy Brown is an awkward icon. To be sure, she flouted the conventional morality of an earlier era. She had sex outside of marriage and then decided to keep and raise her child once she discovered she had unintentionally become pregnant. Although many people in our society still rail against sex other than between married couples, sex outside of marriage has become such a widespread phenomenon that it is generally no longer a stigma. And while it would be easy to chastise Murphy Brown for carelessly getting pregnant, . . . this also is so commonplace that it is barely remarkable any more. Indeed, Murphy Brown might have come in for more censure had she, as a single woman, deliberately become pregnant.

As for deciding to raise her child on her own, this *by itself* does not arouse great public outcry. After all, it is not, as though widowed mothers who make that decision are castigated for choosing not to remarry. As for the unmarried birth mother, shotgun weddings are seen to be less promising than they once were; abortion, while still a right, is hardly thought to be a duty; and while giving a child up for adoption is often commendable, today this is seen primarily as the route for women who do not want to, or cannot afford to, take care of their children themselves.

In short, the strongest objection by those who have assailed Murphy Brown is that she is a bad role model—in particular, that she is a bad role model for *poor* women who, unlike her, cannot provide for their children on their own, but go ahead and have them anyway, planning to turn to the state for financial and other assistance. In many quarters those single mothers are doubly condemned. First, they are seen to be prying money out of the rest of us by trading on our natural sympathy for their innocent children; yet this is said to leave taxpayers both unhappy because they have less money to spend on their own children and with the distasteful feeling that society is condoning, even promoting, the initial irresponsible and self-indulgent behavior by these poor single mothers. Then, these low-income women are rebuked as high-frequency failures as parents—for example, when their children disproportionately drop out of, or are disruptive at, school or turn to criminal behavior. Of course, not everyone disapproves of Murphy Brown or even those poor women who choose to have children on their own knowing that they will have to turn to the state for assistance. Many people believe that every American woman (at least if she is emotionally fit) ought to be able to be a mother if she wants to be.

These various types of single mothers are significant because they raise different issues, and, in turn, they have yielded very different policy solutions and proposals for reform. But before we turn to policy questions, some general demographic information is presented that, among other things, shows single-parent families to differ significantly from some common myths about them. The policy discussion that follows begins with a historical overview that demonstrates how American policies have changed sharply in the past century. . . . The chapter concludes with a call to refocus public attention on the needs of the children in single-parent families.

SINGLE-PARENT FAMILY DEMOGRAPHICS: MYTHS AND REALITIES

Father-Headed Single-Parent Families

In the first place, not all single-parent families are headed by women. In 1970, three-quarters of a million children living in single-parent families lived with their father (about 10 percent of such families); by 2000, more than two million children lived in father-headed single-parent families (an increase to approximately 15 percent) (U.S. Census Bureau, 2001).

These families are not the subject of much policy attention, however. First, most of them are headed by divorced (or separated) men; a few are widowers. It is rare, however, that the father of a child born outside of marriage will gain physical control of that child, and this takes custodial fathers largely outside the most controversial category of single parents. Furthermore, single fathers caring for their children tend to be financially self-supporting and therefore generally beyond the purview of welfare reformers. Finally, they tend to remarry fairly quickly and hence remain heads of single-parent families for only a short time. In fact, the main public policy controversy involving these men today concerns divorce custody law—in what circumstances should fathers be able to become heads of single-parent families in the first place? . . .

Noncustodial fathers are quite another matter—whether divorced from or never married to the mothers of their children. As we will see, they are the subject of a great deal of public attention and concern.

Unmarried as Compared to Divorced and Widowed

Turning back to families headed by single mothers, one myth is that they are predominantly women who have never been married to the father of their child. Yet there are actually more divorced (and separated) single mothers. For example, in 2000, 55 percent of single mothers were divorced or separated, and another 4 percent were widowed (U.S. Census Buresu, 2001). Moreover, because of the predominance of widowed and divorced mothers, large numbers of women become single mothers, not at their child's birth, but later on in their child's life, often not until the child is a teenager. Hence, among the children in single-parent families, living one's entire childhood apart from one's father is by no means the norm.

Cohabitants

Cohabitants with children in their household are a complicated category, and, in turn, they complicate the data. As noted earlier, although the women in these families are decidedly single mothers in a legal sense, in many respects these couples resemble married couples. So, many of these households are better described as two-parent, not single-parent, families. Some demographers have recently suggested that "cohabitation operates primarily as a precursor or a transitional stage to marriage among whites, but more as an alternative form of marriage among blacks" (Manning and Smock, 1995; also see Rindfuss and Vandenheuvel, 1990).

In any case, these cohabiting households come in several varieties. One first thinks of two biological parents not married to each other but living with their child—as exemplified by Susan Saradon and Tim Robbins, and the Swedish model. Cohabiting mothers in this situation still often show up in U.S. surveys as though they were never married mothers living on their own, because survey instruments tend to categorize respondents only as married or single.

A second variety of cohabiting households includes a single mother with her child who is now living with, but not married to, a man who is *not* the child's father. These women are drawn out of the ranks of the never married, the divorced, and the widowed; they, too, are frequently counted in surveys as living on their own. Moreover, in this second category especially, it is often quite unclear to outsiders whether the man is a de facto spouse and stepparent, a casual boyfriend, or something in between.

Yet a third category of cohabiting households contains a homosexual couple (more often two women) in which one of the partners is the legal (usually biological) parent and the other is formally a stranger (although some lesbian couples of late have successfully become dual mothers through adoption)....

Working and Not

Although the myth is that single mothers (especially never-married welfare moms) spend their time lounging around the house, watching TV, doing drugs, and/or entertaining men, this is a wild exaggeration. A large proportion is in the paid labor force. Official data from 1999 show that more than two-thirds of *all* women with children are in the labor force. Married women's rates are about 62 percent where the youngest child is under six and about 77 percent where the youngest child is six or more. Within the ranks of single mothers, divorced women work *more* than married women, whereas never-married women are less likely to *report* working (Committee on Ways and Means, 2000).

Single mothers often feel compelled to work full time even when their children are very young, although the official data again show a difference between divorced and never-married women. According to 1999 figures, of those women with a child under age six, 60 percent of divorced women and 41 percent of never-married women worked *full time;* this was higher than the rate for married women, 39 percent of whom were working full time (Committee on Ways and Means, 2000).

A decade ago, fewer than 10 percent of single mothers who were receiving welfare officially acknowledged earning wages. Research by Kathryn Edin suggests that, in fact, a high proportion of them was actually employed at least part time (Edin and Jencks, 1992). They tended to work for cash in the underground (and sometimes illegal) economy. According to Edin's findings, they did not typically do so to be able to buy drugs or booze, but rather in order to keep their households from utter destitution or to avoid having to live in intolerably dangerous public housing projects. They kept this work a secret from the welfare authorities because if the authorities knew, they would so cut back those women's welfare benefits as to make their wages from work nearly meaningless. Although these women would be viewed by the welfare system as "cheaters," they tended to remain living in fairly impoverished circumstances. As Edin puts it, they felt compelled to break the law by the skimpiness of the welfare benefits they received. As we will see later in this chapter, welfare reform of the 1990s has changed this picture somewhat.

Poor and Nonpoor

Even with the receipt of government assistance, more than a quarter of family households headed by single mothers officially live below the poverty level (as compared with less than 5 percent of families headed by a married couple) (U.S. Census Bureau, 2000). Although this is a distressingly high number, to the extent that the myth is that single mothers are poor and on welfare, the myth is false. A substantial share of single mothers provides a reasonable level of material goods for their children, and well more than half of all single mothers are not on welfare. In 1998, for example, about 30 percent of female-headed households with related children under age eighteen received means-tested cash assistance (Committee on Ways and Means, 2000).

Those who escape poverty for their families have tended to do so primarily through earnings and secondarily through child support and government benefits (or through a combination of these sources)—although typically not by receiving welfare. In 1998, *nonpoor* single mothers received about 81 percent of their income from earnings, 6 percent from child support and alimony, and 7 percent from Social Security, pensions, unemployment compensation and the like, but only 5 percent from welfare, food stamps, and housing assistance (Committee on Ways and Means, 2000). It is not surprising, then, that, in 1998, the poverty rate for single-parent families with children under age eighteen was 43 percent before the receipt of means-tested cash transfers and 37 percent after their receipt, a relatively modest reduction indeed.

White and Nonwhite

The myth is that single mothers primarily come from racial and ethnic minorities. While it is true that these groups are disproportionately represented given their share of the population, in fact, these days more single mothers are white than any other group. For example, in 1999, 40 percent of nonmarital births were to whites, 33 percent to blacks, and 25 percent to Hispanics (Centers for Disease Control, 2000). On a cumulative basis, as of 2000 there were 6.2 million white, mother-headed family groups (including white Hispanics) as compared with three million black, mother-headed family groups (including black Hispanics)—even though 32 percent of all black family groups were headed by mothers and only 10 percent of all white family groups were headed by mothers (U.S. Census Bureau, 2001).

Change over Time

The demography of single-parenting has changed a lot over the twentieth century. There are many more single parents today than there were several generations ago, both in absolute numbers and, more importantly, in terms of the percentage of all children (or all parents) affected.

In 1900, the typical single parent was a widow. Male deaths through industrial and railway accidents were very visible. By contrast, divorce was then scarce (although desertion was a problem). And becoming a single mother by becoming pregnant outside of marriage was not very common, especially because so many who got pregnant promptly married the father. Now, especially since the 1960s, all that is changed. Divorce is more frequent. "Illegitimacy" and cohabitation are also more prevalent than in earlier periods. For example, of women born between 1940 and 1944, only 3 percent had lived with a partner of the opposite sex by age twenty-five; of those born between 1960 and 1964, 37 percent had done so (Hollander, 1996). Moreover, the stigma of bearing a child outside of marriage and/or what some still call "living in sin" is much reduced.

Nonetheless, along with these changing characteristics of the single-parent family has come a change in public empathy. Earlier there was very widespread compassion for single parents and their children when single parents were mainly widows and divorcees, especially in the pre-no-fault era, when divorce usually was triggered (formally at least) by the misbehavior of the husband. Today, at least in some quarters, single mothers are loathed—as those receiving welfare who have borne their children outside of marriage or who are suspected of bringing about the end of their marriages through their own selfishness. As of 1996, the last full year that AFDC ("welfare") was in effect, about 60 percent of those receiving welfare had children outside of marriage as compared with but a trivial share in the 1930s and less than 30 percent as late as 1969 (Committee on Ways and Means, 2000).

CHANGING POLICIES TOWARD SINGLE PARENTS

Widows

In 1909, President Theodore Roosevelt convened a historic first White House Conference on Children, which identified the poverty of widowed mothers and their children as a central policy problem. Then, if states and localities provided any assistance at all, it was too often through the squalid conditions of the "poor house" into which single-parent families might move—something of a counterpart to today's shelters for homeless families. The poor house itself was the successor to an earlier system in which desperate mothers farmed their children out to others, in effect providing young servants to those people who took these semiorphaned children into their homes, farms, and businesses. Reflecting the outlook of the social work profession that was then just getting underway, the White House Conference pushed instead for the adoption of Mothers' Pensions plans. Soon enacted, at least on paper, in most of the states, this new approach envisaged cash payments to single (primarily widowed) mothers who were certified by social workers as capable of providing decent parenting in their own homes if they only had a little more money in their pockets (Bell, 1965).

Mothers' Pensions, the precursor to Aid to Families with Dependent Children (AFDC), reflected both the psychological perspective that it was best for the children to be raised in their own homes and the sociological outlook that it was appropriate for the mothers to stay at home and raise them (perhaps taking in other families' laundry or sewing, but not leaving their children to join the regular paid labor force). As we will see, this benign attitude toward the payment of public assistance to single parents, which was reinforced by the adoption of AFDC in 1935 at the urging of President Franklin D. Roosevelt and maintained at least through the 1960s, has substantially evaporated.

Divorcees

Much earlier in the 1900s, while widows were pitied, marital breakup was broadly frowned upon. Nonetheless, it was increasingly acknowledged that some spouses acted in intolerable ways and should be censured by allowing their spouses to divorce them. Adultery, spousal abuse, and desertion were the main categories of unacceptable marital conduct, and most of it seemed to be engaged in by husbands. As the decades rolled by, however, the divorce law requirement of severe wrongdoing by one spouse and innocence on the part of the complaining spouse soon ill-fit the attitudes of many couples themselves. Especially starting after World War II, and accelerating in the 1960s, many more couples came to realize that their marriages had simply broken down and they both wanted out. Until divorce law changed to reflect this new outlook, couples were prompted to engage in fraudulent charades (often involving the husband pretending to engage in adultery) so as to satisfy domestic relations law judges.

No-fault divorce law first emerged in California in 1970 and was rapidly followed by other states (Kay, 1990). As a practical matter, not only did this reform allow couples amicably to obtain a divorce without having either one of them adjudicated as the wrongdoer but also, in most states, it permitted any dissatisfied spouse to terminate the marriage unilaterally. Whether no-fault divorce actually caused an increase in the divorce rate or merely coincided with (indeed, grew out of) the spiraling demand for divorce is unclear (Peters, 1986). What is clear, however, is that divorce rates today are enormously greater than they were before 1970, thereby contributing to the great increase in single parenting. As we will see, that state of affairs, in turn, has recently generated something of a backlash movement, one that seeks to reintroduce legal barriers to divorce in families with minor children.

Illegitimacy

Public policies toward illegitimacy (and, in turn, toward both abortion and teen pregnancy) have also changed significantly during the 1900s. At an earlier time, children born outside of marriage were pejoratively labeled "bastards" and denied inheritance and other rights connected to their fathers, although their biological fathers did generally have the legal duty to support them. If a single woman became pregnant, a standard solution was to promptly marry the child's father, perhaps pretending that the pregnancy arose during marriage after all. Adoption was available to some, who would be encouraged to go away before their pregnancies began to "show," only to return childless afterwards as though nothing had happened. Pursuing an abortion instead then risked criminal punishment and subjected the woman to grave risks to her life and health.

Rather suddenly, a little more than two-thirds of the way through the twentieth century, policies in these areas turned around dramatically. For those who wanted it, abortion became legal. More important for our purposes, remaining unmarried and then keeping a child born out-of-wedlock became much more acceptable. For example, instead of expelling pregnant teens, schools adopted special programs for them. Fewer women gave up their newborns for adoption—for example, 19 percent of white, unmarried birth mothers did so in the 1960s, but only 3 percent did so in the 1980s (Hollander, 1996). The courts forced states to give many legal rights to illegitimates that had previously been enjoyed only by legitimates; and many legislatures voluntarily expanded the inheritance rights and other entitlements of out-of-wedlock children. Soon, unmarried pregnant women far less often married the biological father during the course of the pregnancy—a drop of from 52 percent to 27 percent between 1960 and 1980 (Hollander, 1996).

Women who had children outside of marriage were no longer casually labeled unsuitable mothers and, as noted previously, soon became the largest category of single mothers receiving welfare. In terms of public acceptability, something of a high-water mark may have been reached in the early 1970s with the conversion of welfare into a "right" by the federal courts, the elimination of welfare's "suitable home" requirement, and the end to one-year waiting periods for newcomers seeking welfare. This ignited an explosion of the welfare rolls, and for the first time in many states, African-American women gained reasonably secure access to benefits. At that time Republican President Richard Nixon proposed turning AFDC from a complex state-federal program into a uniform national scheme.

At the end of the century, however, a policy backlash emerged. Between 1967 and 1997, the proportion of African-American children born outside of marriage skyrocketed from around 25 percent to nearly 70 percent; and the rate for white children was viewed by some as poised for a similar trajectory—and in any case has grown from 8 percent thirty years ago to around 22 percent today (Centers for Disease Control, 2000). Now, curbing illegitimacy, or at least unmarried teen pregnancy, seems to be near the top of many politicians' lists.

Child Support

It has been long understood that fathers have a moral obligation to provide for the financial support of their minor children. In the absent parent context, this means paying "child support." For most of the twentieth century, however, a substantial proportion of men failed to pay the support they might have paid (Chambers, 1979). The default rate by divorced fathers has long been very high, and in out-of-wedlock births the father's paternity often was not even legally determined. (Stepfathers with no legal duties were frequently a more reliable source of support.) Moreover, in many states, even if noncustodial fathers paid all they owed, this was judged to be a pittance when compared with the child's reasonable needs. Deceased fathers were no more reliable, frequently dying with estates of trivial value and without life insurance.

Through the 1930s, AFDC and its predecessors were the main public response to these failures—providing means-tested cash benefits to poor children (and their mothers) deprived of the support of a breadwinner. In 1939, however, special privileged treatment was afforded widows and their children. The Social Security system was expanded so that, upon the death of the working father, "survivor" benefits would be paid to the children and their caretaker mother based upon the father's past wages. This, in effect, created publicly funded life insurance for most widows and their children, with the result that today hardly any widowed mothers find it necessary to apply for welfare.

No comparable "child support insurance" was provided, however, so that divorced and never-married poor mothers have had to continue to turn to the socially less favored means-tested welfare programs instead of Social Security. On behalf of these families, the effort, much enlarged since the mid-1970s, has been to increase the amount of child support an absent father owes and to beef up child support enforcement efforts. Notwithstanding those reforms, it is still estimated that more than ten billion dollars of child support annually goes uncollected, and many custodial mothers are unable to collect any support for their children.

Cohabitation

It appears that American society generally is becoming more accepting of cohabitation, even if it remains frowned upon in many circles. (Clearly, same-sex cohabitation continues to be highly controversial.) So far as public policy is concerned, however, marriage still makes a significant difference. For example, when children are involved and the cohabitants split up, the woman who keeps the children (as is typically the case) continues to be disadvantaged as compared with the woman who had married. Although she is entitled to support for her child, only in very special circumstances can she gain financial support for herself from her former partner. So, too, upon the death of her partner who was the father of her children, while her children can claim Social Security benefits, she does not qualify for the caretaker Social Security benefits that a legal widow would have obtained. . . .

RECENT POLICY REFORMS AND CURRENT INITIATIVES

Child Support

In the past few years, much policy reform has been directed toward getting noncustodial parents to put more money into the hands of single-parent families. Put more simply, the goal has been to force absent fathers (often termed "deadbeat dads") to transfer more of their income to their children and the children's mothers. One reason for these policy changes is that they are among the few solutions on which most liberals and conservatives can agree.

Congress has on several recent occasions prodded states to change their child support regimes in a number of ways. The size of the noncustodial parent's support obligation has been considerably increased in most states. At the same time, the calculation of the sum is now largely determined by formula, instead of being left to the discretion of a local judge in the course of adjudicating a divorce or paternity determination. On the collection side, most of the effort has been directed toward making the process routine, especially through the automatic withholding of support obligations from wages and the direct payment of such obligations over to the custodial parent (or to the welfare authorities if the mother is on welfare).

Nonetheless, child support enthusiasts are by no means satisfied. Although inflation of late has been very low, in the past even moderate inflation has quickly undermined the value of child support awards, necessitating difficult courtroom battles over modifications. Hence efforts are underway to establish a regime of automatic modification based upon changing costs of living. On another front, too many noncustodial fathers remain unidentified, at least formally. In response, some states have posted officials in hospital nurseries on the theory that when unmarried men come in to see their newborns they can be coaxed into admitting paternity on the spot.

An important part of the child support shortfall occurs in the welfare population. There, however, increased support collection generally will benefit the taxpayers, not children in single-parent families. This is because welfare recipients have had to assign their child support rights to the government and have been entitled to keep only fifty dollars a month from what is collected. This helps explain why fathers of children on welfare are not so eager to pay the child support as they might otherwise be. Indeed, a fair proportion of these absent fathers now secretly and informally pay support directly to the mothers of their children, because, if the welfare department managed to capture those funds, the outcome would be the enrichment of the public fisc at the expense of poor children. While redirecting those funds from mothers to the welfare department would strike a blow against what now qualifies as fraud, the result would nonetheless be the further impoverishment of children.

Moreover, one has to be realistic about collecting increased child support from absent fathers. Many of them have new families and new children to support. While some people might find it irresponsible for them to have taken on these new obligations, the practical reality is that we are often talking about shifting money from one set of children to another. In other cases, the nonpaying father is unemployed. Should he be forced to find work, or be placed in a public service job, so that income could be siphoned off to satisfy his child support obligation? Some are urging this very solution. Yet, what is to happen when the men fail to comply with their work obligations? Are we going to imprison thousands of these dads?

In any event, child support policy largely strikes at the single-parent family issue after the fact—even though some men arguably might be deterred from fathering children or abandoning their families if they knew they faced substantial, and nearly certain, collection of child support obligations. Other current policy initiatives are more openly "prevention" oriented.

Divorce Law

Among those who have concluded that it often would be better for the children for the parents to stay together, rather than split up, it is not surprising that no-fault divorce has become a target for reform. The picture these critics present is that some parents selfishly divorce even though they realize they are putting themselves ahead of their children and are likely to harm their children as a result—or else they blithely divorce unaware of the harm they will do to their children. The goal of the critics, they say, is to make divorce more difficult in hopes of helping the children (Galston, 1996).

The problem, however, arises in deciding exactly how to change divorce law. The most sweeping proposal is simply to bar divorce entirely to those with minor children. This solution, however, carries costs that most people would find unacceptable. Suppose one spouse (stereotypically the father) is guilty of domestic violence against the children and/or the other spouse. It seems unimaginable today that, in such circumstances, we would insist that the victim spouse remain married. To be sure, the divorce ban advocates might concede that she would be entitled to a legal separation and/or a protective order keeping him away from the children. But, at that point, to continue to prevent divorce seems gratuitously nasty. Since keeping the parents together for the sake of the children has been abandoned in this case, the only real consequence of the bar would be to prevent the victim spouse from remarrying—and perhaps giving the children a stable new family relationship. So, too, suppose one spouse abandons the family. What good is possibly served by denying the other spouse a divorce and thereby keeping her from remarrying—especially if the alternative is for the abandoned spouse to live with, but not marry, her new love?

These examples make clear that a complete ban on divorce by those with minor children is unsound and unlikely to be adopted. They also demonstrate that even to enact a strong presumption against divorce with special exceptions will inevitably embroil the spouses and the courts in wrangles over individual fault, a prospect that makes most of those familiar with the operation of pre-no-fault divorce law shudder. After all, if you made an exception for physical abandonment, wouldn't you have to make an exception for emotional abandonment, especially if it were combined with extramarital love affairs? And if you made an exception for physical abuse, why not for emotional abuse?

This prospect has caused some of those who want to make divorce law tougher to retreat to the seemingly simple idea that only unilateral divorce would be banned. If one spouse objected, the other could not force a divorce on the one who wanted to remain married. These critics claim that American law seems to have jumped directly from fault-based divorce to unilateral divorce, when it might have stopped in between by allowing divorce only by mutual consent (Glendon, 1989).

Proponents argue that requiring both parents to agree will put an extra roadblock in front of indiscriminate sacrificing of the child's interest. They seem to have in mind the father who gets tired of marriage and family and selfishly wants out—perhaps because he has a new "girlfriend." But, if so, how useful is it really to give his wife a veto? If, as a result, he resentfully stays in the marriage, will this actually be good for the children? Alternatively, what is to prevent him from simply moving out without obtaining a divorce, perhaps taking up housekeeping with another woman? Again the rule really only means that he cannot remarry. Furthermore, this regime is already the law now in the State of New York, and yet we certainly don't hear no-fault divorce critics arguing that everything would be so much better if only the more liberal states tightened up their rules to match New York's.

A final restrictive approach would impose a substantial waiting period (say, two or more years) before one parent could obtain a unilateral divorce and/or insist on marriage counseling before filing for divorce. Although some have argued that either of these measures would benefit the children, here again there is reason to be skeptical. A long waiting period could cause people to file for divorce even more quickly than they now do, or, in any event, simply to treat the rule as a time-hurdle to remarriage. Offering willing parents marriage counseling is probably a good idea, and legislatures might consider making this a mandatory benefit in all health insurance plans (as part of the coverage of mental health services generally). But coerced counseling is likely to have a low payoff.

This analysis suggests that legal change intended to make divorce more difficult to obtain would largely be a symbolic matter and is not a very promising way actually to help children avoid harms that may come from divorce. Perhaps more promising, then, are incentive approaches designed to help cooperative parents who are at risk of divorce to stabilize their marriage. These could include financial support provided through the tax law. For example, in the late 1990s, Republicans pushed through a universal child tax credit of $500 a year, although many Democrats opposed this on the ground that it means spending too much money on families who do not need help. They would rather spend the money through an expansion of the Earned Income Tax Credit, which is better tailored to low-income families. Other reformers would prefer to direct the financial rewards to young families who are first-time home buyers. Yet other intervention strategies designed to help maintain marriages are educationally and psychologically oriented. . . .

REFOCUSING THE POLICY PERSPECTIVE

When it comes to single-parent families, much of our current policy focus is on parents: whether they divorce, whether they pay child support, whether they have children outside of marriage, whether they work, and so on. Suppose instead that policy attention were aimed at the *children* in single-parent families. For example, as we have seen, if a child's breadwinner parent dies, the government ordinarily assures that child far better financial security than it does if that child's breadwinner parent is simply absent from the home: Social Security steps in to satisfy the deceased parent's obligation to have provided life insurance but not the absent parent's duty to provide child support.

Comparable treatment for the latter group implies some sort of publicly funded "child support insurance" scheme. Plans of this sort (including those that would expand Social Security in exactly this way) have in fact been proposed, most notably by Irwin Garfinkel, although so far at least they have not won widespread endorsement (Garfinkel, 1992; Sugarman, 1995, 1993). This sort of scheme could assure all children living in single-parent families of equal financial support—say, up to the poverty level. Or, like Social Security and private child support obligations, the benefits could be related to the absent parent's past wages. In either case, earnings by the custodial parent could supplement, rather than replace, the child support benefit.

Such a plan could be financed by general revenues or Social Security payroll taxes. But it might also be funded, at least in substantial part, by absent parents, thereby making the plan one that guarantees that a suitable level of child support will actually be provided and makes up the shortfall when the collection effort fails. Were this second approach adopted, not only should it dramatically reduce our sense of the cost of the plan but also it should offset any tendency that the plan might otherwise have to increase divorce.

It is important to emphasize that a plan like this would much improve the lot of many children who in the past have been dependent on welfare and large numbers of children with working-class and even middle-class mothers whose absent fathers now default on their child support obligations. It must be conceded, however, that in view of the direction of recent welfare reform, the prospects at present are not favorable for any new initiative to provide cash for children in single-parent families.

A different child-centered approach, therefore, is to try to provide all children with essential goods by means other than providing cash. Ought not all American youths live in decent housing, obtain a quality education, receive adequate food, have access to decent health care, and so on? This is not the place to detail the many alternative mechanisms by which these critical items might be delivered. What needs emphasizing, however, is that any program guaranteeing these sorts of things to all children would have vastly disproportionate benefits to children now living in single-parent families. Moreover, if we can keep the focus on the needy and innocent members of the next generation, perhaps we can escape ideological battles over the worthiness of these children's parents. This is perhaps a naive hope, but one that may be enhanced when the thing delivered to the child's family is other than money: witness the greater public and legislative popularity of the federal food stamps program and federal aid to elementary and secondary education as compared with the now-decimated federal welfare program.

The many policy reforms discussed here are unlikely to have large impacts on people like Jackie Kennedy, Princess Diana, Susan Sarandon, and Murphy Brown. But ordinary single mothers (and their children) who are in analogous situations have a great deal at stake.

SELECTED REFERENCES

Bell, Winifred. 1965. *Aid to Dependent Children.* New York: Columbia University Press, 1965.

Centers for Disease Control and Prevention. 2000. "Nonmarital Childbearing in the United States, 1940–1999," in *National Vital Statistics Reports,* 48:16.

Chambers, David L. 1979. *Making Fathers Pay: The Enforcement of Child Support.* Chicago: University of Chicago Press.

Committee on Ways and Means, U.S. House of Representatives. 2000. *2000 Green Book.* Washington, D.C.: U.S. Government Printing Office.

Edin, Kathryn, and Christopher Jencks. 1992. "Reforming Welfare," in Christopher Jencks, (ed.), *Rethinking Social Policy: Race, Poverty and the Underclass.* Cambridge: Harvard University Press.

Galston, William. 1996. "Divorce American Style." *The Public Interest,* 124:12.

Garfinkel, Irwin. 1992. *Assuring Child Support.* New York: Russell Sage Foundation.

Glendon, Mary Ann. 1989. *The Transformation of Family Law.* Chicago: University of Chicago Press.

Hollander, Dore. 1996. "Nonmarital Childbearing in the United States: A Government Report." *Family Planning Perspectives* 28(1):30, 1996.

Kay, Herma Hill. 1990. "Beyond No-fault: New Directions in Divorce Reform." Pp. 6–36 in Stephen Sugarman and Herma Hill Kay (eds.), *Divorce Reform at the Crossroads.* New Haven: Yale University Press.

Manning, Wendy D., and Pamela J. Smock. 1995. "Why Marry? Race and the Transition to Marriage among Cohabitors." *Demography* 32(4):509.

Peters, Elizabeth. 1986. "Marriage and Divorce: Information Constraints and Private Contracting." *American Economic Review* 76:437.

Rindfuss, Ronald R. and Audrey Vandenheuvel. 1990. "Cohabitation: A Precursor to Marriage or an Alternative to Being Single?" *Population and Development Review* 16:703, 1990.

Sugarman, Stephen D. 1993. "Reforming Welfare through Social Security." *University of Michigan Journal of Law Reform* 26:817–851.

Sugarman, Stephen D. 1995. "Financial Support of Children and the End of Welfare as We Know It." *Virginia Law Review* 81:2523–2573.

U.S. Census Bureau. 2000. "Poverty in the U.S.," in *Current Population Reports* P60–210.

U.S. Census Bureau. 2001. "America's Families and Living Arrangements" in *Current Population Reports* P20–537.

35

Adoptive Families in the United States

CHRISTINE WARD GAILEY,
2004

Looking at adoption on a global scale, the United States has more adopted children than any other country, although Norwegians adopt at a greater rate than any other nation. A 1996 national survey indicated that 6 out of 10 Americans have had some direct connection with adoption: either they, a family member, or a close friend was adopted, had adopted a child, or had placed a child for adoption. A third of the respondents to that national survey said they had "at least somewhat seriously" considered adopting (Lewin 1997:16).

Informal adoption has been part of America from earliest times: most Native American peoples routinely adopted children for a range of reasons, including marking peaceful relations between groups and incorporating children from enemy groups who were orphaned in warfare. In colonial times, European settlers adopted informally because of dramatic death rates among women and children. Among the slaves forcibly brought to the New World, informal adoption sheltered children whose parents were sold separately or who had died from the harsh conditions. In modern times, state-regulated adoption has steadily increased from the 1940s to the present.

We do not have good statistics on adoption, for several reasons; among them the rapidly growing number of so-called independent adoptions, privately arranged transactions involving individual lawyers—a practice that used to be illegal in many states—and the inconsistent manner in which individual states track adoptions and develop adoption data bases. The 2000 Census was the first time adoptive status was tracked at a national level, but it relied on self-reporting. According to the best estimates, between 2% to 4% of U.S. families include an adopted child; about a million children in the U.S. live in an adoptive family (Stolley 1993).

Adoption in the United States today is an arena where issues as controversial as national sovereignty, racial identity, gay and lesbian rights, and single motherhood are contested routinely. It is a terrain that seemingly would challenge what anthropologist David Schneider deemed the dominant ideology of American kinship, namely, that "blood is thicker than water." And yet, even though this kind of family formation flies in the face of a widespread belief that genetic connection is the foundation of kinship, we can see ways in which some adoptive families reassert that dominant ideology, while others model kinship as a life-long process of developing and sustaining networks of people who claim one another and share resources and care-giving.

Based on research I have done over the past decade with adopters and professionals involved in adoption work, I will argue that while some issues affect all adoptive families, the degree to which the parents adhere or reject dominant assumptions about birth being a firmer foundation for kinship than adoption affects the relationships they create. The adopters' motivations to adopt, the kind of adoption they choose, the kinds of children they seek or are willing to accept, and their experiences in an adoptive family shape their attitudes toward their children. Moreover, their relative commitment to dominant ideologies of kinship—itself shaped by the adopters' social class, ethnicity, race, and gender socialization—influences how they relate to their children. In some cases, even though the family is forged through what Judith Modell calls "kinship with strangers", the adopters show a profound commitment to the assumption that birth or genetic connection is a stronger basis for kinship than any other means. In others, the adopters view all kinship as a process of claiming and being claimed, caring, sharing resources, that is no more firmly founded on birth or genetics than any other claim of connection.

While adoption is a growing form of family formation in the U.S., people adopt in a number of different ways. These ways articulate with class and race hierarchies, on the one hand, and, on the other hand, a set of prejudices and policies that favor heterosexual nuclear families and consider single mothers and lesbian and gay parents less than desirable.

TYPES OF ADOPTION IN THE UNITED STATES

In 1992, the last year for which we have any remotely reliable estimates, most adoptions in the United States (42%) involved relatives or stepparents. Independent adoptions—through private agencies or through private lawyers—comprised approximately 37.5% of U.S. adoptions. Also in that year, public agencies arranged only 15.5% of adoptions, and international adoptions accounted for only 5% of reported adoptions. The best statistics available involve international adoptions, as visas issued by the U.S. State Department assure an accurate count. While adoption involves some issues across the different forms, each type also poses special issues for the participants. Let us consider each in turn.

Stepparent Adoption: When Adoption Is Not Seen as Adoption

Stepparent adoptions are not generally considered in the same framework as other adoptions. As an adoption researcher I think many of the problems step families report with parental authority and sibling relationships would be eased if the participants in these families—whether formal adoption is involved or not—recognized some of the dynamics typical of adoption in the U.S. In many cases families could benefit from seeing "older child" adoption as applying to their family formation. Where birth mothers or fathers have joint custody or visitation rights, the dynamics often parallel a form of what Harold Grotevant and Ruth McRoy refer to as "fully disclosed open adoption", that is, a situation where all parties have some contact with one another and the birth and adoptive parents negotiate child rearing. Stepparent adopters could benefit from the work of Patricia Hill Collins and her concept of "othermothers" in a child's life, derived from her research on African American kinship patterns that involve informal adoption and fosterage. Collins describes practices of childrearing where there are mothers and othermothers throughout a person's upbringing in African American communities. These othermothers perform a range of maternal functions without seeking to replace or erase the birth mother, who may be consistently active in the child's life or only intermittently so. These othermothers have authority and respect in their maternal role, and so are not forced to negotiate, as stepmothers so often are, a non-maternal, unauthoritative relationship to children – the husband's wife. There are many parallels in stepparent families to older child open adoption, where the adoptive mother or father must acknowledge the continuing primacy of the birth parent without sacrificing their own, on-site parental involvement. By not acknowledging adoption, the male or female step parent is perceived by children as an interloper, a direct threat to the relationship with the birth parent of the same sex: a dilemma in terms of feelings of loyalty and affection for the children. Appreciating the extraordinarily helpful literature on older child adoption, especially the work of Claudia Jewett, could ease the feelings of grief, loss, and anger typical of children living amid birth and adoptive relatives.

Independent Adoption: From Gray Market to Legitimacy

Private agency or independent adoptions are the second most common form of adoption in the U.S. and constitute one of the least regulated arenas. While international adoptions also involve the work of private agencies and independently contracted lawyers, independent adoptions involve children born in the U.S. Locating a prospective adopted child through newspaper ads targeting pregnant women, through lawyers' channels, or religious agencies' solicitation of pregnant women used to be illegal in many states and was considered by many adoption social workers as a "gray market" until the 1980s. Since that time it has outpaced public agency adoption.

The reasons for this transformation are related to 1) racial preferences of prospective adopters; 2) their preference for infants rather than older children; 3) their reluctance to adopt children with any physical, mental, or development special needs; and 4) changing patterns of childbearing and childrearing among white women. Briefly, the numbers of healthy white infants in the foster care system or coming into adoption through public agencies, have declined precipitously since the early 1970s. While the availability of abortion certainly has played a role in this shift, so has the increasing acceptability of white women keeping children born outside of marriage, and the reduction of gender discrimination in the job market, which has allowed many single mothers to support children on their own. The "white middle class flight" from public agency adoptions has accompanied these cultural and economic changes. It would be less than honest if we did not analytically notice the persistence of racial preferences among these mostly middle class, mostly white adopters. At the same time, after they decide to relinquish their parental rights, birth mothers in this arena of adoption seem to exercise more control than birth mothers in the other arenas over the actual placement of their children, often being allowed by their agencies or lawyers to screen and decide among—and sometimes interview—prospective adopters around issues of education, religion, marital status, sexual orientation, and the like.

Independent adoptions generally involve adopters providing the birth mother with payments to cover prenatal care and living expenses during pregnancy, in addition to the costs for the adopters' home study through the private agency, and legal fees, and in the case of those religious organizations that arrange such private adoptions, often "voluntary donations" to help support agency activities. Direct payments to a birth mother for her child remain illegal as trafficking in children, but judging from the scandals of the past two decades, the arrangements sometimes skitter on the edge of marketing children.

It is in this sector of adoption that we still find some parents who do not openly discuss the circumstance of their children's adoption with them. Urging open discussion of adoption with children throughout their lives is the standard practice in public agency adoption parental training; some private agencies also encourage this. Others, however, do not stress disclosure, or minimize its importance. I have found that those lawyers or agencies that specialize in privately arranged adoption of infants at or near birth, and those that cater to their adopter-clients wishes to "match" the child's appearance to their own—a practice that public agencies have abandoned since the 1970s—are most apt to attract adopters who wish to keep the adoption a secret from their children.[1] Psychologically, parental secrecy about adoption has been shown in study after study to be detrimental to an adoptive child's healthy sense of self, at every stage of development. There literally is no research that does not call for full disclosure and appreciative discussion by parents at a level of depth and sophistication keyed to each stage of the child's development. Precisely because these adoptions are private and only loosely regulated by the state—meaning the clients are the adopters and birth parents, not the children—such practices can be tolerated despite research showing it to be detrimental.

International Adoption

International adoption constitutes only about 5% of all U.S. adoptions, but the numbers are growing steadily, especially over the last ten years. In practice, international adoption in the U.S. constitutes a variant of the private agency/independent adoption type. Prospective adopters must obtain an approved home study, as in independent adoption, and usually work through a private agency or a law firm that specializes in international adoption. As in independent adoption, the adopters are the clients, but unlike domestic independent adoption, the birth mothers or fathers rarely have any say in the placement of their children once they terminate parental rights. Also unlike domestic independent adoptions, and perhaps because of the bulky paper trail that is part of international adoption and perhaps because the children frequently do not resemble the parents, few international adopters I interviewed kept the adoption a secret from their children. Paralleling domestic private adoption, most of international adopters are white and from middle-to-upper income brackets; most whom I interviewed sought acceptably colored, healthy infants or toddlers. Occasionally, as in the Romanian floodgate of international adoptions following the collapse of the Soviet bloc in the 1990s, color trumps age, and older children may be acceptable because they are categorized as white.

A distinct minority of international adopters are workers in non-governmental organizations who know the local language, have sustained relationships in the countries from which their children come, and revisit the area routinely after the adoption. Indeed, some have obtained children from the children's surviving relatives or local community members because of their demonstrated commitment in refugee camps, war zones, and among the urban poor. But the vast majority of international adopters do not have prior ties to their children's countries of origin.

There are situations where prospective adopters are expected to reside in the country for 6 to 8 weeks while the local judiciary processes the paperwork, or examines the situation to be sure it is not a case of child trafficking. In such cases, prospective adopters from the U.S. may reside in the same boarding house or small hotel while they visit with their identified children (still in foster care or in orphanages). These groups on occasion remain in contact after their return to the U.S., but unless they reside near one another, the ties tend to dissolve. Some international adoption agencies sponsor annual reunions or post-adoptive cultural events to which all clients and their families are invited. But unlike most public agency adoptions discussed below, after the adoption is finalized there is no mandatory follow-up visit by a licensed social worker to check on the health and well being of the children.

International adoption is portrayed very positively in the U.S. media as providing a litany of opportunities for the children beyond having a stable and loving family. When there are child-selling scandals, the local country's agents are almost always blamed by the media. In many of the countries from which the children come, governmental regulation is insufficient to guarantee that trafficking in children will not occur, and the situation is compounded by ambiguity in the terms used in international adoption. For example, agencies call all available children "orphans"—a term in international law that simply means a child whose parents have terminated their claims to the child. But the connotation in English, which many of the international adopters believe, is that the child is without living parents. In the vast majority of cases, this is not true, but the legal fiction permits unwary adopters to believe that they are bringing home a child without relatives. Because birth records in many countries are inconsistent, altered to suit the context, or destroyed during wars, it is virtually impossible for adoptees ever to locate their birth parents should they attempt to do so later in life. Indeed, the unlikelihood that any birth relative might find the adopted child is one reason a few international adopters told me that they wanted a child from another country. Here, too, the situation of international adoptive families diverges from domestic independent adoption.

Almost all of the prospective international adopters in my sample considered themselves or their spouses to have fertility problems, but why they sought international adoption required further explanation. Some were deemed too old (over 40) by domestic adoption agencies. Others were single women or women in lesbian partnerships who found domestic agencies hostile to their efforts to become parents. In this regard, a number of countries, such as South Korea, refuse to allow single women to adopt, whether they are from that country or elsewhere; but others sometimes permit it, and sometimes "close down" to single female adopters. Other international adopters in my sample had been rejected by domestic agencies for reasons they found objectionable, such as the couple that had placed an earlier adoptee in a state institution because of the child's significant mental retardation. Still others were convinced that the children in other countries, particularly China, were somehow healthier and from more morally upstanding backgrounds, even if they were poor, than the children of the poor in the U.S. Others simply were seeking a healthy white or white-enough infant without the significant wait and potential for post-adoptive birth mother claims of independent domestic adoption. What all of these people wanted was clear and unchallenged exclusive rights to a child of their choosing.

In an effort to regularize what was threatening in many regions to become simply an international market in children, in the 1990s a range of countries hammered out what became known as The Hague Convention on international adoption. The U.S. signed, but has been very slow to ratify the agreement. A number of countries, such as Haiti, refused to sign, because the kind of adoption permitted under the Convention runs contrary to local practice. Under Haitian law, based on the Napoleonic Code and so, similar to several other Francophone countries, adoption is not plenary, as the Hague Convention specifies. Plenary adoption means the birth parents relinquish all present and future claims to the child in question. Under Haitian law, adoption is recognized, but the child remains responsible in specific ways for the birth parents throughout their lives. As Chantal Collard has noted, the Hague Convention constitutes an imposition of one particular form of adoption on less powerful countries. Similarly, Claudia Fonseca reports from Brazil that overburdened poor mothers often use state orphanages for respite care during crises, intending to reclaim their child or children after the crisis; these illiterate mothers frequently do not realize that in signing adoption relinquishment papers they may never see the child again, as permanent estrangement from adopted children is not the local practice. The implicit cultural imperialism involved in adherence to the Hague Convention—even though provisions of the Convention seek to stop child trafficking—has sparked numerous controversies surrounding international adoption in a range of sending countries. Local media report, often with considerable substantiation, kidnapping of poor children, uninformed consent to relinquishment, or economic pressure by local adoption lawyers or agencies to cede parental rights.

Thus, while international adoption appears to American readers to be a benevolent extension of concern for the well being of the world's children, local communities often disagree. If we stand back and look at where the countries from which most of the U.S. international adoptees come, we can see the clear mark of geopolitical power. Cold War politics and areas of U.S. economic dominion account for most of the children historically: South Korea, Latin America, Vietnam, and the Philippines. In recent years, the "opening" of China to

U.S. influence and the collapse of the U.S.S.R. has led to China and Russia vying for first place among "sending" countries. China's more efficient infrastructure, coupled with its willingness to accept older and until recently, single women as adopters, has given it a competitive edge.

Public Agency Adoptions: Race and Class Issues

Public agency adoption is the most regulated arena and the only type where, consistently, the client is the child, not either the birth parents or adopters. It used to be the most prevalent form of adoption, but because of reasons outlined above now comprises only 15.5% of U.S. adoptions. Because of the tendency of middle class and wealthier prospective adopters to seek independent or international adoptions, those seeking to adopt children in foster care in the U.S. tend to be working class couples from a range of ethnic and racial backgrounds, single women (including middle class professionals), and people currently fostering children they wish to adopt formally. The fee structure tends to be lower than that in private agencies, to the extent that if adopters opt for a child labeled "hard to place" (because of age or background) or "special needs" (because of illness, disabilities, or mental health issues), in many states the fees for the home study and legal processing can be waived and a small post-adoption subsidy arranged. This reduction of financial obstacles to public agency adoption is largely the accomplishment of a series of protests by the National Association of Black Social Workers in the early 1970s, part of its effort to increase the adoptive placement of children in their home communities or communities of similar class and racial or ethnic composition. Before this, adoption was simply unfeasible for most working class families and families of color.

Over the next twenty-five years, the reduction in social services reaching the poor indirectly increased the number of children, especially those from communities of color, being placed in foster care for reasons of neglect or abuse. The "family preservation" approach to foster care implemented during the Reagan and (first) Bush administrations meant that children remained in the foster care system for years on end. Long-term stays in foster care reduce children's chances of placement in permanent families. The longer the child's stay, the more likely he or she was to be moved from foster family to foster family, increasing the risk of physical or sexual abuse in a system where monitoring of families cannot be done adequately because of cutbacks in social worker staffing. To reduce the "backlog" of children in the foster system for longer than 18 months, under the Clinton administration the Metzenbaum adoption reform laws of 1996–1997 banned any agency receiving federal funding from making race or ethnic background a factor in placing a child, and insisted that a permanency plan be developed—including the termination of parental rights if necessary—for any child in state care for 18 months. The Metzenbaum reforms have reduced the number of children moving from foster care to adoption, but at the sametime the indirect impact of welfare reform laws of the mid 1990s has increased the flow of children into the foster care system.

Among the controversies the Metzenbaum reforms reignited was that over transracial adoption. Worried that yet again, Native American children would be preferentially placed with Anglo families, Native American activists, using arguments based on sovereignty, managed to wrest an exemption from the laws regarding jurisdiction over children residing outside reservations. But African American adoption advocates could use no such sovereignty claims, and the issue of group rights in the U.S. is not well supported in legal precedent. As a result, no provisions in the reforms backed or bolstered outreach programs to recruit adopters from communities of color. Considering how deeply racism affects the life chances of black youth, the activists concern about placing children where they can obtain the skills and sense of self to withstand the oppressions of racism is well grounded. The reforms could well return adoption practice in the U.S to the pre-1970s days when little outreach was done to black and other minority communities and transracial placements of black and mixed-race children with white families were done without consideration of whether the child would be isolated from others of similar backgrounds. If housing were not persistently segregated in so many parts of the U.S., this might not be such a concern. Longitudinal studies of transracial adoptive families indicate that if the white parents are aware of racism, make it a consistent concern in their discussions, have friends or regular interaction with peers from their child's background, the children tend to fare well. But this kind of parental engagement with racial issues and friendships across the color line are not common in many parts of the U.S. Transracial adoption issues have centered on public adoptions because most other adoptions are either in-race, or involve children from backgrounds that do not invoke as much consistent racism in the US as do children with African ancestry. Most black children available for adoption are in the foster care system and so, are placed primarily through public agencies.

Among the adopters I studied, public agency adoptive families were the most likely to include birth children as well. Moreover, in the parent training sessions I observed at one public agency in the Northeast and a private agency subcontracted to another public agency, infertility rarely arose as a motivation to adopt. The adopters most often cited a wish to have more children or a second family if their birth children had grown up, or to adopt a child they were fostering or had identified as a foster child in their neighborhood, friendship or kin networks. The married couples among these prospective adopters were the only ones in my sample to show willingness to adopt multiple siblings. Reasons cited by the single women seeking public agency adoptions whom I interviewed included their wish not to birth a child without being married, their "finally" being in a financial position to support a family, and only occasionally their having attempted to get pregnant without success. Whether married or single, the African Americans in the sample were the only ones to express a sense of obligation to the community to "help the generation coming up."

Perhaps because of their willingness to adopt children older than two-to-four years old, the public agency adopters in my sample received children without the dreaded lengthy waiting period after the conclusion of the "home study". Many of the children they adopted—"hard to place" or "special needs"—had experienced severe neglect, physical abuse, or sexual abuse either in their birth home context or in foster care. The emotional issues of children's grief, loss, and disorientation so often associated with adoption transition, thus, were complicated by their histories of violation or inadequate care. For this reason, the agencies I studied provided free or very low-cost post-adoption support groups and workshops on such issues as attachment disorder, post-traumatic stress, and other psychological problems associated with surviving brutality or severe emotional or physical deprivation.

Because single women are rarely considered on a par with married couples, regardless of economic status, they often were paired with children having multiple problems, a situation they expressed to me as "matching" the least desired with the least desired. In the parent training sessions I observed, the preference shown by the social workers conducting the workshops was particularly ironic, since all of the social workers were single mothers of adopted children.

Public agency adoptive families, thus, represented the only consistent effort to redress the situation of so many foster children in the U.S. Because the clients in public agencies are children, the families seeking to adopt tended to be better screened, better prepared to address such complex issues as transracial and interracial family formation, and more aware of post-adoption resources going into the adoption. Compared with the other adoption groups, public agency adopters represented a range of ethnic and racial communities, and cut across working and middle classes.

OPEN ADOPTION AND THE SEALED RECORDS CONTROVERSY

Cutting across the types of adoption is whether or not the adoption is open or closed. While the state of California has had open adoptions as a routine practice for the past fifty years, it is still somewhat unusual in many other states. Open adoption refers to a situation where some kind of contact, ranging from once to routine, from photographs and letters to visitation, exists between the birth parent or parents and the adoptee and adoptive parents. The openness can be minimal: providing the adoptive family with a photograph of the birth mother or father, but no name and address. Among domestic adoptions, greater openness is becoming the practice, as where names and addresses can be obtained when the adoptee reaches the age of 18, where names and addresses are exchanged, or where the adoptive family sends the child's photograph, artwork, a letter, or other age-based material to the birth mother periodically. In a minority of cases openness involves visits to or by the birth mother. In one case I know, because of generational differences, the adoptive parents relate to their child's young birth mother as a kind of older daughter, a role the birth mother told me was "very comfortable" for her. She sees the child frequently, relating to her "more as an older sister than a mom." The relationship is unusually close: for instance, the young woman stayed at the adoptive family's home for a few weeks when she was moving to a nearby city. The child, now six, knows she is his birth mother, but he, too, is more comfortable with her in the role of older sister. Longitudinal studies of open adoptions have found they are especially beneficial to the children's identity formation and sense of continuity.

Closed adoptions, however, remain the norm. Most public agencies have a sealed records policy, meaning identifying information about the birth or adoptive parents cannot be given to the other parties, and the adopted child cannot gain access to the information even after the age of 18. There is much activism by adoptees, adopters, and birth parent organizations to have such policies rescinded. In the interim the Internet has become a vehicle that adoptees and birth parents use to find one another. Virtually all adoption advocacy groups oppose sealed records unless it is at the explicit request of the birth parent and, even then, birth family medical histories and such should be available to adoptive families who want them. Sociologist Katerina Wegar's study of adoption searches delves into the issue of searches and the sealed records controversy. Having grown up knowing her birth mother as all Finnish adoptees do, Wegar finds the preoccupation with privacy peculiarly American.

ADOPTIVE FAMILIES: UPHOLDING OR SUBVERTING AMERICAN KINSHIP IDEOLOGY

Adoptive families in the U.S., then, have to confront the fact that they have made kinship in a way contrary to most Americans' notion of how families should be made: through procreation. In doing so, all parties to the adoption, from birth mother to adoptee, face some kind of public judgment regardless of their conduct. Let us explore some of the ways society marks parties to adoption as unnatural.

Marking Adoption as Unnatural Kinship

People unfamiliar with adoption can be at a loss in figuring out what to call the people involved. Many use terms like "natural" or "biological" child to distinguish birth relations from adoptive ones. This makes the adoptee appear odd, unnatural. As one adoptee remarked when a passer-by asked her mother, "Is that your biological child?", "Why, do I look like an android?" In an effort to reduce such dehumanizing terminology, adopters and writers on adoption have developed the distinction "birth child" and "adoptee". This language of adoption provides a lens through which we can see some of the departures from dominant kinship ideologies, and some instances where uncertainty or ambivalence toward "blood is thicker than water" remains.

Adoptive families and adoption literature often contrast "birth parent" with "adoptive parent" or even "forever parent". While the literature refers to birth parents and adoptive parents, it is clear from the contexts that in most cases, the authors are referring to mothers. Australian researcher Jonathan Telfer has pointed out that the terrain of adoption is heavily gendered where girls are socialized to desire motherhood as a marker of womanhood. Many researchers have noted that the major players in adoption are women: wives almost always initiate adoption proceedings and must persuade their husbands; most adoption social workers are women; and birth mothers figure far larger than birth fathers in adoption imagery and practice. Indeed, the term "birth father" occurs mostly in discussions of terminating parental rights, rather than care giving prior to fosterage or adoption. The term "real father" is rare as a way of describing either the birth or adoptive father. But motherhood is a far more contested matter, and it shows in the terminology: throughout the literature and the interviews I conducted with the parents of 61 adoptees, people sometimes used the term "real mother" to refer to the birth mother and sometimes to the adoptive mother. The ambivalence even among adoptive mothers about the importance of women's procreational role is clear.

General attitudes toward birth mothers who relinquish their children are decidedly harsh, although some understanding is permitted for very young women who do so, viewing relinquishment as "better than abortion". But in general, adoptive parents are viewed as infertile, on the one hand, and, on the other hand, valiant for going ahead and trying to have a family without giving birth. The parties to adoption, as I have argued elsewhere, involve two socially unacceptable "failures": one, a respectable, that is, married woman who has failed in her mission to procreate, the other a disreputable woman who has procreated without rearing her progeny. Most adopters in my sample did not know that the majority of mothers relinquishing parental rights in public adoptions are not teenagers; they harbored an image of birth mothers as a young, unmarried, irresponsible woman. Thus, they generally saw the birth mother as a "bad mother", merging images of youth, carelessness, unacceptable premarital sexuality, and fertility without maternal qualities. The adoptive mother, by way of contrast, while rendered "unnatural" because of her infertility, is somewhat redeemed by her wish to be maternal. Some of the international adopters in my sample did not see their Chinese or Latin American children's birth mothers in a negative light: these couples and single mothers saw their children's birth mothers as poor but honest—married but under economic or social duress. Other exceptions show us the influence of racial and class hierarchies on adoption attitudes. The six African American couples in the study did not condemn their children's birth mothers, arguing that they did not know enough about their lives to cast judgment. The white working class adopters voiced condemnation when they knew the circumstances of their children's early lives, but it was not from a distance: they recognized the circumstances that could lead to abuse and neglect because it had happened to people they grew up with or knew or even some relatives in their own extended families. The most compassionate attitudes toward birth mothers voiced by adoptive mothers came from one African American married woman and several of the white and African American single women, emphasizing either that the birth mother had given their beloved child the gift of life, or, for the single women, that the major difference between themselves and the birth mother was a stable family background, education, and a good job.

Adoption as a way to form a family is pervasively viewed as "second best" to birth-based connections to a child. Among adopters, I found that those who had to grapple with infertility or extensive fertility treatment without success were most apt to see adoption as a "last resort" means of having a child. Other adopters—especially those who had birth children already, single mothers, and gay fathers—viewed adoption positively, as a first choice way of creating a family. National surveys of attitudes toward adoption, however, show that the general public views adoption as a more fragile form of kinship, inferior to adding a child through birth. In such a climate, it should not surprise us to find that adoption language protects adoptive parents if they relinquish the child after adoption. In earlier times, the term used for such adoptions was "failed"; today they are known as "disrupted". In using the term "disrupted", the implication is somehow that the child's problems are what led to terminating the adoption: few adults are viewed as "disruptive", while the term is routinely used to describe children's behavior in schools and other social settings. So the adopters are somehow immunized from any moral condemnation for rejecting adoptive children—a condemnation that relinquishing birth mothers usually receive.

Reproducing Natural Kinship

Among the adopters I interviewed it was clear that the more the parents believed that adoption was inherently more fragile, the more tentative their attachment was to their children. I found this more in the wealthier adopters—mostly international or independent adopters. Where the infants they adopted did not continue to show the "signs of promise" that one of the mothers expressed to me as what she and her husband were seeking, but had developmental or emotional problems, the more likely the parents were to attribute difficulties to "bad breeding" (as one father put it) or other "inherited" characteristics. One of the international adoptive fathers referred to the behavior by his then five-year-old Korean daughter (in my view signs of unattachment) as indicating that she may have been a "bad seed".

Most worrisome to me was a pattern I noted among the wealthier adopters. If the child's behavior or performance did not improve, the parents tended to insert professionals between them and the child, whose duty was to remedy the situation. If the professional intervention did not clear things up over the five years of my study, I noted a deepening emotional distance between parents and child—and sometimes between the parents. One couple ceased to bring their child to family get-togethers at the holidays, for example, for fear he would make a scene. For a child who is grappling with a deep sense of abandonment, such emotional distancing signals another abandonment that can create or exacerbate an attachment disorder. This in turn would make behavior worsen. Yet I was compelled to note that in these families, the relationship between parents also was based on performance. I doubted that a birth child would have been treated differently, although the patterns of blame might have been different (which side of the family was responsible for the problem, rather than who initiated the adoption in the first place). I came to think of the kind of kinship practiced among these wealthy adopters as "contractual": implicitly acceptance and affection were linked with consistent performance up to the level expected. Deviation involved reduction of affection and emotional distancing.

Over the five years two of the international adopter couples divorced. In both cases child custody went, minimally contested, to the wife. Interestingly, both of these suddenly single mothers and fathers saw the adoption as responsible for ruining their marriages. One of the fathers said, "I really did it [the adoption] for her. I never really felt bonded to the kid. He was mine, but not mine, you see?" One of the mothers stated, "I took this on, and I have to see it through. I am her mother; I'm all she's got." The other mother confided, "Frankly, I wish we'd never adopted him: I had a great career, a decent husband, a circle of friends and now—nothing! If I were honest I'd say he's wrecked my life. . . . You think if you get an infant and pay for that prenatal care, everything will be just like it [sic] was yours, but you don't know what's inside, what's waiting for you. A kid that's not yours is like a time bomb. I'm being totally straight with you—and I'll sue if you ever print my name."

Rejecting Natural Kinship

By way of contrast, one of the working class couples, who had adopted three sisters who had been physically and sexually abused by their birth parents, refused to attribute any of the girls' problems (including learning disabilities) to inherited factors. "They're still a mess in some ways," the weary mother said, "but they've been through the wringer, what do you expect? And we love 'em, they're ours." Other working class adopters expressed skepticism at the notion that their children's problems might be genetic in origin. One father told me, "Yeah, they might—so what? He's still our kid and love'll pull you through things that all yer fancy education won't." A well educated, single white mother who had adopted a sexually abused African American girl said, "Some of it [the problems experienced by the daughter] may be hard-wired, but most of it's due to a rotten-awful history. What matters is our relationship, and that's based on acceptance, hope, struggle, love, persistence, and tons of communication." Another of the single mothers of an adolescent girl who'd been adopted at nine, sighed, "Her and me, we've been through hell and back again! If that doesn't make us family, I don't know what family is." Another educated mother, whose son developed significant learning disabilities, shrugged and said, "Look, he could have developed the same thing if he'd been born to me. What difference does it make?"

SUBSTANTIATION: MAKING KINSHIP REAL

What do these experiences of adoptive families tell us about making kinship in America? The parents in the families where children were "their own" rather than "like our own" are the models for how adoption can subvert ideologies of natural kinship, of "blood is thicker than water". Most of the public agency adopters, most of the single mother adopters, about half of the independent adopters, and about a fourth of the international adopters in one way or another had critically examined their own views of family and come to reject notions that genetic connection provides a reliable basis for kinship. As one of the married adoptive mothers expressed, "If I believed that, where would my relationship with my husband be, let alone my relationship with our kid?"

The closely connected adoptive families had invented or adapted rituals that created a kind of supernaturally ordained quality to the family. "Coming Home Day" was celebrated in many of the households. This was the day that the child first moved in. For others the event was "Adoption Day", the day the adoption was legalized. In either case, most of the families made telling the adoption story a central feature of that day. These stories frequently spoke of "fate" or "love at first sight" or "magic" or "joy beyond words" in meeting or seeing the child or the child's picture for the first time.

These families also had special stories about their children's "firsts"—the first time we went to the zoo, the first time we went to the beach, and the like. Most kept a special photograph book about the child's early history, what was known about it, along with the usual family photograph albums. In many of the homes there was a special plaque or certificate from the day the adoption was finalized; several had a framed copy of one of the poems about adoption that circulate on the internet. When they went to pick out a dog at the local animal shelter, one of the families told their daughter that now it was time for her to adopt someone into the family. At home, the six-year-old circled the date on the calendar and announced it was "Lady's Coming Home Day."

"Blood is thicker than water", Schneider states, marks kinship as substance; it exists, it is. Substantiation is the term I use to describe kinship as a process of claiming connection through time. Substantiation can be materialized in shared objects, such as family photographs and home videos. Material claiming can also be seen in gifts that mark the uniqueness of the connection; often these are gender-stereotyped, such as mother-daughter dresses, "World's Greatest Dad" pendants, or birthstone rings for mothers. Substantiation can also take verbal forms: special stories, nicknames that note special attributes of the person, songs made up or dedicated to a specific person or to mark the family itself, and so on. Substantiation can also be through care giving: several mothers mentioned caring for a recent adoptee when severely ill as marking "real family" for both mother and child.

How substantiation is displayed varies with social class and cultural background: photographs in frames all over a mantel or living room wall is expected in many working class contexts, for example. Commonplace in the middle class white households was a refrigerator papered with children's drawings, weekly schedules, and photographs; in wealthier households refrigerators were not an acceptable display space. In Chicano households, the family shrine included all the family members not currently visible, those who were no longer living and those who were far away: among the adopters, a photograph of the child's birth mother or father, if known, was put there. Substantiation is a process of inclusion that must be enacted to maintain people in the net. Michaela di Leonardo has pointed out that "kin work", such as birthday celebrations and sending holiday cards, falls disproportionately on the shoulders of women in many ethnic contexts. It is no different in adoption: mothers in my study exerted far and away the most effort in substantiation.

In sum, appreciating the ways that U.S. adoptive families do kinship gives us some insight into how the range of families in the U.S. work, or don't work. By analyzing how class, gender, and racial hierarchies intersect with beliefs about kinship and ways adoption can occur, we gain insight into the ways in which dominant ideologies of kinship and family can be reproduced or subverted. Some adoptive families are as American as apple pie and motherhood in upholding beliefs that "real" kinship is based on blood connections. Others—in my view the most successful ones—refuse such a defensive posture and celebrate the sharing and caring that pull people in, whether through marriage, adoption, acting as kin, fosterage, or even birth.

REFERENCES

Collard, Chantal. 2000. *Stratified Reproduction: The Politics of Fosterage and International Adoption in Haiti.* Paper presented in the session, "Stratified Reproduction: The Politics of Fosterage and Adoption." American Anthropological Association Annual Meetings. San Francisco, November 15.

Collins, Patricia Hill. 1990. Black Women and Motherhood. *In Black Feminist Thought: Knowledge, Consciousness, and the Politics of Empowerment.* New York, NY: Routledge, pp 115–137.

Di Leonardo, Micaela. 1987. The Female World of Cards and Holidays: Women, Families, and the Work of Kinship. *Signs* 12(3).

Fonseca, Claudia. 1986. Orphanages, Foundlings, and Foster Mothers: The System of Child Circulation in a Brazilian Squatter Settlement. *Anthropological Quarterly* 59(1): 15–27.

Gailey, Christine Ward. 2000a. Ideologies of Motherhood in Adoption. *In Ideologies and Technologies of Motherhood: Race, Class, Sexuality, Nationalism.* Heléna Ragoné and France Winddance Twine, eds. New York, NY: Routledge, pp 11–55.

Gailey, Christine Ward. 2000b. Race, Class, and Gender in Intercountry Adoption in the USA. *In Intercountry Adoption: Developments, Trends, and Perspectives.* Peter Selman, ed. Pp. 295–314. London: Skyline House, for British Agencies for Adoption and Fostering.

Grotevant, Harold, and Ruth McRoy. 1998. *Openness in Adoption: Exploring Family Connections.* Thousand Oaks, CA: Sage Publications.

Jewett, Claudia. 1978. *Adopting the Older Child.* Harvard, MA: Harvard Common Press. 1982. *Helping Children Cope with Grief and Loss.* Harvard, MA: Harvard Common Press.

Modell, Judith Schachter. 1994. *Kinship with Strangers: Adoption and Interpretations of Kinship in American Culture.* Berkeley: University of California Press.

Schneider, David. 1984. The Fundamental Assumption in the Study of Kinship: "Blood Is Thicker than Water."

In A Critique of the Study of Kinship. Ann Arbor: University of Michigan Press, 165–177.

Stolley, K. S. 1993. Statistics on Adoption in the United States. *In The Future of Children.* I. Schulman, ed. Pp. 26–42. Los Altos, CA: Center for the Future of Children.

Telfer, Jonathan. 2000. Pursuing Partnerships: Experiences of Intercountry Adoption in an Australian Setting. *In Intercountry Adoption: Developments, Trends, and Perspectives.* Peter Selman, ed. Pp. 315–345. London: Skyline House, for British Agencies for Adoption and Fostering.

Wegar, Katarina. 1997. *Adoption, Identity, and Kinship: The Debate Over Birth Records.* New Haven, CT: Yale University Press.

CHAPTER 11

Family Violence

Families are often caring and protective places, but they can also harbor some of the most destructive forms of violence ever known. Although we are reluctant to admit it, families are one of the most common contexts for violence in American society and that violence can profoundly affect both adults and children. The likelihood that a man will be assaulted by a family member is more than twenty times greater than the chance he will be assaulted by someone outside his family. For women, the odds are even worse: they are two hundred times more likely to be assaulted by a family member than by an outsider (Straus, 1991). Children, too, are most likely to suffer violence at the hands of family members. Police receive more requests for help with domestic disturbances than with any other single problem, and some researchers claim that police personnel are as likely to be killed trying to settle family fights as in any other line-of-duty activity (Straus, 1991).

These grim facts run counter to our romantic image of families as safe havens from outside threats. Although many families *are* gentle and supportive, we must acknowledge how widespread violence is within them and understand how and why this is so. The selections in this chapter focus attention on the social institutions that promote family violence and provide information to dispel common myths about spouse abuse and child abuse. By focusing attention on the ways that normal cultural practices promote violence against women and children, these selections point the way to ending such violence. Only by highlighting these violent practices and their institutional supports will we be able to understand them, and hopefully, to reduce their incidence in the future.

Historically, violence has been institutionalized within the family. Marriage laws have traditionally given husbands powers over wives and courts have usually supported men's rights to use physical force against their wives. English common law gave husbands the right to "physically chastise" their wives, which was modified by the nineteenth-century "rule of thumb" allowing a husband to beat his wife with a rod no thicker than his thumb. Until the 1870s, wife beating was legal in most of the United States and remained quite common thereafter. The first selection in this chapter, by Charlotte Perkins Gilman, describes how historical norms created families in which men routinely exploited women. Writing near the turn of the century, she echoes the earlier sentiments of John Stuart Mill and Harriet Taylor Mill by equating the role of women in marriage to slavery and promoting an alternative vision of marital equality. Describing "the man-made family" and "the androcentric home" as despotic institutions undermining motherhood and democracy, Perkins Gilman questions whether a man should automatically be treated as the natural head of the family. Instead of embracing the **patriarchal** principles underlying masculine privilege in families, she suggests that "too much manness" detracts from men's potential humanness and harms women and children.

Twentieth-century evidence supports Perkins Gilman's observation that cultural acceptance of male dominance in families contributes to the oppression of women and to violence against them. For example, men who beat their wives tend to regard their actions as proper and justify them as a defense of their natural and traditional rights (Dobash and Dobash, 1979; Dobash, Dobash, Wilson, and Daly, 1992). The larger community has often supported beliefs in men's rights to control "their" women. As late as the 1960s, one in four Americans agreed that it was acceptable for a husband to hit his wife under certain conditions. Law enforcement officers, who have traditionally been unwilling to intervene in what they regarded as "private" domestic affairs, have also supported these beliefs. Until it was changed in 1977, the training manual for domestic-disturbance calls published by the International Association of Chiefs of Police essentially recommended that hitting a spouse be treated as a "private matter" and that arrests should be avoided (Pagelow, 1981; Straus, 1991).

The selection from Neil Jacobson and John Gottman is from their book *When Men Batter Women*. In this excerpt, they describe and debunk ten myths about **battering.** They show how claims about spouse abuse being mutual combat are misleading, how all batterers are not alike, and how drugs and alcohol are often associated with **battery,** even if they do not cause the violence. They also dispel myths that **batterers** cannot control their anger, that battering stops on its own, and that therapy is a better cure than prison. The authors similarly examine common myths about women's role in the violence, rejecting the idea that women provoke men into battering them or that they could stop the violence by changing their own behavior. Finally, Jacobson and Gottman report that most battered women do *not* typically stay in abusive relationships, even though it can be very dangerous to leave them. One of the most interesting parts of the Jacobson and Gottman book is their identification of two types of batterers: "Pit Bulls" who are erratic and explosive but emotionally dependent; and "Cobras" who are antisocial and emotionally abusive, as well as being sadistic in their calculated degradation of female partners. Other researchers have used different terms to describe frequent types of domestic violence. One influential typology discriminates between "common couple violence," which can be mutual and is often embedded in patterns of poverty and substance abuse, and "patriarchal terrorism," which involves violence as one tactic in the implementation of a general pattern of men's power and control over women (Johnson and Ferraro, 2000).

The selection on **corporal punishment** of children by family violence researcher Murray Straus also uses the rhetorical approach of dispelling common myths. Straus draws on research evidence to show that spanking is a relatively ineffective means of child discipline, even though most child-rearing advice at least tacitly accepts it, and even though most American parents do it. Contrary to popular ideals, spanking is not needed as a last resort, it is not harmless, and its infrequent use does not necessarily minimize its impacts on children. The myths that Straus attacks are based on folk beliefs that are typically false, but are nonetheless rarely questioned. Research evidence shows, for example, that children do not need to be spanked to avoid being spoiled, that spanking does not replace verbal abuse, that parents who spank do not typically reserve physical punishment for rare problems, and that parents who rely on corporal punishment do not necessarily stop using it when children get older. Finally, in the interests of ending harmful physical punishment, Straus argues that we should not assume that it is unrealistic for parents to stop spanking and that we should stop assuming that parents need to receive extensive training before they can stop hitting their children. He makes an analogy to workplace abuse (some employers used to hit their employees), and to marital relations (some husbands used to hit their wives and some still do), arguing that hitting children is equally repugnant. In the past it was considered unrealistic to assume that those practices would change, yet they have diminished dramatically in the past few decades.

Straus draws another parallel to the antismoking campaign of the late twentieth century: few expected that Americans would stop smoking, but norms were drastically altered, and a taken-for-granted activity was minimized by redefining it as harmful. Like those exposed to second-hand smoke, Straus hopes that children who are victims of family violence will benefit as new **social norms** and enforcement strategies emerge to limit parental practices that are now known to be harmful.

Although the Straus selection focuses on the relatively common practice of spanking, both mothers and fathers perpetrate many other forms of child maltreatment. More extreme forms of violence used against children range from grabbing and violently shaking to hitting with a fist, cutting or burning, or even killing. In addition, some patterns of emotional abuse do not include physical violence, and if parents fail to meet children's basic needs for food, shelter, and care, it is considered "neglect." Finally, there are cases of child sexual abuse, mostly perpetrated by men, and often including some of the other forms of abuse noted above. Overall, more than three million cases of child abuse or neglect are reported to Child Protective Service Agencies in the United States each year, and experts agree that many times more cases go unreported.

Child abuse is typically hidden in the privacy of the home, and most people are reluctant to admit the full extent of abuse that does occur. On national surveys, however, most parents report that they have slapped or spanked their child in the last year, with about a third admitting that they pushed, grabbed, or shoved their child, and about one in ten admitting hitting a child with an object. Remarkably, one in three parents admit that they continue to hit or physically punish their children even after they become teenagers (Gelles and Straus, 1987). When children of any age are more severely beaten, the episodes are reported as being repeated an average of about once every two months (Gelles and Straus, 1987). Studies also show that most men who abuse their wives also use violence against their children, and wives who are abused by their husbands are more likely than others to use violence against their children. In general, more powerful family members perpetrate violence against weaker members, even though most abusers themselves feel powerless and are attempting to overcome their own insecurity and regain a sense of control over their lives.

Although families vary in the amount of overall violence that occurs between spouses, among siblings, or between parents and children, most violence against children is ordinary physical punishment carried out by loving and concerned parents. However taken-for-granted or motivated by good intentions, such violence teaches children that those who love you hit you, and as Murray Straus (1991, 29) suggests, this generalizes to "those you love are those you can hit." Because these lessons are learned early in life and ritualized in repeated encounters, they tend to carry deep meaning for people. Children who are abused tend to feel that they are bad and unlovable, leading to self-blame, low self-esteem, and difficulties establishing intimate relationships. Although many people who were abused as children never become perpetrators of violence themselves, most studies find that victims of abuse are themselves more likely to use violence against others than those who were not abused as children. Because abusive parents tend to hold unrealistic beliefs about children's abilities and often respond inappropriately to their children's behaviors, instituting educational programs that teach about normal child development hold some promise for decreasing levels of family violence in the future.

REFERENCES

Dobash, R. Emerson, and Russell Dobash. 1979. *Violence against Wives.* New York: Times/Doubleday.

Dobash, Russell, R. Emerson Dobash, Margo Wilson, and Martin Daly. 1992. The myth of sexual symmetry in marital violence. *Social Problems* 39:71–91.

Johnson, Michael P., and Kathleen J. Ferraro. 2000. Research on domestic violence in the 1990s: Making distinctions. *Journal of Marriage and the Family* 62:948–63.

Gelles, Richard J., and Murray A. Straus. 1987. Is violence toward children increasing? A comparison of 1975–1985 national survey rates. *Journal of Interpersonal Violence 2*:212–22.

Pagelow, Mildred D. 1981. *Woman Battering: Victims and Their Experiences.* Beverly Hills, CA: Sage.

Straus, Murray A. 1991. Physical Violence in American Families: Incidence, Rates, Causes, and Trends. In *Abused and Battered,* edited by D. Knudsen and J. Miller. New York: Aldine de Gruyter.

SUGGESTED READINGS

Brownell, Patricia J. 1998. *Family Crimes against the Elderly: Elder Abuse and the Criminal Justice System.* New York: Garland Publishers.

Costin, Lela B., Howard Jacob Karger, and David Stoesz. 1996. *The Politics of Child Abuse in America.* New York: Oxford University Press.

Goetting, Ann. 1999. *Getting Out: Life Stories of Women Who Left Abusive Relationships.* New York: Columbia University Press.

Hamberger, L. Kevin, and Claire Renzetti, eds. 1996. *Domestic Partner Abuse.* New York: Springer Publishing.

Jones, Ann. 2000. *Next Time She'll Be Dead: Battering & How to Stop It.* Boston, MA: Beacon Press.

Leonard, Elizabeth Dermody. 2002. *Convicted Survivors: The Imprisonment of Battered Women Who Kill.* New York: State University of New York Press.

Panter-Brick, Catherine, and Malcolm T. Smith, eds. 2000. *Abandoned Children.* Cambridge, UK: Cambridge University Press.

Peterson Del Mar, David. 1996. *What Trouble I Have Seen: A History of Violence against Wives.* Cambridge, MA: Harvard University Press.

Pelzer, David J. 1995. *A Child Called "It": One Child's Courage to Survive.* Deerfield Beach, FL: Health Communications.

Pleck, Elizabeth. 1987. *Domestic Tyranny: The Making of American Social Policy against Family Violence from Colonial Times to the Present.* New York: Oxford University Press.

Raphael, Jody. 2000. *Saving Bernice: Battered Women, Welfare, and Poverty.* Boston, MA: Northeastern University Press.

Schneider, Elizabeth M. 2000. *Battered Women and Feminist Lawmaking.* New Haven, CT: Yale University Press.

Stark, Evan, and Anne Flitcraft. 1996. *Women at Risk: Domestic Violence and Women's Health.* Thousand Oaks, CA: Sage Publications.

INFOTRAC® COLLEGE EDITION EXERCISES

The exercises that follow allow you to use the InfoTrac® College Edition online database of scholarly articles to explore the sociological implications of the selections in this chapter.

Search Keyword: Marital rape. Is it considered rape when one spouse forces unwanted sex on the other? Until recently, the law in the United States said no, it was not rape. Using InfoTrac® College Edition and the search phrase *marital rape,* search for articles that address this issue. Does the law still condone forced sex between married individuals? How has the law evolved on this issue? How about social norms—does the "person on the street" view marital rape in the same terms as he or she views stranger rape, or is marital rape seen as implicitly involving consent? Discuss these issues in light of the articles that you find. What general attitude about marital rape do you take away after reading these selections? What do the authors of the articles you find believe should be done about marital rape?

Search Keyword: Elder abuse. One outgrowth of the increased longevity of many sectors of the population is the increased prevalence of elder abuse. Search InfoTrac® College Edition to find articles that deal with elder abuse. How is elder abuse defined? What populations are most at risk as victims of this abuse? What populations are most at risk as perpetrators? Are there differences in elder abuse rates and characteristics based on race/ethnicity and/or social class? How can elder abuse be detected? What prevention and/or intervention programs are available to combat the prevalence of elder abuse? Are there any public policy programs proposed? What can be done at the family level to prevent perpetration of elder abuse?

Search Keyword: Child abuse. Use InfoTrac® College Edition to familiarize yourself with the topic of child abuse. What are the different types of child abuse, and what are the telltale signs that a given child is being victimized? Who are the most prevalent perpetrators of this type of abuse? Who are most often the victims? What do the articles you find say about the causes of child abuse? What do they say about solutions to the problem? How do child abuse rates vary between different populations, based on race/ethnicity and/or social class? When are children most at risk, and why? What public policy interventions are available to protect children from family abuse? How effective are these interventions? What can be done to increase their effectiveness and protect children at risk?

Search Keyword: Battered woman syndrome. Look for articles that discuss the syndrome known as *battered woman syndrome.* What are the characteristics of this syndrome? When does it occur? What is the history of this syndrome; i.e., why was this "diagnostic tool" for battered wives needed and, subsequently, developed? Why is this syndrome sometimes viewed as a controversial diagnosis? Discuss the complexities of the issue of battered woman syndrome based on the articles you find in your InfoTrac® College Edition search.

Search Keyword: Dating violence. In many instances, research has shown us that marital violence has been preceded by periods of dating violence. That is, many of the cultural supports for domestic violence may exist in dating relationships, as well. Search for articles that deal with dating violence. What are some of the indicators of dating violence? Has this type of violence increased or decreased over time? Who is at risk for victimization? Who is at risk for becoming a potential perpetrator? What are the cultural supports for this type of violence? What attitudes continue to promote a culture of violence in dating relationships? Why is dating violence likely to continue after the dating couple marries? What is being done to prevent dating violence?

36

The Man-Made Family

CHARLOTTE PERKINS GILMAN

1898

To this day we are living under the influence of the proprietary family. The duty of the wife is held to involve man-service as well as child-service, and indeed far more, as the duty of the wife to the husband quite transcends the duty of the mother to the child.

See for instance the English wife staying with her husband in India and sending the children home to be brought up, because India is bad for children. See our common law that the man decides the place of residence; if the wife refuses to go with him to howsoever unfit a place for her and for the little ones, such refusal on her part constitutes "desertion" and is grounds for divorce.

See again the idea that the wife must remain with the husband though he is a drunkard or diseased, regardless of the sin against the child involved in such a relation. Public feeling on these matters is indeed changing; but as a whole the ideals of the man-made family still obtain.

The effect of this on the woman has been inevitably to weaken and overshadow her sense of the real purpose of the family; of the relentless responsibilities of her duty as a mother. She is first taught duty to her parents, with heavy religious sanction; and then duty to her husband, similarly buttressed; but her duty to her children has been left to instinct. She is not taught in girlhood as to her pre-eminent power and duty as a mother; her young ideals are all of devotion to the lover and husband, with only the vaguest sense of results.

The young girl is reared in what we call "innocence"—poetically described as "bloom"; and this condition is held to be one of her chief "charms." The requisite is wholly androcentric. This "innocence" does not enable her to choose a husband wisely; she does not even know the dangers that possibly confront her. We vaguely imagine that her father and brother, who do know, will protect her. Unfortunately the father and brother, under our current "double standard" of morality, do not judge the applicants as she would if she knew the nature of their offences.

Furthermore, if her heart is set on one of them, no amount of general advice and opposition serves to prevent her marrying him. "I love him!" she says sublimely. "I do not care what he has done. I will forgive him. I will save him!"

This state of mind serves to forward the interests of the lover, but is of no advantage to the children. We have magnified the duties of the wife, and minified the duties of the mother; and this is inevitable in a family relation every law and custom of which is arranged from the masculine view-point.

From this same view-point, equally essential to the proprietary family, comes the requirement that the woman shall serve the man. Her service is not that of the associate and equal, as when she

From Charlotte Perkins Gilman, *The Man-Made World or Our Androcentric Culture,* pp. 37–39, 41–46. London: T. Fisher Un Win. Copyright © 1911.

joins him in his business. It is not that of a beneficial combination, as when she practices another business and they share the profits; it is not even that of the specialist, as the service of a tailor or a barber; it is personal service—the work of a servant.

In large generalisation, the women of the world cook and wash, sweep and dust, sew and mend, for the men.

We are so accustomed to this relation, have held it for so long to be the "natural" relation, that it is difficult indeed to show it to be distinctly unnatural and injurious. The father expects to be served by the daughter, a service quite different from what he expects of the son. This shows at once that such service is no integral part of motherhood, or even of marriage, but is supposed to be the proper industrial position of women, as such.

...The dominant male holding his women as property and fiercely jealous of them, considering them always as *his*—not belonging to themselves, their children, or the world—has hedged them in with restrictions of a thousand sorts—physical, as in the crippled Chinese lady or the imprisoned odalisque; moral, as in the oppressive doctrines of submission taught by all our androcentric religions; mental, as in the enforced ignorance from which women are now so swiftly emerging.

This abnormal restriction of women has necessarily injured motherhood. The man, free, growing in the world's growth, has mounted with the centuries, filling an ever wider range of world activities. The woman, bound, has not so grown; and the child is born to a progressive fatherhood and a stationary motherhood. Thus the man-made family reacts unfavourably upon the child. We rob our children of half their social heredity by keeping the mother in an inferior position; however legalised, hallowed, or ossified by time, the position of domestic servant is inferior.

It is for this reason that child culture is at so low a level, and for the most part utterly unknown. To-day, when the forces of education are steadily working nearer to the cradle, a new sense is wakening of the importance of the period of infancy, and or its wiser treatment; yet those who know of such a movement are few, and of them some are content to earn easy praise—and pay—by belittling right progress to gratify the prejudices of the ignorant.

The whole position is simple and clear, and easily traceable to its root. Given a proprietary family, where the man holds the woman primarily for his satisfaction and service—then necessarily he shuts her up and keeps her for these purposes. Being so kept, she cannot develop humanly, as he has through social contact, social service, true social life. (We may note in passing, her passionate fondness for the child-game called "society" she has been allowed to entertain herself with; that poor simulacrum of real social life, in which people decorate themselves and madly crowd together, chattering, for what is called "entertainment.") Thus checked in social development, we have but a low-grade motherhood to offer our children, reared in the primitive conditions thus artificially maintained, enter life with a false perspective, not only toward men and women, but toward life as a whole.

The child should receive in the family full preparation for his relation to the world at large. His whole life must be spent in the world, serving it well or ill; and youth is the time to learn how. But the androcentric home cannot teach him. We live to-day in a democracy—the man-made family is a despotism. It may be a weak one; the despot may be dethroned and overmastered by his little harem of one; but in that case she becomes the despot—that is all. The male is esteemed "the head of the family"; it belongs to him; he maintains it; and the rest of the world is a wide hunting-ground and battle-field wherein he competes with other males as of old.

The girl child, peering out, sees this forbidden field as belonging wholly to menkind; and her relation to it is to secure a man for herself—not only that she may love, but that she may live. He will feed, clothe, and adorn her—she will serve him; from the subjection of the daughter to that of the wife she steps; from one home to the other, and never enters the world at all—man's world.

The boy, on the other hand, considers the home as a place of women, an inferior place, and longs to grow up and leave it—for the real world. He is quite right. The error is that this great social instinct, calling for full social exercise, exchange, service, is considered masculine, whereas it is human, and belongs to boy and girl alike.

The child is affected first through the retarded development of his mother, then through the arrested conditions of home industry, and further through the wrong ideals which have arisen from these conditions. A normal home, where there was human equality between mother and father, would have a better influence.

We must not overlook the effect of the proprietary family on the proprietor himself.

He, too, has been held back somewhat by this reactionary force. In the process of becoming human we must learn to recognise justice, freedom, human rights; we must learn self-control and to think of others; we must have minds that grow and broaden rationally; we must learn the broad, mutual interservice and unbounded joy of social intercourse and service. The petty despot of the man-made home is hindered in his humanness by too much manness.

For each man to have one whole woman to cook for and wait upon him is a poor education for democracy. The boy with a servile mother, the man with a servile wife, cannot reach the sense of equal rights we need to-day. Too constant consideration of the master's tastes makes the master selfish; and the assault upon his heart direct, or through that proverbial side-avenue, the stomach, which the dependent woman needs must make when she wants anything, is bad for the man as well as for her.

We are slowly forming a nobler type of family,—the union of two, based on love and recognised by law, maintained because of its happiness and use. We are even now approaching a tenderness and permanence of love, high, pure, enduring love, combined with the broad, deep-rooted friendliness and comradeship of equals, which promises us more happiness in marriage than we have yet known. It will be good for all the parties concerned—man, woman, and child; and it will admirably promote our general social progress.

If it needs "a head" it will elect a chairman *pro tem*. Friendship does not need "a head." Love does not need "a head." Why should a family?

37

Basic Facts about Battering:

Myths vs. Realities

NEIL JACOBSON AND JOHN GOTTMAN,
1998

During and following the O. J. Simpson trials the media were full of information about domestic violence. One of the few positive outcomes of the O. J. Simpson murder trial was that the public consciousness about domestic violence greatly increased. Unfortunately, this increased consciousness was a double-edged sword, as misinformation competed with facts, and misinformed opinions often substituted for reality.

Domestic violence is a problem of immense proportions. Unfortunately, there is much that is still not understood about the nature of battering, despite the plethora of opinions and speculative theories. In our presentation of basic information about battering, we hope to begin the process of separating fact from fiction and myth from reality.

MYTH #1: BOTH MEN AND WOMEN BATTER

There has been a backlash against the advocacy movement on behalf of battered women, a backlash that says, "Wait a minute! It is not just women who get battered. It is men too." O. J. Simpson referred to himself repeatedly as a battered husband. There are even those who claim that a huge underground movement of battered husbands refuse to tell their stories because they are reluctant to be identified as "wimps."

In support of these claims, some people cite statistics from two national surveys conducted by sociologists Murray Straus, Richard Gelles, and their colleagues. (Straus, Gelles, and Steinmetz, 1980). These statistics show that the *frequency* of violent acts is about the same in men and women. However, these statistics do not take into account two aspects of violence that are crucial to understanding battering: the impact of the violence and its function. According to statistics from the national surveys of domestic violence. . . . male violence does much more damage than female violence: women are much more likely to be injured, much more likely to enter the hospital after being assaulted by their partner, and much more likely to be in need of medical care (Stets and Straus, 1990; Vivian and Langhinrichsen-Rohling, 1994). Wives are much more likely to be killed by their husbands than the reverse; in fact, women in general are more likely to be killed by their male partners than by all other types of perpetrators combined.

Because men are generally physically stronger than women, and because they are often socialized to use violence as a method of control, it is hard to find women who are even *capable* of battering their husbands. However, battering is not just physical aggression: it is physical aggression with a purpose. The purpose of battering is to control, intimidate, and subjugate one's intimate partner through the use or the threat of physical aggression. Battering often involves injury, and in our sample, it was usually accompanied by fear on the part of the victim. . . . Fear is the force that provides battering with its power. Injuries help sustain the fear. The vast majority of physical assaults reported in the national surveys were pushes, shoves, and other relatively minor acts of violence. They were not the kinds of battering episodes that typically end up in the criminal justice system.

All indications are that in heterosexual relationships, battering is primarily something that men do to women, rather than the reverse. However, as we will show, there are many battered women who are violent, mostly, but not always, in self-defense. Battered women are living in a culture of violence, and they are part of that culture. Some battered women defend themselves: they hit back, and might even hit or push as often as their husbands do. But they are the ones who are beaten up. On a survey that simply totals the frequency of violent acts, they might look equally violent. But there is no question that in most relationships the man is the batterer, and the woman is the one who is being battered.

MYTH #2: ALL BATTERERS ARE ALIKE

Although there is still a tendency for professionals to talk about batterers as if they were all alike, there is growing recognition that there are different types of batterers. There are at least two distinguishable types that have practical consequences for battered women, and perhaps more. Each type seems to have its unique characteristics, its own family history, and perhaps different outcomes when punished by the courts or educated by groups for batterers. Based on our findings of a distinction between the Cobras and the Pit Bulls, and the work of Dr. Amy Holtzworth-Munroe and Gregory Stuart (1994), we think a compelling case can be made for at least two subtypes, roughly corresponding to our distinction between the Cobras and the Pit Bulls.

Cobras

Cobras appear to be criminal types who have engaged in antisocial behavior since adolescence. They are hedonistic and impulsive. They beat their wives and abuse them emotionally, to stop them from interfering with the Cobras' need to get what they want when they want it. Although they may say that they are sorry after a beating, and beg their wives' forgiveness, they are usually not sorry. They feel entitled to whatever they want whenever they want it, and try to get it by whatever means necessary. Some of them are "psychopaths," which means they lack a conscience and are incapable of feeling remorse. In fact, true psychopaths have diminished capacity for experiencing a wide range of emotions and an inability to understand the emotions of others: they lack the ability to sympathize with the plight of others, they do not experience empathy, and even apparent acts of altruism are actually thinly veiled attempts at selfishness. They do not experience soft emotions such as sadness, and rarely experience fear unless it has to do with the perception that something bad is about to happen to *them*.

But not all Cobras are psychopaths. Whether psychopathic or merely antisocial, they are incapable of forming truly intimate relationships with others, and to the extent that they marry, they do so on their terms. Their wives are convenient stepping-stones to gratification: sex, social status, economic benefits, for example. But their commitments are superficial, and their stance in the relationship is a "withdrawing" one. They attempt to keep intimacy to a minimum, and are most likely to be dangerous when their wives attempt to get *more* from them. They do not fear abandonment, but they will not be controlled. Their own family histories are often chaotic, with neither parent providing love or security, and they were often abused themselves as children.

As adults, they can be recognized by their history of antisocial behavior, their high likelihood of drug *and* alcohol abuse, and the severity of their physical and emotional abuse. Their wives fear them, and are often quite depressed. But fear and depression do not completely explain why the women are unlikely to leave the relationship. Nor is it simply that they lack economic and other resources: indeed, Cobras are often economically dependent on their wives. Despite the fact that they are being severely abused, it is often the women rather than the men who continue to fight for the continuance of the relationship. It is these couples where the men exude macabre charisma.

Pit Bulls

The Pit Bulls are more likely to confine their violence to family members, especially their wives. Their fathers were likely to have battered their mothers, and they have learned that battering is an acceptable way to treat women. But they are not as likely as the Cobras to have criminal records, or to have been delinquent adolescents. Moreover, even though they batter their wives and abuse them emotionally, unlike the Cobras the Pit Bulls are emotionally dependent on their wives. What they fear most is abandonment. Their fear of abandonment and the desperate need they have not to be abandoned produce jealous rages and attempts to deprive their partners of an independent life. They can be jealous to the point of paranoia, imagining that their wives are having affairs based on clues that most of us would find ridiculous.

The Pit Bulls dominate their wives in any way they can, and need control as much as the Cobras do, but for different reasons. The Pit Bulls are motivated by fear of being left, while the Cobras are motivated by a desire to get as much immediate gratification as possible. The Pit Bulls, although somewhat less violent in general than the Cobras, are also capable of severe assault and murder, just as the Cobras are. Although one is safer trying to leave a Pit Bull in the short run, Pit Bulls may actually be more dangerous to leave in the long run. Cobras strike swiftly and with great lethality when they feel threatened, but they are also easily distracted after those initial strikes and move on to other targets. In contrast, Pit Bulls sink their teeth into their targets; once they sink their teeth into you, it is hard to get them to let go!

It is not clear how Cobras and Pit Bulls are apportioned within the battering population. In our sample, 20 percent of the batterers were Cobras. Interestingly, Dr. Robert Hare, an internationally renowned expert on psychopaths, estimates that 20 percent of batterers are psychopaths (Hare, 1993). This correspondence is provocative. However, our guess is that Cobras constitute a larger percentage of the clinical or criminal population of batterers than the 20 percent found in our study. The Cobras fit the profile of the type of batterer who comes into contact with the criminal justice system much more than the Pit Bulls do. The profile of the Cobra also describes those referred by judges to treatment groups much better than the profile of the Pit Bull.

MYTH #3: BATTERING IS NEVER CAUSED BY DRUGS AND ALCOHOL

Dr. Kenneth E. Leonard reviewed a body of literature in 1993 suggesting a strong relationship between alcohol use and battering (Leonard, 1993). However, there was at that time a great deal of ambiguity about the extent to which drug and alcohol abuse causes men who would otherwise not be batterers to beat their wives. In 1996, Dr. Leonard conducted the most definitive study to date on the role of alcohol in physical aggression by husbands toward wives (Leonard and Senchak, 1996). In this study of newlyweds, Dr. Leonard reported that alcohol use was one of the strongest indicators that men would be physically aggressive during their first year of marriage. Although the research on drug abuse has been less extensive, it is clear that batterers are a great deal more likely to be drug abusers than are men who do not batter.

None of this research proves that alcohol or drugs "cause" battering. It simply suggests that batterers tend to have drug and alcohol problems. Because of this connection, battered women who haven't given up the dream often see treatment for drugs and alcohol abuse as the ray of hope: "If only he would stop drinking [or shooting up, or snorting coke], everything would be fine." It is easy to understand how it is that battered women develop this belief. Indeed, it may be true that for some batterers, stopping the substance abuse *will* lead to an end to battering.

However, the relationship between substance abuse and battering is an extremely complicated one. First, a substantial portion of batterers are not alcohol or drug abusers. Although battered women married to batterers without drug and alcohol problems would not be inclined to see treatment for substance abuse as relevant, the point is worth making because many people assume that the substance abuse is more connected to battering than it really is. Some batterers use alcohol; some don't. Some batterers abuse illegal substances; others don't.

Second, just because batterers abuse drugs or alcohol doesn't mean that they batter only when they are intoxicated. A battered woman may be just as likely to be beaten when a substance abuser is sober as she is when the batterer is under the influence of these substances.

Third, even when battering episodes typically occur while the batterer is high on drugs, it is not always the intoxication itself that increases the likelihood of battering. The majority of men with drug and alcohol problems do not batter their wives. Alcohol and drug intoxication may lower inhibitions, but they also make for handy rationalizations. Some batterers in our sample got high in order to provide a way of justifying the beating that they had planned before getting high. Many men we talked to attributed their violence to drugs and alcohol: when they did so, it was an attempt to minimize the significance of the battering per se to deny that violence was a problem beyond the other problem of drug abuse, and to distort the cause of the violence, which was the need for control.

Too much focus on drug and alcohol abuse leads us away from the central issue: battering is fundamentally perpetuated by its success in controlling, intimidating, and subjugating the battered woman. The types of men who abuse their partners are also the types of men who abuse drugs and alcohol. Alcohol and drug abuse are part of the lifestyle of the batterer. But it would be a mistake to assume that the drug use causes the violence.

However, it is also true that substance abuse can be one of the causes of violence, and to see it as nothing but an excuse is to oversimplify a complex relationship. We believe that in rejecting alcohol and drug abuse as causal explanations for battering, many theorists in the field of domestic violence have thrown the baby out with the bathwater: in fact, some batterers *do* only batter while intoxicated *because* the state of intoxication transforms them. But these men should not be allowed to use the substance abuse as a justification for the violence. The fact that the batterer was intoxicated should not be grounds for legal exoneration, or a "diminished capacity" defense if the batterer is charged with a battering-related crime. The batterer's accountability should not be affected by whether or not he is sober while battering: a batterer should always be held accountable for battering. Nevertheless, any insistence on separating substance abuse from domestic violence results in lost opportunities for combining what we know about stopping substance abuse with what we know about stopping violence.

As things currently stand, the substance abuse and domestic violence experts typically study one or the other but not both. Rarely do experts from the different fields communicate with one another. If leading figures in the field of substance abuse worked together with leaders in the prevention of domestic violence, it might be possible to develop more effective ways of reducing both problems. And it is entirely possible that if men can be successfully treated for substance abuse, some of them will stop battering. It is perfectly consistent to hold the position that drug and alcohol abuse can be one of many causes of battering on the one hand, and continue to hold the batterer responsible for the violence on the other hand.

MYTH #4: BATTERERS CAN'T CONTROL THEIR ANGER

This is a complicated issue, because there is a sense in which all behavior, even behavior that we think of as voluntary, is actually caused by past and current events in our environment. We are all products of our own unique history, and that history helps to explain how we respond in particular situations.

But voluntary behavior involves a choice, and depending on the outcome one seeks, different choices will be made. Battering is usually voluntary. There are some people with temporal lobe epilepsy whose brains literally trigger violent outbursts that bear no relationship to anything that is going on in the environment. There are some batterers (a small minority) whose battering rampages are truly impulsive and uncontrollable, at least in the early stages of each incident. But in the vast majority of cases, battering is a choice in the same sense that all other voluntary actions are. With Cobras, their physiological responses to conflict are consistent with increased concentration and focused attention. Their lowered heart rates during arguments probably function to focus their attention, to maximize the impact of their aggression. We suspect that Cobras are not only in control, but that they use their control over their own physiology to strike more effectively. But even Pit Bulls, who are highly aroused when they strike, still choose to strike.

Psychologist Donald Dutton and others have written of the "dissociation" often associated with violent episodes (Dutton, 1995). Dutton is talking about the type of batterer whom we would classify as a Pit Bull, and provides some anecdotal evidence that some of them experience an altered state of consciousness during battering episodes. In extreme cases they do not even remember the episodes afterward. Although dissociative states are consistent with the interpretation that they are "out of control" when they batter, few of the batterers in our sample described their episodes of violence in a manner that suggested dissociation. They remembered the episodes but either minimized their significance or denied responsibility for them. Occasionally, they would deny that the violence had occurred. Our interpretation of this denial is that these batterers are lying, using another method of extending their control over their battered wives. The relationship between voluntarily lying and dissociation (truly not remembering) remains an unresolved issue, to be determined by future research. But in the vast majority of violent episodes that occurred among our sample, although the batterers may have been behaving impulsively, they were not out of control.

MYTH #5: BATTERING OFTEN STOPS ON ITS OWN

. . . Battering seldom stops on its own. We found in our research that while many men decrease their level of violence over time, few of them stop completely. And when they do stop, the emotional abuse usually continues.

This is quite important, because most research considers only physical abuse. But emotional abuse can be at least as effective a method of maintaining control, if the physical violence was once there. Once a batterer has achieved dominance through violence and the threat of more violence, emotional abuse often keeps the battered woman in a state of subjugation without the batterer having to use physical force. Since the violence is used in order to obtain control, it is more convenient for the batterer to restrict himself to emotional abuse. That abuse reminds her that the threat of violence is always present, and this threat is often sufficient to retain control. Any intervention which defines success without taking emotional abuse into account will inflate its effectiveness. In our sample, although many batterers *decreased* the frequency and severity of their violence over time, almost none of them stopped completely *and* also ended the emotional abuse.

Frank stopped being violent for eighteen months, and Jane, his wife, was quite happy about it. If we hadn't also studied emotional abuse, they might have looked like a couple whose problems had been solved. But, if anything, we found that Frank was even more emotionally abusive two years later than he had been initially, despite ceasing the physical violence. He was more insulting, more verbally threatening, drove more recklessly when angry, and humiliated Jane at every opportunity. All of these emotionally abusive acts were extremely hurtful and degrading to her. But they served an additional function: they reminded her of the violence, and the tightrope she had to walk to avoid recurrences of that violence. These reminders were all she needed. Frank maintained his control without having to risk breaking the law. Emotional abuse, as destructive as it is, is not against the law.

MYTH #6: PSYCHOTHERAPY IS A MORE EFFECTIVE "TREATMENT" THAN PRISON

Our prisons are overcrowded and judges are constantly looking for alternatives to prison when given discretion in sentencing. Because psychotherapy is available for batterers, judges often find some referrals for court-mandated treatment irresistible as alternatives to imprisonment, especially since domestic assault charges are often misdemeanors rather than felonies.

Unfortunately, what appears at first glance to be an enlightened alternative to imprisonment is often a mistake. There is very little evidence that currently existing treatment programs for batterers are effective, and much reason to be concerned that in their present form, they are unlikely to stop the violence and even less likely to end the emotional abuse. Yet people in our culture believe in psychotherapy, and battered women are no exception. Therefore, when their husbands are "sentenced" to psychotherapy they may be lulled into a false sense of security, thus leading them to return home from a shelter falsely convinced that they are now safe.

As a matter of fact, Roy had been to therapy once, and Helen received glowing reports from the anger-management therapist. Roy received sixteen weeks of group therapy, and his therapist wrote the following during an evaluation at the conclusion of these sixteen weeks:

"In my professional opinion, Roy no longer constitutes a danger to his wife, Helen. He has been a model patient. He has accepted responsibility for being a batterer, has shown no inclination to repeat the violence from the past, and even stands a chance of qualifying for work as a therapist himself, working with batterers who have not yet developed the insight that has changed Roy's life."

Naturally, Helen was quite excited to make things work with the new Roy. But Roy's therapist had been conned. And Roy's therapist had inadvertently misled Helen. Less than two months after the group treatment had ended, Roy came at Helen with a knife, nearly killing her.

Violent criminals who assault strangers are seldom offered psychotherapy as an alternative to prison, in contrast to perpetrators of wife battering. What does this tell us about the criminal justice system and how it views "family violence"? Family violence is still regarded as less serious than violence against strangers, even though most women who are murdered are not killed by strangers but by boyfriends, husbands, ex-husbands, and ex-boyfriends (Koss et al., 1994). But accountability is a prerequisite to decreased violence. We believe that referrals to psychotherapy, in the absence of legal sanctions, send the wrong message to batterers: they have gotten away with a violent crime with nothing but a slap on the wrist. O. J. Simpson is an excellent example. In 1989, he pleaded "no contest" to the charge of misdemeanor assault, after beating his wife Nicole Brown Simpson on New Year's Eve. He received a small fine and was ordered to seek treatment. The treatment ended up being nothing more than a few sessions, some of which were conducted by telephone. We now know just how ineffective this "punishment" and the treatment that was required of him were.

We have no illusions about the rehabilitative power of prison, but at least prison stops the violence temporarily and gives the battered woman time to make plans. It also sends a powerful message to the batterer.

Consider also the ability of the most violent batterers—like Roy and George, both Cobras—to con judges, police, probation officers, and therapists. The Cobra will figure out what the therapist wants him to say, sound contrite, be counted as a success, and yet all that has been accomplished is that the system has been exploited by the Cobra. We believe that some form of treatment, either education or group therapy, should be offered to all convicted batterers on a voluntary basis, but it should never be mandated and it should never be offered as an alternative to the appropriate legal sanctions. We also believe that after the first offense, domestic assault should automatically be a felony. There is no reason to give up on education and treatment. But there is even less reason to allow batterers to use them as additional methods of control: control of the criminal justice system as well as the partner. The message has to be clear and unambiguous: violent crime will not be tolerated, whether the victim is a stranger or a family member.

MYTH #7: WOMEN OFTEN PROVOKE MEN INTO BATTERING THEM

This myth is held by most batterers, many members of the general public, and even by some professionals. But . . . men initiate violence independently of what their wives do or say. One husband came home from work after being criticized by his boss, and as his wife came to greet him at the door, he punched her in the face, knocking her unconscious. Drs. Lenore Walker and Donald Dutton have both described the internal build-up of tension that seems to occur in many batterers, regardless of what the battered woman is doing or saying (Walker, 1984; Dutton, 1995). Ultimately, this tension leads to an explosion of violent rage.

One of the couples in our sample had been in treatment with a family therapist. The husband and wife both quoted the therapist as saying to the wife: "If you would stop using that language, perhaps he wouldn't get so out of control." The husband wore this therapist's comment as a badge of honor and even referred to her use of profanity as "violence." In his view, she started the violence whenever she swore at him. The therapist appeared to be supporting this view. The wife never felt understood by the therapist, but the outcome of treatment was that she blamed herself for the battering and thought that the solution was "to be a better wife."

Holding the husband accountable for using violence, regardless of what the wife does or says, is a necessary step for the violence to stop, but it is not sufficient. The batterer has to "feel" accountable in order for the violence to stop. Feeling the accountability means that in addition to the batterer being punished, he must feel that his punishment is justified. It is the rare batterer who feels this way, which helps to explain why battering infrequently stops on its own. Instead of holding themselves accountable, batterers minimize the severity of their violent actions, deny that they are responsible for them, and distort them to the point where they become trivial, as George did when he said after beating up Vicky, "I didn't think nothing of it because it wasn't important."

In those rare instances where the husband feels accountable, the violence as well as the emotional abuse might stop. One of the few batterers in our sample to stop the abuse entirely told us from the beginning that, "I never felt right about it. Even while I was doing it. When I was arrested I deserved it. No man has the right to hit a woman unless she's trying to kill him."

Thus, even if the husband's violence is a response to remarks made by the wife, it is a mistake to think of these remarks as "provocations." Provocation implies that "she got what she deserved." George believed he was justified in beating Vicky because she had an "attitude" about his being late for dinner. Batterers make choices when they beat their wives. There is no remark or behavior that justifies a violent response unless it is in self-defense.

In our sample, men were rarely if ever defending themselves when they started battering. Nothing a woman says to a man gives him the right to hit her. Therefore, women couldn't possibly precipitate male battering unless they initiated the violence, which they rarely did in our sample. And even if they do initiate physical aggression by pushing their husbands, punching them in the arm out of frustration, or throwing something at them, they haven't provoked a beating. Husbands have a right to defend themselves: when they are punched, they can deflect the blow; when they are pushed, they can hold their wives so that they stop; and when something is thrown at them, they can duck and yell at their wives to stop. None of these physically aggressive behaviors constitute battering. However, they are commonly used by batterers as excuses for battering that would have occurred anyway.

Sam gave Marie a black eye when she shook him in the middle of the night, waking him up. She couldn't sleep because he had been flirting with another woman at a party, and in fact had been dancing with her in a way that was blatantly sexual. He got angry when she woke him up, refused to discuss it with her, and when she hit him with a pillow out of frustration, he punched her in the face.

There is a philosophy of marriage inherent in the view that women provoke men to be violent. It says that the man is the head of the household, the boss. In the old days, being the boss meant having the right to beat and even kill your wife, the way masters had the right to kill their slaves. Now, it means viewing the wife as someone who deserves to be beaten under certain conditions. Wives never deserve to be beaten by their husbands. Battering is a criminal act, and verbal challenges from the wife do not constitute mitigating circumstances.

Harry, one of the husbands in our sample, almost choked his wife Beth to death after she taunted him: "You're probably a fag, just like your father." What she said was wrong, but she didn't deserve to be choked for it. Harry should have been charged with attempted murder.

MYTH #8: WOMEN WHO STAY IN ABUSIVE RELATIONSHIPS MUST BE CRAZY

This myth actually assumes a fact, namely, that most battered women *do* stay in abusive relationships. In our sample, within two years of their first contact with us, 38 percent of the women had left their husbands. When you consider that about 50 percent of couples divorce over the course of their lives, 38 percent in two years is very high indeed! Many battered women *are* getting out of abusive relationships. Abused women ought not to be blamed for not leaving. In fact, they do leave at high rates. And their leaving is often an act of courage because it means having to cope with enormous fear as well as financial insecurity.

But what about those who haven't left yet? Does this mean that there is something wrong or odd about them? The answer is no. It is much easier to get into an abusive relationship than it is to get out of one. Women are often afraid to leave, and with good reason. Their chances of getting seriously hurt or killed increase dramatically for the first two years after they separate from their husbands. Leaving is risky, and often staying is the lesser of two evils. Tracy Thurman went all the way across the country to escape from her abusive husband. But he tracked her down in rural Connecticut, and came after her with a knife, disabling her for life and almost killing her. Police and neighbors watched the slaughter. Her case was a watershed in the quest of advocates for mandatory arrest laws, laws which require that arrests are made when the police find probable cause for a domestic dispute.

Second, women often can't afford to leave, especially if they have children. They are economically dependent on their husbands, and that economic leverage is part of the control exerted by the batterer. Vicky had to wait a long time after she had decided to leave George until she was financially able. So did Clara, who was a homemaker married to a university professor. He controlled all the finances, and whenever he thought she might be contemplating leaving, he would remind her of his resources to hire the best divorce lawyer in town. He assured her that he would settle for nothing less than full custody of the children, and that she would get nothing. Clara was trapped by economic dependency.

Third, after being subjected to physical and emotional abuse for a period of time, women are systematically stripped of their self-esteem, to the point where they falsely believe that they need their husbands in order to survive, despite the violence. This lowering of self-esteem is often part of a constellation of symptoms that are common to survivors of trauma.

Battered women experience trauma similar to that of soldiers in combat, abused children, and rape victims (many battered women *are* also raped by their husbands). These symptoms include depression, anxiety, a sense of being detached from their bodies and numb to the physical world, nightmares, and flashbacks of violent episodes. The syndrome characterized by these symptoms is "post-traumatic stress disorder" (PTSD).

Many battered women suffer from mild to severe versions of PTSD, and as a result are not functioning well. Their parenting is affected, their problem-solving abilities are impaired, and their ability to plan for the future is disrupted. Given this common experience of trauma, it is amazing that battered women manage to be as resourceful and resilient as they are, especially since their lives and the lives of their children are often at stake. The corrosive and cumulative effects of battering go a long way toward explaining how hard it is to get out of an abusive relationship.

Erin had long ago stopped loving her husband, Jack. But Jack had convinced her that she lacked the ability to survive without him. He was constantly telling her that she was stupid, disorganized, ugly, and needed him around in order to get through the day. After years of public insults, severe beatings, and successful attempts at isolating her from the rest of the world, Erin began to believe him. She stayed.

Another thing that keeps some women in violent relationships is that they are holding on to a dream that they have about what life could be like with these men. They love their husbands and they have developed a sympathy for them and their plight in life. They hope that they can help their men become normal husbands and fathers. These dreams can be powerful and are very hard to give up.

Some Cobras exude an inexplicable type of charisma for their partners. Battered women stay in these relationships because they are quite attached to and love their husbands, not because of the violence, but in spite of it. They are afraid of their husbands and want the violence to stop. But to them, violence is not a sufficient reason to leave the relationship. In fact, the wives of Cobras are often *more* committed to maintaining the relationship than their husbands are, despite the severe beatings.

Although as outsiders, it is hard for us to understand this attachment, it helps to remember that violence is normal in many subcultures within North America. When violence is all around you, it tends to be accepted as a fact of life. As hard as it might be to imagine, many battered women assume that violence is part of marriage. They don't like it. They continue to try to change it. But they don't view it as a reason for getting out. They have accepted the culture of violence.

Cobras seem to choose women who are especially vulnerable to their macabre charisma. They also figure out quite quickly where the woman's particular vulnerabilities lie. George, for example, figured out quite quickly that Vicky had a dream, and he altered his presentation of himself so that it would be in accord with that dream. He also happened to meet Vicky at a unique time in her life: she was vulnerable not just because of her dream, but because her self-esteem was at an all-time low, and her life was in shambles. Vicky is no longer vulnerable to the Georges of the world. She has become streetwise. In fact, she would have probably resisted his "charms" at an earlier time. But George met her when she was uniquely available to join with him in the relationship that was to become her worst nightmare.

Another couple with a similar dynamic was Roy and Helen. They had met in prison, where he was serving time for armed robbery, and she was visiting her boyfriend, who had been locked up for sexually abusing Helen's daughter. Helen was one of the most severely battered women in our sample. When they first came into contact with us, Roy and Helen were homeless, although she held a job as a hotel receptionist. He had broken her neck and her back on separate occasions, caused eight miscarriages by beating her whenever he got her pregnant (he refused to use birth control), and he had even stabbed her once or twice, "not to kill her, just to scare her." He was an alcoholic and a heroin addict, and refused to enter treatment for either. He also engaged in frequent extramarital affairs. Yet even though she had made it to a shelter on one occasion, she didn't stay because they "tried to talk me into leaving him." As she put it, "I don't want to leave him; I just want him to stop beating me." She loved him and was committed to him not *because of* the violence, but *in spite of* it. What Vicky once told us about George's mother was probably applicable to Helen: "She never knew that there was anything different; where she was brought up, men beat women. They don't know that they deserve better. They don't even know that there is anything better."

A small percentage of the battered women in our study married to Cobras were themselves antisocial before getting involved in their current abusive relationships, just like their husbands. These women were themselves impulsive and often had criminal records going back to childhood. Helen was one of these women. She and Roy hit it off, immediately decided to live together, and soon after that they were married. Their marriage was both volatile and violent.

Some women in our sample expressed a preference for romance "on the edge," unpredictable and potentially both dangerous and adventurous. They would have it no other way. Even though they did not want to be battered, they found their husbands charismatic and attractive, and had no interest in leaving them. Often they stay against their family's advice and the advice of their friends. By their own account, they had no interest in relationships with boring nice guys.

MYTH #9: BATTERED WOMEN COULD STOP THE BATTERING BY CHANGING THEIR OWN BEHAVIOR

By now, it should be clear that battering cannot be changed through actions on the part of the victim. Battering has little to do with what the women do or don't do, what they say or don't say. It is the batterer's responsibility—and his alone—to stop being abusive. We collected and analyzed data on violent incidents as they unfolded at home, and examined sequences of actions that led up to the violence. We discovered that there were no triggers of the violence on the part of the men, nor were there any switches available for turning it off once it got started.

Countless women from our sample still believed that it was their job to stop the husbands' violence. Helen was a perfect example. She would defend Roy when given the opportunity, usually by blaming herself. "He is a good man who is easily stressed out. I work hard to make it easier for him, but not hard enough. Like when he wants to have sex. I should be there for him more because I know a lot of this has to do with sexual tension. And I like to drink. If there wasn't alcohol around, he probably wouldn't hit me as much."

Because battering has a life of its own, and seems to be unrelated to actions on the part of the woman, couples therapy makes little sense as a first-line treatment. One would not expect couples therapy to stop the violence, since the violence is not about things that the women are doing or saying. Couples therapy has other disadvantages. First, it can increase the risk of violence by forcing couples to deal with conflict on a weekly basis; leaving the batterer in a constant state of readiness to batter. Second, when the couple is seen together, the therapist implies that they are mutually responsible for the violence. This implication is handy for the batterer, since it supports his point of view: "If she would just change her behavior, the violence would stop." The victim ends up being blamed for her own victimization.

Couples therapy can work for couples where there is low-level physical aggression without battering. It might also work in relationships where the husband has demonstrated the ability to stop being abusive for one or two years. But we would never recommend couples therapy as the initial treatment strategy for batterers, even in those instances where psychotherapy of some sort might be appropriate.

MYTH #10: THERE IS ONE ANSWER TO THE QUESTION "WHY DO MEN BATTER WOMEN?"

There are many competing theories among social scientists, legal experts, and advocates about what causes battering. The theories are often pitted against one another as if one is correct and another incorrect. In fact, no one knows what causes battering, and there is in all likelihood no one cause. We are not attempting to answer that question . . . Instead, we are describing what we have learned about the dynamics of battering by looking intensively at the relationships between batterers and battered women, and we recognize that any complete understanding of battering has to include further analyses.

Most important, any complete understanding of battering has to take into account the historical, political, and broad socioeconomic conditions that make battering so common. The subordination of women to men throughout the history of our civilization, and the resulting oppression, is pivotal in this analysis. Our culture has been patriarchal as far back as we can trace it. . . . Patriarchy has sanctioned battering historically and continues to operate to perpetuate battering today: the continued oppression of women provides a context that makes efforts to end violence against women difficult if not impossible.

Battering also is intimately related to social class. It is much more common in lower socioeconomic classes than it is in middle and upper classes. Where violence in general is common, so is violence against women. All of the economic forces which operate to perpetuate class differences, racism, and poverty contribute to high rates of battering. . . .

Finally, it should be noted that our sample was 90 percent Caucasian. Although the couples in our sample were predominantly working class and lower class, there were few African American, Latino, and Native American couples in our sample. Therefore, we are unable to discuss potentially important cross-cultural differences in battering, and are forced to discuss battering without taking into account possibly unique dynamics among ethnic and sexual (gay and lesbian) minorities. This is a crucial area for future research.

In short, battering occurs within a patriarchal culture, and is made possible because such a culture dominates American society. It is further fueled by poverty, racism, and heterosexism. Our focus is on individual differences within the broader society. Not all men are batterers, despite the culture. Therefore, we think our intensive focus on the dynamics of relationships is one crucial area where understanding is needed. But it is only one of many foci that are necessary for a complete understanding of battering.

Although not all or even most men are batterers, most batterers in heterosexual relationships are men. Batterers come in different shapes and sizes. Drug and alcohol abuse are often important components of violent episodes. However, most batterers do not commit their beatings in a state of uncontrollable rage.

. . . Battering can and often does decrease over time, but it seldom completely stops on its own. Until batterers are consistently punished by the criminal justice system in a way that is commensurate with the crime committed, the rates of domestic violence are likely to remain staggeringly high. Women do not, cannot, and should not be implicated, either directly or indirectly, as contributors to the problem of battering, even when they challenge their husbands verbally or stay in abusive relationships. In fact, there is little women *can* do to change the course of a violent episode, or affect its onset.

However, we have personally witnessed many heroic and resourceful steps taken by battered women that have ultimately freed them from their violent relationships, or in other cases turned the relationships in a positive direction. . . .

SELECTED REFERENCES

Dutton, Donald G. 1995. *The Batterer.* New York: Basic Books.

Hare, Robert D. 1993. *Without Conscience.* New York: Pocket Books.

Holtzworth-Munroe, Amy and Gregory L. Stuart. 1994. "Typologies of Male Batterers." *Psychological Bulletin* 116:476–497.

Koss, Mary et al. 1994. *No Safe Haven.* Washington, D.C.: American Psychological Association Press.

Leonard, Kenneth E. 1993. "Drinking Patterns and Intoxication in Marital Violence." In *Alcohol and Interpersonal Violence,* edited by S. E. Martin. National Institute of Alcohol and Alcohol Abuse Monograph No. 24, National Institute of Health Pub. No. 93-3496. Rockville, MD: National Institutes of Health.

Leonard, Kenneth E. and Marilyn Senchak. 1996. "Prospective Prediction of Husband Marital Aggression within Newlywed Couples." *Journal of Abnormal Psychology* 105:369–380.

Stets, Janice E. and Murray A. Straus. 1990. "Gender Differences in Reporting of Marital Violence and Its Medical and Psychological Consequences." pp. 151–165 in *Physical Violence in American Families,* edited by Murray A. Straus and Richard J. Gelles. New Brunswick, NJ: Transaction Publishers.

Straus, Murray A., Richard J. Gelles, and Suzanne K. Steinmetz, 1980. *Behind Closed Doors.* New York: Doubleday Press.

Vivian, Dina and Jennifer Langhinrichsen-Rohling. 1994. "Are Bi-directionality Violent Couples Mutually Victimized? A Gender-sensitive Comparison." *Violence and Victims* 9: 107–123.

Walker, Lenore E. 1984. *The Battered Woman Syndrome.* New York: Springer Publishing Company.

38

Ten Myths That Perpetuate Corporal Punishment

MURRAY STRAUS,
2001

Hitting children is legal in every state of the United States and . . . 84 percent of a survey of Americans agree that it is sometimes necessary to give a child a good hard spanking. . . . Almost all parents of toddlers act on these beliefs. Study after study shows that almost 100 percent of parents with toddlers hit their children. There are many reasons for the strong support of spanking. Most of them are myths.

MYTH 1: SPANKING WORKS BETTER

There has been a huge amount of research on the effectiveness of corporal punishment of animals, but remarkably little on the effectiveness of spanking children. That may be because almost no one, including psychologists, feels a need to study it because it is assumed that spanking is effective. In fact, what little research there is on the effectiveness of corporal punishment of children agrees with the research on animals. Studies of both animals and children show that punishment is not more effective than other methods of teaching and controlling behavior. Some studies show it is less effective. . . .

Day and Roberts (1983) . . . studied three-year-old children who had been given "time out" (sitting in a corner). Half of the mothers were assigned to use spanking as the mode of correction if their child did not comply and left the corner. The other half put their non-complying child behind a low plywood barrier and physically enforced the child staying there. Keeping the child behind the barrier was just as effective as the spanking in correcting the misbehavior that led to the time out.

A study by Larzelere (in press) also found that a combination of *non*-corporal punishment and reasoning was as effective as corporal punishment and reasoning in correcting disobedience.

Crozier and Katz (1979), Patterson (1982), and Webster-Stratton et al. (1988, 1990) all studied children with serious conduct problems. Part of the treatment used in all three experiments was to get parents to stop spanking. In all three, the behavior of the children improved after spanking ended. Of course, many other things in addition to no spanking were part of the intervention. But, as you will see, parents who on their own accord do not spank also do many other things to manage their children's behavior. It is these other things, such as setting clear standards for what is expected, providing lots of love and affection, explaining things to the child, and recognizing and rewarding good behavior, that account for why children of non-spanking parents tend to be easy to manage and well-behaved. What about parents

From Straus, Murray A. 2001. *Beating the Devil out of Them: Corporal Punishment in American Families and Its Effects on Children,* 2nd Edition. New Brunswick, NJ: Transaction Publishers.

who do these things and also spank? Their children also tend to be well-behaved, but it is illogical to attribute that to spanking since the same or better results are achieved without spanking, and also without adverse side effects.

Such experiments are extremely important, but more experiments are needed to really understand what is going on when parents spank. Still, what Day and Roberts found can be observed in almost any household. Let's look at two examples.

In a typical American family there are many instances when a parent might say, "Mary! You did that again! I'm going to have to send you to your room again." This is just one example of a non-spanking method that did *not* work.

The second example is similar: A parent might say, "Mary! You did that again! I'm going to have to spank you again." This is an example of spanking that did *not* work.

The difference between these two examples is that when spanking does not work, parents tend to forget the incident because it contradicts the almost-universal American belief that spanking is something that works when all else fails. On the other hand, they tend to remember when a *non*-spanking method did not work. The reality is that nothing works all the time with a toddler. Parents think that spanking is a magic charm that will cure the child's misbehavior. It is not. There is no magic charm. It takes many interactions and many repetitions to bring up children. Some things work better with some children than with others.

Parents who favor spanking can turn this around and ask, If spanking doesn't work any better, isn't that the same as saying that it works just as well? So what's wrong with a quick slap on the wrist or bottom? There are at least three things that are wrong:

- Spanking becomes less and less effective over time and when children get bigger, it becomes difficult or impossible.
- For some children, the lessons learned through spanking include the idea that they only need to be good if Mommy or Daddy is watching or will know about it.
- . . . There are a number of very harmful side effects, such as a greater chance that the child will grow up to be depressed or violent. Parents don't perceive these side effects because they usually show up only in the long run.

MYTH 2: SPANKING IS NEEDED AS A LAST RESORT

Even parents and social scientists who are opposed to spanking tend to think that it may be needed when all else fails. There is no scientific evidence supporting this belief, however. It is a myth that grows out of our cultural and psychological commitment to corporal punishment. You can prove this to yourself by a simple exercise with two other people. Each of the three should, in turn, think of the most extreme situation where spanking is necessary. The other two should try to think of alternatives. Experience has shown that it is very difficult to come up with a situation for which the alternatives are not as good as spanking. In fact, they are usually better.

Take the example of a child running out into the street. Almost everyone thinks that spanking is appropriate then because of the extreme danger. Although spanking in that situation may help *parents* relieve their own tension and anxiety, it is not necessary or appropriate for teaching the child. It is not necessary because spanking does not work better than other methods, and it is not appropriate because of the harmful side effects of spanking. The only physical force needed is to pick up the child and get him or her out of danger, and, while hugging the child, explain the danger.

Ironically, if spanking is to be done at all, the "last resort" may be the worst. The problem is that parents are usually very angry by that time and act impulsively. Because of their anger, if the child rebels and calls the parent a name or kicks the parent, the episode can escalate into physical abuse. Indeed, most episodes of physical abuse started as physical punishment and got out of hand (. . . Kadushin and Martin, 1981). Of course, the reverse is not true, that is, most instances of spanking do not escalate into abuse. Still, the danger of abuse is there, and so is the risk of psychological harm.

The second problem with spanking as a last resort is that, in addition to teaching that hitting is the way to correct wrongs, hitting a child impulsively teaches another incorrect lesson—that being extremely angry justifies hitting.

MYTH 3: SPANKING IS HARMLESS

When someone says, I was spanked and I'm OK, he or she is arguing that spanking does no harm. This is contrary to almost all the available research. One reason the harmful effects are ignored is because many of us (including those of us who are social scientists) are reluctant to admit that their own parents did something wrong and even more reluctant to admit that we have been doing something wrong with our own children. But the most important reason may be that it is difficult to see the harm. Most of the harmful effects do not become visible right away, often not for years. In addition, only a relatively small percentage of spanked children experience obviously harmful effects.

The delayed reaction and the small proportion seriously hurt are the same reasons the harmful effects of smoking were not perceived for so long. In the case of smoking, the research shows that a third of very heavy smokers die of lung cancer or some other smoking-induced disease. That, of course, means that two-thirds of heavy smokers do *not* die of these diseases (Mattson et al., 1987). So most heavy smokers can say, I've smoked more than a pack a day for 30 years and I'm OK. Similarly, most people who were spanked can say, My parents spanked me, and I'm not a wife beater or depressed.

Another argument in defense of spanking is that it is not harmful if the parents are loving and explain why they are spanking. The research does show that the harmful effects of spanking are reduced if it is done by loving parents who explain their actions. However, . . . a study by Larzelere (1986) shows that although the harmful effects are reduced, they are not eliminated. The . . . harmful side effects include an increased risk of delinquency as a child and crime as an adult, wife beating, depression, masochistic sex, and lowered earnings.

In addition to having harmful psychological effects on children, hitting children also makes life more difficult for parents. Hitting a child to stop misbehavior may be the easy way in the short run, but in the slightly longer run, it makes the job of being a parent more difficult. This is because spanking reduces the ability of parents to influence their children, especially in adolescence when they are too big to control by physical force. Children are more likely to do what the parents want if there is a strong bond of affection with the parent. In short, being able to influence a child depends in considerable part on the bond between parent and child (Hirschi, 1969). An experiment by Redd, Morris, and Martin (1975) shows that children tend to avoid caretaking adults who use punishment. In the natural setting, of course, there are many things that tie children to their parents. I suggest that each spanking chips away at the bond between parent and child. . . .

Contrary to the "spoiled child" myth, children of non-spanking parents are likely to be easier to manage and better behaved than the children of parents who spank. This is partly because they tend to control their own behavior on the basis of what their own conscience tells them is right and wrong rather than to avoid being hit . . . This is ironic because almost everyone thinks that spanking "when necessary" makes for better behavior.

MYTH 4: ONE OR TWO TIMES WON'T CAUSE ANY DAMAGE

The evidence . . . indicates that the greatest risk of harmful effects occurs when spanking is very frequent. However, that does not necessarily mean that spanking just once or twice is harmless. Unfortunately, the connection between spanking once or twice and psychological damage has not been addressed by most of the available research. This is because the studies seem to be based on this myth. They generally cluster children into "low" and "high" groups in terms of the frequency they were hit. This prevents the "once or twice is harmless" myth from being tested scientifically because the low group may include parents who spank once a year or as often as once a month. The few studies that did classify children according to the number of times they were hit by their parents . . . show that even one or two instances of corporal punishment are associated with a slightly higher probability of later physically abusing your own child, slightly more depressive symptoms, and a greater probability of violence and other crime later in life. The increase in these harmful side effects when parents use only moderate corporal punishment (hit only occasionally) may be small, but why run even that small risk when the evidence shows that corporal punishment is no more effective than other forms of discipline in the short run, and less effective in the long run.

MYTH 5: PARENTS CAN'T STOP WITHOUT TRAINING

Although everyone can use additional skills in child management, there is no evidence that it takes some extraordinary training to be able to stop spanking. The most basic step in eliminating corporal punishment is for parent educators, psychologists, and pediatricians to make a simple and unambiguous statement that hitting a child is wrong and that a child *never,* ever, under any circumstances except literal physical self-defense, should be hit.

That idea has been rejected almost without exception everytime I suggest it to parent educators or social scientists. They believe it would turn off parents and it could even be harmful because parents don't know what else to do. I think that belief is an unconscious defense of corporal punishment. I say that because I have never heard a parent educator say that before we can tell parents to never *verbally* attack a child, parents need training in alternatives. Some do need training, but everyone agrees that parents who use *psychological* pain as a method of discipline, such as insulting or demeaning, the child, should stop immediately. But when it comes to causing *physical* pain by spanking, all but a small minority of parent educators say that before parents are told to stop spanking, they need to learn alternative modes of discipline. I believe they should come right out, as they do for verbal attacks, and say without qualification that a child should *never* be hit.

This is not to say that parent education programs are unnecessary, just that they should not be a precondition for ending corporal punishment. Most parents can benefit from parent education programs such as The Nurturing Program (Bavolek, 1983 to 1992), STEP (Dinkmeyer and McKay, 1989), Parent Effectiveness Training (Gordon, 1975), Effective Black Parenting (Alvy and Marigna, 1987), and Los Ninos Bien Educado Program (Tannatt and Alvy, 1989). However, even without such programs, most parents already use a wide range of non-spanking methods, such as explaining, reasoning, and rewarding. The problem is that they also spank.

Given the fact that parents already know and use many methods of teaching and controlling, the solution is amazingly simple. In most cases, parents only need the patience to keep on doing what they were doing to correct misbehavior. Just leave out the spanking! Rather than arguing that parents need to learn certain skills *before* they can stop using corporal punishment, I believe that parents are more likely to use and cultivate those skills if they decide or are required to stop spanking.

This can be illustrated by looking at one situation that almost everyone thinks calls for spanking: when a toddler runs out into the street. A typical parent will scream in terror, rush out and grab the child, and run to safety, telling the child, No! No! and explaining the danger—all of this accompanied by one or more slaps to the legs or behind.

The same sequence is as effective or more effective *without the spanking.* The spanking is not needed because even tiny children can sense the terror in the parent and understand, No! No! Newborn infants can tell the difference between when a mother is relaxed and when she is tense (Stern, 1977). Nevertheless, the fact that a child understands that something is wrong does not guarantee never again running into the street; just as spanking does not guarantee the child will not run into the street again.

If the child runs out again, nonspanking parents should use one of the same strategies as spanking parents—repetition. Just as spanking parents will spank as many times as necessary until the child learns, parents who don't spank should continue to monitor the child, hold the child's hand, and take whatever other means are needed to protect the child until the lesson is learned. Unfortunately, when non-spanking methods do not work, some parents quickly turn to spanking because they lose patience and believe it is more effective: But spanking parents seldom question its effectiveness, they just keep on spanking.

Of course, when the child misbehaves again, most spanking parents do more than just repeat the spanking or spank harder. They usually also do things such as explain the danger to the child before letting the child go out again or warn the child that if it happens again, he or she will have to stay in the house for the afternoon, and so on. The irony is that when the child finally does learn, the parent attributes the success to the spanking, not the explanation.

MYTH 6: IF YOU DON'T SPANK, YOUR CHILDREN WILL BE SPOILED OR RUN WILD

It is true that some non-spanked children run wild. But when that happens it is not because the parent didn't spank. It is because some parents think the alternative to spanking is to ignore a child's misbehavior or to replace spanking with verbal attacks such as, "Only a dummy like you can't learn to keep your toys where I won't trip over them." The best alternative is to take firm action to correct the misbehavior without hitting. Firmly condemning what the child has done and explaining why it is wrong are usually enough. When they are not, there are a host of other things to do, such as requiring a time out or depriving the child of a privilege, neither of which involves hitting the child.

Suppose the child hits another child. Parents need to express outrage at this or the child may think it is acceptable behavior. The expression of outrage and a clear statement explaining why the child should never hit another person, except in self defense, will do the trick in most cases. That does not mean one such warning will do the trick, any more than a single spanking will do the trick. It takes most children a while to learn such things, whatever methods the parents use.

The importance of how parents go about teaching children is clear from a classic study of American parenting—*Patterns of Child Rearing* by Sears, Maccoby, and Levin (1957). This study found two actions by parents that are linked to a high level of aggression by the child: permissiveness of the child's aggression, namely ignoring it when the child hits them or another child, and spanking to correct misbehavior. The most aggressive children. . . are children of parents who permitted aggression by the child and who also hit them for a variety of misbehavior. The least aggressive children. . . are children of parents who clearly condemned acts of aggression and who, by not spanking, acted in a way that demonstrated the principle that hitting is wrong.

There are other reasons why, on the average, the children of parents who do not spank are better behaved than children of parents who spank:

- Non-spanking parents pay more attention to their children's behavior, both good and bad, than parents who spank. Consequently, they are more likely to reward good behavior and less likely to ignore misbehavior.

- Their children have fewer opportunities to get into trouble because they are more likely to child-proof the home. For older children, they have clear rules about where they can go and who they can be with.
- Non-spanking parents tend to do more explaining and reasoning. This teaches the child how to use these essential tools to monitor his or her own behavior, whereas children who are spanked get less training in thinking things through.
- Non-spanking parents treat the child in ways that tend to bond the child to them and avoid acts that weaken the bond. They tend to use more rewards for good behavior, greater warmth and affection, and fewer verbal assaults on the child (see Myth 9). By not spanking, they avoid anger and resentment over spanking. When there is a strong bond, children identify with the parent and want to avoid doing things the parent says are wrong. The child develops a conscience and lets that direct his or her behavior. That is exactly what Sears et al. found. . . .

MYTH 7: PARENTS SPANK RARELY OR ONLY FOR SERIOUS PROBLEMS

Contrary to this myth, parents who spank tend to use this method of discipline for almost any misbehavior. Many do not even give the child a warning. They spank before trying other things. Some advocates of spanking even recommend this. At any supermarket or other public place, you can see examples of a child doing something wrong, such as taking a can of food off the shelf. The parent then slaps the child's hand and puts back the can, sometimes without saying a word to the child. John Rosemond, the author of *Parent Power* (1981), says, "For me, spanking is a first resort. I seldom spank, but when I decide . . . I do it, and that's the end of it."

The high frequency of spanking also shows up among the parents described in this book. The typical parent of a toddler told us of about 15 instances in which he or she had hit the child during the previous 12 months. That is surely a minimum estimate because spanking a child is generally such a routine and unremarkable event that most instances are forgotten. Other studies, such as Newson and Newson (1963), report much more chronic hitting of children. My tabulations for mothers of three- to five-year-old children in the National Longitudinal Study of Youth found that almost two-thirds hit their children during the week of the interview, and they did it more then three times in just that one week. As high as that figure may seem, I think that daily spanking is not at all uncommon. It has not been documented because the parents who do it usually don't realize how often they are hitting their children.

MYTH 8: BY THE TIME A CHILD IS A TEENAGER, PARENTS HAVE STOPPED

. . . Parents of children in their early teens are also heavy users of corporal punishment, although at that age it is more likely to be a slap on the face than on the behind. . . . More than half of the parents of 13– to 14-year-old children in our two national surveys hit their children in the previous 12 months. The percentage drops each year as children get older, but even at age 17, one out of five parents is still hitting. To make matters worse, these are minimum estimates.

Of the parents of teenagers who told us about using corporal punishment, 84 percent did it more than once in the previous 12 months. For boys, the average was seven times and for girls, five times. These are minimum figures because we interviewed the mother in half the families and the father in the other half. The number of times would be greater if we had information on what the parent who was not interviewed did.

MYTH 9: IF PARENTS DON'T SPANK, THEY WILL VERBALLY ABUSE THEIR CHILD

The scientific evidence is exactly the opposite. Among the nationally representative samples of parents in this book, those who did the least spanking also engaged in the least verbal aggression.

It must be pointed out that non-spanking parents are an exceptional minority. They are defying the cultural prescription that says a good parent should spank if necessary. The depth of their involvement with their children probably results from the same underlying characteristics that led them to reject spanking. There is a danger that if more ordinary parents are told to never spank, they might replace spanking by ignoring misbehavior or by verbal attacks. Consequently, a campaign to end spanking must also stress the importance of avoiding verbal attacks as well as physical attacks, and also the importance of paying attention to misbehavior.

MYTH 10: IT IS UNREALISTIC TO EXPECT PARENTS TO NEVER SPANK

It is no more unrealistic to expect parents to never hit a child than to expect that husbands should never hit their wives, or that no one should go through a stop sign, or that a supervisor should never hit an employee. Despite the legal prohibition, some husbands hit their wives, just as some drivers go through stop signs, and a supervisor occasionally may hit an employee.

If we were to prohibit spanking, as is the law in Sweden (. . . Deley, 1988; Haeuser, 1990), there still would be parents who would continue to spank. But that is not a reason to avoid passing such a law here. Some people kill even though murder has been a crime since the dawn of history. Some husbands continue to hit their wives even though it has been more than a century since the courts stopped recognizing the common law right of a husband to "physically chastise an errant wife" (Calvert, 1974).

A law prohibiting spanking is unrealistic only because spanking is such an accepted part of American culture. That also was true of smoking. Yet in less than a generation we have made tremendous progress toward eliminating smoking. We can make similar progress toward eliminating spanking by showing parents that spanking is dangerous, that their children will be easier to bring up if they do not spank, and by clearly saying that a child should *never,* under any circumstances, be spanked.

SELECTED REFERENCES

Alvy, Kirby T., and Marilyn Marigna. 1987. *Effective Black Parenting.* Studio City, CA: Center For the Improvement of Child Caring.

Bavolek, Stephen J. 1992. *The Nurturing Programs.* Park City, Utah: Family Development Resources.

Calvert, Robert. 1974. "Criminal and Civil Liability in Husband-Wife Assaults." Chapter 9 in *Violence in the Family,* edited by S. K. Steinmetz and M. A. Straus. New York: Harper and Row.

Crozier, Jill, and Roger C. Katz. 1979. "Social Learning Treatment of Child Abuse." *Journal of Behavioral Therapy and Psychiatry* 10:213–20.

Day, Dan E., and Mark W. Roberts. 1983. "An Analysis of the Physical Punishment Component of a Parent Training Program." *Journal of Abnormal Child Psychology* 11:141–52.

Deley, Warren W. 1988. "Physical Punishment of Children: Sweden and the USA." *Journal of Comparative Family Studies* 19:419–31.

Dinkmeyer Sr., Don, and Gary D. McKay. 1989. *Systematic Training for Effective Parenting.* Circle Pines, MN: American Guidance Service.

Gordon, Thomas. 1975. *Parent Effectiveness Training.* New York: New American Library.

Haeuser, Adrienne A. 1990. "Banning Parental Use of Physical Punishment: Success in Sweden." Presented at the Eighth International Congress on Child Abuse and Neglect, Hamburg, Germany, 2–6 September.

Hirschi, Travis. 1969. *The Causes of Delinquency.* Berkeley and Los Angeles: University of California Press.

Kadushin, Alfred, and Judith A. Martin. 1981. *Child Abuse: An Interactional Event.* New York: Columbia University Press.

Larzelere, Robert E. 1986. "Moderate Spanking: Model or Deterrent of Children's Aggression in the Family?" *Journal of Family Violence* 1:27–36.

Larzelere, Robert E. 1994. "Should Corporal Punishment by Parents Be Considered Abusive?—No." In *Children and Adolescents: Controversial Issues,* edited by E. Gambrill and M. A. Mason. Newbury Park, CA: Sage Publications.

Mattson, Margaret E., Earl S. Pollack, and Joseph W. Cullen. 1987. "What Are the Odds that Smoking Will Kill You?" *American Journal of Public Health* 77:425–31.

Newson, John, and Elizabeth Newson. 1963. *Patterns of Infant Care in an Urban Community.* Baltimore: Penguin Books.

Patterson, Gerald R. 1982. "A Social Learning Approach to Family Intervention: III." *Coercive Family Process.* Eugene, OR: Castalia.

Redd, William H., Edward K. Morris, and Jerry A. Martin. 1975. "Effects of Positive and Negative Adult-Child Interactions on Children's Social Preference." *Journal of Experimental Child Psychology* 19:153–164.

Rosemond, John K. 1981. *Parent Power, A Common Sense Approach to Raising Your Children in the '80s.* Charlotte, NC: East Woods Press.

Sears, Robert R., Eleanor C. Maccoby, and Harry Levin. 1957. *Patterns of Child Rearing.* Evanston, IL: Row, Peterson, and Company.

Stern, Daniel. 1977. *The First Relationship: Mother and Infant.* Cambridge, MA: Harvard University Press.

Tannatt, Lupita Montoya, and Kirby T. Alvy. 1989. *Los Ninos Bien Educados Program.* Studio City, CA: Center for the Improvement of Child Caring.

Webster-Stratton, Carolyn, Mary Kolpacoff, and Terri Hollinsworth. 1988. "Self-Administered Videotape Therapy for Families with Conduct-Problem Children: Comparison with Two Cost-Effective Treatments and a Control Group." *Journal of Consulting and Clinical Psychology* 56:558–66.

Webster-Stratton, Carolyn. 1990. "Enhancing the Effectiveness of Self-Administered Videotape Parent Training for Families with Conduct-Problem Children." *Journal of Abnormal Child Psychology* 18:479–92.

CHAPTER 12

Families over the Life Course

Although researchers and the general public tend to think about and symbolize families as containing parents and young children, this phase of family life is relatively short. The bulk of most people's family experiences occur after the early child-rearing years. Most theories and research focus on the importance of early childhood, but social scientists are beginning to understand that other phases of family life are also important. Much research now focuses on **adolescence** and the various transitions associated with becoming an adult. Other research focuses on later critical family transitions of **family formation** (cohabitation, marriage, having children, etc.); **family disruption** (separation, divorce, death, etc.); **family reconstitution** (shared custody/visitation, launching children, new partner cohabitation, remarriage, stepparenting, etc.); and **aging family transitions** (becoming grandparents, caring for aging parents, dealing with death, surviving a spouse, etc.).

Although many topics could be covered in a chapter like this one, the concept that ties them all together is the **life course.** Earlier researchers recognized that human development does not happen all at once and is not completed when one reaches adulthood (Erikson, 1959). Adulthood is marked by various peaks, valleys, and turning points that tend to follow general patterns according to culture and era, even though individuals' life paths may vary. The term *life cycle* suggests that everyone must pass through fixed sequential stages of development. More complex formulations based on the concept of **life course** suggest that life transitions do not occur in the same way or at the same time for everyone. Individuals in the same culture tend to differ somewhat not only in the onset of various stages of adult development, but also in the length, sequence, and composition of those transitions. Earlier chapters used historical and comparative data to show how different cultural and economic conditions organized daily family life in starkly contrasting ways and how ideals about childhood, motherhood, or fatherhood change over time. This is not to suggest that we cannot observe some general patterns about family transitions and how they change over time. For example, in the United States, we are returning to timing patterns for getting married and having children that were more common at the end of the nineteenth century than they were in the middle of the twentieth century. Between 1890 and the 1960s we witnessed a gradual quickening of the pace of life course transitions, with people tending to get married, become parents, leave home, and become grandparents much sooner chronologically than their own parents. Since the 1960s, the trend has reversed, with both men and women marrying later (and less often), having children later than their own parents did, and becoming grandparents at a comparatively later age. As discussed in previous chapters, women's increasing likelihood of attending college and getting jobs is the major engine driving these changes in life course timing.

Another important demographic factor explaining emergent life course patterns is death—more specifically, the timing of death. Because of improvements in sanitation, nutrition, and medicine, since 1900 average life expectancy increased a whopping twenty-seven years. The average woman in the United States can now expect to live to be about eighty years old, and the average man can expect to live to be about seventy-four years old. This increasing **longevity** has dramatically altered opportunities for family interaction and multigenerational relationships. For example, the average woman born in 1960 who becomes a mother can expect to spend about forty years as a grandmother. Compared to the early nineteenth century, American women spend four times as many years as a daughter with both parents alive. At the beginning of the nineteenth century, they were likely to spend about seven years with a parent over the age of sixty-five, but today they are likely to spend almost twenty years with a parent over that age. And although we symbolize families as containing young children, the average years spent with children under the age of eighteen is now less than the average number of years spent with a parent over the age of sixty-five. In other words, even though they may not be living together, adult Americans are spending more years in family relationships than ever before. The selections in this chapter introduce some challenging questions about those relationships and how they might be changing.

The first article in this chapter is based on a lecture Kingsley Davis presented to the American Sociological Society in 1939. In "The Sociology of Parent-Youth Conflict," Davis asks why Western civilization has such high levels of parent–adolescent conflict. He addresses this question with reference to "constants" (what he sees as universal features of parent–youth relations), and "variables" (the things that differ across societies). The rate of social change is one variable that strongly influences parent–youth relations, because when a culture is changing rapidly, parents have less access to knowledge that will be guaranteed to help the young-adult child to succeed. This undermines the authority of the parent and the trust of the child that the parent knows best. Davis suggests that most societies avoid the potential clash of old and young by assigning them different social positions, but in complex societies like ours that allow for quick vertical mobility, this built-in buffer breaks down. Western society, according to Davis, is characterized by small families, open competition for economic position, conflicting **norms** between young and old, a wealth of competing authoritative knowledge sources, sexual tensions, and few guidelines for readjusting authority between parent and child as the adolescent makes the transition to adulthood. All these variables tend to increase the potential for parent–youth conflict. Although he wrote decades before the youth rebellion of the late 1960s and 1970s and well before the emergence of a youth-oriented consumer culture in the 1980s and 1990s, Davis's predictions seem prophetic. His central point is a provocative one: traditional societies based on family production can maintain stable parental authority over youth, but urban **industrial societies** like our own will inevitably produce parent–youth conflict. Do you agree?

Drawing on the theories of Dorothy Smith (see chapter 1), Karen Pyke explores one of the premises of Kingsley Davis's observations: that immigrant families exemplify rapid social change and provide evidence of conflict and tension in parent–youth relationships. Pyke focuses on how grown children of Korean and Vietnamese immigrants draw on images of "the Normal American Family" to discuss family and filial obligation. Popular images of the normal American family provide an ideological code through which people evaluate their lives. The family ideal contains notions about the appropriate values, norms, and beliefs that guide family members' interactions and relationships. Pyke focuses on the tension between modern American family images emphasizing sensitivity, openness, communication, flexibility, and forgiveness (generally coded white and middle-class), and images predicated on duty, responsibility, obedience, and commitment to the family collective (generally coded as racial-ethnic). She finds that the adult children of Asian immigrants tended to criticize their parents for lacking American values emphasizing psychological well-being and expressive love. When talking about the care they expected to provide to their parents, however, respondents switched to an interpretive lens that denigrated Western individualism and valued ethnic family solidarity. This ideological code switching is not uncommon among people who live in conflicting social worlds, and Pyke shows how these young adults were constructing positive self-identities that both acknowledged and downplayed parent–youth conflict, but did not fundamentally challenge mainstream ethnocentric family imagery. Can you think of other groups for whom such family conflicts and symbolic processes might be common?

In an excerpt from the book *Family Ties* entitled "Extending the Family," authors John Logan and Glenna Spitze discuss relationships between parents and their grown children. By focusing on parent–child relations in their adult stage, they consider various phases in the life course, including later career trajectories, grandparenthood, retirement, widowhood, disability, and death. Their central finding is that family ties are strong and enduring, and that they remain centrally important to both parents and adult children in modern America. Contrary to popular conceptions, most exchanges (until parents reach very old ages) flow from parent to child. They find that female adult children take more responsibility for some aspects of these kin relationships than males, but discover fewer gender differences than previous research or theories have predicted. For example, between one-third and one-half of male adult children help their aging mothers and fathers with routine **household labor.** Whether measured as contact, visiting, help given, or help received, family ties between parents and adult children remain preferred network ties, especially when parents and children continue to live in the same neighborhood. Logan and Spitze suggest that even though kinship relations are becoming increasingly complicated by divorce, remarriage, and the like, people are adopting expanded family structures because they are living longer and because latent kinship bonds are more easily activated than other support systems. Instead of finding that the family is in crisis, Logan and Spitze discover that multigenerational family ties endure throughout adulthood, in spite of the fact that norms governing marriage and living arrangements are more fluid than in past times. How do you think that the large size of the aging baby boom cohort will influence the processes that these authors discuss?

REFERENCES

Erikson, Eric H. 1959. "Identity and the life cycle." *Psychological Issues 1*(1).

SUGGESTED READINGS

Bengtson, Vern L., and Robert A. Harootyan. 1994. *Intergenerational Linkages: Hidden Connections in American Society.* New York: Springer Publishing.

Bengtson, Vern L., and Joan F. Robertson. 1985. *Grandparenthood.* Beverly Hills, CA: Sage Publications.

Csikszentmihalyi, Mihaly, and Barbara Schneider. 2000. *Becoming Adult: How Teenagers Prepare for the World of Work.* New York: Basic Books.

Elder, Glen H., Jr., John Modell, and Ross D. Parke. 1993. *Children in Time and Place: Developmental and Historical Insights.* Cambridge; New York: Cambridge University Press.

Fass, Paula S., and Mary Ann Mason, eds. 2000. *Childhood in America.* New York: New York University Press.

Furstenberg, Frank F., Thomas D. Cook, Jacquelynne Eccles, Glen H. Elder, Jr., and Arnold J. Sameroff. 1999. *Managing to Make It: Urban Families and Adolescent Success.* Chicago, IL: University of Chicago Press.

Giele, Janet Z., and Glen H. Elder, Jr. 1998. *Methods of Life Course Research: Qualitative and Quantitative Approaches.* Thousand Oaks, CA: Sage Publications.

Jenks, Chris. 1996. *Childhood.* London, UK: Routledge.

Lopata, Helena Z. 1996. *Current Widowhood: Myths and Realities.* Thousand Oaks, CA: Sage Publications.

Snarey, J. 1993. *How Fathers Care for the Next Generation: A Four-Decade Study.* Cambridge, MA: Harvard University Press.

INFOTRAC® COLLEGE EDITION EXERCISES

The exercises that follow allow you to use the InfoTrac® College Edition online database of scholarly articles to explore the sociological implications of the selections in this chapter.

Search Keyword: Adolescence. Using InfoTrac® College Edition, search for articles discussing the stage of life known as *adolescence.* Why is this stage considered a transitional one? What changes (social, emotional, physical, for instance) do teens experience that create tensions in their family relationships? In the articles that you find, what ways are proposed to address the family tensions associated with adolescence? What institutional supports are available to the adolescent, and to her/his family, for this life course transition? What supports are lacking?

Search Keyword: Widowhood. Death of a spouse often occurs as part of the life course, most often, although not always, during an individual's later years. Look up articles that deal with the marital status of widowhood. How do men and women experience widowhood differently? How is widowhood experienced differently based on age? How do social class and race/ethnicity affect the likelihood of experiencing widowhood, and how do they affect the way widowhood is experienced? What social and institutional support is available (and what is unavailable) to widow/ers?

Search Keyword: Gen Y. Some of you probably belong to the Generation Y (or Gen Y) cohort, having been born sometime between 1977 and 1994. Use InfoTrac® College Edition and the search keyword *Gen Y* to find articles that deal with this population. What are some of the characteristics of the Gen Y cohort? What cultural images are associated with this particular group? What historical events have occurred to shape the life course of the Gen Y population? What sociological influences and demographic trends have affected the life course trajectory of Gen Y'ers? How do these influences, trends, and events cause the Gen Y cohort to look similar to, or different from, earlier cohorts such as the Baby Boom generation born between roughly 1946 and 1964?

Search Keyword: Midlife crisis. It is generally believed that people, particularly men, suffer from a midlife crisis, occurring any time between ages thirty-five and fifty. Search for articles that deal with midlife crisis. How do these articles characterize the phenomenon? Does the midlife crisis appear to be biological, or is it a socially constructed event? Is this crisis specific to men? If not, are there differences in how men and women are alleged to experience midlife crises? According to the articles you find, what are the common triggers for the onset of the midlife crisis life transition? What are the implications of midlife crises for family life and family relationships?

Search Keyword: Graying of the population. As longevity has increased and fertility decreased, the population of the United States is increasingly older. Use InfoTrac® College Edition to search for articles that address the graying of the population. What are the implications of this demographic trend? How will the graying of the population affect family life in the twenty-first century? What public policies are in effect, or are proposed, that would increase support for this aging population? What policies are needed to support families faced with caring for aging parents and grandparents?

39

The Sociology of Parent-Youth Conflict

KINGSLEY DAVIS,
1940

It is in sociological terms that this paper attempts to frame and solve the sole question with which it deals, namely: Why does contemporary western civilization manifest an extraordinary amount of parent-adolescent conflict? In other cultures, the outstanding fact is generally not the rebelliousness of youth, but its docility. There is practically no custom, no matter how tedious or painful, to which youth in primitive tribes or archaic civilizations will not willingly submit. What, then, are the peculiar features of our society which give us one of the extremist examples of endemic filial friction in human history?

Our answer to this question makes use of constants and variables, the constants being the universal factors in the parent-youth relation, the variables being the factors which differ from one society to another. Though one's attention, in explaining the parent-youth relations of a given milieu, is focused on the variables, one cannot comprehend the action of the variables without also understanding the constants, for the latter constitute the structural and functional basis of the family as a part of society.

The Rate of Social Change. The first important variable is the rate of social change. Extremely rapid change in modern civilization, in contrast to most societies, tends to increase parent-youth conflict, for within a fast-changing social order the time-interval between generations, ordinarily but a mere moment in the life of a social system, become historically significant, thereby creating a hiatus between one generation and the next. Inevitably, under such a condition, youth is reared in a milieu different from that of the parents; hence the parents become old-fashioned, youth rebellious, and clashes occur which, in the closely confined circle of the immediate family, generate sharp emotion.

That rapidity of change is a significant variable can be demonstrated by three lines of evidence: a comparison of stable and nonstable societies; a consideration of immigrant families; and an analysis of revolutionary epochs. If, for example, the conflict is sharper in the immigrant household, this can be due to one thing only, that the immigrant family generally undergoes the most rapid social change of any type of family in a given society. Similarly, a revolution (an abrupt form of societal alteration), by concentrating great change in a short span, catapults the younger generation into power—a generation which has absorbed and pushed the new ideas, acquired the habit of force, and which, accordingly, dominates those hangovers from the old regime, its parents.

*Presented to the American Sociological Society, Philadelphia, Dec. 28, 1939.

The Birth-Cycle, Decelerating Socialization, and Parent-Child Differences. Note, however, that rapid social change would have no power to produce conflict were it not for two universal factors: first, the family's duration; and second, the decelerating rate of socialization in the development of personality. "A family" is not a static entity but a process in time, a process ordinarily so brief compared with historical time that it is unimportant, but which, when history is "full" (i.e., marked by rapid social change), strongly influences the mutual adjustment of the generations. This "span" is basically the birth-cycle—the length of time between the birth of one person and his procreation of another. It is biological and inescapable. It would, however, have no effect in producing parent-youth conflict, even with social change, if it were not for the additional fact, intimately related and equally universal, that the sequential development of personality involves a constantly decelerating rate of socialization. This deceleration is due both to organic factors (age—which ties it to the birth-cycle) and to social factors (the cumulative character of social experience). Its effect is to make the birth-cycle interval, which is the period of youth, the time of major socialization, subsequent periods of socialization being subsidiary.

Given these constant features, rapid social change creates conflict because *to* the intrinsic (universal, inescapable) differences between parents and children it adds an extrinsic (variable) difference derived from the acquisition, at the same stage of life, of differential cultural content by each successive generation. Not only are parent and child, at any given moment, in different stages of development, but the content which the parent acquired at the stage where the child now is, was a different content from that which the child is now acquiring. Since the parent is supposed to socialize the child, he tends to apply the erstwhile but now inappropriate content.

. . . He makes this mistake, and cannot remedy it, because, due to the logic of personality growth, his basic orientation was formed by the experiences of his own childhood. He cannot "modernize" his point of view, because *he* is the product of those experiences. He can change in superficial ways, such as learning a new tune, but he cannot change (or *want* to change) the initial modes of thinking upon which his subsequent social experience has been built. To change the basic conceptions by which he has learned to judge the rightness and reality of all specific situations would be to render subsequent experience meaningless, to make an empty caricature of what had been his life. . . .

Although, in the birth-cycle gap between parent and offspring, astronomical time constitutes the basic point of disparity, the actual sequences, and hence the actual differences significant for us, are physiological, psychosocial, and sociological—each with an acceleration of its own within, but to some degree independent of, sidereal time, and each containing a divergence between parent and child which must be taken into account in explaining parent-youth conflict.

Physiological Differences. Though the disparity in chronological age remains constant through life, the precise physiological differences between parent and offspring vary radically from one period to another. The organic contrasts between parent and *infant,* for example, are far different from those between parent and adolescent. Yet whatever the period, the organic differences produce contrasts (as between young and old) in those desires which, at least in part, are organically determined. Thus, at the time of adolescence the contrast is between an organism which is just reaching its full powers and one which is just losing them. The physiological need of the latter is for security and conservation, because as the superabundance of energy diminishes, the organism seems to hoard what remains.

Such differences, often alleged (under the heading of "disturbing physiological changes accompanying adolescence") as the primary cause of parent-adolescent strife, are undoubtedly a factor in such conflict, but, like other universal differences to be discussed, they form a constant factor present in every community, and therefore cannot in themselves explain the peculiar heightening of parent-youth conflict in our culture.

The fact is that most societies avoid the potential clash of old and young by using sociological position as a neutralizing agent. They assign definite and separate positions to persons of different ages, thereby eliminating competition between them for the same position and avoiding the competitive emotions of jealousy and envy. Also, since the expected behavior of old and young is thus made complementary rather than identical, the performance of cooperative functions as accomplished by different but mutually related activities suited to the disparate organic needs of each, with no coercion to behave in a manner unsuited to one's organic age. In our culture, where most positions are *theoretically* based on accomplishment rather than age, interage competition arises, superior organic propensities lead to a high evaluation of youth (the so-called "accent on youth"), a disproportionate lack of opportunity for youth manifests itself, and consequently, arrogance and frustration appear in the young, fear and envy, in the old.

Psychosocial Differences: Adult Realism versus Youthful Idealism. The decelerating rate of socialization (an outgrowth both of the human being's organic development, from infant plasticity to senile rigidity, and of his cumulative cultural and social development), when taken with rapid social change and other conditions of our society, tends to produce certain differences of orientation between parent and youth. Though lack of space makes it impossible to discuss all of these ramifications, we shall attempt to delineate at least one sector of difference in terms of the conflict between adult realism (or pragmatism) and youthful idealism.

Though both youth and age claim to see the truth, the old are more conservatively realistic than the young, because on the one hand they take Utopian ideals less seriously and on the other hand take what may be called operating ideals, if not more seriously, at least more for granted. Thus, middle-aged people notoriously forget the poetic ideals of a new social order which they cherished when young. In their place, they put simply the working ideals current in the society. There is, in short, a persistent tendency for the ideology of a person as he grows older to gravitate more and more toward the status quo ideology, unless other facts (such as a social crisis or hypnotic suggestion) intervene. With advancing age, he becomes less and less bothered by inconsistencies in ideals. He tends to judge ideals according to whether they are widespread and hence effective in thinking about practical life, not according to whether they are logically consistent. Furthermore, he gradually ceases to bother about the *untruth* of his ideals, in the sense of their failure to correspond to reality. He assumes through long habit that, though they do not correspond perfectly, the discrepancy is not significant. The reality of an ideal is defined for him in terms of how many people accept it rather than how completely it is mirrored in actual behavior. Thus, we call him, as he approaches middle age, a realist.

The young, however, are idealists, partly because they take working ideals literally and partly because they acquire ideals not fully operative in the social organization. Those in authority over children are obligated as a requirement of their status to inculcate ideals as a part of the official culture given the new generation. The children are receptive because they have little social experience—experience being systematically kept from them (by such means as censorship, for example, a large part of which is to "protect" children). Consequently, young people possess little ballast for their acquired ideals, which therefore soar to the sky, whereas the middle-aged, by contrast, have plenty of ballast.

This relatively unchecked idealism in youth is eventually complicated by the fact that young people possess keen reasoning ability. The mind, simply as a logical machine, works as well at sixteen as at thirty-six. Such logical capacity, combined with high ideals and an initial lack of experience, means that youth soon discovers with increasing age that the ideals it has been taught are true and consistent are not so in fact. Mental conflict thereupon ensues, for the young person has not learned that ideals may be useful without being true and consistent. As a solution, youth is likely to take action designed to remove inconsistencies or force actual conduct into line with ideals, such action assuming one of several typical adolescent forms—from religious withdrawal to the militant support of some Utopian scheme—but in any case consisting essentially in serious allegiance to one or more of the ideal moral systems present in the culture.

A different, usually later reaction to disillusionment is the cynical or sophomoric attitude; for, if the ideals one has imbibed cannot be reconciled and do not fit reality, then why not dismiss them as worthless? Cynicism has the advantage of giving justification for behavior that young organisms crave anyway. It might be mistaken for genuine realism if it were not for two things. The first is the emotional strain behind the "don't care" attitude. The cynic, in his judgment that the world is bad because of inconsistency and untruth of ideals, clearly implies that he still values the ideals. The true realist sees the inconsistency and untruth, but without emotion; he uses either ideals or reality whenever it suits his purpose. The second is the early disappearance of the cynical attitude. Increased experience usually teaches the adolescent that overt cynicism is unpopular and unworkable, that to deny and deride all beliefs which fail to cohere or to correspond to facts, and to act in opposition to them, is to alienate oneself from any group, because these beliefs, however unreal, are precisely what makes group unity possible. Soon, therefore, the youthful cynic finds himself bound up with some group having a system of working ideals, and becomes merely another conformist, cynical only about the beliefs of other groups.

While the germ of this contrast between youthful idealism and adult realism may spring from the universal logic of personality development, it receives in our culture a peculiar exaggeration. Social change, complexity, and specialization (by compartmentalizing different aspect of life) segregate ideals from fact and throw together incompatible ideologies while at the same time providing the intellectual tools for discerning logical inconsistencies and empirical errors. Our highly elaborated burden of culture, correlated with a variegated system of achieved vertical mobility, necessitates long years of formal education which separate youth from adulthood, theory from practice, school from life. Insofar, then, as youth's reformist zeal or cynical negativism produces conflict with parents, the peculiar conditions of our culture are responsible.

Sociological Differences: Parental Authority. Since social status and office are everywhere partly distributed on the basis of age, personality development is intimately linked with the network of social positions successively occupied during life. Western society, in spite of an unusual amount of interage competition, maintains differences of social position between parent and child, the developmental gap between them being too clearcut, the symbiotic needs too fundamental, to escape being made a basis of social organization. Hence, parent and child, in a variety of ways, find themselves enmeshed in different social contexts and possessed of different outlooks. The much publicized critical attitude of youth toward established ways, for example, is partly a matter of being on the outside looking in. The "established ways" under criticism are usually institutions (such as property, marriage, profession) which the adolescent has not yet entered. He looks at them from the point of view of the outsider (especially since they affect him in a restrictive manner), either failing to imagine himself finding satisfaction in such patterns or else feeling resentful that the old have in them a vested interest from which he is excluded.

Not only is there differential position, but also *mutually* differential position, status being in many ways specific for and reciprocal between parent and child. Some of these differences, relating to the birth-cycle and constituting part of the family structure, are universal. This is particularly true of the super- and subordination summed up in the term *parental authority.*

Since sociological differences between parent and child are inherent in family organization, they constitute a universal factor potentially capable of producing conflict. Like the biological differences, however, they do not in themselves produce such conflict. In fact, they may help to avoid it. To understand how our society brings to expression the potentiality for conflict, indeed to deal realistically with the relation between the generations, we must do so not in generalized terms but in terms of the specific "power situation." Therefore, the remainder of our discussion will center upon the nature of parental authority and its vicissitudes in our society.

Because of his strategic position with reference to the new-born child (at least in the familial type of reproductive institution), the parent is given considerable authority. Charged by his social group with the responsibility of controlling and training the child in conformity with the mores and thereby insuring the maintenance of the cultural structure, the parent, to fulfill his duties, must have the privileges as well as the obligations of authority, and the surrounding community ordinarily guarantees both.

The first thing to note about parental authority, in addition to its function in socialization, is that it is a case of authority within a primary group. Simmel has pointed out that authority is bearable for the subordinate because it touches only one aspect of life. Impersonal and objective, it permits all other aspects to be free from its particularistic dominance. This escape, however, is lacking in parental authority, for since the family includes most aspects of life, its authority is not limited, specific, or impersonal. What, then, can make this authority bearable? Three factors associated with the familial primary group help to give the answer: (1) the child is socialized within the family, and therefore knowing nothing else and being utterly dependent, the authority of the parent is internalized, accepted; (2) the family, like other primary groups, implies identification, in such sense that one person understands and responds emphatically to the sentiments of the other, so that the harshness of authority is ameliorated; (3) in the intimate interaction of the primary group control can never be purely one-sided; there are too many ways in which the subordinated can exert the pressure of his will. When, therefore, the family system is a going concern, parental authority, however, inclusive, is not felt as despotic.

A second thing to note about parental authority is that while its duration is variable (lasting in some societies a few years and in others a lifetime), it inevitably involves a change, a progressive readjustment, in the respective positions of parent and child—in some cases an almost complete reversal of roles, in others at least a cumulative allowance for the fact of maturity in the subordinated offspring. Age is a unique basis for social stratification. Unlike birth, sex, wealth, or occupation, it implies that the stratification is temporary, that the person, if he lives a full life, will eventually traverse all of the strata having it as a basis. Therefore, there is a peculiar ambivalence attached to this kind of differentiation, as well as a constant directional movement. On the one hand, the young person, in the stage of maximum socialization, is, so to speak, *moving into* the social organization. His social personality is expanding, i.e., acquiring an increased amount of the cultural heritage, filling more powerful and numerous positions. His future is before him, in what the older person is leaving behind. The latter, on the other hand, has a future before him only in the sense that the offspring represents it. Therefore, there is a disparity of interest, the young person placing his thoughts upon a future which, once the first stages of dependence are passed, does not include the parent, the old person placing his hopes vicariously upon the young. This situation, representing a *tendency* in every society, is avoided in many places by a system of respect for the aged and an imaginary projection of life beyond the grave. In the absence of such a religio-ancestral system, the role of the aged is a tragic one.

Let us now take up, point by point, the manner in which western civilization has affected this *gemeinschaftliche* and processual form of authority.

1. *Conflicting Norms.* To begin with, rapid change has, as we saw, given old and young a different social content, so that they possess conflicting norms. There is a loss of mutual identification, and the parent will not "catch up" with the child's point of view, because he is supposed to dominate rather than follow. More than this, social complexity has confused the standards *within* the generations. Faced with conflicting goals, parents become inconsistent and confused in their own minds in rearing their children. The children, for example, acquire an argument against discipline by being able to point to some family wherein discipline is less severe, while the parent can retaliate by pointing to still other families wherein it is firmer. The acceptance of parental attitudes is less complete than formerly.

2. *Competing Authorities.* We took it for granted, when discussing rapid social change, that youth acquires new ideas, but we did not ask how. The truth is that, in a specialized and complex culture, they learn from competing authorities. Today, for example, education is largely in the hands of professional specialists, some of whom, as college professors, resemble the sophists of ancient Athens by virtue of their work of accumulating and purveying knowledge, and who consequently have ideas in advance of the populace at large (i.e., the parents). By giving the younger generation these advanced ideas, they (and many other extrafamilial agencies, including youth's contemporaries) widen the intellectual gap between parent and child.

3. *Little Explicit Institutionalization of Steps in Parental Authority.* Our society provides little explicit institutionalization of the progressive readjustments of authority as between parent and child. We are intermediate between the extreme of virtually permanent parental authority and the extreme of very early emancipation, because we encourage release in late adolescence. Unfortunately, this is a time of enhanced sexual desire, so that the problem of sex and the problem of emancipation occur simultaneously and complicate each other. Yet even this would doubtless be satisfactory if it were not for the fact that among us the exact time when authority is relinquished, the exact amount, and the proper ceremonial behavior are not clearly defined. Not only do different groups and families have conflicting patterns, and new situations arise to which old definitions will not apply, but the different spheres of life (legal, economic, religious, intellectual) do not synchronize, maturity in one sphere and immaturity in another often coexisting. The readjustment of authority between individuals is always a ticklish process, and when it is a matter of such close authority as that between parent and child it is apt to be still more ticklish. The failure of our culture to institutionalize this readjustment by a series of well-defined, well-publicized steps is undoubtedly a cause of much parent-youth dissension. The adolescent's sociological exit from his family, via education, work, marriage, and change of residence, is fraught with potential conflicts of interest which only a definite system of institutional controls can neutralize. The parents have a vital stake in what the offspring will do. Because his acquisition of independence will free the parents of many obligations, they are willing to relinquish their authority; yet, precisely because their own status is socially identified with that of their offspring, they wish to insure satisfactory conduct on the latter's part and are tempted to prolong their authority by making the decisions themselves. In the absence of institutional prescriptions, the conflict of interest may lead to a struggle for power, the parents fighting to keep control in matters of importance to themselves, the son or daughter clinging to personally indispensable family services while seeking to evade the concomitant control.

4. *Concentration within the Small Family.* Our family system is peculiar in that it manifests a paradoxical combination of concentration and dispersion. On the one hand, the unusual smallness of the family unit makes for a strange intensity of family

feeling, while on the other, the fact that most pursuits take place outside the home makes for a dispersion of activities. Though apparently contradictory, the two phenomena are really interrelated and traceable ultimately to the same factors in our social structure. Since the first refers to that type of affection and antagonism found between relatives, and the second to activities, it can be seen that the second (dispersion) isolates and increases the intensity of the affectional element by sheering away common activities and the extended kin. Whereas ordinarily the sentiments of kinship are organically related to a number of common activities and spread over a wide circle of relatives, in our mobile society they are associated with only a few common activities and concentrated within only the immediate family. This makes them at once more instable (because ungrounded) and more intense. With the diminishing birth rate, our family is the world's smallest kinship unit, a tiny closed circle. Consequently, a great deal of family sentiment is directed toward a few individuals, who are so important to the emotional life that complexes easily develop. This emotional intensity and situational instability increase both the probability and severity of conflict.

In a familistic society, where there are several adult male and female relatives within the effective kinship group to whom the child turns for affection and aid, and many members of the younger generation in whom the parents have a paternal interest, there appears to be less intensity of emotion for any particular kinsman and consequently less chance for severe conflict. Also, if conflict between any two relatives does arise, it may be handled by shifting mutual rights and obligations to another relative.

5. *Open Competition for Socioeconomic Position.* Our emphasis upon individual initiative and vertical mobility, in contrast to rural-stable regimes, means that one's future occupation and destiny are determined more at adolescence than at birth, the adolescent himself (as well as the parents) having some part in the decision. Before him spread a panorama of possible occupations and avenues of advancement, all of them fraught with the uncertainties of competitive vicissitude. The youth is ignorant of most of the facts. So is the parent, but less so. Both attempt to collaborate on the future, but because of previously mentioned sources of friction, the collaboration is frequently stormy. They evaluate future possibilities differently, and since the decision is uncertain yet important, a clash of wills results. The necessity of choice at adolescence extends beyond the occupational field to practically every phase of life, the parents having an interest in each decision. A culture in which more of the choices of life were settled beforehand by ascription, where the possibilities were fewer and the responsibilities of choice less urgent, would have much less parent-youth conflict.

6. *Sex Tension.* If until now we have ignored sex taboos, the omission has represented a deliberate attempt to place them in their proper context with other factors, rather than in the unduly prominent place usually given them. Undoubtedly, because of a constellation of cultural conditions, sex looms as an important bone of parent-youth contention. Our morality, for instance, demands both premarital chastity and postponement of marriage, thus creating a long period of desperate eagerness when young persons practically at the peak of their sexual capacity are forbidden to enjoy it. Naturally, tensions arise—tensions which adolescents try to relieve, and adults hope they will relieve, in some socially acceptable form. Such tensions not only make the adolescent intractable and capricious, but create a genuine conflict of interest between the two generations. The parent, with respect to the child's behavior, represents morality, while the offspring reflects morality *plus* his organic cravings. The stage is thereby set for conflict, evasion, and deceit. For the mass of parents, toleration is never possible. For the mass of adolescents, sublimation is never sufficient. Given our system of morality, conflict seems well nigh inevitable.

Yet it is not sex itself but the way it is handled that causes conflict. If sex patterns were carefully, definitely, and uniformly geared with nonsexual patterns in the social structure, there would be no parent-youth conflict over sex. As it is, rapid change has opposed the sex standards of different groups and generations, leaving impulse only chaotically controlled.

The extraordinary preoccupation of modern parents with the sex life of their adolescent offspring is easily understandable. First, our morality is sex-centered. The strength of the impulse which it seeks to control, the consequent stringency of its rules, and the importance of reproductive institutions for society, make sex so morally important that being moral and being sexually discreet are synonymous. Small wonder, then, that parents, charged with responsibility for their children and fearful of their own status in the eyes of the moral community, are preoccupied with what their offspring will do in this matter. Moreover, sex is intrinsically involved in the family structure and is therefore of unusual significance to family members *qua* family members. Offspring and parent are not simply two persons who happen to live together; they are two persons who happen to live together because of past sex relations between the parents. Also, between parent and child there stand strong incest taboos, and doubtless the unvoiced possibility of violating these unconsciously intensifies the interest of each in the other's sexual conduct. In addition, since sexual behavior is connected with the offspring's formation of a new family of his own, it is naturally of concern to the parent. Finally, these factors taken in combination with the delicacy of the authoritarian relation, the emotional intensity within the small family, and the confusion of sex standards, make it easy to explain the parental interest in adolescent sexuality. Yet because sex is a tabooed topic between parent and child, parental control must be indirect and devious, which creates additional possibilities of conflict.

Summary and Conclusion. Our parent-youth conflict thus results from the interaction of certain universals of the parent-child relation and certain variables the values of which are peculiar to modern culture. The universals are (1) the basic age or birth-cycle differential between parent and child, (2) the decelerating rate of socialization with advancing age, and (3) the resulting intrinsic differences between old and young on the physiological, psychosocial, and sociological planes.

Though these universal factors *tend* to produce conflict between parent and child, whether or not they do so depends upon the variables. We have seen that the distinctive general features of our society are responsible for our excessive parent-adolescent friction. Indeed, they are the same features which are affecting *all* family relations. The delineation of these variables has not been systematic, because the scientific classification of whole societies has not yet been accomplished; and it has been difficult, in view of the interrelated character of societal traits, to seize upon certain features and ignore others. Yet certainly the following four complex variables are important: (1) the rate of social change; (2) the extent of complexity in the social structure; (3) the degree of integration in the culture; and (4) the velocity of movement (e.g., vertical mobility) within the structure and its relation to the cultural values.

Our rapid social change, for example, has crowded historical meaning into the family time-span, has thereby given the offspring a different social content from that which the parent acquired, and consequently has added to the already existent intrinsic differences between parent and youth, a set of extrinsic ones which double the chance of alienation. Moreover, our great societal complexity, our evident cultural conflict, and our emphasis upon open competition for socioeconomic status have all added to this initial effect. We have seen, for instance, that they have disorganized the important relation of parental authority by confusing the goals of child control, setting up competing authorities, creating a small family system, making necessary certain significant choices at the time of adolescence, and leading to an absence of definite institutional mechanisms to symbolize and enforce the progressively changing stages of parental power.

If ours were a simple rural-stable society, mainly familistic, the emancipation from parental authority being gradual and marked by definite institutionalized steps, with no great postponement of marriage, sex taboo, or open competition for status, parents and youth would not be in conflict. Hence, the presence of parent-youth conflict in our civilization is one more specific manifestation of the incompatibility between an urban-industrial-mobile social system and the familial type of reproductive institutions.

40

"The Normal American Family" as an Interpretive Structure of Family Life among Grown Children of Korean and Vietnamese Immigrants

KAREN PYKE,
2000

The use of monolithic images of the "Normal American Family" as a stick against which all families are measured is pervasive in the family wars currently raging in political and scholarly discourses (Holstein & Gubrium, 1995). The hotly contested nature of these images—consisting almost exclusively of White middle-class heterosexuals—attests to their importance as resources in national debates. Many scholars express concern that hegemonic images of the Normal American Family are ethnocentric and that they denigrate the styles and beliefs of racial-ethnic, immigrant, gay–lesbian, and single-parent families while encouraging negative self-images among those who do not come from the ideal family type (Bernades, 1993; Dilworth-Anderson, Burton, & Turner, 1993; Smith, 1993; Stacey, 1998; Zinn, 1994). Yet we still know little about how the Family ideology shapes the consciousness and expectations of those growing up in the margins of the mainstream. This study examines the accounts that grown children of Korean and Vietnamese immigrants provide of their family life and filial obligations. The findings suggest that public images of the Normal American Family constitute an ideological template that shapes respondents' familial perspectives and desires as new racial-ethnic Americans.

FAMILY IDEOLOGY AS AN INTERPRETIVE STRUCTURE

Images of the Normal American Family (also referred to as the Family) are pervasive in the dominant culture—part of a " 'large-scale' public rhetoric" (Holstein & Miller, 1993, p. 152). They are found in the discourse of politicians, social commentators, and moral leaders; in the talk of everyday interactions; and in movies, television shows, and books. Smith (1993, p. 63) describes these ubiquitous images as an "ideological code" that subtly "inserts an implicit evaluation into accounts of ways of living together." Such images serve as instruments of control, prescribing how families ought to look and behave (Bernades, 1985). Most scholarly concern centers on how this ideology glorifies and presents as normative that family headed by a breadwinning husband with a wife who, even if she works for pay, is devoted primarily to the care of the home and children. The concern is that families of diverse

From Karen Pyke. 2000. " 'The Normal American Family' as an Interpretive Structure of Family Life Among Grown Children of Korean and Vietnamese Immigrants." *Journal of Marriage and the Family,* 62:240–255. Reprinted by permission.

structural forms, most notably divorced and female-headed families, are comparatively viewed as deficient and dysfunctional (Fineman, 1995; Kurz, 1995; Stacey, 1998). Scholars concerned about the impact of such images point to those who blame family structures that deviate from this norm for many of society's problems and who suggest policies that ignore or punish families that don't fit the construct (e.g., Blankenhorn, 1995; Popenoe, 1993, 1996).

In addition to prescribing the structure of families, the Family ideal contains notions about the appropriate values, norms, and beliefs that guide the way family members relate to one another. The cultural values of "other" families, such as racial–ethnic families, are largely excluded. For example, prevailing family images emphasize sensitivity, open honest communication, flexibility, and forgiveness (Greeley, 1987). Such traits are less important in many cultures that stress duty, responsibility, obedience, and a commitment to the family collective that supercedes self-interests (Chung, 1992; Freeman, 1989). In further contrast to the traditional family systems of many cultures, contemporary American family ideals stress democratic rather than authoritarian relations, individual autonomy, psychological well-being, and emotional expressiveness (Bellah, Madsen, Sullivan, Swidler, & Tipton, 1985; Bernades, 1985; Cancian, 1987; Coontz, 1992; Skolnick, 1991). Family affection, intimacy, and sentimentality have grown in importance in the United States over time (Coontz, 1992), as evident in new ideals of fatherhood that stress emotional involvement (Coltrane, 1996).

These mainstream family values are evident in the therapeutic ethic, guiding the ways that those who seek professional advice are counseled and creating particular therapeutic barriers in treating immigrant Asian Americans (Bellah et al., 1985; Cancian, 1987; Tsui and Schultz, 1985). Family values are also widely disseminated and glorified in the popular culture, as in television shows like *Ozzie and Harriet, Leave It To Beaver, The Brady Bunch, Family Ties,* and *The Cosby Show,* many of which are rerun on local stations and cable networks (Coontz, 1992). Parents in these middle-class, mostly White, television families are emotionally nurturing and supportive, understanding, and forgiving (Shaner, 1982; Skill, 1994). Indeed, such shows tend to focus on the successful resolution of relatively minor family problems, which the characters accomplish through open communication and the expression of loving concern. Children in the United States grow up vicariously experiencing life in these television families, including children of immigrants who rely on television to learn about American culture. With 98% of all U.S. households having at least one television set, Rumbaut (1997, p. 949) views TV as an immense "assimilative" force for today's children of immigrants. Yet, he continues, it remains to be studied how their world views are shaped by such "cultural propaganda." The images seen on television serve as powerful symbols of the "normal" family or the "good" parent—and they often eclipse our appreciation of diverse family types (Brown & Bryant, 1990; Greenberg, Hines, Buerkel-Rothfuss, & Atkin, 1980). As the authors of one study on media images note, "The seductively realistic portrayals of family life in the media may be the basis for our most common and pervasive conceptions and beliefs about what is natural and what is right" (Gerbner, Gross, Morgan, & Signorielli, 1980, p. 3). Family scholars have rarely displayed analytic concern about the emphasis on emotional expressiveness and affective sentimentality that pervades much of the Family ideology, probably because the majority—who as middle-class, well-educated Whites live in the heartland of such values—do not regard them as problematic. As a result, this Western value orientation can seep imperceptibly into the interpretive framework of family research (Bernades, 1993; Dilworth-Anderson et al., 1993; Fineman, 1995; Smith, 1993; Thorne & Yalom, 1992).

The theoretical literature on the social construction of experience is an orienting framework for this study (Berger & Luckmann, 1966; Holstein & Gubrium, 1995). According to this view, cultural ideologies and symbols are integral components of the way individuals subjectively experience their lives and construct reality. The images we carry in our heads of how family life is supposed to be frame our interpretation of our own domestic relations. This is evident in the different ways that Korean and Korean American children perceived their parents' childrearing behavior in a series of studies. In Korea, children were found to associate parental strictness with warmth and concern and its absence as a sign of neglect (Rohner & Pettengill, 1985). These children were drawing on Korean family ideology, which emphasizes strong parental control and parental responsibility for children's failings. In this interpretive framework, parental strictness is a positive characteristic of family life and signifies love and concern. Children of Korean immigrants living in the United States, on the other hand, viewed their parents' strictness in negative terms and associated it with a lack of warmth—as did American children in general (Pettengill & Rohner, 1985). Korean American children drew on American family ideology, with its emphasis on independence and autonomy, and this cast a negative shadow on their parents' strict practices.

Although pervasive images of the Normal American Family subtly construct Asian family patterns of interaction as "deviant," countervailing images of Asians as a "model minority" are also widely disseminated. News stories and scholarly accounts that profile the tremendous academic success among some immigrant Asian children or describe the upward economic mobility observed among segments of the Asian immigrant population credit the cultural traditions of collectivist family values, hard work, and a strong emphasis on education. Such images exaggerate the success of Asian immigrants and mask intraethnic diversity (Caplan, Choy, & Whitmore, 1991; Kibria, 1993; Min, 1995; Zhou & Bankston, 1998). Meanwhile, conservative leaders use model minority images as evidence of the need to return to more traditional family structures and values, and they blame the cultural deficiency of other racial minority groups for their lack of similar success, particularly African Americans and Latinos (Kibria, 1993; Min, 1995; Zhou & Bankston, 1998): The model minority construct thus diverts attention from racism and poverty while reaffirming the Family ideology. In the analysis of the accounts that children of immigrants provided of their family life, references to such cultural images and values emerged repeatedly as a mechanism by which respondents gave meaning to their own family lives.

KOREAN AND VIETNAMESE IMMIGRANT FAMILIES

This study focuses on children of Vietnamese and Korean immigrants because both groups constitute relatively new ethnic groups in the United States. Few Vietnamese and Koreans immigrated to the United States before 1965. However, from 1981 to 1990, Korea and Vietnam were two of the top five countries from which immigrants arrived (*Statistical Yearbook,* 1995, table 2, pp. 29–30). Thus adaptation to the United States is a relatively new process for large groups of Koreans and Vietnamese, one that is unassisted by earlier generations of coethnic immigrants. The children of these immigrants, located at the crossroads of two cultural worlds, offer a good opportunity to examine the familial perspectives and desires of new racial–ethnic Americans. . . .

Although more research is needed that closely examines Asian ethnic differences in family practices, the existing literature reveals patterns of similarities among the family systems of Koreans and Vietnamese that differentiate them from American family patterns. The role prescriptions, family obligations, hierarchal relations, lack of emotional expressiveness, and collectivist values associated with the traditional family systems of Korea and Vietnam contrast sharply with the emphasis on individualism, self-sufficiency, egalitarianism, expressiveness, and self-development in mainstream U.S. culture (Bellah et al., 1985; Cancian, 1987; Chung, 1992; Hurh, 1998; Kim & Choi, 1994; Min, 1998; Pyke & Bengtson, 1996; Tran, 1988; Uba, 1994). Immigrant children tend to quickly adopt American values and standards, creating generational schisms and challenges to parental control and authority. That parent-child conflict and cultural gaps exist in many Asian immigrant families is well documented (Gold, 1993; Freeman, 1989; Kibria, 1993; Min, 1998; Rumbaut, 1994; Zhou & Bankston, 1998; Wolfe, 1997). However, no study to date has closely examined the cultural mechanisms at play in this process. This study begins that task.

METHOD

The data are from an interview study of the family and social experiences of grown children of Korean and Vietnamese immigrants. Respondents were either located at a California university where 47% of all undergraduates are of Asian descent (Maharaj, 1997) or were referred by students from that university. In-depth interviews were conducted with 73 respondents consisting of 34 Korean Americans (24 women, 10 men) and 39 Vietnamese Americans (23 women, 16 men). Both parents of each respondent were Korean or Vietnamese, except for one respondent, whose parents were both Sino-Vietnamese. Respondents ranged in age from 18–26 and averaged 21 years. Only one respondent was married, and none had children. . . .

RESULTS

I examine two ways in which respondents commonly used the typification of American family life as a contrast structure against which behavior in immigrant Asian families was juxtaposed and interpreted (Gubrium & Holstein, 1997). When describing relations with their parents, most respondents provided negative accounts of at least one aspect of their relationship, and they criticized their parents for lacking American values that emphasize psychological well-being and expressive love. Recurring references to a narrow Americanized notion of what families ought to look like were woven throughout many such accounts. However, when respondents described the kinds of filial care they planned to provide for their parents, the respondents switched to an interpretive lens that values ethnic family solidarity. In this context, respondents' references to notions of the Normal American Family became a negative point of comparison that cast their own immigrant families, and Asian families in general, in positive terms.

Viewing Parental Relations through an Americanized Lens

Respondents were asked to fantasize about how they would change their parents if they could change anything about them that they wanted to change. The three areas of desired change that respondents mentioned most often reveal their adoption of many mainstream American values. They wished for parents who: (a) were less strict and gave them more freedom; (b) were more liberal, more open-minded, more Americanized, and less traditional; (c) were emotionally closer, more communicative, more expressive, and more affectionate. These three areas are interrelated. For example, being more Americanized and less traditional translates into being more lenient and expressive. A small minority of respondents presented a striking contrast to the dominant pattern by describing, in terms both positive and grateful, parents who had liberal attitudes or Americanized values and parenting styles.

The communication most respondents described with parents focused on day-to-day practical concerns, such as whether the child had eaten, and about performance in school and college, a major area of concern among parents. Conversations were often limited to parental directives or lectures. For the most part, respondents were critical of the emotional distance and heavy emphasis on obedience that marked their relations with their parents. Chang-Hee, an 18-year-old who immigrated from Korea at 8, provided a typical case. When asked about communication, she disparaged her parents for not talking more openly, which she attributed to their being Asian. Respondents typically linked parental styles with race and not with other factors such as age or personality. Like many other respondents, Chang-Hee constructed an account not only of her family relations, but also of Asian families in general.

> To tell you the truth, in Asian families you don't have conversations. You just are told to do something and you do it. . . . You never talk about problems, even in the home. You just kind of forget about it and you kind of go on like nothing happened. Problems never really get solved. That's why I think people in my generation, I consider myself 1.5 generation, we have such a hard time because I like to verbalize my emotions. . . . [My parents] never allowed themselves to verbalize their emotions. They've been repressed so much [that] they expect the same out of me, which is the hardest thing to do because I have so many different things to say and I'm just not allowed.

Some respondents volunteered that their parents never asked them about their well-being, even when their distress was apparent. Chang-Hee observed, "If I'm sad, [my mom] doesn't want to hear it. She doesn't want to know why. . . . She's never asked me, 'So what do you feel?' " This lack of expressed interest in children's emotional well-being, along with the mundane level of communication, was especially upsetting to respondents because, interpreted through the lens of American family ideology, it defined their parents as emotionally uncaring and distant.

Several respondents longed for closer, more caring relationships with their parents that included expressive displays of affection. Thanh, a married 22-year-old college student who left Vietnam when she was 6, said, "I'd probably make them more loving and understanding, showing a bit more affection. . . . A lot of times I just want to go up and hug my parents, but no, you don't do that sort of thing."

Research indicates that the desire for greater intimacy is more common among women than men (Cancian, 1987). Thus it was surprising that many male respondents also expressed strong desires for more caring and close talk—especially from their fathers, who were often described as harsh and judgmental. Ralph, 20, a Korean American man born in the United States, said:

> My dad, he's not open. He is not the emotional type. So he talks . . . and I would listen and do it. It's a one-way conversation, rather than asking for my opinions. . . . I would think it'd be nicer if he was . . . much more compassionate, caring, because it seems like he doesn't care.

Similarly, Dat, a 22-year-old biology major who left Vietnam when he was 5, said:

> I would fantasize about sitting down with my dad and shooting the breeze. Talk about anything and he would smile and he would say, "Okay, that's fine, Dat." Instead of, you know, judge you and tell me I'm a loser. . . .

A definition of love that emphasizes emotional expression and close talk predominates in U.S. culture (Cancian, 1986). Instrumental aspects of love, like practical help, are ignored or devalued in this definition. In Korean and Vietnamese cultures, on the other hand, the predominant definitions of love emphasize instrumental help and support. The great divide between immigrant parents who emphasize instrumental forms of love and children who crave open displays of affection was evident in the following conversation, which occurred between Dat and his father when Dat was 7 or 8 years old. Dat recalled, "I tried saying 'I love you' one time and he looked at me and said, 'Are you American now? You think this is *The Brady Bunch?* You don't love me. You love me when you can support me.' " These different cultural definitions of love contributed to respondents' constructions of immigrant parents as unloving and cold.

The Family as a Contrast Structure in the Negative Accounts of Family Life

Many of the images of normal family life that respondents brought to their descriptions came in the form of references to television families or the families of non-Asian American friends. Although these monolithic images do not reflect the reality of American family life, they nevertheless provided the basis by which respondents learned how to be American, and they served as the interpretive frame of their own family experiences. By contrasting behavior in their immigrant families with mainstream images of normalcy, the respondents interpreted Asian family life as lacking or deficient. Dat referred to images of normal family life in America, as revealed on television and among friends, as the basis for his desire for more affection and closeness with his father:

> Sometimes when I had problems in school, all I wanted was my dad to listen to me, of all people. I guess that's the American way and I was raised American. . . . That's what I see on TV and in my friends' family. And I expected him to be that way too. But it didn't happen. . . . I would like to talk to him or, you know, say "I love you," and he would look at me and say, "Okay." That's my ultimate goal, to say, "I love you." It's real hard. Sometimes when I'm in a good mood, the way I show him love is to put my hands over his shoulders and squeeze it a little bit. That would already irritate him a little. . . . You could tell. He's like, "What the fuck's he doing?" But I do it because I want to show him love somehow. Affection. I'm an affectionate person.

Similarly, Hoa, a 23-year-old Vietnamese American man who immigrated at age 2, referred to television in describing his own family: "We aren't as close as I would like. . . We aren't as close as the dream family, you know, what you see on TV. Kind of like . . . . *Leave It To Beaver.* You know, stuff I grew up on."

Paul, a 21-year-old Korean American born in the United States, also criticizes his father, and Asian fathers in general, in relation to the fathers of friends and those on television:

> I think there is somewhat of a culture clash between myself and my parents. They are very set on rules—at least my father is. He is very strict and demanding and very much falls into that typical Asian father standard. I don't like that too much and I think it is because . . . as a child, I was always watching television and watching other friends' fathers. All the relationships seemed so much different from me and my father's relationship. . . . I guess it's pretty cheesy but I can remember watching *The Brady Bunch* returns and thinking Mike Brady would be a wonderful dad to have. He was always so supportive. He always knew when something was wrong with one of his boys. Whenever one of his sons had a problem, they would have no problem telling their dad anything and the dad would always be nice and give them advice and stuff. Basically I used what I saw on television as a picture of what a typical family should be like in the United States. . . .

Sometimes respondents simply made assumptive references to normal or American families, against which they critically juxtaposed Asian immigrant families. Being American meant that one was a member of the Normal American Family and enjoyed family relations that were warm, close, and harmonious. Being Asian, on the other hand, meant living outside such normality. Thuy, a 20-year-old Vietnamese American woman who had immigrated when 13, said:

> If I could, I would have a more emotional relationship with my parents. I know they love me, but they never tell me they love me. They also are not very affectionate. This is how I've always grown up. It wasn't really

> until we came to the United States that I really noticed what a lack of love my parents show. *American* kids are so lucky. They don't know what it's like to not really feel that you can show emotion with your own parents.

Similarly, Cora, 20, a Korean American woman born in the United States, remarked:

> I would probably want [my parents] to be more open, more understanding so I could be more open with them, 'cause there's a lot of things that I can't share with them because they're not as open-minded as *American* parents. . . . 'Cause I have friends and stuff. They talk to their parents about everything, you know?

When asked how he was raised, Josh, 21, a Vietnamese American man who immigrated when 2, responded by calling up a construction of the "good" American Family and the "deficient" Asian Family. He said, "I'm sure that for all *Asian* people, if they think back to [their] childhood, they'll remember a time they got hit. *American* people, they don't get hit."

Respondents repeatedly constructed American families as loving, harmonious, egalitarian, and normal. Using this ideal as their measuring stick, Asian families were constructed as distant, overly strict, uncaring, and not normal. In fact, respondents sometimes used the word "normal" in place of "American." For example, Hoa, who previously contrasted his family with the one depicted in *Leave It To Beaver,* said, "I love my dad but we never got to play catch. He didn't teach me how to play football. All the stuff a *normal* dad does for their kids. We missed out on that." Thomas, 20, a Korean American who arrived in the United States at age 8, said, "I always felt like, maybe we are not so normal. Like in the real America, like Brady Bunch normal. . . . I always felt like . . . there was something irregular about me." Similarly, after describing a childhood where she spoke very little to her parents, Van, 24, who immigrated to the United States from Vietnam at 10, began crying and noted, "I guess I didn't have a *normal* childhood." To be a normal parent is to be an American parent. Asian immigrant parents are by this definition deficient. Such constructions ignore diversity within family types, and they selectively bypass the social problems, such as child abuse, that plague many non-Asian American families. It is interesting, for example, that respondents did not refer to the high divorce rate of non-Asian Americans (Sweet & Bumpass, 1987) to construct positive images of family stability among Asian Americans. This may be because, applying an Americanized definition of love, many respondents described their parents' marriage as unloving and some thought their parents ought to divorce.

Respondents relied on the Family not only as an interpretive framework, but also as a contrast structure by which to differentiate Asian and American families. This juxtaposition of American and Asian ignores that most of the respondents and the coethnics they describe are Americans. "American" is used to refer to non-Asian Americans, particularly Whites. The words "White" and "Caucasian" were sometimes used interchangeably with "American." Indeed, the Normal American Family *is* White. This Eurocentric imagery excludes from view other racial minority families such as African Americans and Latino Americans. It is therefore not surprising that racial-ethnic families were not referenced as American in these interviews. In fact, respondents appeared to use the term "American" as a code word denoting not only cultural differences but also racial differences. For example, Paul, who was born in the United States, noted, "I look Korean but I think I associate myself more with the *American race.*" The oppositional constructions of Asian and American families as monolithic and without internal variation imply that these family types are racialized. That is, the differences are constructed as not only cultural but also racially essential and therefore immutable (Omi & Winant, 1994). . . .

These data illustrate how Eurocentric images of normal family relationships promulgated in the larger society served as an ideological template in the negative accounts that respondents provided of their immigrant parents. However, as described next, when respondents discussed their plans for filial care, they presented positive accounts of their immigrant families.

Maintaining Ethnic Values of Filial Obligation

Respondents were not consistent in their individual constructions of Asian and American families as revealed in their interviews. When discussing future plans for filial care, most respondents positively evaluated their family's collectivist commitment to care. Such an interpretation is supported by model minority stereotypes in mainstream U.S. culture that attribute the success enjoyed by some Asian immigrants to their strong family values and collectivist practices (Kibria, 1993; Zhou & Bankston, 1998).

The majority of respondents valued and planned to maintain their ethnic tradition of filial care. For example, Josh, who criticized his parents (and Asian parents in general) for using physical forms of punishment, nonetheless plans to care for his parents in their old age. He said, "I'm the oldest son, and in Vietnamese culture the oldest son cares for the parents. That is one of the things that I carry from my culture. I would not put my parents in a [nursing] home. That's terrible." In contrast to White Americans who condition their level of filial commitment on intergenerational compatibility (Pyke, 1999), respondents displayed a strong desire to fulfill their filial obligation and—especially among daughters—were often undeterred by distant and even conflict-ridden relations with parents. For example, after describing a strained relationship with her parents, Kimberly, 20, who came to this country from Vietnam when 7, added, "I would still take care of them whether I could talk to them or not. It doesn't matter as long as I could take care of them."...

Most respondents expected to begin financially supporting their parents prior to their elderly years, with parents in their 50s often regarded as old. A few respondents had already begun to help out their parents financially. Many planned on living with parents. Others spoke of living near their parents, often as neighbors, rather than in the same house, as a means of maintaining some autonomy. The tradition of assigning responsibility for the care of parents to the eldest son was not automatically anticipated for many of these families, especially those from Vietnam. Respondents most often indicated that responsibility would be pooled among siblings or would fall exclusively to the daughters. Several said that parents preferred such arrangements, because they felt closer to daughters. Although the tendency for daughters to assume responsibility for aging parents is similar to the pattern of caregiving common in mainstream American families (Pyke & Bengston, 1996), several respondents noted that such patterns are also emerging among relatives in their ethnic homeland.

The Importance of Collectivism as an Expression of Love

Respondents typically attributed their future caregiving to reciprocation for parental care in the past and a cultural emphasis on filial respect and support. Yet the enthusiasm and strong commitment that pervades their accounts suggests that they are motivated by more than obligation. For example, Vinh, 26, a graduate student who immigrated from Vietnam at age 5, said about his parents:

> They are my life. They will never be alone. I will always be with them. When I was growing [up] as a child, my parents were always with me. And I believe . . . when you grow up, you should be with them; meaning, I will take care of them, in my house, everything. Your parents didn't abandon you when you were a kid. They did not abandon you when you [were] pooping in your diaper. Then when they do, I will not abandon them. . . . Whatever it takes to make them comfortable, I will provide it. There is no limit.

Unable to express love via open displays of affection and close talk, filial assistance becomes a very important way for adult children to symbolically demonstrate their affection for their parents and to reaffirm family bonds. Blossom, 21, who immigrated to the United States from Korea when 6, described the symbolic value of the financial assistance her father expects. She said, "Money is not really important, but it's more about our heart that [my dad] looks at. Through money, my dad will know how we feel and how we appreciate him." Remember that Dat's father told him, "You love me when you can support me." Because instrumental assistance is the primary venue for expressing love and affection in these immigrant families, adult children often placed no limits on what they were willing to do. For example, John, 20, who immigrated from Vietnam when 3, remarked "I'm willing to do anything (for my parents), that's how much I care."

Parental financial independence was not always welcomed by those children who gave great weight to their role of parental caregivers. For example, it was very important to Sean, 19, an only child, to care for his Sino-Vietnamese parents. Sean, who planned to become a doctor, commuted from his home to a local university. He said, "I want my parents to stay with me. I want to support them.... I'll always have room for my parents.... When I get my first paycheck, I want to support them financially." ...

The Family as a Contrast Structure in the Positive Accounts of Filial Obligation

Many respondents distinguished their ethnic collectivist tradition of filial obligation from practices in mainstream American families, which they described as abandoning elderly parents in retirement or nursing homes. The belief that the elderly are abandoned by their families is widespread in U.S. society and very much a part of everyday discourse. Media accounts of nursing home atrocities bolster such views. Yet most eldercare in this country is not provided in formal caregiving settings but by family members (Abel, 1991). Nonetheless, respondents used this tenacious myth as a point of contrast in constructing Asian American families as more instrumentally caring. For example, Thuy, who previously described wanting a "more emotional relationship" with her parents, like "American kids" have, explained:

> With the American culture, it's ... not much frowned upon to put your parents in a home when they grow old. In our culture, it is a definite no-no. To do anything like that would be disrespectful.... If they need help, my brother and I will take care of them, just like my mom is taking care of her parents right now.

Similarly, Hien, 21, a Vietnamese American woman who arrived in this country as an infant, noted, "I know a lot of non-Asians have their parents go to the nursing homes ... but I personally prefer to find a way of trying to keep them at home."

Mike, who wished he could talk to his parents the way his friends do to theirs, plans to care for his Americanized parents even though they have told him they do not want him to. He was not alone in remaining more committed to filial care than his parents required him to be. He said:

> They tell me to just succeed for yourself and take care of your own family. But [referring to filial care] that's just how the Vietnamese culture is. Here is America, once your parents are old, you put them in a retirement home. But not in my family. When the parents get old you take care of them. It doesn't matter if they can't walk, if they can't function anymore. You still take care of them.

When discussing relationships with their parents, respondents used the Family as the ideological raw material out of which they negatively constructed their parents as unloving and distant. However, when the topic changed to filial care, respondents switched to an ethnic definition of love that emphasizes instrumental support and that casts a positive and loving light on their families. . . .

In describing their plans for parental care, respondents turned their previous construction of Asian and American families on its head. In this context the Family was constructed as deficient and uncaring, while the families of respondents—and Asian families in general—were described as more instrumentally caring and closer. Respondents' view of American families as uncaring should not be interpreted as a departure from mainstream family ideology. There has been much concern in the public discourse that today's families lack a commitment to the care of their elders and children (e.g., Popenoe, 1993; see Coontz, 1992, pp. 189–191). Indeed, the pervasive criticism that "individualism has gone haywire" in mainstream families—bolstered by references to the solidarity of model minority families—provides ideological support for ethnic traditions of filial care. That is, children of immigrants do not face ideological pressure from the dominant society to alter such practices; rather, they are given an interpretive template by which to view such practices as evidence of love and care in their families. In fact, U.S. legislative attempts to withdraw social services from legal immigrants without citizenship, with the expectation that family sponsors will provide such support, structurally mandate collectivist systems of caregiving in immigrant families (Huber & Espenshade, 1997). In other words, the dominant society ideologically endorses and, in some ways, structurally requires ethnic immigrant practices of filial care. Filial obligation thus serves as a site where children of Korean and Vietnamese immigrants can maintain their ethnic identity and family ties without countervailing pressure from the mainstream.

DISCUSSION

Interweaving respondents' accounts with an analysis of the interpretive structure from which those accounts are constructed suggests that the Family ideology subtly yet powerfully influences the children of immigrants, infiltrating their subjective understandings of and desires for family life. Respondents relied on American family images in two ways. When discussing their relations with parents and their upbringing, respondents used the Family ideology as a standard of normal families and good parents, leading them to view their immigrant parents as unloving, deficient, and not normal. However, when respondents discussed filial care, a complete reversal occurred. Respondents referred to negative images of rampant individualism among mainstream American families, specifically in regard to eldercare, to bolster their positive portrayals of the instrumental care and filial piety associated with their ethnic families. Thus the Family ideology was called upon in contradictory ways in these accounts—in the denigration of traditional ethnic parenting practices and in the glorification of ethnic practices of filial obligation.

Findings from this study illustrate how a narrow, ethnocentric family ideology that is widely promulgated throughout the larger culture and quickly internalized by children of immigrants creates an interpretive framework that derogates many of the ethnic practices of immigrant families. As others have argued, the cultural imposition of dominant group values in this form of "controlling images" can lead minorities to internalize negative self-images (Espiritu, 1997). That is, racial-ethnic immigrants can adopt a sense of inferiority and a desire to conform with those values and expectations that are glorified in the mainstream society as normal. Indeed, many respondents explicitly expressed a desire to have families that were like White or so-called American families, and they criticized their own family dynamics for being different. Rather than resist and challenge the ethnocentric family imagery of the mainstream, respondents' accounts reaffirmed the Normal American Family and the centrality of White native-born Americans in this imagery. This research thus reveals a subtle yet powerful mechanism of internalized oppression by which the racial-ethnic power dynamics in the larger society are reproduced. This is a particularly important finding in that racial-ethnic families will soon constitute a majority in several states, causing scholars to ponder the challenge of such a demographic transformation of the cultural and political hegemony of White native-born Americans (Maharidge, 1996). This study describes an ideological mechanism that could undermine challenges to that hegemony.

This research also uncovered an uncontested site of ethnic pride among the second-generation respondents who drew on mainstream images of elder neglect in their positive interpretation of ethnic filial commitment. As previously discussed, the belief that mainstream American families abandon their elders is tenacious and widespread in the dominant society, despite its empirical inaccuracy. This negative myth has been widely used in popular discourse as an example of the breakdown of American family commitment, and it sometimes serves as a rallying cry for stronger "family values." Such cries are often accompanied by references to the family solidarity and filial piety celebrated in the model minority stereotype. Thus the mainstream glorification of ethnic filial obligation, as contrasted with negative images of abandoned White American elders, provided respondents with a positive template for giving meaning to ethnic practices of filial care. The mainstream endorsement of filial obligation marks it as a locale where respondents can maintain family ties and simultaneously produce a positive self-identity in both cultural worlds. This might explain why some respondents were steadfastly committed to filial care despite parental requests to the contrary.

It remains to be seen, however, whether these young adults will be able to carry out their plans of filial obligation. It is likely that many will confront barriers in the form of demanding jobs, childrearing obligations, geographic moves, unsupportive spouses, competing demands from elderly in-laws, and financial difficulties. Furthermore, parents' access to alternative sources of support such as Social Security and retirement funds could diminish the need for their children's assistance. Some research already finds that elderly Korean immigrants prefer to live on their own and are moving out of the homes of their immigrant children despite the protests of children, who see it as a public accusation that they did not care for their parents (Hurh, 1998). Although this research examined first-generation immigrant adults and their aging parents, it suggests a rapid breakdown in traditional patterns of coresidential filial care that will likely be reiterated in the next generation. Future research is needed to examine these dynamics among second-generation immigrants, to look at how they will cope with inabilities to fulfill ethnic and model minority expectations of filial obligation, and to assess the impact of any such inabilities on their ethnic identity.

SELECTED REFERENCES

Abel, E. K. (1991). *Who cares for the elderly?* Philadelphia: Temple University Press.

Bellah, R. N., Madsen, R., Sullivan, W. M., Swidler, A., & Tipton, S. (1985). *Habits of the heart.* San Francisco: Harper & Row.

Berger, P. L., & Luckmann, T. (1966). *The social construction of reality.* New York: Doubleday.

Bernades, J. (1985). "Family ideology": Identification and exploration. *Sociological Review, 33,* 275–297.

Bernades, J. (1993). Responsibilities in studying post-modern families. *Journal of Family Issues, 14,* 35–49.

Blankenhorn, D. (1995). *Fatherless America.* New York: Basic Books.

Brown, D., & Bryant, J. (1990). Effects of television on family values and selected attitudes and behaviors. In J. Bryant (Ed.), *Television and the American family* (pp. 253–274). Hillsdale, NJ: Erlbaum.

Cancian, F. M. (1986). The feminization of love. *Signs, II,* 692–708.

Cancian, F. M. (1987). *Love in America.* New York: Cambridge University Press.

Caplan, N., Choy, M. H., & Whitmore, J. K. (1991). *Children of the boat people: A study of educational success.* Ann Arbor: University of Michigan Press.

Chung, D. K. (1992). Asian cultural commonalities: A comparison with mainstream American culture. In S. Furuto, R. Biswas, D. Chung, K. Murase, F. Ross-

Sheriff (Eds.), *Social work practice with Asian Americans* (pp. 27–44). Newbury Park, CA: Sage.

Coltrane, S. (1996). *Family man.* New York: Oxford University Press.

Coontz, S. (1992). *The way we never were.* New York: Basic Books.

Dilworth-Anderson, P., Burton, L. M., & Turner, W. L. (1993). The importance of values in the study of culturally diverse families. *Family Relations, 42,* 238–242.

Espiritu, Y. L. (1997). *Asian American women and men.* Thousand Oaks, CA: Sage.

Fineman, M. A. (1995). *The neutered mother, the sexual family, and other twentieth century tragedies.* New York: Routledge.

Freeman, J. M. (1989). *Hearts of sorrow: Vietnamese-American lives.* Stanford, CA: Stanford University Press.

Gerbner, G., Gross, L., Morgan, M., & Signorielli, N. (1980). *Media and the family: Images and impact.* Washington, DC: White House Conference on the Family, National Research Forum on Family Issues. (ERIC Document Reproduction Service No. ED 198 919).

Gold, S. J. (1993). Migration and family adjustment: Continuity and change among Vietnamese in the United States. In H. P. McAdoo (Ed.), *Family ethnicity* (pp. 300–314). Newbury Park, CA: Sage.

Greeley, A. (1987, May 17). Today's morality play: The sitcom. *New York Times,* p. HI.

Greenberg, B. S., Hines, M., Buerkel-Rothfuss, N., & Atkin, C. K. (1980). Family role structures and interactions on commercial television. In B. S. Greenberg (Ed.), *Life on television: Content analyses of U.S. TV drama* (pp. 149–160). Norwood, NJ: Ablex.

Gubrium, J. E, & Holstein, J. A. (1997). *The new language of qualitative method.* New York: Oxford University Press.

Holstein, J. A., & Gubrium, J. F. (1995). Deprivatization and the construction of domestic life. *Journal of Marriage and The Family, 57,* 894–908.

Holstein, J. A., & Miller, G. (1993). Social constructionism and social problems work. In J. A. Holstein & G. Miller (Eds.), *Reconsidering social constructionism* (pp. 151–172). New York: Aldine De Gruyter.

Huber, G. A., & Espenshade, T. J. (1997). Neo-isolationism, balanced-budget conservatism, and the fiscal impacts of immigrants. *International Migration Review, 31,* 1031–1054.

Hurh, W. M. (1998). *The Korean Americans.* Westport, CN: Greenwood Press.

Kibna, N. (1993). *The family tightrope: The changing lives of Vietnamese Americans.* Princeton, NJ: Princeton University Press.

Kim, U., & Choi, S. (1994). Individualism, collectivism, and child development: A Korean perspective. In P. Greenfield & R. Cocking (Eds.), *Cross-cultural roots of minority child development* (pp. 227–57). Hillsdale, NJ: Erlbaum.

Kurz, D. (1995). *For richer, for poorer.* New York: Routledge.

Maharidge, D. (1996). *The coming white minority: California eruptions and America's future.* New York: New York Times Books.

Maharaj, D. (1997, July 8). E-mail hate case tests free speech protections. *Los Angeles Times,* pp. A1, A16.

Min, P. G. (1995). Major issues relating to Asian American experiences. In P. G. Min (Ed.), *Asian Americans* (pp. 38–57). Thousand Oaks, CA: Sage.

Min, P. G. (1998). *Changes and conflicts: Korean immigrant families in New York.* New York: Allyn and Bacon.

Omi, M., & Winant, H. (1994). *Racial formation in the United States.* New York: Routledge.

Pettengill, S. M., & Rohner, R. P. (1985). Korean-American adolescents' perceptions of parental control, parental acceptance-rejection and parent-adolescent conflict. In I. R. Lagunes & Y. H. Poortinga (Eds.), *From a different perspective: Studies of behavior across culture* (pp. 241–249). Berwyn, IL: Swets North America.

Popenoe, D. (1993). American family decline, 1960–1990: A review and appraisal. *Journal of Marriage and the Family, 55,* 527–555.

Popenoe, D. (1996). *Life without father: Compelling new evidence that fatherhood and marriage are indispensable and for the good of the children and society.* New York: Martin Kessler/Free Press.

Pyke, K. D. (1999). The micropolitics of care in relationships between aging parents and adult children: Individualism, collectivism, and power. *Journal of Marriage and the Family, 61,* 661–672.

Pyke, K. D., & Bengtson, V. L. (1996). Caring more or less: Individualistic and collectivist systems of family eldercare. *Journal of Marriage and the Family, 58,* 379–392.

Rohner, R. P., & Pettingill, S. M. (1985). Perceived parental acceptance-rejection and parental control among Korean adolescents. *Child Development, 56,* 524–528.

Rumbaut, R. G. (1997). Assimilation and its discontents: Between rhetoric and reality. *International Migration Review, 31,* 923–960.

Shaner, J. (1982). Parental empathy and family role interactions as portrayed on commercial television. *Dissertation Abstracts International, 42,* 3473A.

Skill, T. (1994). Family images and family actions as presented in the media: Where we've been and what we've found. In D. Zillmann, J. Bryant, & A. C. Huston (Eds.), *Media, children, and the family* (pp. 37–50). Hillsdale, NJ: Erlbaum.

Skolnick, A. (1991). *Embattled paradise: The American family in an age of uncertainty.* New York: Basic Books.

Smith, D. E. (1993). The standard North American family: SNAF as an ideological code. *Journal of Family Issues, 14,* 50–65.

Stacey, J. (1998). The right family values. In K. Hansen & A. Garey (Eds.), *Families in the U.S.* (pp. 859–880). Philadelphia: Temple University Press.

Sweet, J. A., & Bumpass, L. (1987). *American families and households.* New York: Russell Sage Foundation.

Thorne, B., & Yalorn, M. (1992). *Rethinking the family: Some feminist questions.* Boston: Northeastern University.

Tran, T. V. (1988). The Vietnamese American family. In C. H. Mmdel, R. W. Habenstein, & R. Wright, Jr. (Eds.), *Ethnic families in America: Patterns and variations* (pp. 276–299). New York: Elsevier.

Tsui, P., & Schultz, G. (1985). Failure of rapport: Why psychotherapeutic engagement fails in the treatment of Asian clients. *American Journal of Orthopsychiatry, 55,* 561–569.

Uba, L. (1994). *Asian Americans: Personality patterns, identity, and mental health.* New York: The Guilford Press.

U.S. Immigration and Naturalization Service (1997). *Statistical yearbook of the immigration and naturalization service, 1995.* Washington, DC: U.S. Government Printing Office.

Wolf, D. (1997). Family secrets: Transnational struggles among children of Filipino immigrants. *Sociological Perspectives, 40,* 457–482.

Zhou, M., & Bankston, III, C. (1998). *Growing up American: How Vietnamese children adapt to life in the United States.* New York: Russell Sage Foundation.

Zinn, M. B. (1994). Feminist rethinking of racial-ethnic families. In M. B. Zinn & B. T. Dill (Eds.), *Women of color in U.S. society* (pp. 303–314). Philadelphia: Temple University Press.

41

Extending the Family

JOHN LOGAN AND GLENNA SPITZE,
1996

The older family is important in itself and also as part of the larger "family matrix" that includes young children, their parents, and their grandparents. Other researchers have demonstrated the potential contributions of the older generation to the well-being of young children. We find that ties between the middle and older generations are strong and durable under conditions of demographic change. We suggest that discussions of crisis in the modern family should take into account the whole life course of the family.

Probably no day passes without a reminder in the mass media that the family is in trouble. Americans have come to accept this view without question. Commentators and politicians bemoan the crisis of the family and the failure of family values without having to explain what they are talking about. These understandings have become part of our common folklore at the close of the twentieth century, part of what we take for granted as an issue of our time. Of course there is a crisis, and it sells newspapers and garners votes.

Social scientists have been more cautious about the prospects for the family, but academic writing echoes the concerns found in the media. These issues are highly politicized. For example, in their discussion of single-parent families, McLanahan and Booth (1989, p. 569) argue that "the position taken by analysts on this issue is shaped by their values regarding women's traditional family roles and whether they view single motherhood as a cause or consequence of economic insecurity." At the extreme, analysts who see the traditional two-parent family as a primary civilizing influence, "view women's employment and the subsequent decline in the nuclear family as disasters for society" (McLanahan and Booth 1989, p. 571).

As we acknowledged at the very beginning, we also started with the presumption that the family as a core set of social relations was headed for trouble. Our focus has been exclusively on parent-child relations in their adult stage, the years when "children" have come of age. This is an important period in the family life cycle because it encompasses so many critical events in the lives of many members of the older generation: the peak of working years, grandparenthood, retirement, widowhood, disability, death. It now includes the larger part of the family life cycle: one is typically a dependent minor child for only eighteen or twenty years (or at the outside, twenty-five years), but potential relations with parents take up as much as thirty or forty or even fifty years of one's adult life.

We identified several sources of apprehension about the resilience of these intergenerational ties, especially as these trends are played out in the future:

- the aging of the population, accompanied by a sharpening consciousness of generational conflicts as a growing older generation demands that its needs be met;
- the shrinking of average family sizes, associated with rising standards and costs of living, that reduces children's chances of being able to share the duty of caring for aging parents;
- the growth of external demands on young parents, particularly women who are increasingly involved in employment outside the home, cutting their involvement with parents;
- the rising rate of divorce, depleting the financial and time resources of younger adults, and making them both more reliant on parents and less able to provide help to them;
- the special burden on women as the traditional kinkeepers in the American family, caught more often in the tangle of conflicting family roles, especially between their own parents and minor children;
- and finally, and more broadly, the replacement of traditional social networks, both family and local community, by secondary relationships and public institutions.

The underlying population trends listed here have been continuing for decades, and they show little sign of reversal. The question is how they are affecting intergenerational relations. One by one, . . . we have laid out the theoretical assumptions on which each expectation is based, reviewed the empirical literature, and offered our own new evidence. And in every case, the reality, as best we can determine it, is different from expectations.

Despite speculation to the contrary, we find that the subjective boundaries between generations are not at all clear. In terms of value orientations—in expectations about family relationships, as well as about public programs and policies—people in the older generation express greater concern for the younger generation than for their own self-interest. Older people with fewer children are less involved in intergenerational exchanges (though in partial compensation, children with fewer siblings are more involved with parents). But in fact most exchanges, until parents reach very old ages, flow from the parent to the child. Thus the trend toward smaller families actually relieves parents from children's demands for assistance. Neither employment nor divorce has a negative impact on an employed or divorced daughter's relations with parents. Though women clearly still have disproportionate responsibility for household labor, child care, and some forms of kinkeeping, there are important dimensions of intergenerational ties that do not exhibit gender differences. It is indeed the case that many middle-aged people simultaneously have the responsibilities of being a parent and being a child (as well as a spouse and a paid worker), but few devote more than three hours per week to both their parents and their adult children. And if either women or men experience difficulty with juggling multiple roles, this effect is not measurable in terms of reported burden, stress, or life satisfaction.

We have been surprised to find fewer gender differences than we expected, not only in intergenerational relations but also in such issues as age identity. Others, including Rossi and Rossi (1990), have argued that gender is a major organizing factor in intergenerational relations. They conclude that "gender of parent and child is a highly salient axis of family life and intergenerational relations. On topic after topic, we have found that ties among women were stronger, more frequent, more reciprocal, and less contingent on circumstances than those of men. Women's ties to women, as mothers, daughters, sisters, or grandmothers, provide social and emotional connecting links among members of a family and lineage" (p. 495).

Our findings regarding gender are less clearcut. We think these seeming discrepancies can be explained in several ways. First, we do find some differences between men and women. Most prominently, there are clearcut gender differences in frequency of telephoning, the most common form of interaction between parents and adult children.

Second, there are some commonly gender-identified activities that we did not tap, such as kinkeeping and arranging family visits. Much of this may occur during telephone contacts. We think it is possible that some of the gendered patterns in kin contact are more subtle than can be measured in a structured interview survey, and some would require different kinds of questions. Just as surveys about household labor for years measured task performance without tapping task responsibility and management (e.g., Mederer 1993), studies of intergenerational relations need to measure both frequency of contact and who arranges contact.

Further, we think part of the difference between our own conclusions and those of others depends on one's emphasis. For example, Rossi and Rossi find, as we do, that there is no significant difference between frequency of visiting with sons and daughters in multivariate analysis (though there *is* a difference in telephone contact). They also find that daughters help with certain tasks—specifically household chores—more than sons do. However, even among sons, 48 percent help mothers and 36 percent help fathers with household chores (compared to 59 percent and 48 percent of daughters). Obviously there is a difference here, but it is also striking that between one-third and one-half of sons are providing this kind of help. Patterns for "fixing or making something for you" show the opposite pattern, with sons helping more than daughters. We measured help in slightly different categories, and also found that some tasks are gender-typed. But when we measured total weekly hours of help (as suggested by Rossi and Rossi 1990, p. 507) we found much smaller gender differences. There are strong differences in types of help, but weak differences in the overall total.

Finally, it is probably worth repeating that much of the literature on gender and aging families focuses on intensive caregiving situations. We did not study people in these kinds of situations in very large numbers, and such caregivers are predominantly female.

CENTRALITY OF FAMILY TIES

Our analyses are remarkable for the apparent resilience of parent-child ties in later life. Even more noteworthy is the importance of these ties compared to other parts of people's social networks. Whether measured as contact, visiting, help given, or help received, these family ties are at the very core of people's networks outside the home. And the most traditional of them—family neighbors, that is, parents or children living within the same neighborhood—retain a special weight in social relationships. Rather than being displaced by secondary associations, by friends or coworkers living at a distance or by paid helpers or public services, we find that family ties—when they are present—are people's preferred source of routine assistance. Others play distinctly subordinate roles. These conclusions are parallel to those of Rossi and Rossi regarding normative obligations. They describe a pattern of normative social relations as a set of concentric circles, with parent and child in the center.

No doubt the family is being stretched today, as could only be expected under conditions of growing social inequality and poverty, weakening job security, and reduced public assistance. Even the most affluent households bear the emotional scars of divorce, the strains of remarriage, and the pressures of long working hours. Family patterns including high incidence of female-headed households, resulting from divorce and out-of-wedlock childbirth, have extended well beyond the confines of the inner city and the minority community. Yet there are reasons to believe that the family is also being regenerated, and the task of sociologists is not only to identify the failures of the old family but also to perceive the emergence of the new one. Vital institutions don't just cease to function; they may be battered and punished by change, but people must and do find ways to adapt to their conditions.

It is not a Pollyana but a realist who, like Stacey (1991, p. 16), looks for the ways in which people "have been creatively remaking American family life." Stacey describes the current period as one where little is taken for granted about family form and people struggle to adapt existing institutions to new needs. No single family structure has emerged to replace the modern family: this is "a transitional and contested period of family history, a period after the modern family order, but before what we cannot foretell" (1991, p. 18).

Our own view is that less is changing than meets the eye. Any interpretation reflects the particular blinders of the theorist. Our interpretation may reflect the emphases of the gerontological literature in which we have worked together for nearly ten years. Let us be clear about what we mean. Sociologists are now more inclined than ever before to study family issues in the context of people's whole life histories, from childhood to retirement and widowhood (Elder 1985). Yet the core of family studies has been, and perhaps still is, that phase of family life that runs from marriage to nest leaving: the family, the nuclear family or the modern family, is a childbearing and childraising institution. Social gerontologists, grounded in a medical and social work tradition and specializing in the problems of older people, have been separated from this theoretical core. The linkage between aging and family studies was established by the recognition that adult children were often the primary caregivers for the frail elderly. But the framing of the issues at that time—the concern about whether these caregivers would continue to be available because of divorce, women's employment, etc.—simply reflected the questions that family theorists were already asking about the younger family. The crisis of the younger family became a crisis in support for older family members.

EXTENDING THE FAMILY

Thus our tendency has been to study the younger family only insofar as it is mirrored in intergenerational relations in the older family. The corresponding bias of most family researchers has been to assume that the extended family died out long ago, and that only the younger nuclear family has survived. For example, a single-minded focus on the young family is found in the recent essay on the decline of the family by David Popenoe (1993). . . . Though he recognizes that family change is not necessarily family decline, he regards recent changes as "both unique and alarming." This is because, in his view, "it is not the extended family that is breaking up but the nuclear family. The nuclear family can be thought of as the last vestige of the traditional family unit; all other adult members have been stripped away, leaving but two—the husband and the wife. The nuclear unit—man, woman, and child—is called that for good reason: It is the fundamental and most basic unit of the family. Breaking up the nucleus of anything is a serious matter. . . . Adults for their own good purposes, most recently self-fulfillment, have stripped the family down to its nucleus. But any further reduction—either in functions or in number of its members—will likely have adverse consequences for children, and thus for generations to come" (Popenoe 1993, pp. 539–40).

We do not dispute the importance of this younger nuclear family. The family of parents and minor children is the site of many key activities. It is first of all the setting where most child-raising issues arise. It is where divorce has its greatest incidence and perhaps its greatest impact. And it is arguably where the greatest gender-related tensions are found, particularly because the tasks of economic support, homemaking, and child-rearing are typically most stressful for young parents. Perhaps for these same reasons it is the least secure component of the American family.

But it is not the whole family. On the contrary, as others have argued persuasively (Rossi and Rossi 1990), the nuclear family is a lifelong and multigenerational event. It is a network of two sorts of relationships of central importance in people's lives: with their parents, lasting until their parents' death (when, in many cases, they themselves are well into middle age), and with their own children, usually extending through their own lifetime. We all acknowledge this point, when confronted by it, but family sociologists have focused particularly on the childhood years.

The research literature on young families and on grandparenting demonstrates the relevance for children of relationships with older kin. The well-known ethnographic accounts of resilient kin and neighbor networks of poor single mothers by Stack (1974) offer at least the possibility that there are workable alternatives to the two-parent family, even in the worst conditions. There is evidence that intergenerational ties do play a strong role in coping with difficult family situations. Only a small minority of the grandparents studied by Cherlin and Furstenberg (1986) were "involved" in a parent-like relationship with grandchildren, but others were viewed by the authors as "sources of support in reserve. This latent support may never be activated; but, like a good insurance policy, it is important nevertheless" (p. 206). They predict that demographic trends including the increase in divorce will work to make this support more salient in the future.

The great majority of teenage mothers live in a larger family environment (Voydanoff and Donnelly 1990), typically in the grandmother's home, and this may occur even when the mother is married. This is particularly true for black teen mothers (married or unmarried), who are twice as likely to live with other adult kin than are white teen mothers (Hogan et al. 1990; see also Taylor et al. 1993). White mothers, though less likely to coreside with kin, are more likely to receive other forms of assistance (Hofferth 1984).

Coresidence may create problems for grandparents, who may feel burdened by and conflicted with the contradictory norms of noninterference and obligation to help if needed, but grandchildren and their mothers may benefit from these arrangements (Aldous 1995). Generally, mothers who live with kin benefit from higher levels of assistance in child care, psychological support, and help in returning to school or work (Hogan et al. 1990). And the payoff of this intergenerational presence is evident in the performance of the children: the negative effects of having been born to a teenage mother are partly offset by the presence of other adults in the household (Voydanoff and Donnelly 1990, p. 90). McLanahan and Sandefur (1994) note that studies of younger families have found that "being raised by a mother and grandmother is just as good as being raised by two parents, at least in terms of children's psychological well-being" (p. 74; see also Cherlin and Furstenberg 1986). Their own study of teenage grandchildren found no positive effect.

Adopting a multigenerational view of the family leads in a different direction than the reports of its decline. A recent essay by Matilda White Riley and John Riley (1993) offers an alternate perspective. Riley and Riley argue that the "simple kinship structure" of a young nuclear family that only sometimes included a surviving older generation has already been supplanted by an "expanded" structure. Because people live longer, the number of generations surviving jointly has increased. As a result, "linkages among family members have been prolonged. Thus parents and their offspring now survive jointly for so many years that a mother and her daughter are only briefly in the traditional relationship of parent and little children; during the several remaining decades of their lives they have become status equals" (1993, p. 163).

This expanded family, extended in generations and in duration, points the way toward an even more flexible type of kinship structure, a "latent matrix" of kin connections. Riley and Riley (1993, p. 169) point to four features of this new family:

"First, numerous social and cultural changes—especially cohort increases in longevity—are yielding a large and complex network of kin relationships. Second, these many relationships are flexible . . . increasingly matters of choice rather than obligation. A plethora of options is potentially at hand. Third, these relationships are not constrained by age or generation; people of any age, within or across generations, may opt to support, love, or confide in one another. Fourth, many of these kinship bonds remain latent until called upon. They form a safety net of significant connections to choose from in case of need."

The "latent matrix" loses "the sharp boundaries set by generation or age or geographical proximity . . . the boundaries of the family network have been widened to encompass many diverse relationships, including several degrees of stepkin and in-laws, single-parent families, adopted and other 'relatives' chosen from outside the family." (1993, p. 174). Further, it adds the element of greater choice rather than obligation to family relationships—even for motherhood itself.

THE FUTURE OF THE FAMILY

So we can't equate the future of the family with the future of the nuclear family, or the "modern" nuclear family, as Stacey calls it. It is the family as a latent matrix or multigenerational network that has to be evaluated, the family that endures throughout adulthood and that only sometimes is characterized by living together. This is a family structure in which people are relatively independent and self-sufficient, but nonetheless they maintain core relationships.

In this family form (the "extended" family, by which we mean the nuclear family extended through time and generations), we certainly find consequences of the changing nuclear family, though in most respects we do not share a sense of crisis. Perhaps the most often discussed set of family issues is associated with divorce and single-parent families. The harshest impacts here are probably on children in their younger years, most commonly including poverty and weakened or absent ties to their fathers. Our own research bears only on intergenerational relations at a later stage in the life cycle. Goldscheider (1990; see also Bumpass 1990) argues that there are special risks for people, especially men, who will become old in the next century, due to their weak ties to children (already visible in their relationships to these children today). But it is hard to predict the future impact of the single-parent or blended families of today.

We cannot be sure of the degree to which the multigenerational family network can substitute for the two-parent nuclear family that became our cultural model in the 1950s. Commentators have often pointed out that divorce and remarriage simultaneously undermine a child's links to one parent while expanding the universe of potential kin in the form of stepparents, stepsiblings, and stepgrandparents. But as Furstenberg and Cherlin (1991, pp. 91–95) point out, little is known about the long-term consequences of exchanging one very close tie for a wider network of weaker ties (see also Bumpass 1990). This is largely uncharted territory. The evidence so far, however, is that intergenerational ties in adulthood remain strong in spite of the social currents that have been expected to disrupt them. These ties are a firm foundation for the family of the future.

SELECTED REFERENCES

Aldous, Joan. 1995. "New Views of Grandparents in Intergenerational Context." *Journal of Family Issues* 16: 104–22.

Bumpass, Larry L. 1990. "What's Happening to the Family? Interactions between Demographic and Institutional Change." *Demography* 27: 483–98.

Cherlin, Andrew, and Frank Furstenberg Jr. 1986. *The New American Grandparent: A Place in the Family, A Life Apart.* New York: Basic Books.

Elder, Glen H., Jr. 1985. *Life Course Dynamics: Trajectories and Transitions, 1968–1980.* Ithaca: Cornell University Press.

Furstenberg, Frank, Jr. and Andrew Cherlin. 1991. *Divided Families.* Cambridge, MA: Harvard University Press.

Goldscheider, Frances K. 1990. "The Aging of the Gender Revolution: What Do We Know and What Do We Need to Know?" *Research on Aging* 12 (December): 531–45.

Hofferth, Sandra. 1984. "Kin Networks, Race, and Family Structure." *Journal of Marriage and the Family* 46: 791–806.

Hogan, Dennis P. Ling-Xin Hao, and William Paris. 1990. "Race, Kin Networks, and Assistance to Mother-Headed Families." *Social Forces* 68: 797–812.

McLanahan, Sara, and Karen Booth. 1989. "Mother-Only Families: Problems, Prospects, and Politics." *Journal of Marriage and the Family* 51: 557–80.

McLanahan, Sara, and Gary Sandefur. 1994. *Growing Up with a Single Parent: What Hurts, What Helps.* Cambridge, MA: Harvard University Press.

Mederer, Helen J. 1993. "Division of Labor in Two-Earner Homes: Task Accomplishment Versus Household Management as Critical Variables in Perceptions about Family Work." *Journal of Marriage and the Family* 55: 133–45.

Popenoe, David. 1993. "American Family Decline, 1960–1990: A Review and Appraisal." *Journal of Marriage and the Family* 55: 527–55.

Riley, Matilda White, and John W. Riley Jr. 1993. "Connections: Kin and Cohort." Pp. 169–89 in *The Changing Contract across Generations,* eds. Vern Bengston and W. Andrew Achenbaum. New York: Aldine de Gruyter.

Rossi, Alice S., and Peter H. Rossi. 1990. *Of Human Bonding: Parent-Child Relations Across the Life Course.* New York: Aldine de Gruyter.

Stacey, Judith. 1991. *Brave New Families: Stories of Domestic Upheaval in Late Twentieth Century America.* New York: Basic Books.

Stack, Carol. 1974. *All Our Kin: Strategies for Survival in a Black Community.* New York: Harper and Row.

Taylor, Robert Joseph, Linda M. Chatters, and James S. Jackson. 1993. "A Profile of Familial Relations among Three-Generation Black Families." *Family Relations* 42: 332–41.

Voydanoff, Patricia, and Brenda Donnelly. 1990. *Adolescent Sexuality and Pregnancy.* Newbury Park, CA: Sage.

CHAPTER 13

Family Policy

Some scholars suggest that family policy does not really exist in the United States, because no set of laws or administrative orders is labeled "family policy" and because no high-level offices in federal or state governments are responsible for directly overseeing government policies that affect families (Bane and Jargowsky, 1989). Many industrialized nations in the world today *do* have specific agencies devoted to family well-being, but the U.S. government has typically taken an official hands-off policy toward families. That may be changing, as recent government policies from welfare reform to covenant marriage have attempted to articulate specific family-related objectives, though in a way that tends to reinforce traditional marriages rather than focusing on the needs of children.

When politicians and journalists use the term *family policy,* they mean government actions that are specifically addressed to families—such as guaranteeing that you will still have your job after you take time off to have a baby, or allowing you to take a tax deduction if you pay for childcare or send your children to college. But *family policy* is a vague term that leaves out much of what governments do to shape family life. When legislators or government agencies mandate minimum wages, limit immigration, subsidize farm crops, build highways, set interest rates, enact trade embargoes, regulate health care, alter criminal sentencing rules, adjust taxes, or declare war, they are influencing how families are able to operate in society. In so doing, they also indirectly shape how husbands and wives act toward their children and toward each other. How people form and sustain sexual and marital relationships are also directly affected when laws and policies governing rape, prostitution, pornography, homosexuality, abortion, birth control, or adoption are reformed. And when lawmakers and the courts alter the formal ways they treat marriage, divorce, custody, child support, schools, or welfare, families are affected even more directly (Coltrane, 1998; Kamerman and Kahn, 1997).

In an effort to improve the life chances of children, many social scientists and child advocates argue that family policies should be defined in terms of the well-being of children (Hernandez, 1993). For example, Marian Wright Edelman (1987), former president of the Children's Defense Fund, suggested that instead of formulating something called family policy, we should fund health, nutrition, education, and housing programs aimed at directly benefiting children (virtually all of whom live in some sort of family). She has publicized the plight of children living in single-parent families, many of whom live at or near the poverty level. She focuses especially on the plight of African American children, whose life chances have been declining in recent years. When compared to white children, black children are "*twice as likely to* die in the first year of life, be born prematurely, suffer low birth weight, have mothers who received late or no prenatal care, see a parent die, live in substandard housing, be suspended from school or suffer corporal punishment, be unemployed as teenagers, have no parent employed, live in institutions; three *times as likely to* be poor, have their mothers die in childbirth, live with a parent who has separated, live in a female headed family, be placed in an educable mentally retarded class, be murdered between five and nine years of age, be in foster care, die of known child abuse; [and] *four times as likely to* live with neither parent and be supervised by a child welfare agency, be murdered before one year of age or as a teenager, be incarcerated between fifteen and nineteen years of age" (Edelman, 1989). Similar declining life chances are evident for Latino youth (Hernandez, 1993).

The harsh conditions faced by many American families, coupled with the lack of government attempts to alleviate them, put poor children in the United States at a distinct disadvantage relative to children in other developed countries. Studies comparing child poverty in the United States to other industrial nations show that things are not getting much better for the majority of poor American children (Rainwater and Smeeding, 1995). Compared with seventeen other industrial democracies in Europe, Australia, and North America, only the poverty rates for children in Ireland and England begin higher than the child poverty rate in the United States. In those countries, however, government assistance programs lift about half of their impoverished children out of poverty. The United States, in contrast, lifts only about 17 percent of its poor children out of poverty, a record that lags behind all the other comparison nations. After figuring in government assistance, the level of child poverty in the United States is 50 percent higher than the next highest country (Australia), twice as high as in Ireland or England, three times higher than in Germany or France, and seven times higher than in Scandinavian welfare sates like Denmark, Sweden, or Finland. Even though the majority of children are better off in those countries to begin with, these governments contribute greater proportionate resources to lift children who are poor up to a minimal standard of living (O'Hare, 1996; Rainwater and Smeeding, 1995).

Most observers agree that the U.S. government could do much more to enhance the well-being of children and families, but various political impediments have blocked efforts over the years. The selections in this chapter focus on what governments in the United States have done in the past, what they are currently doing, and what they might do in the future to assist families and children. In the first selection, Nancy Hooyman and Judith Gonyea define family policy and explain why the United States has been reluctant to adopt an explicit family policy. Using an historical and comparative analysis, these authors observe that countries defining the family as two parents (one of whom is the predominant **breadwinner**) typically lack an explicit family policy. They introduce two concepts that help explain this tendency: an "ideology of familism" and "residualism." An ideology of **familism** assumes that the private family should care for its own—both young and old. In turn, following a principle of residualism, governments become involved in serving families only after a private family has assumed maximum responsibility for the care of family members. Tracing the development of family policy up through the mid-1990s, Hooyman and Gonyea suggest that familism and residualism have produced family policies that reflect and reinforce the traditional male-headed family. They observe that this ideal is used to avoid creating a national family policy with consequent devastating effects for many mothers and children. In presenting feminist alternatives to such approaches, Hooyman and Gonyea argue that gender equality should be used as the primary organizing principle of family policy.

In an excerpt from *Flat Broke with Children,* Sharon Hays continues the analysis of family policy by exploring the cultural logic of welfare reform. Focusing on the historical development and implementation of welfare in the United States from the New Deal's 1935 Aid to Dependent Children through and beyond the 1996 Personal Responsibility and Work Opportunity Act, Hays documents how such legislation reflects contradictory values. The new law replaced the "old" welfare (Aid to Families with Dependent Children—AFDC) with "new" welfare (Temporary Assistance to Needy Families—TANF) and sent a strong message that all mothers must be prepared to leave the home and find paying jobs that will support themselves and their children. Earlier welfare policies assumed that mothers should stay home and take care of their children, but the new law suggests that they should be treated as individual workers rather than as dependents of men in families. At the same time that the law promotes individual self-reliance through paid work, however, it also champions an older set of family values. As Hays notes, the act begins: "Marriage is the foundation of a successful society. . . . Promotion of responsible fatherhood and motherhood is integral to successful child rearing and the well-being of children." The "family values" version of welfare reform has a different set of edicts, including promoting marriage, discouraging divorce, and preventing and reducing the incidence of out-of-wedlock pregnancies. Hays notes how the two goals are meant to work together to discourage women from choosing divorce or single parenthood: "removing the safety net and forcing welfare mothers to work is actually a way to reinforce all women's proper commitment to marriage and family." She calls the two distinct visions of work and family life "the Work Plan" and "the Family Plan." Hays documents how these competing visions are connected to a broader set of dichotomies underlying our culture and politics, and explains how they were transformed into simplified slogans and stereotypes that obscured more difficult dilemmas and disturbing social inequities. Finally, Hays explains why welfare reform should be seen as a response to the widespread employment of mothers and highlights some of the new law's unintended negative consequences for women and children.

The next selection by Sandra Hofferth, Deborah Phillips and Natasha Cabrera turns our attention to childcare, something the Work Plan part of welfare reform encourages, albeit with insufficient funding. These authors begin by reviewing how and why economic hardship and related family stressors negatively influence child development. They then consider research on nonparental childcare, examining its quality, cost, and availability and summarizing the contexts in which parents must make choices about caring for their children while they are working. Mothers are increasingly likely to return to work while their children are still young: About half of all mothers in the United States are back at work within six months of their first birth. Although the Family and Medical Leave Act (FMLA) ensured that parents who took time off could have their jobs back, the law covers only large employers and mandates only twelve weeks of *unpaid* leave. As the authors note, in many European nations, mothers of infants receive paid leave during most of the first year of life and keep their jobs (though they are not working during that period). The high level of employment for U.S. mothers with infants, preschoolers, and school-aged children creates a great demand for organized childcare. Unfortunately, the demand far exceeds the supply, especially for affordable, high-quality childcare. Hofferth, Phillips, and Cabrera discuss various childcare policies and programs, including direct subsidies and tax credits, concluding that families with higher incomes receive higher quality care (on average) than the working poor or middle-income families, with poor families who are eligible for directly subsidized preschool programs like Head Start also receiving higher quality, center-based childcare.

In the final selection, Evelyn Nakano Glenn asks questions about how to create a caring society. She focuses on the relegation of care to the private sphere and to women and discusses how such arrangements are linked to the devaluation of caring work and caring relationships. Glenn draws on political philosophy and feminist theory to suggest that caring work has been excluded from the concept of citizenship, but that it ought to be considered as a public societal contribution comparable to paid employment. By exploring alternate definitions of care, she argues that it should be valued as "real work" and carry the rights and responsibilities of citizenship. Glenn reviews literature on family care and paid care, encouraging us to rethink our taken-for-granted assumptions about who should do the work and whether it should be well paid. With reference to various family policy options, Glenn and other care work proponents call for transforming care work into a publicly valued activity for the good of all men, women, and children.

REFERENCES

Bane, Mary Jo, and P. Jargowsky. 1989. The Links between Government Policy and Family Structure: What Matters and What Doesn't. In *The Changing American Family and Public Policy,* edited by A. J. Cherlin. Washington, DC: Urban Institute Press.

Coltrane, Scott. 1998. Gender and Families. Thousand Oaks, CA: Pine Forge Press.

Edelman, Marion Wright. 1987. *Families in Peril: An Agenda for Social Change.* Cambridge, MA: Harvard University Press.

———. 1989. Black Children in America. In J. Dewart (Ed.) *The State of Black America.* (pp. 63–76). New York: National Urban League.

Hernandez, Donald J. 1993. *America's Children: Resources from Family, Government, and the Economy.* New York: Russell Sage Foundation.

Kamerman, Sheila, and Alfred Kahn. 1997. Family change and Family Policy in Great Britain, Canada, New Zealand, and the United States. New York: Oxford University Press.

O'Hare, William P. 1996. *A New Look at Poverty in America.* Population Bulletin, 51, 2. Washington, DC: Population Reference Bureau.

Rainwater, L., and T. M. Smeeding. 1995. *Doing Poorly: The Real Income of American Children in a Comparative Perspective.* Working Paper No. 127. Luxembourg Income Study, Maxwell School of Citizenship and Public Affairs. Syracuse, NY: Syracuse University.

SUGGESTED READINGS

Bottoms, Bette L., Margaret Bull Kovera, and Bradley D. McAuliff, eds. 2002. *Children, Social Science, and the Law.* Cambridge, UK: Cambridge University Press.

Crotty, Patricia McGee. 1999. *Family Law in the United States: Changing Perspectives.* New York: Peter Lang.

Eichler, Margrit. 1997. *Family Shifts: Families, Policies, and Gender Equality.* Toronto: Oxford University Press.

Folbre, Nancy. 1994. *Who Pays for the Kids?: Gender and the Structures of Constraint.* London: Routledge.

Folbre, Nancy. 2001. *The Invisible Heart: Economics and Family Values.* New York: The New Press.

Harrington, Mona. 1999. *Care and Equality: Inventing a New Family Politics.* New York: Knopf.

Helburn, Suzanne W., and Barbara R. Bergmann. 2002. *America's Child Care Problem: The Way Out.* New York: Palgrave.

Joseph, Lawrence B., ed. 1999. *Families, Poverty, and Welfare Reform: Confronting a New Policy Era.* Chicago: University of Illinois Press.

Koven, Seth, and Sonya Michel, eds. 1993. *Mothers of a New World: Maternalist Politics and the Origins of Welfare States.* New York: Routledge.

Mazur, Amy G. 2002. *Theorizing Feminist Policy.* Oxford, UK: Oxford University Press.

Michel, Sonya. 1999. *Children's Interests/Mothers' Rights: The Shaping of America's Child Care Policy.* New Haven, CT: Yale University Press.

Roberts, Dorothy E. 2002. *Shattered Bonds: The Color of Child Welfare.* New York: Basic Books.

Seccombe, Karen. 1999. *So You Think I Drive a Cadillac? Welfare Recipients' Perspectives on the System and Its Reform.* Boston, MA: Allyn and Bacon.

Wadlington, Walter, and Raymond C. O'Brien. 2001. *Family Law in Perspective.* New York: Foundation Press.

INFOTRAC® COLLEGE EDITION EXERCISES

The exercises that follow allow you to use the InfoTrac® College Edition online database of scholarly articles to explore the sociological implications of the selections in this chapter.

Search Keyword: Family policy. Using InfoTrac® College Edition, find articles that focus on current family policy issues. What are some of the family policy concerns that are being addressed at this time? What proposals for policy change are emerging from present family policy debates? How do these proposals attempt to address the expressed concerns? Are the proposals controversial? If so, what controversies surround the family issues being debated?

Search Keyword: Childcare policy. Use the search phrase *childcare policy* to find articles that address childcare policy proposals in the United States. What is the current policy of the federal government regarding childcare? What controversies surround this issue? Why is childcare seen as a pressing issue at this time? What proposals for childcare policy appear to be most popular? Which appear to be least popular? How are childcare proposals related to welfare reform?

Search Keyword: Family friendly workplace. Look for articles that discuss the need for family friendly workplace policies and laws. What does it mean for a workplace to be "family friendly"? What policies and laws are proposed that might create this type of workplace environment? Are there already any policies in place that contribute to a family friendly workplace environment? If so, how are they working? What are the controversies that surround such policies? Do the articles that you find propose any solutions to this issue? If so, what are they?

Search Keyword: Family Medical Leave Act (FMLA). Use InfoTrac® College Edition to search for articles that address the FMLA. What does this policy entail? When was it enacted? What businesses and employees are affected by this act? What family concerns was it supposed to alleviate? Has it been successful in alleviating those concerns? What concerns does it not address? What changes to the act are proposed? Would these changes make it more effective?

Search Keyword: Welfare reform. Search for articles, using InfoTrac® College Edition, that deal with welfare reform. How has social welfare evolved over the course of the late twentieth and early twenty-first centuries? What is the current state of welfare provision in the United States? What welfare provisions are targeted for revision? What welfare recipients are targeted for changes in funding? What effects will welfare reform be likely to have on these recipients? How do the proposed revisions reflect the present state of family values rhetoric? What implications does the proposed welfare reform have for different family types in the United States?

42

Defining Family Policy

NANCY HOOYMAN AND JUDITH GONYEA,
1995

Defining *family policy* is a difficult task. No country represents family policy by a single policy; instead, a collection of policies is directed at the family (Kamerman & Kahn, 1989). Moreover, unlike many public policy areas such as education or health care, family policies lack clear boundaries. The family is the central institution in people's lives; thus, in the broadest sense, anything the government does that affects families can potentially be construed as family policy. For example, transportation and environmental policies, such as where communities build roads or locate waste sites or how government subsidizes public transportation services, directly affect the quality of families' lives. In this chapter, however, our focus will be restricted to government actions specifically designed to affect families' lives. In defining family policy, we draw heavily on the work of two leading pioneers in this arena, Sheila Kamerman and Alfred Kahn (1978), who distinguish between *explicit* and *implicit* family policy and offer three alternative definitions.

Countries that develop explicit family policies view the family as the key unit of analysis. Countries that develop implicit family policies institute policies and programs affecting families, but the individual members of the family are the typical unit of analysis (i.e., children, the aged, the physically disabled, the juvenile delinquent). Kamerman and Kahn (1978) note the following:

> The explicit-implicit dimension clearly is interesting, reflecting as it does whether a society has enough internal homogeneity to announce objectives for this most intimate of institutions, enough power to do something about such objects, and a value system which supports such actions. (p. 477)

Countries such as the United States and the United Kingdom, in which rhetoric about the traditional family (two parents, one breadwinner) predominates, are more likely to reject a national family policy, as compared to countries such as France, Sweden, or Norway that support the concept of family diversity.

Kamerman and Kahn (1978, 1989) also distinguish among three dimensions of family policy. Family policy as a field, they argue, focuses on defining certain objectives regarding the family, such as achieving healthier children, less financial burden attached to raising children, or greater equality for women. Based on the specified objectives, the parameters of family policy may include such domains as income-transfer programs, tax policy, health, housing, personal social services, and education. Family policy as a social instrument encompasses population policies, labor market policies, and social control policies (Kamerman & Kahn, 1978). Recent government interest in developing policies to support families in providing care to relatives with disabilities represents an example of family policy as an instrument of social control (Kamerman & Kahn, 1989). Despite women's changing social roles, Kamerman and Kahn (1978, p. 7) suggest that the "assumption here is that family care is cheaper and may be more humane." Thus, government should actively encourage family caregiving (usually meaning women's caring) to prevent the transfer of responsibility to the state. Finally, family policy as a perspective assumes that debate on all public policies should include an exploration of their potential consequences or effects on families' well-being. This type of investigation, similar to the environmental field, is often referred to as "family impact analysis" (Kamerman & Kahn, 1989).

For our discussion, it is also important to emphasize that, across countries, family policy is largely defined as benefits directed toward families and children. Although Wilensky (1985, p. 56) views family policy as "a wide umbrella of providing shelter across the life span," most policy makers, family advocates, and social scientists exclude adult and elderly dependents from family policy discussions. Analysts often justify this perspective by arguing that the family's economic situation is most likely to be strained when children are young (Kamerman & Kahn, 1989). They tend to ignore the extent and cost of caring for dependent adults, particularly the cost to women. Moreover, this socially constructed dichotomy may add fuel to the current conflict presumed to exist between generations (Binney & Estes, 1988; Quadagno 1989; Wisensale, 1988). The focus on the nuclear family obscures the widespread phenomenon of intergenerational support—support flowing in both directions across generations. It also ignores commonalities in the caring experience across the life span, promotes special-interest mentality (i.e., "kids versus canes"), and discourages the adoption of "a principle of life course entitlement" to basic human needs (Binney & Estes, 1988).

UNDERSTANDING AMERICA'S LACK OF AN EXPLICIT FAMILY POLICY

Cultural ideologies and values provide the context both for how countries define problems and how they seek solutions. As Clark (1993, p. 33) underscores, "Ethical principles guide our development and implementation of policies and programs. . . . The moral dimension of public policy . . . influences our perception of acts, our loyalties, and our assumptions about human nature." Public policy is inevitably tied to empirical data; it is our interpretation of these data that is key. Empirical data become the instrument to advance a particular social agenda based on an ideological understanding of the nature of family and state.

In attempting to understand the United States' lack of an explicit, coherent, and comprehensive family policy, analysts have often used a cross-national perspective. As Meyers, Ramirez, Walker, Langton, and O'Connor (1988, p. 139) note, "modern societies differ in their construction of the public, and in particular the way in which the state becomes linked to private life." They distinguish between three forms of modernization: organic corporatism, communal corporatism, and liberal individualism (for a detailed discussion of this typology of modernization, see Meyers et al., 1988). Many Latin American countries are models of organic corporatist societies, in which the traditions of natural law and the church dominate, and individualism is deemphasized. The family is viewed as a natural entity worthy of public protection. The state does not penetrate the internal side of family life and therefore intrafamilial conflicts are invisible to public organizations. "The internal arrangements of family life are determined by what is conceived to be natural or religious. . . . The conflicts between women and children as putative individuals are handled in more traditional ways . . . and do not much enter the public agenda" (Meyers et al., 1988, p. 144). The state concerns itself mainly with women and children who are not in families.

Northern European communities offer examples of communal corporatism. Within these societies, "modernization is built on a perception of the state as a national community, rather than a nation of free individuals" (Meyers et al., 1988, p. 144). Instead of being hidden within the family, individuals are viewed as having rights and responsibilities in the public sphere based on one's own personal status as women, men, children, the aged, homosexuals, etc. Public organizations recognize the natural characteristic of conflicts between groups (based on differing capacities and needs) and expect that they "may be called upon to clarify or redefine mutual rights and obligations" (Meyers et al., 1988, p. 145).

The United States represents the liberal individualistic society. The state, the economy, culture, and religion all emphasize the concept of individual choice and freedom. Ironically, this society overlooks women as individuals by presuming that women will assume traditional roles and caregiving burdens, thereby limiting their choices. Meyers and his associates argue that "liberal society is built on the myth of the individual who is free to choose and act—free beyond age, gender and family relations" (Meyers et al., 1988, p. 146). They contend that familial conflicts are redefined as public issues when persons assert their individual rights against situations that constrain them, such as marriage, fatherhood, motherhood, and filial obligation.

In his comparison of the United States' and Canada's domestic policies toward older families, Clark (1993) notes that both countries possess an ideology of familism, that is, an assumption of the primacy of families in meeting the care needs of their members. Canada's emphasis on collectiveness or a sense of community, however, tempers familism; whereas America's stress on individualism heightens familism (and reduces options for women as individuals). Clark stresses that this preeminence of the individual is ingrained in our national identity. America's founding document, the Declaration of Independence, emphasizes "life, liberty and the pursuit of happiness." In contrast, Canada's founding legislation, the British North American Act, stresses the importance of "peace, order and government." He argues the following:

> [The United States] particularizes and compartmentalizes social policies along lines of individual or static group-based need, rather than seeing public programs as responding to changing life course needs across the entire society . . . [Thus] the United States has spawned the generational equity debate. (Clark, 1993, p. 34)

Canada's emphasis on a sense of community responsibility—a concern for the welfare of others versus a concern directed mainly toward self—makes the development of more universal policies such as a national health care system much more possible.

As we have previously noted, it is these values of individualism and familism that form the basis for the residual approach to public policy in the United States. Residualism—meaning the state becomes involved only after the family has assumed as much responsibility as possible—especially serves the current federal government's goal of cost containment. Faced with an increasing federal deficit and growing interest in cost effectiveness, "familism offers a convenient justification to cloak the real reasons for withholding support" (Clark, 1993, p. 27). As feminists stress, those most affected by such withholding of government support are women. It is women who, in the great majority of cases, are the caregivers; therefore, legislation strengthening family obligation restricts women's lives.

History, social and cultural traditions, and economic and demographic circumstances shape each country's approach to family policy. The interest of Western European countries in family policy came about in part due to the heavy loss of life during World War I. Governments created policies and programs that made having children less of a financial burden to families. Following World War II, more financial aid for families was seen as "part of a push for social equity by socialist and church-related parties represented in post-war governments" (Aldous, 1980, p. x.). Thus, in Europe the concept of family policy has an extensive history of being employed as both income redistribution policy and population policy—a means to promote higher fertility rates and larger families. Resurgence of interest in family policy in Europe reflects the recent "demographic panic" about declining birth rates in many European nations (Kamerman & Kahn, 1989). Current European fertility rates are significantly lower than the United States' 2.1 births for each woman. Only one European country, Ireland, has a similar fertility. Spain and Italy (1.2 births) presently have the lowest fertility rate of the European Union (Le Monde de l'education, 1994). Thus, European family policy is viewed as offering families incentives for childbearing and child rearing.

This pronatalistic value underlies most countries' family policies that feminists have argued are restrictive of women's choices. Most European family policies were instituted to promote reproduction. "They do not give people choices of whether or not to have children. They seek to provide supports for the family roles of women as mothers and men as fathers" (Spakes, 1989a, p. 612). Of European nations, only France includes a couple without children in its definition of family. All other countries with an explicit family policy have in their definition of family the presence of at least one parent and at least one child (Kamerman & Kahn, 1989).

Although the United States has not experienced concern with fertility decline or population size as has Europe, the pronatalistic value has been incorporated into many of our social policies. "AFDC (Aid to Families with Dependent Children) provides an example. . . . The program is clearly pro-natalistic. No forms of financial support are available for low-income women without children" (Spakes, 1989a, p. 616). In the United States, interest in family policy emerged primarily during the mid-1970s in response to growing public concern about the disturbing number of children living in poverty.

THE EVOLUTION OF THE FAMILY POLICY DEBATE IN THE UNITED STATES

The concept of a national family policy was not broadly promoted in the United States until the Carter-Mondale administration (1976 to 1980). Drawing on the experiences of European and Nordic nations, the Carter administration suggested that the federal government should assume a more proactive role through the development of comprehensive policies and programs to support all families in performing their functions. Increasing public concern about the state of the American family focused primarily on three issues (Bane & Jargowsky, 1988):

- The large number of children who live in poverty. The poverty rate for children is greater than any other age group in this country. More than 20% of all children, and nearly 25% of children under 6 years of age, live in poverty in the United States. In fact, children are worse off than they were two decades ago when America undertook its "war on poverty."
- Children living in single-parent families are at greatest risk for poverty, and the percentage of children who spend at least part of their lives in single-parent families continues to increase.
- Rising divorce rates and greater acceptance of heterosexual and homosexual cohabitation are leading many adults to fail either to form, or to stay in, "families" in the traditional definition. This situation may reflect the declining importance of the institution of family in our society.

Bane and Jargowsky (1988) note that there is general agreement on the first point, that poverty among children is a problem, but less agreement on the second two points: whether single-parenthood is a problem or whether the family is in decline. Feminists argue that the single-parent family structure per se is not the problem. Instead, the association of poverty with single-parent families reflects the following: (a) only with both parents working are families able to enjoy the same standard of living as single-earner families a generation ago; (b) the inability of the majority of women to earn an adequate family wage due to gender-based occupational segregation and pay inequity; and (c) in many cases, fathers' failure to assume economic responsibility for their children. Feminists also emphasize that the plurality of family forms does not mean that essential family functions are less important than they were a generation ago. Critical functions of the family remain: economic security; nurturance, affiliation, and emotional support; socialization and education; and procreation.

These ideological disputes dominated the 1980 White House Conference on Families. Convened by President Carter to address the growing problems of children and families in America, the Conference is best remembered for the polemics between conservatives and liberals on the definition of the family, the passage of the Equal Rights Amendment, and the issue of abortion and reproductive choice for women. Whitehead (1993, p. 48) notes that "no president since has tried to hold a national family conference." During the 1980s, as the Reagan administration entered the White House and the Republican party took control of the U.S. Senate, the notion of a national family policy disappeared from debate. The Right carried the dual messages of being "profamily" and "antigovernment" (Langley, 1991). Under Reagan, the emphasis was on eliminating government programs viewed as "undercutting the family" (Seaberg, 1990). For instance, Reagan appointed administrators to the Department of Health and Human Services who were antiabortion and opposed to recognition of adolescent sexuality. As a result of these appointments, new federal regulations were introduced under Title X requiring federally funded family planning agencies to notify both parents or legal guardians within ten days of giving birth control technology to any minor.

The Reagan administration also instituted major cutbacks in the levels of benefits and the coverage of income maintenance programs. Especially hard hit were the means-tested programs for those in poverty, such as AFDC, food stamps, and Medicaid. During Reagan's first term, the total reduction in income maintenance programs was $57 billion, representing 33% of all federal cuts, in programs that accounted for only 10% of the federal budget (Abbott & Wallace, 1992). The changes in AFDC that resulted from the Omnibus Budget Reconciliation Act (OBRA) of 1981 made life worse for most women and children on welfare. Particularly devastating were the diminution of the work-incentive provisions and the imposition of limits of $75 per month for work expenses and $160 per month for child care. As noted by Miller (1990, pp. 36–37), "restricting [AFDC] eligibility to the third trimester of pregnancy is a punitive measure directed toward single mothers, and is ironic in view of the Reagan administration's position that a fetus is a person from the moment of conception."

A similar philosophy continued with the Bush administration. Although seeking a "kinder, gentler" society, Bush's message of "a thousand points of light" emphasized the role of family, church, charitable organizations, volunteers, and business—as opposed to government's responsibility—for the well-being of citizens. For example, although the Family Support Act of 1988 mandated (a) the establishment of procedures to assure that center-based child care met health and safety standards and (b) the development of guidelines for family day care, the Bush administration's Department of Health and Human Services' regulations emphasized, instead, the use of free informal sources of care (Miller, 1990).

The . . . Clinton administration stands apart from the previous two administrations in suggesting that the federal government should play a greater role in safeguarding and promoting the well-being of American families. There is symbolic importance in the fact that the first Act signed into law by President Clinton was the Family and Medical Leave Act (FMLA), which offered job protection to workers requiring short-term leaves from their jobs to care for ill or disabled family members. Moreover, the number one priority of the Clinton administration has been health care reform, an issue of great concern to many American families. It is important to note, however, that Clinton has chosen not to structure his reforms, whether in health, housing, welfare, or education, around a family policy framework. This decision may reflect a desire to avoid some of the intense debate that surrounds the changing American family and the role of government in family life.

> Family policy in liberal circles is understood to mean economic assistance and social services that will put a floor under family income and lead the way to self-sufficiency. . . . For [conservatives], family policy appears to involve the use of national resources to legitimize behavior not concomitant with behavior typical of the American family. Right-minded national policy should reinforce traditional American patterns, but not abide deviations that smack of irresponsibility. (Steiner, 1981, p. 17)

Such "deviations that smack of irresponsibility" presumably include nontraditional forms of families as well as any choices women make away from traditional caring roles.

THE NEW RIGHT'S PERSPECTIVE OF FAMILY AND SOCIAL POLICY

The 1980s represented a decade in which the appropriate role of government was being reevaluated, not just in the United States, but globally (Hula, 1991). Eastern Europe and the Soviet Union experienced the collapse of socialist and communist governments, and in United Kingdom and the United States, dominant themes were government retrenchment and privatization. Hula (1991) suggests the following:

> Given the importance of family in the framing of social welfare policy, it is hardly surprising that current political debate often focuses on the effects of social policy on the family. . . . It is apparent that family is a symbol of fundamental importance in American politics and it is equally clear it offers no guide to action. (p. 3)

Both those who argue for less federal intrusion into families lives and those who advocate for increased public spending to strengthen the family view themselves as "profamily." Central to the debate about social policy is whether the traditional nuclear family should serve as the guiding symbol.

The New Right idealizes the traditional patriarchal nuclear family, a family composed of a male breadwinner, an economically dependent female homemaker, and socially and economically dependent children. The appeal of the New Right is one of nostalgia for a time when marriage was for a lifetime, children obeyed their parents, crime was low, and families felt safe in their neighborhoods. Feminists argue that what is ignored by the New Right is that for many families, especially working-class families, immigrant families, and families of color, this image was never a reality. Throughout U.S. history, working-class women and children were frequently in the labor force, often employed long hours in unsafe or unhealthy environments. Working-class women and women of color have seldom had the luxury of opting out of paid employment. These families have often lived in substandard housing, in neighborhoods lacking adequate sanitation, or in high-crime areas.

Gender roles, defined by the ideology of separate spheres, are central to the Right's view of the family:

> A man's responsibility to his family is best met by his success in the market, his ability as a wage earner to support his wife and children; a woman's worth is measured by her

> dedication to her role as wife and mother. (Cohen & Katzenstein, 1988, p. 26)

Neoconservatives Brigitte and Peter Berger (1983), as well as New Right author George Gilder (1981), emphasize the importance of the bourgeois or traditional heterosexual nuclear family to both capitalism and democratic order. Gilder argues that married men contribute more to society by virtue of their stable work patterns than do bachelors, who dissipate their energies in nonproductive sexual and economic concerns. The Bergers suggest that the bourgeois family's promotion of individualism of male members is conducive to entrepreneurial capitalism and democracy. Although they maintain that women will need to decide on their own priorities, the Bergers express a strong wish that women come to recognize that "life is more than a career and that this 'more' is above all to be found in the family." They warn women that "they should not expect public policy to underwrite and subsidize their life plan" (Berger & Berger, 1983, p. 205).

Understanding the New Right's views on gender relations and the family is crucial to understanding their argument that reducing the role of the state is the way to solve economic and social problems.

> The family is seen as the linchpin of New Right economic and social policies because, within this discourse, men are seen as the "individuals" of economic liberal thought whereas women are seen as outside the market place, a part of the dependent family, not citizens in their own right. (Abbott & Wallace, 1992)

The Right sees the family as the cornerstone of society, crucial to maintaining stability. It views the diversification of family forms as the moral and economic decline of the family and, in turn, defines this as leading to the moral and economic decline of society. Similarly, the Right sees legislation allowing abortion, instituting easier divorce procedures, permitting sex education in the schools, and mandating equal opportunity in the labor market, along with welfare programs that offer benefits to never-married or divorced single parents, as sabotaging the traditional nuclear family and validating alternative family forms and lifestyles. The New Right views the breakdown of family life in its idealized form as the cause of the high rates of criminal activity, substance abuse, school drop-outs, and unemployment.

Whitehead (1993) begins her *Atlantic Monthly* article, "Dan Quayle Was Right" with this statement:

> The social-science evidence is in: though it may benefit the adults involved, the dissolution of intact two-parent families is harmful to large numbers of children. . . . Family diversity in the form of increasing numbers of single-parent and stepparent families does not strengthen the social fabric but, rather, dramatically weakens and undermines society. (p. 47)

For evidence, she cites poverty statistics such as that children in single-parent families are 6 times more likely to be poor and they are also likely to stay poor longer. The empirical data are correct, but Whitehead and other conservatives err in their interpretation. They falsely assume that correlation equals causality (Cowin, 1993). It is beyond dispute that single-parent families are at greater risk for poverty, but the Right chose to "blame the victim" and ignore the synergy of social forces that place these families at risk. The Right fails to consider alternative causal hypotheses for the economic and social stresses experienced by these families, such as the worldwide economic upheaval, the decline in real wages for American workers, gender inequality in the labor market, and social policies that penalize nontraditional families. They do not see the connection between women's traditional caring roles and the poverty of women and children.

Within New Right thinking, Abbott and Wallace (1992) identify two conflicting ideologies: *economic liberalism* and *traditional authoritarian conservatism*. (For an in-depth feminist analysis of the New Right, see Abbott & Wallace, 1992.)

Economic liberalism argues a laissez-faire position, in which government should not intervene in the economy nor in individuals' lives unless they endanger the rights of others. In contrast, traditional authoritarian conservatism emphasizes the need for government to reinforce traditional patterns of authority (i.e., institutions of family and church) in order to protect a strong central state. They note that these contrasting ideologies do not conflict in practice. For example the economic liberal objectives of reducing public spending for welfare implies traditional roles for women and the family (Abbott & Wallace, 1992).

Much of the writing of New Right intellectuals, George Gilder, Charles Murray, Martin Anderson, and Lawrence Mead, has criticized income maintenance programs, particularly AFDC. Common to most of their writings is the view that the welfare programs of the Great Society "undermine the moral responsibility of individuals by removing their incentives to work and support themselves and by discouraging marriage" (Abbott & Wallace, 1992, p. 97). They reject the liberal view that poverty is derived from structural disadvantage and argue instead that it arises from individuals' own behaviors. In establishing a behavioral basis for poverty, Murray and Mead distinguish between the deserving or respectable poor and the undeserving or underclass of poor. They view the underclass as complacent about illegitimacy, inadequate parents, and lacking a strong work ethic (Mead, 1986; Murray, 1984, 1988). Moreover, they assert that government programs have fostered dependency in the underclass. Mead (1986, p. 38) suggests that "especially since the late 1960's on, millions of female-headed families and other low-income groups signed up for AFDC, food stamps, and other programs rather than continue to struggle to support themselves without assistance." Interestingly, Abbott and Wallace (1992) note that Murray and Mead view the problem similarly but they would seek different solutions. Murray sees the solution as getting, or keeping, women married and supported by a male wage-earner. Mead proposes welfare mothers enter the workforce, although he advocates that they should furnish their own child care arrangements rather than rely on government to do so. Abbott and Wallace (1992, p. 96) conclude that "neither position is particularly helpful to women, especially poor women."

The New Right's attitude toward care of America's elderly is also based on a mythical past in which families supposedly lived in multigenerational households (Ford, 1991). Gerontologists have consistently found, however, that within the United States, multigenerational households have never been the standard. In the past, fewer persons survived to old age, and those who did most often lived in a two-person household with their spouse. It is the normative preference of older persons, even in widowhood, to maintain a household independent from their children (Daniels, 1982; Treas, 1975; Troll, Miller, & Atchley, 1979). Members of the New Right argue that both the responsibilities and costs of long-term care should be shifted back to the family rather than incurred by the state. "Family responsibility, therefore, has become the central underpinning of current long-term care policy implementation and formulation. . . . The ideological constructs of the New Right exert tremendous influence on current policy development" (Ford, 1991, p. 100). "Family," as used here, is essentially a euphemism for women, who carry most of the burdens and costs of caregiving.

Filial responsibility—the concept that the state is not responsible until children (in practice, female children) have made the maximum effort—dominates the New Right's philosophy regarding long-term care. Evidence suggests, however, that family responsibility laws regarding elder care are neither beneficial nor cost-effective. Forty states currently have, but do not enforce, laws requiring financial support of elderly parents (Hagestad, 1987). In general, these family responsibilities have not produced substantial revenues for the states, have been found to be time-consuming and difficult to administer, and have entailed substantial administrative and judicial costs (Bulcroft, Leynseele, & Borgotta, 1989; Schorr, 1980). Guilliland (1986, p. 33) posits that, "the impact of filial responsibilities will be greatest on adult children who, although not denying their filial responsibility, must impoverish themselves and their children to make a financial contribution to their parents' care." She also notes that elder abuse is more common among families who are forced into elder care, and suggests that, on a societal level, mandated filial responsibility will only heighten intergenerational tensions and conflicts (Guilliland, 1986).

FEMINISTS' PERSPECTIVES OF FAMILY AND FAMILY POLICY

Feminists argue that the traditional nuclear family should not be the guiding metaphor for American social policy. For more and more families, the traditional nuclear family is simply no longer a reality. Focusing only on family structure (and not family roles), a 1994 U.S. Census Bureau Report confirms that only 50% of American children now live in the traditional nuclear family and that there is considerable variation by race and ethnicity. As we have seen, approximately 56% of Caucasian children live in a traditional nuclear family, but only 38% of Hispanic children (who can be of any race), and 26% of Black children do so (*Boston Globe,* 1994).

Despite demographic and social changes such as the diversity of family structures and women's entry into the labor force, policy makers have been reluctant to modify existing policies or to create new ones because such proposals are often perceived as being "antifamily."

> While there is no doubt that the traditional family has become less common, it continues to have enormous normative power. That is, such families are seen as being the ideal, the model to which reasonable citizens strive. Such an argument underlies the demands that social policy continue its traditional posture toward families even as its central model becomes a statistical rarity. (Hula, 1991, p. 3)

It is questionable whether public policy can be used to strengthen the traditional nuclear family. Bane and Jargowsky (1988) contend that although family support and welfare policies may substantially affect quality of life, the evidence suggests such policies will not bring about much change in family structure. Their analysis of European countries' pronatalistic policies and fertility rates reveals that even substantial policies had minimal effect. They suggest that "both conservatives and liberals should find this analysis, if they believe it, troubling," but also insist that "this conclusion does not imply that policies for families are of no importance or should not be pursued. . . . Improving the well-being of families and the general environment to support families also is important" (Bane & Jargowsky, 1988, p. 245).

Feminists maintain that if policies are to truly support women, they cannot be based on the system of patriarchy that currently underlies our social welfare programs—patriarchy being a system of sexual hierarchical relations maintained by law, culture, and societal norms in which masculine dominance is upheld (Miller, 1990). Although first-wave feminists of the 19th century fought for women's emancipation, women's rights to choose marriage or a career, and to vote, they did not challenge women's place in the family. Second-wave feminists of the 1960s and 1970s, however, focused on sexual politics in both the public and private spheres. Rapp (1978) makes this assertion:

> One of the more valuable contributions of feminist theory has been its effort to "deconstruct" the family as a natural unit, and to reconstruct it as a social unit, as ideology, as an institutional nexus of social relationships and cultural meanings. (p. 280)

Women's roles in the family and the consequences of a male-dominated power structure for women are central to these discussions. Simone de Beauvoir (1968) conceptualized woman as "Other" to describe women's second-class status in a patriarchal society in which man is "Subject" or "Absolute." She noted that within the economic sphere, women's lower standing resulted in men holding the better jobs, getting higher wages, and having greater opportunity for success than their female competitors.

Miller (1990, p. 23) uses the concept of *patriarchal necessity*—"the need among the collectivity of men to separate the sexes and devalue and control women"—as the driving force of the social welfare system's treatment of women. The Social Security Act of 1935, the foundation of our national social welfare system, is predicated on a notion of family in which a breadwinner father and a homemaker mother are committed to each other in a lifelong marriage. Thus, women who find themselves living outside of marriage—whether from personal choice, divorce, or widowhood—are often penalized for these circumstances. Women who are not under the protection or control of men through marriage are generally viewed as falling under the protection or control of the state.

AFDC (originally entitled Aid to Dependent Children), for instance, was enacted as part of the Social Security Act to be a small, temporary program for widows with young children. For the first 20 years, the majority of recipients were widows; however, by the 1960s the vast majority of recipients were divorced or never-married women with children (Segal, 1989). The shift in the AFDC population led to a change in public opinion. Rather than being seen as the "deserving poor," AFDC mothers were blamed for their own condition and stigmatized as amoral, lazy, and unsuitable parents. AFDC is now regarded as, at best, a subsistence program that keeps women and children in poverty.

Social Security's Old Age and Survivor Insurance Program (OASI, but commonly referred to as Social Security) also may penalize older women who are alone. Because women usually earn lower wages and have more interrupted work histories (due to caregiving) than men, the majority of older women choose a Social Security benefit based on their spouses' work career rather than their own in order to receive a larger benefit amount. Divorced women, however, must have been married to their ex-spouse for a minimum of 10 years in order to be eligible for a spousal benefit. Moreover, a divorced woman cannot receive payments until her ex-partner (if living) begins to draw his benefits. Even widows, who are often viewed as more deserving than the divorced, may suffer penalties for outliving their spouses and being unattached. For instance, widows whose husbands die prior to their intended retirement (usually age 65) do not have those "lost-income years" computed into their spousal benefit rates and, thus, receive smaller payments. In essence, women of all ages—but especially poor women—who live outside marriage may face negative consequences for their choices or circumstances.

Because women provide most of the caring in the family—caring that has been undervalued and relatively invisible—it would seem at first glance that the concept of a national family policy would be readily embraced by feminists. Would not women have the most to gain from a comprehensive and coherent national family agenda? There is, however, considerable caution among feminists. Dornbusch and Strober (1988) note the following:

> Among proponents of feminism, there are some who are suspicious of the motives of those who have introduced familial language into the national agenda. Is concern for the family only a ruse, a device for transmuting the terms of national debate in ways that result in a vote for stability in the relations between men and women? (p. 4)

Within the family policy debate, the issue that most divides feminists and conservatives is the matter of adult roles—specifically women's autonomy—not children's needs (Cohen & Katzenstein, 1988).

Spakes (1991, p. 23) suggests "the idea of national family policy and what it seems to promise is in ways seductive, particularly to working mothers. However, proposals that have been offered thus far have varied considerably." A number of American feminists have looked toward Sweden, with its national priorities of family welfare and gender equity, as a model for the United States

(Acker & Hallock, 1992; Rosenthal, 1994). Sweden's family policy links both family and labor objectives. Concerned with its declining population in the 1930s, Sweden developed policies specifically to boost its birth rate. Similarly faced with a national need for more workers after World War II, Sweden had two choices—either seek immigrants or recruit women. Because of a concern for cultural homogeneity, the former strategy was viewed as undesirable. Instead, policies were instituted to encourage Swedish women into the labor force (Spakes, 1992).

> Thus, what is commonly thought of as the Swedish "family policy" is actually a combination of universal, pro-natalistic and economic goals. These policies include: (1) tax-free children allowances; (2) parental insurance which entitles parents to a leave of absence up to 12 months for childbirth, plus leave to care for a sick child (60 days per year), and 2 days leave per year for daycare or school visits; and (3) day care, subsidized by municipal and national government and provided for an income-based fee. (Spakes, 1992, p. 47)

These policies were viewed as encouraging women both to bear children and to participate in the labor force. It was assumed that by providing women with these public supports for their family responsibilities, they would no longer be disadvantaged in the labor market relative to their male counterparts. Has that occurred—has gender equity in the labor market been achieved in Sweden?

Swedish women have moved into the labor force in slightly greater numbers than their American counterparts. Swedish women represent 48% of their country's entire workforce compared to America's 40% figure. Yet Sweden's labor market is even more gender-segregated than the United States. As in America, Swedish women have primarily transferred their reproductive functions or caring work into the paid labor market and are predominately employed as nurses, teachers, child caretakers, home helpers, kitchen staff, cleaners, secretaries, and shop assistants (Rosenthal, 1994; Spakes, 1989a). Moreover, 43% of Swedish women (as compared to only 6% of Swedish men) work part-time, an average of 26 hours per week. Swedish women who work full-time earn 78% of men's wages, but because so many work part-time, they earn only 37% of total wages (Rosenthal, 1994). Despite women's high level of labor force participation and the presence of universal social welfare benefits, such as the children's allowance, because of their pronounced concentration in low-paying female occupations coupled with part-time employment, 40% of all single-mother families receive means-tested public assistance. Spakes (1992, p. 50) concludes "these supports have not created economic independence for women . . . because many women are now dependent on the public, rather than the private patriarchy."

Still, women and children fare better in Sweden than in the United States. Sweden's poverty rate among children is 5% compared to approximately 20% in the United States (Smeeding, Torrey, & Rein, 1988). This low poverty rate is achieved through a combination of universal programs such as health care, pension rights, children allowances, and state-advanced child support payments (for noncustodial parents) along with means-tested programs such as public assistance and housing subsidies. Single-parent families are more likely to receive public assistance and housing subsidies as well as priority for publicly supported child care, for which they pay lower fees (Rosenthal, 1994). Public assistance in Sweden is required to provide recipients with a "reasonable standard of living"; they do not live in poverty as do their American counterparts on public welfare.

PROMOTING A FEMINIST FAMILY POLICY

Feminists maintain that, contrary to the alarm of conservatives and the New Right, the family is not declining; rather, it is changing (Stacey, 1993). Feminists argue that the family as an institution is resilient, precisely because variation allows it to adjust to changing social conditions (Cohen & Katzenstein, 1988). Social policies must address the needs of all families, recognize the diversity of family forms, and become more flexible in application. The traditional family can no longer be the central model for social policy, nor can the traditional assignment of women to private and unpaid labor in the household go unquestioned. These assumptions have only served to constrain women to marginal, second-class roles in the labor market (Hula, 1991). Feminists caution against the politically popular term "family values," used by conservatives to imply that (a) families should care for their own and can do so best with women doing the unpaid caregiving, (b) the "good family" is a traditional family, and families who do not fit this model are the cause of current social ills, and (c) that upholding family values means maintaining a separation between family and state, private and public (Rosenthal & Hendricks, 1993).

Feminists contend that family policies in the United States should not be based on pronatalistic and patriarchal concepts of the family. Spakes (1989a) emphasizes the following:

> The best family policy for all families would, in fact, be a woman's policy: one that is not paternalistic, patriarchal, or exclusively pronatalistic, one that is not class biased or preferential, and one that gives all women real choices about their lives. (p. 616)

How do we achieve this? First, as noted earlier, we must acknowledge that the personal is political. It is not possible to separate the public and private spheres of life or of responsibility; the home and the state are intertwined and interdependent (Spakes, 1991). We must, therefore, strive for gender equality both in the family and in broader social institutions.

Scanzoni (1982, 1991) argues that family policy can be used to promote egalitarian relationships in the family and the worth of androgyny in sex roles. He maintains that men should be expected to be efficient house-holders and effective parents if they choose parenthood.

> The conventional family model relieves men of primary responsibility to households and for children. That model is reflected . . . in part by child custody awards that generally go to women and also by the difficulties of collecting child support payments from men who sense minimal involvement in, and influence over, the offspring they have sired. (Scanzoni, 1991, p. 21)

He criticizes conservatives who "argue that children are so extraordinarily vital and then fail to assign shared parenting to men by retaining women as primary parenting agents" (Scanzoni, 1991, p. 21).

The pronatalistic values that underlie family policy must be removed to promote gender equality in the family. Family policy must advance women's right to reproductive choice and freedom of sexual expression. Reproductive choice means not only when but also whether they choose to parent. Freedom of sexual expression notes women's right to choose heterosexual or homosexual relationships, and to parent within lesbian relationships (Spakes, 1989b).

A life course perspective must also be incorporated into family policy, recognizing that the caring role extends beyond parenting to include care for such family members as adult children, siblings, spouses, parents, and even grandparents. Children's advocates have successfully argued that children's needs and interests can best be understood within the context of the family or through a family policy approach, but similar gains have not been made by advocates for adults and elders with disabilities. Thus, within the policy arena, younger families continue to be pitted against older families, fueling intergenerational tensions. Feminists note that it is society's extension of the culturally appointed nurturing role of mother that leads women to care for other family members. Moreover, it is women's family obligations that restrict their opportunities in the market-place and promote their economic dependence across the life span, and in turn, contribute to their poverty in old age.

A key objective of family policy must be promoting women's economic autonomy. "It makes little sense to speak of the well-being of women and children (as well as men) unless and until gender makes no difference with respect to adult economic autonomy" (Scanzoni, 1991, p. 21). The fundamental patriarchal nature of industrial society that promotes gender inequality must be changed. Money and power are distributed through the production system, which is dominated by men, whereas women remain tied to the reproductive system (Spakes, 1989a). Despite the remarkable rise in women's labor force participation, they have primarily transferred their reproductive or caregiving roles into the paid sector. Gender-based occupational segregation and pay inequity continue to dominate the labor market. In order for women to achieve economic autonomy, family policy must encompass the issues of equal opportunity in education and employment, pay equity, and comparable worth.

In conclusion, feminists have been charged by conservatives as being antifamily. Yet, as Cohen and Katzenstein (1988, p. 29) emphasize, "this antifamily charge is basically a slogan that muddles rather than clarifies the true political issue. The real debate is over women's autonomy within and outside the family." Feminism and family policy are not intrinsically incompatible. Rather, gender equality must be an organizing principle of family policy.

SELECTED REFERENCES

Abbott, P., & Wallace, C. (1992). *The family and the new right.* Boulder, CO: Pluto Press.

Acker, J., & Hallock, M. (1992, March). *Economic restructuring and women's wages: Equity issues in the U. S. and Sweden.* Paper presented at the Women, Power and Strategies for Change Seminar, New York University, NY.

Aldous, J. (1980). Introduction. In J. Aldous & W. Dumon (Eds.), *The politics and programs of family policy* (pp. ix–xix). Notre Dame, IN: University of Notre Dame Press.

Bane, M. J., & Jargowsky, P. A. (1988). The links between government policy and family structure: What matters and what doesn't. In A. J. Cherlin (Ed.), *The changing American family and public policy* (pp. 219–261). Washington, DC: Urban Institute.

Berger, B., & Berger, P. L. (1983). *The war over the family: Capturing the middle ground.* New York: Anchor/Doubleday.

Binney, E., & Estes, C. (1988). The retreat of the state and its transfer of responsibility: The intergenerational war. *International Journal of Health Services, 18*(1), 83–96.

Bulcroft, K., Leynseele, J.V., & Borgotta, E. F. (1989). Filial responsibility laws. *Research on Aging, 7,* 374–393.

Clark, P. (1993). Public policy in the United States and Canada: Individualism, familial obligation and collective responsibility in the care of the elderly. In J. Hendricks & C. Rosenthal (Eds.), *The remainder of their days: Domestic policy and older families in the United States and Canada* (pp. 13–49). New York: Garland.

Cohen, S., & Katzenstein, M. F. (1988). The war over the family is not over the family. In S. M. Dornbusch & M. H. Strober (Eds.), *Feminism, children, and the new families* (pp. 25–46). New York: Guilford.

Cowin, P. A. (1993). The sky is falling, but Popenoe's analysis won't help us do anything about it. *Journal of Marriage and the Family, 55,* 548–553.

Daniels, N. (1982). *Am I my parent's keeper*? New York: Oxford University Press.

de Beauvoir, S. (1968). *The second sex.* New York: Modern Library.

Dornbusch, S. M., & Strober, M. H. (1988). Our perspective. In S. M. Dornbusch & M. H. Strober (Eds.), *Feminism, children, and the new families* (pp. 3–24). New York: Guilford.

Ford, D. E. D. (1991). Translating the problems of the elderly into effective policies. In E. A. Anderson & R. C. Hula (Eds.), *The reconstruction of family policy* (pp. 91–108). Westport, CT: Greenwood.

Gilder, G. (1981). *Wealth and poverty.* New York: Basic Books.

Guilliland, N. (1986). Mandating family responsibility for elderly members, costs and benefits. *Journal of Applied Gerontology, 5,* 26–36.

Hagestad, G. O. (1987). Family. In G. L. Maddox (Ed.), *The encyclopedia of aging* (pp. 247–249). New York: Springer.

Hula, R. C. (1991). Introduction: Thinking about family policy. In E. A. Anderson & R. C. Hula (Eds.), *The reconstruction of family policy* (pp. 1–7). Westport, CT: Greenwood.

Kamerman, S. B., & Kahn, A. J. (1978). *Family policy: Government and families in fourteen countries.* New York: Columbia University Press.

Kamerman, S. B., & Kahn, A. J. (1989). *The responsive workplace: Employers and a changing labor force.* New York: Columbia University Press.

Langley, P. A. (1991). The coming of age of family policy. *Families in Society: The Journal of Contemporary Human Services, 72,* 116–120.

Mead, L. (1986). *Beyond entitlement: The social obligations of citizenship.* New York: Free Press.

Meyers, J. W., Ramirez, F. O., Walker, H. A., Langton, N., & O'Connor, S. M. (1988). The state and the institutionalization of the relations between women and children. In S. M. Dornbusch & M. H. Strober (Eds.), *Feminism, children, and the new families* (pp. 137–158). New York: Guilford.

Miller, D. (1990). *Women and social welfare: A feminist analysis.* New York: Praeger.

Murray, C. (1984). *Losing ground, American social policy 1950–1980.* New York: Basic Books.

Murray, C. (1988). *In pursuit of happiness and good government.* New York: Simon & Schuster.

Quadagno, J. (1989). Generational equity and the politics of the welfare state. *Politics and Society, 17*(3):353–376.

Rapp, R. (1978). Family and class in contemporary America: Notes toward an understanding of ideology. *Science and Society, 42,* 278–300.

Rosenthal, C. J., & Hendricks, J. (1993). Conclusion. In J. Hendricks & C. J. Rosenthal (Eds.), *The remainder of their days: Domestic policy and older families in the United States and Canada* (pp. 223–227). New York: Garland.

Rosenthal, M. G. (1994). Single mothers in Sweden: Work and welfare in the welfare state. *Social Work, 39,* 270–278.

Scanzoni, J. (1982). Reconsidering family policy: Status quo or force for change? *Journal of Family Issues, 3,* 277–300.

Scanzoni, J. (1991). Balancing the policy interests of children and adults. In E. A. Anderson & R. C. Hula (Eds.), *The reconstruction of family policy* (pp. 11–22). Westport, CT: Greenwood.

Schorr, A. L. (1980). *Thy father and thy mother: A second look at filial responsibility and family policy.* Washington, DC: Department of Health and Human Services.

Seaberg, J. R. (1990). Family policy revisited: Are we there yet? *Social Work, 35,* 548–554.

Segal, E. A. (1989). Welfare reform: Help for poor women and children? *Affilia: Journal of Women and Social Work, 4,* 42–50.

Smeeding, T., Torrey, B., & Rein, M. (1988). Patterns of income and poverty: The economic status of children and the elderly in eight countries. In J. L. Palmer, T. Smeeding, & B. B. Torrey (Eds.), *The vulnerable* (pp. 89–119). Washington, DC: Urban Institute.

Spakes, P. (1989a). A feminist case against national family policy: A view to the future. *Policy Studies Review, 8,* 610–621.

Spakes, P. (1989b). Reshaping the goals of family policy: Sexual equality, not protection. *Affilia: Journal of Women and Social Work, 4,* 7–24.

Spakes, P. (1992). National family policy: Sweden versus the United States. *Affilia: Journal of Women and Social Work, 7,* 44–60.

Stacey, J. (1993). Good riddance to "The Family": A response to David Popenoe. *Journal of Marriage and the Family, 55,* 545–547.

Steiner, G. (1981). *The futility of family policy.* Washington, DC: Brookings Institute.

Treas, J. (1975). Aging and the family. In D. S. Woodruff & J. E. Birren (Eds.), *Aging, scientific perspectives, and social issues* (pp. 92–108). New York: Van Nostrand.

Troll, L. E., Miller, S., & Atchley, R. (1979). *Families in later life.* Belmont, CA: Wadsworth.

Whitehead, B. D. (1993). Dan Quayle was right. *Atlantic Monthly, 271,* 47–84.

Wilensky, H. L. (1985). *Comparative social policy: Theories, methods, and findings.* Berkeley, CA: University of California, Berkeley, Institute of International Studies.

Wisensale, S. K. (1988). Generational equity and intergenerational policies. *The Gerontologist, 28,* 773–778.

43

Work, the Family, and Welfare

SHARON HAYS,
2003

THE CULTURAL LOGIC OF WELFARE REFORM

A nation's laws reflect a nation's values.

Like all laws, the law reforming welfare operates as a mechanism of social control to deter would-be transgressors and to discipline those who are measured as deviant according to its standards. By punishing those who break a society's moral code and supporting those considered worthy, laws can also serve to strengthen and affirm the values prescribed (Abramovitz, 1996; Fraser and Gordon, 1997; Mink, 1998). This is no less true of the belief that people should stop at stoplights than it is of Americans' affirmation that older citizens deserve Social Security and our collective condemnation of murder as the highest crime. Thus, the Personal Responsibility Act is much more than a set of policies aimed at managing the poor, it also provides a reflected image of American culture and reinforces a system of beliefs about how *all* of us should behave.

Of course our laws are an imperfect reflection of our values. In the case of welfare reform, for instance, it is clearly important to consider the power and financial resources of the politicians primarily responsible for designing the law relative to those who are its central targets. The content and form of laws are also constrained and shaped by the organizational practices and procedures of political and social life—the partisan bickering, the vote mongering, the committee negotiations, the language and structure of the legal system, the existing mechanisms for implementation and enforcement. And it certainly cannot be assumed that the values legislated are fully shared, let alone regularly practiced, by all members of the society. Still, in modern democracies where politicians are charged with representing the interests of their constituents, it can be said without much doubt that there is a strong relationship between the nation's laws and more widespread cultural norms, beliefs, and values. . . .

WORK, THE FAMILY, AND WELFARE

Welfare policy in the United States has long been closely connected to the nation's cultural vision of the appropriate commitment to work. Nineteenth-century poor laws established the moral distinction between the "deserving" and "undeserving" poor—providing aid to those who were out of work through no fault of their own and punishing the "intemperate," "immoral," "idle" undeserving with placement in poorhouses, miserly aid, and forced work (Katz, 1986;

Gans, 1995). From the start there were concerns about the (innocent) children of the undeserving poor, and some provisions were established to place such children in good homes where they could be properly cared for and trained as workers. But these anxieties regarding children did not, at first, translate into concern for protecting mothers or for maintaining family cohesion.

What we have come to understand as "welfare" today, however, was firmly connected to our values regarding family life from its inception. Its roots are in early twentieth-century state laws providing Mother's Pensions, specifically aimed at protecting widows so that they might care for their children at home (Skocpol, 1992; Abramovitz, 1996). These laws were expanded and made more inclusive when New Deal legislation instituted the program of Aid to Dependent Children in 1935.

The 1935 federal law establishing welfare followed directly from the American family ideal of a breadwinning husband and a domestic wife—if the husband was absent, the state would step in to take his place in the support of mother and children. The history of welfare makes it clear that, in practice, aid was denied to many women who were understood as not "virtuous" enough to be worthy of the family ideal (Boris, 1999; Goodwin, 1995; Gordon, 1994). Yet the cultural message clearly asserted that *good* women should stay at home with their children. And by the late 1960s, increasing numbers of poor single mothers were using welfare for precisely the purposes for which it was originally intended—they were staying at home to care for their young children, just as the ideal of appropriate family life prescribed.

The ensuing rise in the welfare rolls that began in the 1960s and continued into the 1990s was a major propellant for the Personal Responsibility Act, and was also directly connected to a broad range of changes in American society, including the feminization of poverty, the increase in single parenting, and the changing shape of the workforce and economy. The size of the welfare rolls was also impacted by the 1960s focus on aid to the poor. That decade was marked by the War on Poverty, the formation of the National Welfare Rights Organization, a series of court challenges and legislative changes that equalized the giving of aid, and the creation of the federal poverty programs of food stamps and Medicaid. But even at that moment in history, when the nation seemed collectively dedicated to providing support for the poor and vulnerable, there was fierce disagreement about the best approach. And the rise of the welfare rolls, alongside the continuing rise in single parenting, caused increasing concern among politicians and the public (Mink, 1995; Bone and Ellwood, 1994; O'Connor, 2001).

Numerous successive welfare "reforms" were enacted in hopes of stemming the tide. State welfare programs began decreasing their benefit amounts to make welfare less attractive. New federal rules were enacted throughout the 1970s and 1980s. Welfare mothers were encouraged to get training and go to work, a limited number of two-parent families were allowed to receive welfare benefits under stringent criteria, and the system of child support enforcement was linked to welfare. Relative to the changes wrought by the Personal Responsibility Act, however, these reforms were mere tinkering, leaving lots of loopholes and exemptions. Less than 10 percent of welfare recipients actually participated in the work programs by the 1990s, very few two-parent families qualified for aid, and only a small proportion of welfare clients actually received any child support. And throughout all these changes, the national guarantee of a familial safety net remained solidly in place (Katz, 1986; U.S. House of Representatives, 1998).

Both the cultural logic and the practical reality of welfare changed dramatically with the passage of the 1996 legislation renaming welfare as Temporary Assistance to Needy Families (TANF). The Personal Responsibility Act firmly established the absolute demand that mothers participate in the paid labor force, offering no exceptions to the more "virtuous" or more vulnerable women among them. The only indication of concern for the fate of ("innocent") children within it was the provision of temporary subsidies for paid childcare. Most significantly, by ending the entitlement to welfare benefits, this law suggested that the nation no longer believed that women and children deserved any form of special protection.

From the moment I recognized this logic, it seemed to me a rather one-sided reflection of the nation's values. Many people, after all, still think that children, at least, deserve some form of protection. And welfare reform's demand that mothers take paying jobs occurs at a time when society as a whole is still expressing tremendous ambivalence regarding the labor force participation of mothers. Although 73 percent of mothers are now employed, and 59 percent of women with infants now work for pay, many Americans are still worried about the consequences of this change (Bachu and O'Connell, 2000; Hays, 1996). In fact, the very same politicians who signed the law reforming welfare have continued to busily espouse the "family values" that had previously relied on women staying at home to maintain the warm hearth, provide for the emotional sustenance of family members, and shore up family ties. Similarly, scholars of family life continue to debate the problems of paid child care, women's second shift, the time crunch at home, and the declining commitment to family and children said to be signaled by rising rates of divorce and single parenthood (Blankenhorn, 1995; Coontz, 1997; Hochschild 1989; Popenoe, 1988). The tensions between the values of home and the values of paid work are apparent in these debates. But the most widely hailed message sent by welfare reform appears straightforward—all mothers must be prepared to leave the home to find paying jobs that will support themselves and their children.

From this perspective, welfare reform might be said to represent the triumph of classical liberal individualism. That is, women are no longer seen as the dependents of men, properly embedded in family life. Instead, women are treated as genderless individuals and, just like men, they are understood as competent, rational, independent beings who can be held responsible for their own lives and their own breadwinning. Similarly, welfare reform could be interpreted as representing the success of liberal feminist goals in constructing a new vision of family life. Earlier welfare policies followed the logic of difference feminism, assuming that mothers should stay at home and practice a distinctively female ethic of nurturing care. The work requirements of welfare reform, on the other hand, seem to signal the expectation that women can and should join men in the public sphere of paid work, operating according to an individualistic ethic of "personal responsibility."

All welfare mothers are now required to work, including those with infants and toddlers. (Gallagher et al., 1998; National Governors' Association, 1999). From the moment they enter the welfare office, they must be looking for a job, training for a job, or in a job. If they can't find a paying job or suitable short-term training, they are assigned to work full-time for a state-appointed agency in return for their welfare checks. But the provision of welfare reform that gives work requirements real teeth, and the provision that is in some respects even more harsh than nineteenth-century policies, is the federal time limits on benefits. After five years, all welfare recipients are expected to be self sufficient—and no matter how destitute they might be, they will remain ineligible to receive welfare assistance for the rest of their lives (U.S. House of Representatives, 1998, 2000; U.S. Congress, 1996). Many states have chosen even shorter time limits, as is true of Arbordale and Sunbelt City. In both, after two years of aid, single mothers are barred from welfare receipt for two years; when that period is complete, they may return to the welfare office and repeat the cycle until their five-year limit is reached.

Given the power of the work requirements and the virtually airtight enforcement mechanism of time limits, should we understand the law as saying that male breadwinners are a thing of the past, and women should be seen as perfectly able to care for themselves and their children on their own? Has the cultural championing of individualism won out over the concern for children? Is the old family ideal dead?

As it turns out, the promotion of perfected individual self-reliance is not the only message sent by reform. Although the attention paid to state efforts at placing welfare recipients in jobs has led many to believe that work requirements are the centerpiece of this legislation, a reading of the Personal Responsibility Act makes it appear that the intent of lawmakers was to champion family values above all else. It begins, "*Marriage is the foundation of a successful society.* Marriage is an essential institution of a successful society which promotes the interests of children. Promotion of responsible fatherhood and motherhood is integral to successful child rearing and the well-being of children" (U.S. Congress, 1996).

The law goes on to describe the problems of teenage pregnancy, out-of-wedlock births, children raised in single-parent homes, and fathers who fail to pay child support. Indeed, a reading of this statement of the law's intent would lead one to believe that the problem of poverty itself is the direct result of failures to live up to the family ideal. Congress emphasizes the close connection between the rising number of births to unmarried women and the growing number of people receiving welfare benefits. We are told that these single-parent homes not only create dependence on welfare, they also foster higher rates of violent crime and produce children with low cognitive skills, lower educational aspirations, and a greater likelihood of becoming teen parents—who will then produce children prone to repeat the cycle and foster ever-higher rates of crime, poor educational attainment, teen pregnancy, and welfare receipt.

These problems, Congress proclaims, are responsible for "a crisis in our Nation." To solve this crisis, the Act sets forth the following four goals:

1. provide assistance to needy families so that children may be cared for in their own homes or in the homes of relatives;
2. end the dependence of needy parents on government benefits by promoting job preparation, work, and marriage;
3. prevent and reduce the incidence of out-of-wedlock pregnancies and establish annual numerical goals for preventing and reducing the incidence of these pregnancies; and
4. encourage the formation and maintenance of two-parent families. (U.S. Congress, 1996).

It should be noted that only one of these goals is directed at paid work. And even in this case it is set alongside marriage as one of the two proper paths leading away from welfare.

Read in this way, work requirements can be understood as a method of enforcing family values through their deterrent effect—as measures meant to discourage women from choosing divorce or single parenthood (Gillespie and Schellhas, 1994). Single mothers on welfare are effectively punished for having children out of wedlock or for getting divorced. The punishment they face is being forced to manage on their own with low-wage work. But in this argument, the punishment of current welfare mothers is less important than the training of other poor, working-class, and middle-class women who, when they contemplate divorce or out-of-wedlock childbearing, will learn to think twice before they decide to raise children without the help of men. Hence, removing the safety net and forcing welfare mothers to work is actually a way to reinforce all women's proper commitment to marriage and family (Mink, 1995, 1998; Abramovitz, 1996, 1988; Gordon, 1994, 1988).

This, the "family values" version of welfare reform, is also the basis for a second set of edicts contained in the Personal Responsibility Act. Congress offers, for instance, financial incentives to states for the development and promotion of programs of sexual abstinence education. It calls on the nation to "aggressively" enforce statutory rape laws. Our lawmakers insist that teenage welfare mothers live in adult-supervised arrangements in order to receive benefits. And above all, this version of welfare reform absolutely requires that all welfare mothers establish the identity of their children's fathers and work with child-support enforcement officials in demanding that fathers provide financial support (U.S. Congress,

1996). These edicts are designed not only to regulate the reproductive behavior of poor men and women but also to convince men everywhere that if they should consider divorce or unwed parenting, they will be held responsible for the financial support of their progeny—and should therefore reassess their behavior and their options.

How, then, are we to interpret the message of welfare reform? Are marriage and family commitment the central concern? Or is the importance of individual self-sufficiency so great that the care of children can take a back seat to mothers' paid work? Are we reasserting the portrait of a nurturing mom and a breadwinning husband, or are we pressing for a world full of breadwinners?

There are, in fact, two distinct (and contradictory) visions of work and family life embedded in this legislation. For shorthand purposes (and to emphasize the disjunction between them), I have come to call these two visions the Work Plan and the Family Plan. In the Work Plan, work requirements are a way of *rehabilitating* mothers, transforming women who would otherwise "merely" stay at home and care for their children into women who are self-sufficient, independent, productive members of society. The Family Plan, on the other hand, uses work requirements as a way of *punishing* mothers for their failure to get married and stay married. In the Work Plan we offer women lots of temporary subsidies, for childcare, transportation, and training, to make it possible for them to climb a career ladder that will allow them to support themselves and, presumably, their children. No longer dependent on men or the state, these women will make their own choices about marriage and children. According to the Family Plan, work requirements will teach women a lesson; they'll come to know better than to get divorced or to have children out of wedlock. They will learn that their duty is to control their fertility, to get married, to stay married, and to dedicate themselves to the care of others.

If the Work Plan follows the logic of classical liberal individualism, imagining all women and men as equally competent individuals capable of competing in the market, achieving self-sufficiency, and utilizing market-based solutions to the problem of caring for children, then the Family Plan can be said to follow the logic of a certain form of classical conservatism. According to this model, systems of social connection, obligation, and commitment, epitomized by the operations of traditional family life, are essential to the maintenance of social order. Also crucial to social stability is the requirement that people conform to their proper roles within the social hierarchy. (Burke, 1910; Beecher, 1841; Blankenhorn, 1995; Popenoe, 1988). Hence, the Family Plan can solve the "problem" of where children fit by relying on an image of family life where women are subservient nurturers and men are financially successful heads of households.

As you might have noticed, however, welfare reform represents something more than a simple disagreement between liberals and conservatives. Conservatives are certainly not the only ones who worry about the fate of children and community and familial ties, and liberals are not the only ones who think that people should be able to achieve self-sufficiency. The two competing visions embedded in welfare reform are directly connected to a much broader set of cultural dichotomies that haunt us all in our attempts to construct a shared vision of the good society—independence and dependence, paid work and caregiving, competitive self-interest and obligations to others, the value of the work ethic and financial success versus the value of personal connection, familial bonding, and community ties. These cultural oppositions also inform debates between liberals and communitarians regarding the primacy of individual freedom versus the centrality of moral community, and arguments among feminists over whether to stress women's independence or valorize women's caregiving. These oppositions also mirror the uncertainties we find in public concern over women's labor force participation, the costs and quality of childcare, the time pressures faced by dual-earner couples, and the problems of divorce and single parenting. And the tensions in the Work and Family Plans are tightly connected to a whole series of issues often treated as the result of declining family and community values—including latchkey kids, unsupervised teens, deadbeat dads, abortion, gang violence, drug abuse, rising rates of crime, declining civic engagement, and ever-lower levels of social trust (Blankenhorn et al., 1990; Etzioni, 1993; Popenoe, 1988; Putnam, 2000, 1995).

As a reflection of our nation's values, welfare reform thus represents a powerful tug-of-war taking place in a society that is uncertain about the proper path. Rather than offering a single, coherent, and inclusive solution to problems of work and family life today, welfare reform offers us two narrow and opposing visions. It thereby simultaneously promises to solve all our problems, and promises to solve none of them.

HIGHER VALUES, MULTIPLE MEANINGS, CULTURAL DISTORTIONS, AND THE POLITICS OF EXCLUSION

Part of the reason that welfare reform has been so widely affirmed is surely because its two competing messages are able to satisfy two distinct constituencies. Depending upon one's angle of vision, welfare reform can be seen as a valorization of independence, self-sufficiency, and the work ethic, as well as the promotion of a certain form of gender equality. On the other hand, it can serve as a condemnation of single parenting, a codification of the appropriate preeminence of lasting family ties and the commitment to others, and a reaffirmation that women's place is in the home.

Further, it is certainly no accident that the primary guinea pigs in this national experiment in family values and the work ethic are a group of social subordinates—overwhelmingly women, disproportionately non-white, single parents, and of course, very poor. Politicians and welfare critics have labeled them "wolves," "alligators," "reckless breeders," and "welfare queens." They have become throwaway people. And those powerful stereotypes have made them readily identifiable symbols of societal failures in family and work life.

But there is more. The popularity of welfare reform has also followed from its ability to satisfy even more numerous constituencies, at the same time it arose from a much more widely shared set of higher moral ideals.

If you scrub off all the controversy and contradiction of welfare reform, at bottom you can find a set of honorable moral principles. The worthy ideals implicitly championed by this reform represent collective and long-standing commitments to the values of independence, productivity, conscientious citizenship, family togetherness, social connection, community, and the well-being of children. There is nothing *inherently* contradictory in these principles. The reasons they emerge as contradictory and even punitive relative to welfare reform is that this legislation takes place in the context of massive changes in family and work life, deepening levels of social distrust, rising social inequalities, and an increasingly competitive and global capitalist marketplace. In a connected way, these higher principles were refashioned and debased through processes of cultural distortion and exclusion—processes that have translated social and moral complexity into simplified slogans and stereotypes that obscure the more difficult dilemmas and the more disturbing social inequities involved.

By the time our nation's ideals were codified into the law reforming welfare, they had been passed through so many hands and been sifted through so many (often conflicting) interests, beliefs, and experiences that they were transformed almost beyond recognition. They had been tossed about by politicians seeking votes, policymakers hoping that their bright ideas would win out over others, states trying to trim their budgets, and scholars and pundits who sell books and make it big on the lecture circuit by providing simple, provocative, one-sided portraits of complex issues.

The values of independence, citizenship, connection, and community were similarly reinterpreted by a populace that includes members of the working and middle classes who have become increasingly worried about their chances of achieving or sustaining the American dream. Their concerns may differ depending on where they sit in the class hierarchy, but Americans have good reason to be worried about global competition, trade wars, corporate downsizing, the technological revolution, declining real wages, rising home prices, the volatile stock market, the decline of trade unions, a precarious Social Security system, and the rising split between educated professionals and less-educated blue-collar and service workers, just to name a few (Blau, 1999; Hochschild, 1989; Levy, 1998). When it comes to welfare reform, these worries get funneled into the condemnation of *those* people who are spending the nation's tax dollars while avoiding work and remaining apparently immune from all the economic woes faced by the rest of the country.

The principles of independence and commitment also took on different meanings for the growing numbers of working mothers who are struggling to juggle the demands of work and home. That welfare reform is a response to the widespread employment of mothers is a fact that is hard to miss. Working mothers today face not only glass ceilings, a sex-segregated labor force, and the "mommy track," they also face intense demands on their time and energy, especially with regard to child-rearing (Hays, 1996; England, 1992). If mothers choose to stay at home to avoid such pressures, like welfare recipients, they are often devalued and dismissed and their work at home is treated as inconsequential. Nearly all women recognize this. And like some mothers who work in the welfare offices of Arbordale and Sunbelt City, many working mothers imagine that welfare recipients have been spared from the intense demands and difficult choices they face. To many women, this seems unfair.

Long-standing national values took on a distinct significance for the many employers who have had trouble finding and keeping workers for the lowest paid jobs. Such employers couldn't help noticing the benefits of welfare reform. Work requirements and time limits throw millions of desperate women into the labor market and put them in a position where they must accept low wages, the most menial work, the poorest hours, with no benefits, and little flexibility. Thus, low-wage employers gain not only the benefit of this large pool of "eager" new workers, they arguably also gain greater control over their existing workers—who must now fear that if they don't accept their current working conditions, they can be replaced by former welfare recipients (Ehrenreich, 2001; Piven, 1999).

A further, and central, source of exclusionary images of welfare mothers is persistent racial tensions and continuing discrimination against nonwhite and immigrant groups. Race is so powerful in shaping negative images of welfare recipients that when Charles Murray wrote his famous book attacking the welfare system, *Losing Ground,* he focused almost exclusively on blacks—ignoring the other two-thirds of welfare recipients. When Ronald Reagan immortalized the image of a Cadillac-driving "welfare queen," it was not by chance that the story of fraud he chose to (grossly) exaggerate was a story of a black woman. In *The Color of Welfare,* sociologist Jill Quadagno forcefully argues that a central reason that U.S. welfare policies have long been less generous and inclusive than those of other Western industrial nations is precisely because of this country's history of racism. And as Martin Gilens demonstrates in *Why Americans Hate Welfare,* the racial coding of welfare recipients shows up in opinion polls as a primary feature of Americans' disdain for the welfare system (Murray, 1984; Zucchino, 1997; Quadagno, 1994; Hochschild 1995; Gilens, 1999). Most people recognize that welfare has come to be associated with blacks, even though they have never been a majority of welfare recipients. Laying all the problems associated with poverty at their doorstep allows many people to feel smugly superior, and it also helps to perpetuate the cultural and economic under-pinnings of racial inequality.

Finally, the process of distortion continues as our higher moral principles are continually tattered and corrupted by all these groups—that is, all of us—through the propagation of stereotypes and mythologies and slogans that speak to our particular interests and concerns and offer neat and tidy responses to complicated social problems.

The Work and Family Plans of welfare reform are both examples of distortion and exclusion. In the Work Plan, values of independence and productivity, once grounded in ideals of democratic citizenship and notions of collective progress, are reduced to a vision of calculating, self-interested individuals competing in the "free" market. This image provides no answer to who will care for the children, and leaves us to wonder just how we will care for one another or how we might be convinced to work together to build a better society. It implicitly suggests that we conceptualize children as relatively meaningless appendages and view our fellow citizens as merely potential rivals in the quest for success. And it ultimately excludes from full social membership all those people who fail to achieve middle-class economic stability.

The Family Plan, on the other hand, implicitly transforms the values of community ties and commitment to others—values that have long served to temper the rampant self-interest described in the work model—with an extremely narrow and rigid vision of the "traditional" family. It thus excludes all those people whose families diverge from the 1950s Leave-It-to-Beaver model. And it implies that we simply turn back the clock on women's movement into the paid labor force, failing to notice, apparently, that this movement is not only connected to the changing shape of the contemporary family but has also been crucial to women's greater independence and to their claim to productive social membership.

By the time our worthy moral principles have made it through state policymakers and local welfare offices into the lives of welfare clients, they have taken further twists and turns. . . . Enforced at the local level, our collective commitment to healthy family life can, for instance, have the effect of pressuring women to enter into or maintain relationships with physically abusive, drug-abusing, or law-breaking men. Independence, in this context, often looks like a job on the graveyard shift at Burger King or Dunkin' Donuts, a job that forces you to spend half your wages on substandard childcare and leaves you unable to buy winter coats for the kids.

SELECTED REFERENCES

Abramovitz, Mimi. 1996. *Under Attack, Fighting Back: Women and Welfare in the United States.* New York: Monthly Review Press.

Abramovitz, Mimi. 1988 [1996]. *Regulating the Lives of Women: Social Welfare Policy from Colonial Times to the Present.* Revised edition. Boston, MA: South End Press.

Bachu, Amara, and Martin O'Connell. 2000. *Fertility of American Women, 1998.* U.S. Census Bureau, Current Population Reports, P20–526. Washington, DC: U.S. Government Printing Office.

Bane, Mary Jo, and David T. Ellwood. 1994. *Welfare Realities: From Rhetoric to Reform.* Cambridge, MA: Harvard University Press.

Beecher, Catharine Esther. 1841 [1988]. *A Treatise on Domestic Economy: For the Use of Young Ladies at Home and at School.* New York: Harper and Brothers.

Blankenhorn, David. 1995. *Fatherless America: Confronting Our Most Urgent Social Problem.* New York: Basic Books.

Blankenhorn, David, Jean Bethke Elshtain, and Steven Bayme, editors. 1990. *Rebuilding the Nest: A New Commitment to the American Family.* Milwaukee: Family Service America.

Blau, Joel. 1999. *Illusions of Prosperity: Americas Working Families in an Age of Economic Insecurity.* New York: Oxford University Press.

Boris, Eileen. 1999. "When Work Is Slavery," pp. 36–55 in *Whose Welfare?* edited by Gwendolyn Mink. Ithaca, NY: Cornell University Press.

Burke, Edmund. 1910 [1951]. *Reflections on the French Revolution.* New York: E. P. Dutton.

Coontz, Stephanie. 1997. *The Way We Really Are: Coming To Terms with America's Changing Families.* New York: Basic Books.

Ehrenreich, Barbara. 2001. *Nickel and Dimed: On (Not) Getting By in America.* New York: Metropolitan Books.

England, Paula. 1992. *Comparable Worth: Theories and Evidence.* New York: Aldine de Gruyter.

Etzioni, Amitai. 1993. *The Spirit of Community: The Reinvention of American Society.* New York: Touchstone.

Fraser, Nancy, and Linda Gordon. 1997. "A Genealogy of 'Dependency': Tracing a Keyword of the U.S. Welfare State," pp. 121–149 in *Justice Interruptus: Critical Reflections on the "Postsocialist" Condition,* edited by Nancy Fraser. New York: Routledge.

Gallagher, L. Jerome, Megan Gallagher, Kevin Perese, Susan Schreiber, and Keith Watson. 1998. *One Year After Federal Welfare Reform: A Description of State Temporary Assistance for Needy Families (TANF) Decisions as of October 1997.* Washington, DC: Urban Institute.

Gans, Herbert J. 1995. *The War Against the Poor: The Underclass and Antipoverty Policy.* New York: Basic Books.

Gilens, Martin. 1999. *Why Americans Hate Welfare: Race, Media, and the Politics of Antipoverty Policy.* Chicago: University of Chicago Press.

Gillespie, Ed, and Bob Schellhas, editors. 1994. *Contract with America: The Bold Plan by Rep. Newt Gingrich, Rep. Dick Armey and the House Republicans to Change the Nation.* New York: Times Books.

Goodwin, Joanne L. 1995. " 'Employable Mothers' and 'Suitable Work': A Re-evaluation of Welfare and Wage-Earning for Women in the Twentieth-Century United States," *Journal of Social History* 29 (Winter): 253–274.

Gordon, Linda. 1988. *Heroes of Their Own Lives: The Politics and History of Family Violence.* New York: Penguin.

Gordon, Linda. 1994. *Pitied but not Entitled: Single Mothers and the History of Welfare.* Cambridge, MA: Harvard University Press.

Hays, Sharon. 1996. *The Cultural Contradictions of Motherhood.* New Haven, CT: Yale-University Press.

Hochschild, Arlie Russell with Anne Machung. 1989. *The Second Shift: Working Parents and the Revolution at Home.* New York: Viking.

Hochschild, Jennifer L. 1995. *Facing Up to the American Dream: Race, Class, and the Soul of the Nation.* Princeton, NJ: Princeton University Press.

Katz, Michael B. 1986 [1996]. *In the Shadow of the Poorhouse: A Social History of Welfare in America.* Revised edition. New York: Basic Books.

Levy, Frank. 1998. *The New Dollars and Dreams: American Incomes and Economic Change.* New York: Russell Sage.

Mink, Gwendolyn. 1995. *The Wages of Motherhood: Inequality in the Welfare State, 1917–1942.* Ithaca, NY: Cornell University Press.

Mink, Gwendolyn. 1998. *Welfare's End.* Ithaca, NY: Cornell University Press.

Murray, Charles. 1984. *Losing Ground: American Social Policy, 1950–1980.* New York: Basic Books.

National Governors' Association for Best Practices. 1999. *Round Two Summary of Selected Elements of State Programs for Temporary Assistance for Needy Families.* Washington, DC: National Governors' Association.

O'Connor, Alice. 2001. *Poverty Knowledge: Social Science, Social Policy, and the Poor Twentieth-Century U.S. History.* Princeton, NJ: Princeton University Press.

Piven, Frances Fox. 1999. "Welfare and Work," pp. 83–99 in *Whose Welfare*? edited by Gwendolyn Mink. Ithaca, NY: Cornell University Press.

Popenoe, David. 1988. *Disturbing the Nest: Family Change and Decline in Modern Societies.* New York: Aldine de Gruyter.

Putnam, Robert D. 1995. "Bowling Alone: America's Declining Social Capital." *Journal of Democracy 6*(1): 65–78.

Putnam, Robert D. 2000. *Bowling Alone: The Collapse and Revival of American Community.* New York: Simon and Schuster.

Quadagno, Jill. 1994. *The Color of Welfare: How Racism Undermined the War on Poverty.* New York: Oxford University Press.

Skocpol, Theda. 1992. *Protecting Mothers and Soldiers: The Political Origins of Social Policy in the United States.* Cambridge, MA: Harvard University Press.

U.S. Congress. 1996. *Personal Responsibility and Work Opportunity Reconciliation Act of 1996.* Public Law 104-193, H. R. 3734.

U.S. House of Representatives, Committee on Ways and Means. 1998. *Green Book: Overview of Entitlement Programs.* Washington, DC: U.S. Government Printing Office.

U.S. House of Representatives, Committee on Ways and Means. 2000. *Green Book: Overview of Entitlement Programs.* Washington DC: U.S. Government Printing Office.

Zucchino, David. 1997. *Myth of the Welfare Queen.* New York: Touchstone.

44

Public Policy and Family and Child Well-Being

SANDRA HOFFERTH, DEBORAH PHILLIPS, AND NATASHA CABRERA,
2001

Public policy has a major effect on families because it affects the resources and constraints families have to work with in their personal lives, which include their decisions regarding children. Families make many choices. However, their choices are constrained by the information and values they have, by financial resources available to them, or by their residential environment. Helping families make appropriate decisions in an informed way and reducing constraints on decisions are appropriate goals of public policy.

The extent to which personal behavior can be changed independent of changing a family's resources and opportunities is a major issue in public policy today. Policy changes restricting smoking in public places have resulted in private behavior changes; in other areas, such as liberalization of divorce laws, behavior changes preceded policy changes. Recent legislative changes in public assistance incorporated stricter work requirements for single mothers, increasing the numbers of recipients who are employed (DeParle 1997). Health and child-care services are provided to facilitate the transition, but the policy assumes that single mothers' willingness to leave their small children will change as well. That attitudes and behavior do not change easily is indicated by mothers leaving public assistance rather than meet these new requirements (DeParle 1997).

Another key issue is whether the absolute level or the relative level of resources affects behavior. Douglas Massey has hypothesized that the increased poverty and affluence of the past two decades, combined with persistent racial residential segregation and increased geographic concentration of the poor and the affluent "will cause an acute sense of relative deprivation among the poor and heightened fears among the rich, yielding rising social tension and growing conflict between the haves and the have-nots" (1996, 395). Even if such social upheaval does not materialize, however, relative deprivation may have harmful effects on children. The harmful effects of financial hardship on children are well-documented (Sherman 1994; Duncan and Brooks-Gunn 1997). Relative to the typical U.S. child, economically disadvantaged children are more likely to experience physical and mental illnesses, to have reduced access to medical and mental health services, to live in environments that are more violent and hazardous to their safety, to attend inferior schools, and to have less access to higher education.

Economic hardship is also positively related to adult psychological distress. As a result, adverse changes in the labor market not only increase financial hardship, but also contribute to increased physical and mental health problems for parents. Increased hardship affects parenting stress, parenting practices, and family structure. All of these, in

From Sandra Hofferth, Deborah Phillips, and Natasha Cabrera, "Public Policy and Family and Child Well-Being," from *The Well-Being of Children and Families,* edited by Arland Thorton, pp 384–408 (Ann Arbor: The University of Michigan Press, 2001). Used by permission.

turn, contribute to child physical and mental health problems, school performance problems, nonmarital childbearing, and other young adult outcomes. Taken together, these cumulative disadvantages reduce the subsequent socioeconomic attainment of disadvantaged children and extend the cycle of risk into the next generation (for a review of this literature, see Danziger et al. 1997)....

EFFECTS OF EARLY ENVIRONMENTS ON CHILD DEVELOPMENT

Home Environments

Extensive research has demonstrated that how parents react to inadequate financial resources structures the consequences of poverty for children. Economic disadvantage exerts its influence on development from infancy through adolescence directly by affecting the adequacy of resources available to the child at home (and outside the home), as well as indirectly through its detrimental effects on parents and parenting (Bradley and Whiteside-Mansell 1997; Chase-Lansdale and Brooks-Gunn 1995; Duncan and Brooks-Gunn 1997; Huston et al. 1994; McLoyd 1997)....

Variation in income manifests itself in virtually every facet of children's home environments. Children living in poverty are substantially less likely to have access to the material resources, opportunities, or interactive experiences that are associated with cognitive and language development, school readiness, and achievement. Their parents are also significantly more likely to experience mental health problems associated with the stresses of poverty and low-wage work and to provide less effective parenting and monitoring that, in turn, can compromise children's social and emotional development.

Child-Care Environments

Developmental Effects. Research on the developmental consequences of nonparental child-care environments has shifted away from a focus on whether the use of child care per se poses a risk to early development to whether and how variation in the quality, timing, and quantity of child care that children receive affects their development. Research in the 1990s has also looked at child care in the context of the family environment and found that, in comparison to child care, children's parents exert a much stronger influence on development. This is not to say, however, that child care does not matter.

The timing and amount of child care have few consistent effects on development, particularly when the quality of care experienced by children is taken into consideration. Significant and sometimes sizable associations between quality of child care and children's development are, however, among the consistent findings in developmental science (see reviews by Lamb 1997; Love et al. 1996; Hayes et al. 1990). Children who are cared for by providers who are warm and responsive and in centers where there are low staff-to-child ratios, small group sizes, high levels of training, low staff turnover, and higher staff salaries tend to be socially competent and score higher than children who experience low-quality care on a variety of child development measures (Phillips 1987). The training of the provider and the number of children for whom she or he is responsible also appear to be significantly associated with the quality of care in family day-care homes and more informal home-based arrangements (NICHD Early Child Care Research Network 1996). Regardless of maternal education, family income, or child gender and ethnicity, children's cognitive, language, and social-emotional development are associated with the quality of their child-care experience.

The emerging research focuses on quality, cost, and availability of care, and their impact on development as children prepare to enter school. Conceptions of quality of nonparental care vary somewhat according to who is defining quality—parents, policymakers, or researchers. The early child development community identifies three interrelated types of quality: structural, process, and developmental. Some of the attributes of structural quality (e.g., staffing ratio, group size) are easily observable, measured, and state-regulated. Process qualities, on the other hand, include the general environment and the social relations in the center and are time-consuming to measure because they require on-site observation. While structural characteristics per se do not "cause" good outcomes for children, they influence process variables, especially caregiving and developmentally appropriate activities (NICHD Early Child Care Research Network 1996). The impact of high-quality care on child development is measured as observable cognitive and social functioning of the children, as well as their success in school. . . .

The most recent findings from the Cost, Quality, and Child Outcomes Study suggest that child-care quality has long-term implications for all children, but especially for children who are most at risk of poor school outcomes (Peisner-Feinberg et al. 1999). Child-care experiences predict children's academic and social competence in second grade, even after adjusting for family characteristics and a variety of intervening school experiences. In accord with the NICHD child-care study, family characteristics are important predictors of school success, but child-care quality has a significant independent effect on later school success.

Levels of Quality. A series of studies have documented the lack of high-quality affordable child care. The National Child Care Staffing Study (NCCSS) concluded that the needs of child-care staff are so poorly met that staff and children's well-being are being jeopardized (Whitebook et al. 1990). The Profile of Child Care Settings (PCS) study concluded that while centers serving preschool-age children generally meet standards of quality, programs that serve infants and toddlers often fail to meet basic structural standards of quality care (Kisker et al. 1991). The Study of Children in Family Child Care and Relative Care found generally poor levels of quality in these non-center-based arrangements used by most families (Galinsky et al. 1994). Finally, the Cost, Quality, and Child Outcomes Study of center-based arrangements reported that "most child care—especially for infants and toddlers—is mediocre in quality and sufficiently poor to interfere with children's emotional and intellectual development" (Helburn et al. 1995). While the NICHD Study of Early Child Care, which encompassed all forms of care, found fewer than 20 percent of infants received insensitive caregiving, close to three-quarters received care that was relatively unstimulating of cognitive development (NICHD Early Child Care Research Network 1996). Variation in measurement, interpretation, or sampling may account for differences in study findings. Moreover, the NICHD study reveals the importance of looking in a more differentiated way at the qualities of the interactions that transpire between caregivers and children.

One of the greatest issues of concern for policymakers is that child-care centers and family child care meet minimum standards for health and safety. Most parents assume that operating facilities meet these minimum standards. However, children's environments may violate these basic standards (U.S. General Accounting Office 1992). . . .

PARENTAL DECISIONS ABOUT CHILDREN'S ENVIRONMENTS

Available options and resources affect parental choices. In a simple economic model, at any point in time parents select from a set of options, given their resources and characteristics, the option that has the highest value to them (Ben-Akiva and Lerman 1985). To the extent that access to both opportunities and resources depends on economic circumstances, we would expect differences by financial status in the choices of options as well. Given this framework, public policies can affect parental choices by changing available opportunities and providing resources. How have changes in economic circumstances and the policy environment affected parental choices and, hence, the quality of the child's experience?

We have chosen to focus on children's access to nonparental child-care programs while mothers are employed. Employment is an important aspect of family economic self-sufficiency. Important decisions parents make include whether or not to work and, if they do work, how to care for children while they are working. Because there is no single child-care system, parents face a set of private markets with many choices. Once children turn school-age, parental choices are more limited, and most still choose to have their children attend public schools. Schooling options have become more fragmented and privatized, however, with some public provision of vouchers for attending private schools, public funding of charter schools, and a growing number of parents choosing to educate their children at home (Holloway and Fuller 1992). Decisions about after-school care continue to occur in the context of a highly fragmented set of options.

Constraints on parental choices concerning early child care can be characterized from the point of view of (1) availability, (2) cost, and (3) information.

Availability

First, given the supply of programs at any time in any given area, certain options are less available than others. For example, child-care centers may be few but family child-care homes may be plentiful. Second, only options with certain characteristics may be available. For example, even if the supply of group programs may be large, all of them may have high child-teacher ratios.

Geography. Geographic areas vary in the availability and characteristics of early childhood programs. While centers are distributed across regions proportional to the number of children, family child-care homes are relatively rare in the Northeast and South and relatively more common in the Midwest and West (Kisker et al. 1991). Data from the National Child Care Survey 1990 (NCCS) (Hofferth et al. 1991) and A Profile of Child Care Settings (PCS) (Kisker et al. 1991) show substantial geographic differences in number of programs, in fees, in average child-staff ratios, and in teacher training. A recent study found child-care availability to be lower in nonmetropolitan than in metropolitan areas (Gordon and Chase-Lansdale 1999).

Location also affects parental choices. The child-care market is quite localized. Focus group accounts suggest that parents primarily consider alternatives close to their home or place of work (EDK Associates 1992; Mitchell et al. 1992). Large national samples show similar findings—availability within about 30 minutes time is a key factor influencing child-care choices (Hofferth and Collins 2000). Additionally, some neighborhoods do not lend themselves well to certain types of programs, such as family day care or preschool programs (Gordon and Chase Lansdale 1999). Houses may be far apart, neighbors may not know one other, or there may be few young children. Just as in preschool programs, location is a key factor in choice of public school. Even when families can choose a school, many still select one in their own neighborhood (Boyer 1992). Only half of families are within five miles of the next closest public school with their child's grade level (Boyer 1992).

Family Members as Caregivers. The availability of a family member as a caregiver clearly affects parental demand for centers and family child care. Relative care is the first choice for many families, particularly low-income families and those with very young children (Hofferth 1995). In 1993; 41 percent of the children of employed mothers were cared for by a relative (including fathers), a level that has been remarkably stable over the past decade. Single-parent families are more constrained because they lack a partner to share the care. Other families lack a nearby relative who can assist with child care. In 1990, only one out of three parents reported that a relative was available who could provide child care, whereas three out of five said that a center-based program was available (Hofferth et al. 1991).

Neighborhood and Community Economic Conditions. We know little about the relationship between the socioeconomic characteristics of a community and the number and characteristics of the early childhood environments available. Fronstin and Wissoker (1994) found that in areas with low per capita income, fees paid by parents were significantly lower, the number of family child-care providers was smaller, and the number of public school, Head Start, and center-based programs was slightly higher. This makes sense for public school programs and Head Start, since public policy places compensatory programs in areas in which the need is greater. Another study found no significant difference in the number of center-based programs in low-income areas, but substantially fewer spaces for children (Fuller and Liang 1996). This raises concerns that even though the number of programs does not differ, programs are smaller and meet area care needs less adequately.

Neighborhood and Community Culture/Values. Community or cultural norms and values may constrain the use of certain arrangements as well as the supply of programs. For example, Zinsser (1990) emphasizes cultural attitudes in a working-class neighborhood that lead parents to choose relatives and close friends as child-care providers. Latino families also appear less likely to use centers compared to blacks and whites, controlling for a variety of other socioeconomic and demographic factors (Hofferth et al. 1994; Fuller et al. 1996).

Parental Work Hours. Some families need care for nonstandard hours, such as during the evening or on a weekend. According to the National Child Care Survey, one out of eight mothers and one out of seven fathers worked one weekend day, and about the same proportion worked a nonday shift (Willer et al. 1991). Few child-care programs operate during nonstandard hours. About 10 percent of centers and 6 percent of family day-care homes offered care on weekends, while 13 percent of family day-care homes, 20 percent of nonregulated family day-care homes, and only 3 percent of centers provided care in the evenings (Willer et al. 1991). Low-income families are more likely to have nonstandard schedules (Hofferth 1995; Presser 1989), but it is not clear whether they are choosing these schedules or if they have no other work options. One study found that the flexibility of the child-care program affected child-care preferences and parental ability to work (Emlen 1998).

Employers. While employers have greatly increased the benefits to help mothers manage their work and family lives, low-income mothers have less access to most employer benefits, including flextime, work at home, unpaid leave, and flexible spending accounts (Hofferth et al. 1991). Additionally, in the United States few workers have access to paid maternity or paternity leave. The availability of unpaid leave increased with the passage of the Family and Medical Leave Act (FMLA) in 1994. The Commission on Family and Medical Leave (1996) found that only 3.4 percent of those who needed leave for a reason covered by the FMLA did not take it. Of those who did not take it, the main reason was that they could not afford it. Research has shown that mothers who have access to part-time work, a flexible spending account, and liberal unpaid leave return to work sooner than mothers who do not have access to such policies (Hofferth 1996a). These results are consistent with research showing the importance of flexible child-care arrangements to parental decisions (Emlen 1998).

Selectivity of Residential Location Decision. Given that most children attend schools where they live, residential location decisions determine the quality of schools children attend. Race and socioeconomic status are the major factors influencing this decision (Massey et al. 1994; South and Deane 1993; Duncan and Newman 1976; Yinger 1996). "Relationships among schools, families, and students are greatly complicated by the fact that many families select the schools that their children attend . . . There is reason to believe, then, that schools whose students are enrolled by choice will presumably have an easier time producing a given level of achievement than schools whose students are not self-selected" (Chubb and Moe 1990, 111).

Low-income families are constrained in their ability to access high-quality educational programs without substantial public intervention. While public schools are financed by local property taxes, expenditures per pupil vary directly with housing values in their school districts. In many states, per pupil spending in affluent districts is two or three times that in low-income districts (Burtless 1996). While child care is still generally privatized, the ability of providers to charge fees that permit high-quality programming will be dependent upon the incomes of their clientele. In addition, as public schools incorporate younger and younger children through prekindergarten programs, the same issues of resources will arise for preschool as for school-age children.

Cost

Even if there are many alternative suppliers of child care and many available options in terms of quality, parents may still be constrained by income and cost as they choose among options.

Maternal Employment. A family's total income affects maternal employment in the first place. First, the higher the income from other family members, the less likely a woman is to work during pregnancy. Thus, mothers with fewer resources work longer in pregnancy. Mothers with fewer financial resources also return to work sooner after childbirth (NICHD Early Child Care Research Network 1997a; Hofferth 1996a). Working mothers, of course, need nonparental child care more than those who do not work outside the home. Other sources of income have a substantial influence on maternal employment, before and after children are born. U.S. mothers now return to work very soon after childbirth—about half of all mothers are back at work within 6 months of their first birth (Hofferth 1996a). The NICHD Study of Early Child Care found very high employment rates and use of substitute care by mothers soon after birth. Almost three out of four infants experienced some nonparental child care during the first year of life, with an average age at entry of 3.3 months (NICHD Early Child Care Research Network 1997a). The high level of employment of mothers of infants in the United States differs dramatically from many European nations, which provide paid leave during much of the first year of life; mothers have jobs, but are not working during that period (Hofferth and Deich 1994).

Economic Factors. Family income and the price of care are key factors in parental choice of early childhood environments once the mother is employed. The higher the price, the lower the use of a type of care; parents with higher incomes use higher priced care; those with lower incomes use lower priced and informal care (Hofferth and Wissoker 1992; Chaplin et al. 1998). Families are likely to exclude from their choices options that they view as too expensive, but it is difficult to define this subjective concept. While fees overall do not seem unreasonably high, and many low-income families do not have to pay anything because of public subsidies, child care can account for a large share of a family's budget. Low-income mothers who pay for care spend about one-quarter of their income on child care, compared with an average of 10 percent for all families and 6 percent for high-income families (Hofferth et al. 1991).

Not all families pay the full price for care. Low-income children from nonworking poor families are not enrolled in centers at levels approaching high-income children, but they are enrolled at levels higher than working poor and working-class children (Hofferth 1995; Phillips et al. 1994; NICHD Early Child Care Research Network 1997b). This is likely the result of government subsidies (e.g., Head Start) for low-income children that in the past were disproportionately provided through center-based programs.

Information, Preferences, and Values

What parents include in their choices depends on what they know about. Families using center-based care, in-home providers, or family day-care providers were asked how they first learned about their care arrangement. The majority (66 percent) reported their source as a friend, neighbor, or relative. Advertisements were a distant second at 13 percent (Hofferth et al. 1991), while only 9 percent used a resource and referral service. There was no variation in source of information by employment status of the mother, age of child, or family income.

A substantial proportion of parents not using a type of care have no idea what it would cost (Hofferth et al. 1991). This is particularly the case for relatives and for center-based programs, for whom one out of three parents said they did not know, than for in-home providers and family day-care providers, with 18 and 24 percent saying they did not know. It is possible that better information on the price of center-based child-care arrangements might lead to increased use of services.

Research examining the implementation of the Family Support Act in 1988 suggests that low-income parents are constrained by lack of full information on their options (Mitchell et al. 1992). Caseworkers may not provide the information on other options if they think parents already have one child-care option available to them. In addition, some parents are suspicious of nonfamily child care. Extensive discussion and information are required to convince them (particularly young parents) otherwise (Kisker and Silverberg 1991). If anything, parents are most comfortable with child-care centers, which they view as the safest and most desirable for their children (Porter 1991).

Preferences or Values. Finally, parental choices depend upon their preferences. What aspects of children's environments are important to parents? Children's environments can be characterized in two major ways: (1) parent-focused attributes, which include convenience, cost, location, and hours; and (2) child-focused characteristics, which include child-staff ratio, group size, training of teachers, teacher turnover, warmth and sensitivity of provider, and other attributes affecting children, such as facilities.

Mason and Kuhlthau (1991) suggest that the age of the child strongly influences choices. Most mothers of preschool children believed that a parent is the ideal caregiver until their child enters school. For working mothers, another relative, such as the grandmother or father, was the ideal caregiver. Many parents share the care of children when they can by working different shifts (Presser 1989; Brayfield 1995). While almost all say that the mother or another parent is ideal for infants and toddlers, the proportion who say formal care is ideal increases to close to half as their child ages from 3 to 5 years of age. This suggests considerable parental support for the enormous expansion in early childhood programs over the past several decades, even among younger children (Hofferth 1996b).

The attributes that are important to parents may vary across time and across income and cultural groups based upon tastes and preferences, values, cultural influences, and professional advice. A comparison of 1975 and 1990 parent data (Hofferth et al. 1991) suggests that there was an increase in the importance of quality (from about 50 to 60 percent) and a decline in the importance of cost (from 12 to 4 percent) as reasons for changing child-care arrangements. Recent surveys also document that quality is an important criterion. Parents are more likely to report that having a small number of children and a trained provider were important than to say that cost and convenience were important (Hofferth et al. 1998). In fact, their children's programs reflect these preferences. Large increases have been reported in center-based enrollments relative to increases in all early education and child-care programs (Willer et al. 1991). Parents are more likely today to report that their program has an educational component (over 90 percent) than in the past, blurring the distinction between nursery school and child-care centers. Finally, there are few income differences in preferences. Low-income parents are just as concerned about the quality of care as high-income parents (Sonenstein and Wolf 1991). . . .

Direct Child-Care Subsidies

Child care is one underrated success story of payoff from federal investments. Federal expenditures on early education and care for low-income families have tripled since 1980 (Hofferth 1993), and the supply of center-based care rose threefold as well (Willer et al. 1991). What was the result of these investments? Most child-care programs are privately run; Head Start centers and public-school-based centers made up only about 17 percent of all centers in 1990 (Willer et al. 1991), though Head Start doubled its enrollment between 1990 and 1996 to 752,077 children (U.S. Department of Health and Human Services 1998). However, even privately funded centers may receive some public funding. One study found the highest quality center-based programs were those that received substantial public funding (Helburn et al. 1995). Global ratings of developmentally appropriate care for three different age groups of children, by the income level of the families using the center, showed that children from low- and high-income families were in higher quality care than those from working-poor families (Phillips et al. 1994). Another study found children from both low- and high-income families to be in centers with higher quality ratings than those just above the poverty line (NICHD Early Child Care Research Network 1997b). In all other forms of care, however, the lower the family income the poorer the quality of care. While we expect children of high-income families to be in high-quality centers—their parents can afford them—we do not expect this of children from low-income families. One explanation for the better care received by the latter than children from middle-income families is their eligibility for subsidized programs such as Head Start and publicly subsidized preschool programs (Hofferth 1995), thus demonstrating the difference such subsidies make to low-income children's lives.

Children from working-poor families are still at a disadvantage because of their inability to take advantage of nonrefundable tax credits such as the Child and Dependent Care Tax Credit. The federal government increased its child-care expenditures by 100 percent between 1980 and 1995 (to about $10 billion). Still, in 1995 only about 60 percent of federal expenditures for child care went to low-income families; the rest went to middle-class families, primarily through the Child and Dependent Care Tax Credit. Of families relying on supplemental arrangements, 18 percent of working-poor families with a preschool child reported receiving direct financial assistance in paying for child care, compared with 12 percent of working-class and 3 percent of middle-class families (Hofferth 1995). Thirty-seven percent of non–working-poor families with a preschool child received such financial assistance. In contrast, 12 percent of working-poor families, 24 percent of working-class, and 34 percent of middle-class families claimed the Child and Dependent Care Tax Credit in 1988. When direct assistance and tax relief are summed, only 30 percent of working-poor families receive subsidies at the present time (compared with 36 to 37 percent of other families) because they are least likely to take the tax credit. . . .

CONCLUSION

In sum, a lot has been learned about how parents select programs for their children and how this may influence the relationship between program and outcomes. Future data collection efforts by developmental psychologists should incorporate such factors as the availability of programs, community characteristics, and public policies into their studies; those by sociologists and economists should include child assessments and observations in programs. One major gap at the national level is a regularly scheduled survey of children's programs comparable to the Profile of Child Care Settings conducted in 1990 and the National Study of Before- and After-School Programs conducted in 1991. Federal agencies have focused more on the demand (parent) side than on the supply (program) side in supporting research on early childhood. Both are critical to understanding the effects of public policy on family decisions and, ultimately, the well-being of children.

SELECTED REFERENCES

Ben-Akiva, Moshe, and Steven R. Lerman. 1985. *Discrete Choice Analysis: Theory and Application to Travel Demand.* Cambridge, MA: MIT Press.

Boyer, Ernest L. 1992. *School Choice.* Princeton, NJ: Carnegie Foundation for the Advancement of Teaching.

Bradley, Robert H., and L. Whiteside-Mansell. 1997. "Children in Poverty." In *Handbook of Prevention and Treatment with Children and Adolescents,* ed. R. T. Ammerman and M. Hersen, 13–58. New York: Wiley.

Brayfield, April. 1995. "Juggling Jobs and Kids: The Impact of Work Schedules on Fathers' Caring for Children." *Journal of Marriage and the Family* 57:321–32.

Burtless, Gary. 1996. *Does Money Matter? The Effect of School Resources on Student Achievement and Adult Success.* Washington, DC: Brookings.

Chaplin, Duncan, Sandra Hofferth, Douglas Wissoker, and Paul Fronstin. 1998. "The Price Elasticity of Child Care Demand: A Sensitivity Analysis." Washington, DC: Urban Institute.

Chase-Lansdale, P. Lindsay, and Jeanne Brooks-Gunn. 1995. *Escape from Poverty: What Makes a Difference for Children?* New York: Cambridge University Press.

Chubb, John E., and Terry M. Moe. 1990. *Politics, Markets, and America's Schools.* Washington, DC: Brookings Institution.

Danziger, Sheldon, Sandra Danziger, and J. Stern. 1997. "The American Paradox: High Income and High Child Poverty." In *Child Poverty and Deprivation in Industrialized Countries,* ed. G. A. Cornia and S. Danziger. Oxford, England: Oxford University Press.

DeParle, Jason. 1997. "Success, and Frustration, as Welfare Rules Change." *New York Times* 117:1, 12–13.

Duncan, Greg, and Jeanne Brooks-Gunn. 1997. *Consequences of Growing Up Poor.* New York: Russell Sage Foundation.

Duncan, Greg J., and Sandra Newman. 1976. "Expected and Actual Residential Mobility." *Journal of the American Institute of Planners* 42:174–86.

EDK Associates. 1992. *Choosing Quality Child Care: A Qualitative Study.* New York: EDK Associates.

Emlen, Arthur C. 1998. "From a Parent's Point of View: Flexibility, Income, and Quality of Child Care." Paper presented at a Conference on Child Care in the New Policy Context. Bethesda, MD, April 30, May 1.

Fronstin, Paul, and Doug Wissoker. 1994. "The Effects of the Availability of Low-Cost Child Care on the Labor Supply of Low-Income Women." Manuscript. Urban Institute, Washington, DC.

Fuller, Bruce, and Xiaoyan Liang. 1996. "Market Failure? Estimating Inequality in Preschool Availability." *Educational Evaluation and Policy Analysis* 18:31–9.

Fuller, Bruce, Susan Holloway, and Xiaoyan Liang. 1996. "Family Selection of Child-Care Centers: The Influence of Household Support, Ethnicity, and Parental Practices." *Child Development* 67:3320–37.

Galinsky, Ellen, Carollee Howes, Susan Kontos, and Marybeth Shinn. 1994. *The Study of Children in Family Child Care and Relative Care.* New York: Families and Work Institute.

Gordon, Rachel, and P. Lindsay Chase-Lansdale. 1999. *Women's Participation in Market Work and the Availability of Child Care in the United States.* Working Paper. Alfred P. Sloan Working Families Center, vol. 99–05. Chicago: University of Chicago.

Hayes, Cheryl, John Palmer, and Martha Zaslow. 1990. *Who Cares for America's Children? Child Care Policy for the 1990s.* Washington, DC: National Academy Press.

Helburn, Suzanne, Mary Culkin, Carollee Howes, M. Cryer, and E. Peisner-Fein-berg. 1995. *Cost, Quality and Child Outcomes in Child Care Centers.* Denver: University of Colorado at Denver.

Hofferth, Sandra L. 1993. "The 101st Congress: An Emerging Agenda for Children in Poverty." In *Child Poverty and Public Policy,* ed. J. Chafel. 203–43. Washington, DC: Urban Institute.

Hofferth, Sandra L. 1995. "Caring for Children at the Poverty Line." *Children and Youth Services Review* 17 (1–3): 61–90.

Hofferth, Sandra L. 1996a. "Effects of Public and Private Policies on Working after Childbirth." *Work and Occupations* 23 (4):378–404.

Hofferth, Sandra L. 1996b. "Child Care in the United States Today." *The Future of Children* 6(2): 41–61.

Hofferth, Sandra, April Brayfield, Sharon Deich, and Pamela Holcomb. 1990. *National Child Care Survey 1990.* Washington, DC: Urban Institute.

Hofferth, Sandra, and Nancy Collins. 2000. "Child Care and Employment Turnover." *Population Research and Policy Review* 19(4): 1–32.

Hofferth, Sandra L., and Sharon Gennis Deich. 1994. "Recent U.S. Child Care and Family Legislation in Comparative Perspective." *Journal of Family Issues* 15 (3): 424–48.

Hofferth, Sandra, Kimberlee Shauman, Robin Henke, and Jerry West. 1998. *Characteristics of Children's Early Care and Education Programs.* Washington, DC: National Center for Educational Statistics.

Hofferth, Sandra L., Jerry West, Robin Henke, and Phillip Kaufman. 1994. *Access to Early Childhood Programs for Children at Risk.* Washington, DC: National Center for Educational Statistics.

Hofferth, Sandra, and Douglas Wissoker. 1992. "Price, Quality, and Income in Child Care Choice." *Journal of Human Resources* 27(1): 70–111.

Holloway, Susan, and Bruce Fuller. 1992. "The Great Child-Care Experiment: What Are the Lessons for School Improvement?" *Educational Researcher* 21:12–19.

Huston, Aletha C., Vonnie C. McLoyd, and Cynthia T. Garcia Coll. 1994. "Children and Poverty: Issues in Contemporary Research." *Child Development* 65:275–82.

Kisker, Ellen E., and Marsha Silverberg. 1991. "Child Care Utilization by Disadvantaged Teenage Mothers." *Journal of Social Issues* 47:159–77.

Kisker, Ellen E., Sandra Hofferth, Deborah Phillips, and Elizabeth Farquhar. 1991. *A Profile of Child Care Settings: Early Education and Care in 1990.* Washington, DC: U.S. Government Printing Office.

Lamb, Michael E., ed. 1997. *The Role of the Father in Child Development.* New York: John Wiley.

Love, John, Peter Schochet, and Alicia Meckstroth. 1996. *Are They in Any Real Danger? What Research Does—and Doesn't—Tell Us about Child Care Quality and Children's Well-Being.* Princeton, NJ: Mathematica Policy Research.

Mason, Karen O., and Karen Kuhlthau. 1991. "Determinants of Child Care Ideals among Mothers of Preschool-Aged Children." *Journal of Marriage and the Family* 51:593–603.

Massey, Douglas. 1996. "The Age of Extremes: Concentrated Affluence and Poverty in the Twenty-First Century." *Demography* 33:395–412.

Massey, Douglas S., Andrew B. Gross, and Kumiko Shibuya. 1994. "Migration, Segregation and the Geographic Concentration of Poverty." *American Sociological Review* 59:425–45.

McLloyd, Vonnie C. 1997. "Children in Poverty: Development, Public Policy, and Practice." In *Handbook of Child Psychology,* 5th ed., ed. I. Sigel and K. A. Renninger, 135–210. New York: Wiley.

Mitchell, Anne, Emily Cooperstein, and Mary Larner. 1992. *Child Care Choices, Consumer Education and Low-Income Families.* New York: National Center for Children in Poverty.

NICHD Early Child Care Research Network. 1996. "Characteristics of Infant Child Care: Factors Contributing to Positive Caregiving." *Early Childhood Research Quarterly* 11:269–306.

NICHD Early Child Care Research Network. 1997a. "Child Care in the First Year of Life." *Merrill-Palmer Quarterly* 43(3): 340–60.

NICHD Early Child Care Research Network. 1997b. "Poverty and Patterns of Child Care." In *Consequences of Growing Up Poor,* ed. Greg J. Duncan and Jeanne Brooks-Gunn, 100–131. New York: Russell Sage Foundation.

Peisner-Feinberg, Ellen, M. Burchinal, R. Clifford, M. Culkin, C. Howes, and S. L. Kagan. 1999. "The Children of the Cost, Quality, and Outcomes Study Go to School." <www.fpg.unc.edu/~NCEDL/PAGES/cqes.htm> (16 June 1999).

Phillips, Deborah. 1987. *Quality in Child Care: What Does Research Tell Us?* Washington, DC: National Association for the Education of Young Children.

Phillips, Deborah, M. Voran, and Ellen Kisker. 1994. "Child Care for Children in Poverty: Opportunity or Inequity." *Child Development* 65:472–92.

Porter, Toni. 1991. *Just Like Any Parent: The Child Care Choices of Welfare Mothers in New Jersey.* New York: Bank Street College of Education.

Presser, Harriet. 1989. "Can We Make Time for Children? The Economy, Work Schedules, and Child Care" *Demography* 26(4): 523–43.

Sherman, Arloc. 1994. *Wasting America's Future.* Washington, DC: Children's Defense Fund.

Sonenstein, Freya, and Douglas Wolf. 1991. "Satisfaction with Child Care: Perspectives of Welfare Mothers." *Journal of Social Issues* 47(2): 15–31.

South, Scott, and Glenn Deane. 1993. "Race and Residential Mobility: Individual Determinants and Structural Constraints." *Social Forces* 12(1): 147–67.

U.S. Department of Health and Human Services. 1998. *The Green Book (Overview of Entitlement Programs)*. Washington, DC: U.S. Department of Health and Human Services.

U.S. General Accounting Office. 1992. "Child Care: States Face Difficulties Enforcing Standards and Promoting Quality." GAO/HRD Report, vol. 93–13. Washington, DC: U.S. General Accounting Office.

Whitebook, Marcy, Carollee Howes, and Deborah Phillips. 1990. *Who Cares? Child Care Teachers and the Quality of Care in America.* Oakland, CA: Child Care Employee Project.

Willer, Barbara, S. L. Hofferth, E. E. Kisker, P. Divine-Hawkins, E. Farquhar, and F. B. Glantz. 1991. *The Demand and Supply of Child Care in 1990.* Washington, DC: National Association for the Education of Young Children.

Yinger, John. 1996. *Closed Doors, Opportunities Lost: The Continuing Costs of Housing Discrimination.* New York: Russell Sage Foundation.

Zinsser. Caroline. 1990. *Born and Raised in East Urban: A Community Study of Informal and Unregulated Child Care.* New York: Center for Public Advocacy Research.

45

Creating a Caring Society

EVELYN NAKANO GLENN,
2000

Why is it important to achieve a society that values caring and caring relationships? The answer might appear obvious: It seems inherent in the definition of a good society that those who cannot care for themselves are cared for; that those who can care for themselves can trust that, should they become dependent, they will be cared for; and that people will be supported in their efforts to care for those they care about. But even more is at stake. Currently we are caught in a nasty circle. To the extent that caring is devalued, invisible, underpaid, and penalized, it is relegated to those who lack economic, political, and social power and status. And to the extent that those who engage in caring are drawn disproportionately from among disadvantaged groups (women, people of color, and immigrants), their activity—that of caring—is further degraded. In short, the devaluing of caring contributes to the marginalization, exploitation, and dependency of care givers. Conversely, valuing and recognizing caring would raise the status and rewards of those who engage in it and also increase the incentives for other groups to engage in caring. Thus, a society that values care and caring relationships would be not only nicer and kinder, but also more egalitarian and just.

In addressing the question of how to create a society in which caring is valued, I first give a brief account of the contemporary "crisis" in care which stems from its being defined as a privatized, feminized, and therefore devalued domain. In the next section I review recent feminist attempts to rethink the concept of care in ways that open it up to critical analysis. I then define some desirable goals for a society that values care. In the final section I outline four major directions for change in social citizenship rights, family responsibility, organization of paid care, and employment policies and practices.

THE CONTEMPORARY PROBLEM OF CARE

A spate of popular books and articles in the last decade has sounded an alarm about a new "crisis in care," a crisis occasioned by the exodus of women from the home into the work force. The need for care of children, the elderly, and the chronically ill and disabled has not diminished, and may have grown because of increased longevity and medical advances that keep people with serious injuries or illnesses alive. Yet traditional caretakers—stay-at-home wives and mothers—are now less available to provide care on a full-time basis.

Dual-worker families—and more concretely, employed women—are said to be increasingly

From Evelyn Nakano Glenn, "Creating a Caring Society," from *Contemporary Sociology*, Vol. 29, No. 1, January 2000. Used with permission of the American Sociology Association.

overburdened and strained by the need to meet both earning and care responsibilities. At the same time, most families don't have the economic means to purchase care, and state services are grossly inadequate. As Mona Harrington (1999: 17) says in a recent popular treatment, "we have patchwork systems, but we have come nowhere near replacing the hours or quality of care that the at home women of previous generations provided for the country." The question of how care is to get done without substantial numbers of nonemployed women to do it has become the subject of research and policy initiatives. For example, the Alfred P. Sloan Foundation has funded several university research centers on work and family life, including one at my campus devoted to "Cultures of Care."

The "crisis in care" is just one impetus for recent critical examinations of the concept and organization of care in modern political democracies. Feminist theorists and researchers for some time have been examining care in its gendered dimensions. Their work makes it clear that the current crisis is a product of a privatized and gendered caring regime in which families, rather than the larger society, are responsible for caring and in which women (and other subordinate groups) are assigned primary responsibility for care giving.

The relegation of care to the private sphere and to women has had two further corollaries: the devaluation of caring work and caring relationships, and the exclusion of both from the arena of equality and rights. As feminist critics of liberal political philosophy have explained, the very concept of citizenship (i.e., full membership in the community, including reciprocal rights and responsibilities) has been premised on two conceptual dichotomies. First has been a split between the public and private, with the private realm of concrete relations of care defined not only as separate from, but also in opposition to citizenship. The private realm encompasses emotion, particularity, subjectivity, and the meeting of bodily needs, while the public arena of citizenship is ruled by thought, universality, objectivity, and the ability to act on abstract principles. Those relegated to the private sphere and associated with its values—women, servants, and children—were long excluded from full citizenship. Second has been a dichotomy between independence and dependence, with the ideal citizen defined as an autonomous individual who can make choices freely in the market and in the political realm. Within the liberal polity, citizenship supposedly created a realm of equality in which independent individuals had identical rights and responsibilities, regardless of differences in economic standing and other attributes. Those deemed dependent, whether categorically (as in the case of women, slaves, and children) or by reason of condition (as in the case of mental or physical disability) lacked standing and therefore were defined as outside the realm of equality (Okin 1979; Pateman 1988). The fiction of liberal philosophy that independent and autonomous actors exist also obscures the actual interdependence among people and the need for care that even "independent" people have.

Historically, then, in the United States caring work within the family has not been recognized as a public societal contribution comparable to paid employment. As Judith Shklar (1991) has pointed out, earning has always been seen as a responsibility of citizenship because it is the basis for independence. In this view, earners fulfill citizenship responsibilities and therefore deserve certain entitlements, such as old age pensions, unemployment insurance, and health and safety protection. In contrast, unpaid family caregivers perform strictly private responsibilities and do not fulfill broader citizenship responsibilities. Hence, they are not accorded entitlements comparable to those of wage earners.

Moreover, the dominant family model assumes that support for dependents and caregivers comes from the male breadwinner. Historically, the United States has provided little support for care giving, compared to other Western nations where paid parental leave, family allowances, child care services, housing subsidies, and health care coverage have been common (Fraser and Gordon 1993). During the World War I era, Progressive reformers pushed though maternalist programs, such as the Mothers' Pension program, to allow widowed women to keep their children rather than sending them to orphanages. But pensions were so low that single mothers were forced to work as well as care for their children. The Mothers' Pension was quickly phased out. New Deal-era social welfare policies institutionalized a two-tier system based on a male breadwinner-female caregiver model. The upper tier consisted of safety net entitlements for male breadwinners, which provided relatively generous, non-means-tested benefits such as unemployment insurance, social security retirement, and disability payments. Dependents of male breadwinners, including female caregivers, received indirect benefits through their relationship to a male earner, via provisions such as social security survivor benefits. The lower tier for women without connections to male breadwinners provided relatively ungenerous, means-tested "welfare" as in the original Aid to Dependent Children (ADC) and in the later Aid to Families with Dependent Children (AFDC). These benefits were considered a response to the neediness of children, not as an entitlement for mothers' caring labor (Nelson 1990; Gordon 1994; Abramovitz 1996). These programs were not only gendered, they were also raced. Black single mothers in the South and Mexican single mothers in the Southwest were routinely denied relief on the grounds that they were "employable." Thus, these women were not seen as "dependent" caregivers in the same way that white women were (Mink 1994).

Yet despite the prevailing ideology of the family as the realm of care, the growing need for care has generated a demand for paid care giving as an alternative or supplement to unpaid family care. Some of the demand has been met by institutions and services administered by the state and non-profit organizations. The greatest growth, however, has been in institutions and services organized by for-profit corporate entities formed to take advantage of payments available through (industry-backed) government medical insurance. Overall, then, there has been a shift of some portion of caring to publicly organized settings, whether administered by state, non-profit, or for-profit entities.

In these settings, the actual work of caring is done by "strangers"—paid workers, sometimes supplemented by unpaid volunteer workers. When caring is done as paid work, it not only remains gendered, it also becomes conspicuously racialized. In institutional settings such as hospitals, nursing homes, and group homes, nursing aides and other workers who actually do the day-to-day work of caring are overwhelmingly women of color, many of them recent immigrants. Home care workers also are drawn disproportionately from the ranks of women of color (Glenn 1992).

When care work is done by people who are accorded little status and respect in the society by reason of race, class, or immigrant status, it further reinforces the view of caring as low-skilled "dirty" work. This dual devaluation—of care work and care workers—rationalizes the low wages and lack of benefits that characterize care work. From her analysis of national wage data, Paula England (1992: 182) concluded that "being in a job requiring nurturing carries a net wage penalty of between $.24/hour and $1.70/hour." Taking into account such factors as workers' education, service jobs involving care giving paid less than comparable jobs not involving care giving. Thus child care workers earned less than manicurists; nursing aides and orderlies earned less than janitors; and psychiatric aides earned less than elevator operators. One ironic result is that those who care for others usually have to give up caring for their own dependents, yet cannot afford to pay anyone to care for them. Caring work is considered low-skilled and largely physical in nature, despite the importance of emotional and psychological aspects of caring.

Care in institutional settings is compromised by a combination of factors: pressures to cut costs, government regulations, medicalization, and bureaucratization (Foner 1994). Deborah Stone (1999) notes that cost-containment pressures affect both private for-profit care and public nonprofit and taxpayer-supported facilities. Efforts to reduce or control costs have resulted in inadequate training and chronic understaffing. Government regulations, reflected in institutional procedures, also require caregivers to spend time on extensive paper work. As workers are stretched thin, they experience stress and frustration, leading to burnout and high turnover. Bureaucratic structures and regulations, which are designed to both keep down costs and protect care receivers, nonetheless often restrict the caring activities of caregivers. For example, because of Medicare regulations, health care institutions try to limit staff to performing strictly medical and medical-related tasks such as changing dressings, and not getting involved in social and emotional caregiving. All of these pressures directly affect the care relationship. Caregivers complain about the lack of time and autonomy to respond to individual needs. Care receivers may be subject to controls that maintain "order" under conditions of understaffing (e.g., through use of sedation or physical restraints). Care receivers may not receive the kind of individualized and time-consuming care that would allow them maximum dignity and autonomy.

RETHINKING CARE

To develop alternatives to the present situation, we need to rethink the concept of care. Because care is so closely associated with womanhood, feminist philosophers and social theorists have subjected care to close analysis. My reading of several theorists of care, including Joan Tronto (1993), Diemut Bubeck (1995), Emily Abel and Margaret Nelson (1990), and Sara Ruddick (1998), suggests the usefulness of defining care as a practice that encompasses an ethic (caring about) and an activity (caring for). "Caring about" engages both thought and feeling, including awareness and attentiveness, concern about and feelings of responsibility for meeting another's needs. "Caring for" refers to the varied activities of providing for the needs or well-being of another person. These activities include physical care (e.g., bathing, feeding), emotional care (e.g., reassuring, sympathetic listening), and direct services (e.g., driving a person to the doctor, running errands). The definition is not free of ambiguity, but it does establish some boundaries. For example, defining caring in terms of direct meeting of needs differentiates caring from other activities that may foster survival. Thus, economic provision would not be included, even though it may help support care giving. Men are often said to be "taking care of their family" when they earn and bring money into the household. Despite the use of the term *care* in this phrase, breadwinning would not be considered "caring." In fact, economic support has historically been seen as men's contribution in lieu of actual care giving; simultaneously, care giving has been viewed as women's responsibility, an exchange for being supported by the primary breadwinner.

Within this definition of care as a practice, three features are important. First, this definition recognizes that everyone needs care, not just those we consider incapable of caring for themselves. Often only children, the elderly, the disabled, or the chronically ill are seen as requiring care, while the need for care and receiving of care by so-called independent adults is suppressed or denied. As Sara Ruddick (1998: 11) notes, "most recipients of care are only partially 'dependent' and often becoming less so; most of their 'needs,' even those clearly physical, cannot be separated from more elusive emotional requirements for respect, affection, and cheer." At the same time, even those we see as fully independent—that is, able to care for themselves in terms of "activities of daily living"—may for reasons of time or energy or temporary condition need care to maintain their physical, psychological, and emotional well-being. They may turn to a family member, friends, a servant, or a service provider for hot meals, physical touch, or a sympathetic ear. The difference is that "independent adults" may preserve their sense of independence if they have sufficient resources, economic or social, to "command" care from others, rather than being beholden to relatives or charity.

A second aspect of defining care as practice is that care is seen as creating a relationship; as Ruddick (1998: 14) puts it, "[caring] work is constituted in and through the relationship of those who give and receive care." The relationship is one of interdependence. Generally we think of the caregiver as having the power in the relationship; but the care receiver, even if subordinate or dependent, also has agency/power in the relationship. Focusing on relationships brings into relief the influence of the recipients of care on caring work. Tronto (1993) notes that for the work of care to be successful, its recipients have to respond appropriately—e.g., a screaming child betokens failure. In some situations where the care receiver employs the caregiver or has social authority (e.g., due to the norm of respect toward elders), the care receiver may have more power than the caregiver.

Third, the definition of care as practice recognizes that caring can be organized in a myriad of ways. The paradigmatic care relationship is the mother-child dyad, which often serves as the template for thinking about caring. In this model, caring (mothering) is viewed as natural and instinctive—women's natural vocation. However, this idealized model is deceptive in that it ignores the actual diversity in the ways mothering/caring is actually carried out within and across cultures. Caring can take place in the household or in publicly organized institutions, and can be carried out individually or collectively and as paid or unpaid labor. Much caring takes place in the family, usually as the unpaid work of women, but it is also done as paid work (e.g., by babysitters, home health aides, and the like). It also takes place in the community as unpaid volunteer work, as in the case of church or charitable organizations that run day care or senior activity centers. It also takes place in institutions organized by the state, corporations, or individuals as commodified services using paid caregivers.

Care can also be "fragmented," divided among several caregivers and between "private" and "public" settings. Thus, a parent may take ultimate responsibility for ensuring that a child has care after school but delegate the actual work of care giving to a babysitter, a relative, a paid home care worker, and/or an after-school program. Barrie Thome (1999) found in her study of childhoods in an urban multicultural community that parents often have to patch together several of these arrangements.

WHAT SHOULD OUR GOALS BE?

To achieve a society in which caring is valued in all spheres of social life, all of the elements—the work of care giving and the people involved (care receivers and caregivers)—would have to be recognized and valued. Hence, a society in which caring is valued would be one in which:

- Caring is recognized as "real work" and as a social contribution on a par with other activities that are valued, such as working, military service, or community service, regardless of whether caring takes place in the family or elsewhere or as paid or unpaid labor.
- Those who need care (including children, the elderly, disabled, and chronically ill) are recognized as full members of the society and accorded corresponding rights, social standing, and the voice of citizens. This would mean that care receivers are empowered to have influence over the type of care, the setting, and the caregivers, and that they have access to sufficient material resources to obtain adequate care.
- Those who do caring work are accorded social recognition and entitlements for their efforts similar to those who contribute through paid employment or military service. These entitlements include working conditions and supports that enable them to do their work well and an appropriate level of economic return, whether in wages or social entitlements.

For each of these ideals to be achieved, additional specific conditions would have to be fulfilled;

these conditions are also desirable for reasons of equity and social justice.

- Caring is legitimated as a collective (public) responsibility rather purely a family or private responsibility.
- Access to care is relatively equally distributed and not dependent on economic or social status. Ultimately, the ideal would be a society in which there is an adequate amount and quality of care for all who need it—i.e., care that is individualized, culturally appropriate, and responsive to the preferences of those who are cared for.
- The responsibility and actual work of caring is shared equitably so that the burden of care does not fall disproportionately, as it now does, on disadvantaged groups—women, racialized minorities, and immigrants.

SOME DIRECTIONS FOR CHANGE

Rethinking social citizenship: One important step is to redefine social citizenship to make care central to the rights and entitlements of citizens. This would involve a radical reversal of the present situation, in which care is defined as a private responsibility and therefore outside the realm of citizenship. Making care central to citizenship would entail three elements: establishing a right to care as a core right of citizens; establishing care giving as a public social responsibility; and according caregivers recognition for carrying out a public social responsibility. These three elements are interrelated. If citizens have a right to care, then there is a corresponding responsibility on the part of the community to ensure that those who need care get it. Further, if care giving is a public social responsibility, then those who do care giving fulfill an obligation of citizenship and thus are entitled to societal benefits comparable to those accorded for those fulfilling the obligation to earn—for example, social security, seniority, and retirement benefits.

Additionally, a constraint that is specific to caring (in contrast to earning) and that needs to be addressed is what Kittay (1995) has called the "secondary dependence" of the caregiver. By taking on the care of a dependent and foregoing earning, unpaid caregivers become dependent on a third party—a breadwinner or the state—for resources to sustain both those they care for (primary dependents) and themselves (secondary dependents). Historically, U.S. welfare policy has been premised on the assumption that support for care giving belonged to the male breadwinner, and that the state should assume responsibility for support of caring only in the absence of a male breadwinner. Sometimes, as in the case of black single mothers, the lack of a male breadwinner was not seen as adequate grounds for the state to step in. Instead, black single mothers were deemed to be "employable mothers" who should support themselves and their dependents. In a step backward from recognizing caregivers' need for support, the U.S. Congress passed the Personal Responsibility Act in 1996, which abolished AFDC, devolved welfare back to individual states, and restricted the amount of lifetime benefits; most states have mandated stringent "workfare" to get single mothers off welfare.

In contrast to the U.S. welfare system, European welfare states have all provided some forms of family allowance for citizens with children. Most countries have supported caregivers with child allowances, and some even give small pensions to those who engage in unpaid care work. In conservative welfare regimes, such as France and Germany, the rationale for maternal allowances typically has been framed in terms of child welfare and promoting natalism, to ensure the size and well-being of the future population, rather than in terms of the value of caring and social citizenship rights and responsibilities in caring. Nonetheless, the allowances have been designed as universal entitlements not tied to income or means testing, unlike U.S. welfare programs. In more progressive social democratic welfare regimes support for care giving is extensive, including allowances, subsidies, and direct services, such as child care and home aides (Pederson 1993; Sainsbury 1996).

Transforming citizenship in the United States to make care central to rights and entitlements would require us to challenge the linked ideologies of individual independence and family responsibility that I have described above. The United States for the most part has not even recognized mothering/parenting as a contribution to the national welfare, nor has it assumed a larger societal responsibility for supporting caregivers. As with previous historic changes in the boundaries and meanings of citizenship, it would require concerted struggle. Political citizenship, in the form of suffrage, was gradually extended to include previously excluded groups: nonpropertied white men in the early nineteenth century, black men after the Civil War, and, finally, women in 1920. The democratization of the vote was achieved only after concerted struggles by each of the groups in the course of over 100 years. Social citizenship rights of the welfare state, including social security, unemployment relief, minimum wage, and job creation were responses to the political mobilization of millions of Americans displaced by the Great Depression. In the second half of the twentieth century, the second civil rights movement and second-wave feminism impelled legal, political, and social changes that dramatically expanded employment, education, and legal rights for racial minorities and women.

An important recent example of expanding citizenship is the success of the disability rights movement in establishing federal laws and policies that require schools and universities, employers, and public programs to provide facilities and activities that enable differently-abled citizens to work, study, travel, and otherwise participate in the social and cultural life of the society. The latter movement comes closest to addressing the issues central to caring and social citizenship. It addresses the rights of citizens who have physical and mental conditions that limit their physical and economic independence to receive services and accommodations that allow them to achieve social and political independence. There is thus a precedent for claiming the right to care as essential for meaningful citizenship.

Rethinking the family as the primary site of care: The previous discussion about state policies on social citizenship and care has assumed that most care takes place within the family and is carried out as part of unpaid labor of family members. However, if we take seriously the notion that caring is a public social responsibility, we also need to examine critically the conception of the family as the institution of first resort for caring. Indeed, one can argue that keeping the family as the "natural" unit for caring relationships helps anchor the gender division of caring labor. Seeing family and women's caring as "natural" disguises the material relationships of dependence that undergird the arrangement. But as those who care for others know, love is not enough: Care requires material resources. We need therefore to consider "defamilializing" care in order to relieve women of disproportionate responsibility for care giving and also to free both care receivers and care givers from economic dependence on a male breadwinner.

Utopian societies in the past, ranging from communes to the Kibbutz movement, have attempted to transform care, especially infant and child care, into a public or communal responsibility by collectivizing child care. Theoretically, communal arrangements in which child care is treated as a form of "public" labor equal to other forms of labor free those who engage in caring from dependence on a breadwinner and also free children from dependence on (and therefore subordination to) biological parents. In practice, collectivized care has not eliminated the gendered division of caring labor, since it was still women who were the principal caregivers in publicly organized child care. Moreover, collectivized care generally has arisen in homogeneous religious and socialist communities where members shared fundamental cultural and political values. Completely collectivized care would be unlikely and perhaps undesirable in large-scale multicultural societies in which people maintain divergent cultural and political values. Family remains the main institutional nexus for anchoring distinctive cultural and social identities.

Thus, for both practical and ideological reasons it seems likely that families (broadly defined) will continue to value caring, and that family members will feel responsible for caring for children and, to a lesser extent, elderly and disabled members and will choose to do so. This does not mean that the family should be defined in the traditional way as the conjugal heterosexual household or that it should be the first resort for care in all cases. The states' and employers' care policies currently recognize dependency and caring relationships in rather traditional terms of parents and children (whether biological or adoptive) and spouses (defined through legal marriage). However, there are many other types of family relations that generate relationships of care, including cohabiting couples, gay and lesbian couples, extended kin such as grandparents and siblings, and sometimes "fictive kin" who participate in mutual support. As Carol Stack and Linda Burton (1994) point out in relation to their study of African American families, men, women, and children may be "kin-scripted" to care for the children of siblings, grandparents, grandchildren, aunts, and uncles when there is no one else able to do so. To the extent that caring in the "family" is valued, the notion of "family" must be extended to encompass diverse kin relations, including "voluntary" or "fictive" relationships.

Regarding the knotty question of the primacy of family vs. the larger community in care giving: In a survey conducted in England by Janet Finch (1996), respondents affirmed the importance of kin ties; they indicated that "rallying around in times of crisis" was what defined a functional family. The actual degree of responsibility that respondents felt in particular situations and toward particular relatives varied, however, depending on prior relationship and current circumstances. (I would also add that in a diverse society, there is considerable cultural difference in degree of obligation and in who is included in the net of obligation.) In general, Finch's respondents emphasized that relatives should not expect or take for granted assistance from other family members. Another British researcher, Jenny Morris, found that, in turn, people requiring care often prefer not to rely on family. Many of the disabled adult women Morris interviewed said they preferred paid helpers or helpers provided by social service to help from family members, because it allowed them more independence (cited in Cancian and Oliker 2000: 99).

Finch (1996: 207) argues that the moral reasoning of people in her survey suggests the principle that people should have the right *not* to have to rely on their families for help: "To point in another way, the family should not be seen as the option of first resort for giving assistance to its adult members, either financial or practical."

Finch is careful to say that her point is not to deprecate generosity, care, and support within families, but only to see these as "optional, voluntary, freely given" (1996: 207).

Taken together, the findings from Finch's and Morris's studies support the case that the community, as represented by the state, has primary responsibility for care of its citizens, and that citizens in turn have the right to nonfamily care. Public policy would thus be that all persons are entitled to publicly organized care or to allowances or vouchers to pay for care, regardless of whether or not family members are available to provide it.

RETHINKING PAID CARE

As noted in the introduction, the sheer demand for care, the inability of families to provide all care, and economic incentives to commodity care have brought about significant shift of caring to paid caregivers. This is especially the case for those needing physically demanding, round-the-clock care, such as children or adults with severe mental and physical disabilities, and elderly with dementia or Alzheimer's. Much of the latter care takes place in institutional settings, nursing homes, hospitals, and residential facilities, where the intensive face-to-face caring is done by nursing aides and other nonprofessional workers under the supervision and authority of administrators and medical and nursing professionals.

Thus, any scheme to create a society in which caring is valued in all spheres must address the growing commodification and defamilization of care. We need to think about the changes that occur when caring is made into a public rather than private function, when "strangers" rather than family members provide care, when care giving is paid rather than unpaid, and most importantly when caring is regulated and controlled by bureaucratic rules and hierarchy.

Transferring caring from private household into publicly organized settings inserts "third parties" into the caring relationship. Both caregivers and care receivers are hemmed in by rules and regulations about time spend and kinds of care that are covered (e.g., shopping). Foner (1994) and others have argued that the "iron cage" of bureaucracy that constrains people in organizations creates fundamental dilemmas for care workers who are caught between conflicting ideals. Whereas bureaucracies operate according to principles of standardization, impersonal rules, and efficiency, care relationships encourage individual treatment, personal ties, and patience.

Bureaucratic rules and control were instituted because of publicity about widespread abuse and neglect of patients. Having done an ethnographic study of a nursing home in New York, Foner (1994) agrees that bureaucratic rules and oversight are necessary to protect elderly patients, and that nursing aides, who do the actual physical care, feeding, cleaning, bathing, and so on, cannot be allowed to act autonomously. However, the rules and the way they are administered emphasize "efficiency" in getting physical care tasks done, meeting time deadlines, and maintaining records. Yet, as Tim Diamond (1988: 48) found in his ethnography of a nursing home, emotional care is essential to the nursing aide's job: "holding someone trying to gasp for breath" or talking to residents to "help them hold on to memories of their past." Diamond observed that these kinds of emotional support were not listed in the aides' job descriptions, nor were the aides rewarded for these activities. In the nursing home she studied, Foner found that Ana, a nursing aide who regularly took time to talk to patients, and comfort or reassure them while bathing them or changing them, was constantly reprimanded for being inefficient, while Ms. James, an aide who never spoke to patients and handled them roughly to get them through their routines, was praised by supervisors as a model aide.

Deborah Stone (2000) found that home care workers also faced a conflict between bureaucratic rules and principles and their own ethic of care. Thus, they often stretched or evaded rules and supervisors to provide personal care, or spent off-work time or money to provide extra services.

The various ethnographic studies reveal that many care workers do provide quality emotional care, but they do so "around the fringes" so that their skills and effort are unrecognized or they do so in direct defiance of the rules. These studies point to the existence of "oppositional cultures" in which workers cooperate to provide the kind of care that the bureaucratic structure does not recognize or disallows. One case study of a psychiatric hospital (Lundgren et al. 1990) found the quality of care was excellent because psychiatric technicians who did the daily care carved out areas of autonomy in which they could act in accordance with an ethos of care. Because the psychiatric technicians had opportunities to interact freely when residents were in classes, they developed camaraderie. Workers supported one another to go beyond the policies they considered unreasonable or against the interests of the residents. They developed customs, such as "time out" to leave the unit when they were about to lose control. These kinds of practices that workers themselves develop could be incorporated into organizational practices. Encouraging a team approach in which workers model and support each other for sensitive caring would be one such salutary practice. Procedures could also be reformed to build in more opportunity and recognition for aides who show kindness and go out of their way for patients. Organizations could offer more regular training in sensitivity and emotional aspects of care, include emotional caring work in job descriptions and worker evaluations, and provide a reward system for caregivers who go beyond the call of duty to help patients.

At the professional level, the bureaucratic and chart-keeping imperatives of caring institutions could be harnessed to build in accountability for the social and emotional well-being of care receivers. Foner (1994) notes that one reform that has been adopted in many institutions is the psychosocial model of care, which pays attention to the emotional and social as well as the physical aspects of caring. The psychosocial model involves a case management approach that includes both health and social service needs of care receivers. Cancian and Oliker (2000) describe a "Clinical Practice Model" of nursing that Bonnie Wesorick has developed and introduced in several hospitals. This model challenges the medical model by emphasizing "holistic caring." It does so by such methods as keeping a record on each patient that includes personal histories, religious orientation, family situation, and individual concerns. Importantly, it calls for writing a plan of care that documents the patient's needs, concerns, and problems and an individualized approach to reach desired outcomes.

Encouraging caregivers to focus on social and emotional aspects of care may be salutary in some respects. Yet there is an inherent pitfall to empowering caregivers: It may exacerbate the already unequal relationship between caregivers and care receivers. Caregivers may feel that they understand the needs of care receivers and that they are acting in their best interests. However, care receivers might have different values and priorities. To the extent that care receivers depend emotionally and physically on their caregivers, they may feel they have no choice but to defer to the caregiver's judgment.

Thus, an additional concern should be to ensure that care receivers are given voice and influence over their care. In the case of mentally competent adults requiring home care assistance, for example, it would be preferable for them to be given grants or vouchers to hire their own caregivers rather than being assigned a helper by a social service agency. Several of the 50 disabled women interviewed by Jenny Morris in England said they especially valued helpers they hired and paid for themselves rather than those sent by government social services, because they had greater control. One woman said that only when she started employing her own helper did she feel she could pay attention to her own appearance. She had her paid helper assist her with clothing and makeup, which she felt justified in doing because "They need to be patient and I'm paying for that patience so I feel OK about expecting it." (quoted in Cancian and Oliker 2000: 99). One group already has direct access to government funds for paid care. The Department of Veterans' Affairs has a program for Universal Aid and Attendance Allowance that gives direct unrestricted cash payments to 220,000 veterans to pay for homeworkers or attendants (Cancian and Oliker 2000: 155). The right of veterans to state supported paid care is acknowledged because of their "service to the country." What is needed is a more universal approach that extends entitlements to nonfamilial paid care to all citizens.

In short, both paid caregivers and receivers of paid care need to be empowered. Sometimes, when the interests of caregivers and care receivers intersect, it makes sense for them to organize together. For example, when social service agency budgets are cut and home care and other services are reduced, caregivers may be forced to serve more clients less well and clients don't get the care they need. During the 1980s and '90s, coalitions of home health care workers, care receivers, and community leaders have formed to improve wages and benefits for care workers. Since services are paid from Medicaid or other public funds, care receivers will support wage increases for care workers, especially if it means that their caregivers will continue rather than leaving for higher-paying jobs in other fields (Cobble 1996; SEIU 1999).

RETHINKING EMPLOYMENT PRACTICES

Changes in employment practices are also needed to make it possible for people to integrate work and care and so that care giving is not penalized. A small proportion of citizens currently benefit from private-sector initiatives by corporations that recognize the caring responsibilities of their employees. Some of these corporate employers provide child care and unpaid leaves to care for children or elderly relatives. Model programs include those by CitiBank, Stride Rite, and Campbell's Soups, which provide child care on or near their premises. Bristol Myers-Squibb has a family leave policy for employees that covers care for elderly relatives (Cancian and Oliker 2000: 75, 155).

The passage of the 1993 Family and Medical Leave Act marked a first step in developing a national policy that supports combining work and care. The act recognizes care responsibilities for those engaged in paid work and accepts public responsibility so that dependents can receive adequate care. As in many European countries, the stated goal of the legislation was the development of children and promotion of the family unit rather than recognition of care giving as a social responsibility. The preamble to the Act recognizes job security and parenting as important for citizens' well-being and acknowledges the role of the state in supporting both. However, coverage is extremely limited. By mandating only unpaid leave, the government accommodates care rather than fully supporting it, since few parents can afford to use the unpaid leave. Moreover, by exempting employers with fewer than 50 employees, the law leaves an estimated half of the workforce uncovered—56 percent of women and 48 percent of men, according to Spalter-Roth and Hartmann (1990). Ultimately, when employer interests are at stake, employer needs are allowed to trump care needs. Finally, the Act recognizes dependency only within traditional conjugal family relationships—spouse, children, and parents (Kittay 1995). It thereby "refamilizes" care by excluding other types of voluntary relations of dependency and care.

Besides parental and care-giving leave and child care, employment policy must consider the sheer number of hours needed for care. A national survey of a representative sample of 1509 English-speaking households found an average of 17.9 hours of care giving per week per household, while several other specialized surveys found a much higher number of unpaid caregivers hours in households with persons having specific medical conditions or disabilities (Arno et al. 1999). At the same time, work hours of employed Americans have become the longest of those in all industrialized nations, according to a 1997 United Nations survey. The survey found that U.S. workers averaged 40 percent more hours than Norwegians and 25 percent more than the French (calculated from figures in the *San Francisco Chronicle,* September 6, 1999).

In combination with lack of state support for nonemployed caregivers, long work hours increase the strain on U.S. workers who have care responsibilities. Comparisons of worker productivity suggest that the longer hours of U.S. workers have not produced comparable increases in productivity. Thus reduction of work hours can be justified on economic as well as social welfare grounds. The 40-hour week was the goal of labor movements starting after the Civil War, but it was only when organized labor acquired sufficient political power in the 1930s that it became the standard. It involved the recognition of workers' rights for a life apart from the job. It is now time to recognize the reality of workers' multiple responsibilities for earning and caring by reducing work hours through a combination of reducing the standard for "full-time work" and increasing vacation and leave time.

CLOSING THOUGHTS

I have focused on specific ideological and structural constructions of caring. But ideas about and structures of caring are tied to other ideologies and structures that they support and are supported by. Achieving the kinds of changes needed to create a society that values caring will require transforming the ways we think about ourselves, our relationships with others, the family, civil society, the state, and the political economy. Ultimately, the transformation of caring must be linked to major changes in political-economic structures and relationships. Perhaps most fundamentally, the liberal concept of "society" as made up of discrete, independent, and freely choosing individuals will have to be discarded in favor of notions of interdependence among not wholly autonomous members of a society.

SELECTED REFERENCES

Abel, Emily, and Margaret Nelson, eds. 1990. *Circles of Care.* Albany: State University of New York Press.

Abramovitz, Mimi. 1996. *Under Attack: Fighting Back.* New York: Monthly Review Press.

Amo, Peter S., Carol Levine, and Margaret M. Memmott. 1999. "The Economic Value of Informal Caregiving." *Health Affairs* 18(2): 182–87.

Bubeck, Diemut Elisabet. 1995. *Care, Gender, and Justice.* Oxford: Clarendon Press.

Cancian, Francesca M., and Stacey J. Oliker. 2000. *Caring and Gender.* Thousand Oaks, CA: Pine Forge Press.

Cobble, Dorothy Sue. 1996. "The Prospects for Unionization in a Service Economy." Pp. 333–58 in *Working in a Service Economy,* edited by Cameron MacDonald and Carmen Siriani. Philadelphia: Temple University Press.

Diamond, Timothy. 1988. "Social Policy and Everyday Life in Nursing Homes: A Critical Ethnography." Pp. 39–55 in *The Worth of Women's Work: A Qualitative Synthesis,* edited by Anne Statham, Eleanor M. Miller, and Hans O. Mauksch. Albany: State University of New York Press.

England, Paula. 1992. *Comparable. Worth: Theories and Evidence.* Hawthorne, NY: Aldine de Gruyter.

Finch, Janet. 1996. "Family Rights and Responsibilities." Pp. 193–208 in *Citizenship Today: The Contemporary Relevance of T. H. Marshall,* edited by Martin Bulmer and Anthony M. Rees. London: UCL Press.

Foner, Nancy. 1994. *The Caregiving Dilemma: Work in the American Nursing Home.* Berkeley: University of California Press.

Fraser, Nancy, and Linda Gordon. 1993. "Contract versus Charity: Why Is There No Social Citizenship in the United States?" *Socialist Review* 212(3): 45–68.

Glenn, Evelyn Nakano. 1992. "From Servitude to Service Work: Historical Continuities in the Racial Division of Paid Reproductive Labor." *Signs* 18: 1–43.

Gordon, Linda. 1994. *Pitied but Not Entitled: Single Mothers and the History of Welfare 1890–1935.* New York: Free Press.

Harrington, Mona. 1999. *Care and Equality: Inventing a New Family Politics.* New York: Knopf.

Kittay, Eva Feder. 1995. "Taking Dependence Seriously: The Family and Medical Leave Act Considered in Light of the Social Organization of Dependency Work and Gender Equality." *Hypatia* 10:8–29.

Lundgren, Rebecka Inga, and Carole H. Browner. 1990. "Caring for the Institutionalized Mentally Retarded: Work Culture and Work-Based Social Supports." Pp. 150–172 in *Circles of Care,* edited by Emily Abel and Margaret Nelson. Albany: State University of New York Press.

Mink, Gwendolyn. 1994. *The Wages of Motherhood.* Ithaca, NY: Cornell University Press.

Nelson, Barbara. 1990. "The Origins of the Two-Channel Welfare State: Workmen's Compensation and Mothers' Aid." Pp. 123–51 in *Women, The State and Welfare,* edited by Linda Gordon. Madison: University of Wisconsin Press.

Okin, Susan. 1979. *Women in Western Political Thought.* Princeton, NJ: Princeton University Press.

Pateman, Carole. 1988. *The Sexual Contract.* Stanford, CA: Stanford University Press.

Pedersen, Susan. 1993. *Family, Dependence, and the Origins of The Welfare State: Britain and France, 1914–1945.* Cambridge: Cambridge University Press.

Ruddick, Sara. 1998. "Care as Labor and Relationship." Pp. 3–25 in *Norms and Values: Essays on the Work of Virginia Held,* edited by Joram G. Haber and Mark S. Halfon. Lanham, MD: Lanham, Rowman & Littlefield.

Sainsbury, Diane. 1996. *Gender, Equality and Welfare States.* Cambridge: Cambridge University Press.

San Francisco Chronicle. 1999. "UN Says Americans Are the Hardest Workers." September 6, p. 3.

Service Employees' International Union (SEIU). 1999. "Drive to Improve L.A. Homecare Takes Big Step Forward." Press Release.

Shklar, Judith. 1991. *American Citizenship: The Quest for Inclusion.* Cambridge, MA: Harvard University Press.

Spalter-Roth, Roberta M., and Heidi I. Hartmann. 1990. *Unnecessary Losses: Cost to Americans of the Lack of Family and Medical Leave.* Washington, DC: Institute for Women's Policy Research.

Stack, Carol, and Linda Burton. 1994. "Kinscripts: Reflections on Family, Generation, and Culture." Pp. 33–44 in *Mothering: Ideology, Experience and Agency,* edited by Evelyn Nakano Glenn, Grace Chang, and Linda Forcey. New York: Routledge.

Stone, Deborah. 1999. "Care and Trembling." *The American Prospect* 43 (March-April): 61–67.

———. 2000. "Care as We Give It, Work as We Know It." In *Care Work: Gender, Labor and the Welfare State,* edited by Madonna Harrington-Meyer. New York: Routledge.

Thome, Barrie. 1999. "Pick-Up Time at Oakdale Elementary School: Work and Family from the Vantage Points of Children." Working Paper No. 2. Center for Working Families, University of California, Berkeley.

Tronto, Joan. 1993. *Moral Boundaries: A Political Argument for an Ethic of Care.* New York: Routledge.

Glossary

abolitionist One who advocates the abolition (end) of the slave trade.

abstinence-only sex education programs Teachings that focus on not having sexual intercourse as the only acceptable means of contraception.

adolescence (Adolescent) Period of life in between childhood and adulthood.

aging family transitions The behaviors associated with later life changes in family composition or interaction, such as becoming grandparents, caring for aging parents, dealing with death, surviving a spouse, etc.

agrarian societies Civilizations based on agriculture. Agrarian state societies were based on technological innovations like the plow and irrigation, and they were characterized by the growth of cities and high levels of stratification.

battering (Battery, Batterer) The act of physically hitting or beating another person.

birth control Contraception; any method of preventing conception or pregnancy.

birthrate The ratio of births to population within a given time period.

breadwinner (Good Provider) Ideal of providing for the material needs of family members through one's paid labor; usually assumed to apply to fathers. Compare to Homemaker.

child support Income paid to a former spouse for support of dependent children following a divorce or separation.

corporal punishment Punishing using physical means (e.g., spanking, hitting, flogging).

cohabitation Residence of a couple in a shared household without being married.

compulsory motherhood The idea that all women should be mothers and that they should gain intense satisfaction from motherhood.

contraception Birth control; any method of preventing conception or pregnancy.

courtship A process of acquaintance, selection, and attachment between potential mates that leads to the formation of strong sexual ties and possibly marriage.

coverture Legal doctrine stipulating that the identity of a wife is subsumed by her husband.

custody (Child Custody) Guardianship; usually refers to a legal award following divorce, including responsibility for residence, care, and control of children.

divorce The legal and formal dissolution of a marriage.

divorce rate The number of divorces over a specified period per specified population. The divorce rate is often calculated per 1,000 population or by estimating the proportion of all marriages that are expected to end in divorce.

economic class Groups within a society distinguished and often stratified according to economic, occupational, and social factors.

emotional property The idea or belief that one's emotional attachment to another is owned exclusively.

entitlement The feeling that one has a right to certain privileges and that specific rewards should be forthcoming by virtue of what they have done or who they are.

exogamy Marriage to a partner outside one's group, clan, tribe, or ethnicity.

familism The idea that family members should give priority to family (versus individual) interests, welfare, and survival, hence family members should cleave to one another and support one another under all circumstances.

family Two or more persons who are related by birth, marriage, or adoption who live together as one household (Census Definition).

family disruption The behaviors associated with altering or terminating family and pseudo-family units; separation, annulment, divorce, disownment, death, etc.

family formation The behaviors associated with creating family and pseudo-family units; cohabitation, engagement, marriage, giving birth, adoption, etc.

family reconstitution The behaviors associated with reforming family and pseudo-family units; shared custody/visitation, launching children, cohabitation with a new partner, remarriage, stepparenting, etc.

feminism The policy, practice, or advocacy of political, economic, and social equality for women.

feminist An advocate of feminism.

functionalism Type of social theory that explains social institutions by the contribution they make to the functioning and survival of society.

gender Socially defined "acceptable" behavior and characteristics of men and women.

good provider (Breadwinner) Ideal of providing for the material needs of family members through one's paid labor; usually assumed to apply to fathers. Compare to Homemaker.

homemaker (Homemaking) Ideal of caring for the emotional and physical needs of family members and specializing in unpaid work in the home; usually assumed to apply to mothers. Compare to Breadwinner.

homogamy The tendency for people to marry those with similar social characteristics.

household A residential unit in which coresidents typically share some resources. Households vary in membership and composition. Some households are families (parents and children or husbands and wives), but many families are not households (some family members may not live in the same residence).

household labor The labor involved in maintaining and operating a household, including shopping, cooking, cleaning, laundry, repairs, yard work, and childcare.

ideology of separate spheres (see Separate Spheres)

industrial societies Large-scale, complex societies whose economic basis derives from the use of inanimate energy sources such as coal, oil, and electricity.

infanticide Killing an infant.

infant mortality rate The number of deaths per 1,000 live births in a given population.

interaction ritual Any activity in everyday life that brings people together, focuses their attention on a common activity or set of symbols, and builds up a shared emotion or mood, resulting in a feeling of group membership.

institution A set of roles graded in authority that have been embodied in consistent patterns of actions that have been legitimated and sanctioned by society or segments of that society; whose purpose is to carry out certain activities or prescribed needs of that society or segments of that society.

joint legal custody Legal award following divorce in which divorced parents share decisions about such things as a child's education and medical treatment.

joint physical custody Legal award following divorce in which child resides approximately equally with each parent for alternating time periods.

kinship groups (Kinship Structure) Individuals related to each other by marriage or descent who have reciprocal obligations.

life course The chronological unfolding of an individual's development, from infancy through adulthood to death, patterned by physiological capacities, individual differences, social circumstances, and cultural ideals.

life expectancy The number of years an individual is expected to live.

longevity Length of time (years) that people live.

matrilineal Descent system in which family membership and inheritance are traced through the female line, from a mother to her children.

norm (Social Norm) A cultural rule or standard.

nuclear family Family group consisting of a mother, a father, and their children.

patriarchy (Patriarchal) Male-dominated social system in which the chief authority is the father or eldest male member of the family or clan (i.e., the patriarch).